THE COMPLETE ILLUSTRATED ENCYCLOPEDIA OF

DINOSAURS
& PREHISTORIC CREATURES

THE COMPLETE ILLUSTRATED ENCYCLOPEDIA OF
DINOSAURS
& PREHISTORIC CREATURES

The ultimate illustrated reference guide to 1000 dinosaurs and prehistoric creatures, with 2000 specially commissioned illustrations, maps and photographs

DOUGAL DIXON

HERMES
HOUSE

CONTENTS

INTRODUCTION

The science of palaeontology (the study of fossil animals) is developing at an exciting speed. New discoveries are being made so quickly, that before this book is on the bookshelves there will have been an overwhelming number of new finds and developments in the understanding of the subject. Each week there is something new to report, whether it be a skeleton that constitutes an entirely new branch of the evolutionary tree, or some indication of life gleaned through new finds of footprints or feeding traces. Microscopic analysis of fossilized dung provides ancient information about diet eaten by prehistoric creatures. Footprints tell us about the lifestyles of extinct animals.

By far the greatest number of fossils that have been found are those of invertebrate animals that lived in the sea. It is easy to understand why. Fossils are found in sedimentary rocks. Most sedimentary rocks are formed from deposits laid down on the sea bed, and so animals that lived on the sea bed would have a better chance of being buried there. There are also far more invertebrates than vertebrate species. The bodies of animals that live and die on land tend to be broken up and scavenged before they have a chance to be buried.

Our knowledge of land animals is weighted towards those that we know existed close to rivers, or in deserts, or on the banks of lakes or lagoons – places where dead bodies are likely to have been buried quickly and fossilized. We do not have direct evidence yet of the animals that lived on mountains or highland forests, or other places that were far from quick burial sites.

The scope of this book encompasses the fossil tetrapods. Tetrapod means, literally, "four-footed". But to zoologists and palaeontologists it has a more technical meaning. It covers all the vertebrates except those that we would regard as fish. It ranges from the most primitive of the early amphibians to everything that has evolved from them since – the various lizard-like and crocodile-like and mammal-like early reptiles, the dinosaurs and their contemporary reptiles in the ocean and the air, the rich array of mammal and bird life that succeeded the dinosaurs, and even, eventually, ourselves. As such it would appear to be something of a misnomer, since it covers birds, that have only two legs and a pair of wings, whales, that have no legs but paddles instead, and snakes, that have no limbs whatsoever. In fact all of these evolved from four-footed ancestors and so belong in this classification.

In general, tetrapods are land-living animals, and so fall into the categories of animals that are rarely fossilized or at least fossilized less frequently than the marine invertebrates. Sometimes we are lucky and find complete skeletons of animals that have been buried suddenly in sandstorms, or have been preserved in the sterile sediments of poisonous lake beds. More often, however, the bones are broken up and scattered, and it takes a great deal of analysis and interpretation to work out what the living animal originally looked like.

The general introduction provides an overview of the ancient world, introducing the key areas of research that have helped paleontologists to paint a picture of what the world was like in remote times.

Below: Ichthyostega, *one of the earliest Devonian tetrapods.*

Below: Eogyrinus, *a crocodile-like amphibian from the Carboniferous.*

Below: Estemmenosuchus, *a Permian plant-eating reptile.*

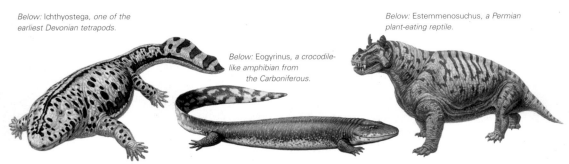

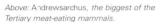

Left: Sinodelphis, *an early marsupial mammal from the Cretaceous.*

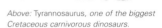

Above: Andrewsarchus, *the biggest of the Tertiary meat-eating mammals.*

Above: Eucladoceros, *a spectacular Tertiary deer.*

Above: Tyrannosaurus, *one of the biggest Cretaceous carnivorous dinosaurs.*

From fossil evidence, we can say with certainty much about how the animals lived, what food they ate, whether they lived in groups, had family networks, and what the landscape looked like.

The second section of the book is an encyclopedia of all the representative groups of fossil tetrapods. The 1000 entries are arranged chronologically, from the late Palaeozoic, when vertebrates first began to leave the water and live on land, through the Mesozoic, the time of dinosaurs, to the Tertiary when the mammals spread throughout the world, and culminating in the Quaternary, which covers the great Ice Age and the tiny sliver of geological time that represents the present day.

Fascinating information about each animal entry is provided and is accompanied by a description of the features that make the animal distinctive. A fact box lists some of the technical data, such as the period of history when the animal lived, its dimensions and its discoverer. Each is illustrated with a watercolour that shows what scientists think that the animal looked like. The appearance is based on the evidence available, using studies of related animals to make

the best attempt possible of a restoration of the living beast. Skin colours and textures are often assumptions. Finally a map pinpoints the site at which the fossils have been found.

In science an animal is known by its scientific name, or its "binomial". For example, humans are scientifically known as *Homo sapiens*. *Homo* is the genus name and *sapiens* is the species name. The names, usually derived from Latin or classical Greek, are always italicized with only the genus name capitalized. For dinosaurs it is customary to use only the genus name in popular literature. *Tyrannosaurus rex*, however, is so evocative that often both are used. Once the genus name has been introduced, it can then be referred to by its genus initial along with its species name. Hence *T. rex*. In many instances a particular dinosaur genus has several species. This should help to explain the use of names in the encyclopedia.

For anyone interested in prehistoric life, this rich and colourful volume presents creatures that have never been illustrated before.

Below: Peteinosaurus, *one of the first pterosaurs from the Triassic.*

Left: Tanystropheus, *a bizarre long-necked Triassic reptile.*

Right: Proceratosaurus, *a small Jurassic carnivorous dinosaur.*

Below: Liopleurodon, *the biggest Jurassic pliosaur.*

THE GEOLOGICAL TIMESCALE

Geological time is an unbelievably massive concept to grasp. Millions of years, tens of millions of years, hundreds of millions of years. These unfathomable stretches of time are often referred to as "deep time". This is the scale that palaeontologists and anyone interested in dinosaurs must use.

When did dinosaurs appear? About 220,000,000 years ago. And when did they die out? About 65,000,000 years ago. It is easier when we say 220 million and 65 million, but we could use a better system.

To make the concept easier, geologists split geological time into named periods. It is the same when we talk about human history. We can say 150 years ago, or 200 years ago, or 600 years ago, but it gives a clearer idea of the time if we say Victorian London, or Napoleonic Europe or Pre-Columbian North America – then we can put events into their chronological context.

Below: The three periods of the Mesozoic era, the Triassic, the Jurassic and the Cretaceous, are the periods in which the dinosaurs lived. They evolved in the latter part of the Triassic and died out at the end of the Cretaceous. These periods are further divided into stages.

When geologists refer to different parts of a period, they talk about "upper" Cretaceous or "lower" Jurassic. This is a reference to the rock sequence in which the rocks formed. When we talk about the events that took place at these times, we use the terms "late" Cretaceous or "early" Jurassic instead.

Geological time periods are named after the rock sequences that were formed at that time, and the names were given by the scientists (mostly Victorian, about 150 years ago) who first studied them in the regions in which they typically outcrop.

Each period encompasses tens of millions of years, so is subdivided into stages for ease of reference. The stages, usually named after the places where they were first studied, are given at the bottom of these pages, along with the actual number of years that they entailed, so that they can be referred to when necessary. The stages are then divided into zones, usually named after a distinctive fossil, but these time divisions are too small to be of any interest to us here.

There are two ways in which geological events are dated. The first is "relative dating" – placing events on the geological time scale in relation to each other. This is the principle involved in most studies of the past. Fossil A is found in rocks that lie above those in which fossil B is found. That means that fossil B is older than

Above: A fossil forms when an animal dies and its body falls into sediment that is accumulating at the time. The body is buried and the organic matter of the hard parts is transformed into mineral at the same time as the sediment is transformed into sedimentary rock. The Pterosaur shown must have died while it was flying over a shallow lagoon in late Jurassic times. It sank to the bottom of the lagoon, where it was buried by contemporary sediment.

fossil A; in an undisturbed sequence of sedimentary rocks the oldest is always on the bottom. If a fossil is found in a rock on another continent from which that fossil is usually found, then the two rocks will be of the same age, even if there are no other clues to the age of the rocks. The fossil dates the rock.

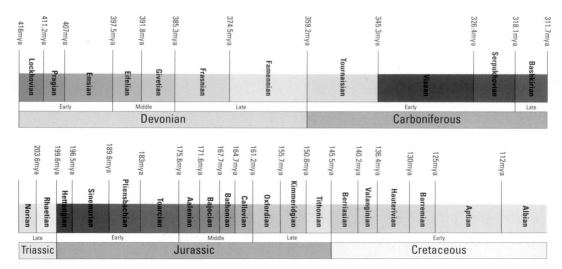

416mya	411.2mya	407mya		397.5mya	391.8mya	385.3mya		374.5mya		359.2mya		345.3mya		326.4mya	318.1mya	311.7mya
Lochkovian	Pragian	Emsian		Eifelian	Givetian	Frasnian		Famennian		Tournaisian		Viséan			Serpukhovian	Bashkirian
Early				Middle		Late						Early				Late
		Devonian										Carboniferous				

203.6mya	199.6mya	196.5mya		189.6mya		183mya		175.6mya	171.6mya	167.7mya	164.7mya	161.2mya	157.7mya	150.8mya		145.5mya	140.2mya	136.4mya	130mya	125mya		112mya	
Norian	Rhaetian	Hettangian	Sinemurian		Pliensbachian		Toarcian	Aalenian	Bajocian	Bathonian	Callovian	Oxfordian	Kimmeridgian	Tithonian		Berriasian	Valanginian	Hauterivian	Barremian		Aptian		Albian
Late		Early						Middle				Late				Early							
Triassic		Jurassic														Cretaceous							

The second type of geological dating is absolute dating. This is much more tricky, and involves studying the decay of radioactive minerals in a particular rock. A radioactive mineral breaks down at a particular known rate. If we can measure the amount of that mineral remaining, and compare it with the amount of what is called the "decay residue", we can tell how long it has been decaying and how long ago it formed. Several radioactive elements are used in this method.

One disconcerting aspect about geological time, however, is that the absolute dates keep changing. This is because the science used to determine them becomes increasingly sophisticated and precise with developments in technology and understanding. A century ago we were talking in terms of tens of millions of years, whereas nowadays the same periods are talked of in hundreds of millions of years. This is why dates may differ in various dinosaur books.

Above: Geological periods are defined by the fossils found in the sedimentary rocks formed at that particular time. Sedimentary rocks are those that are built up from layers of mud and sand, and have been compressed and solidified over time. Those shown here represent an angular unconformity between two rock formations: Triassic rocks are the horizontal ones lying above Devonian rocks, which are inclined at 40 degrees. These were laid down horizontally, but have been tilted by movements under the surface.

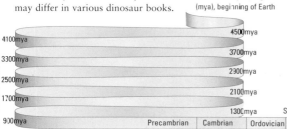

4,600 million years ago
(mya), beginning of Earth

Left: The age of the Earth is so immense that it can only be shown diagramatically in some kind of distorted image. The Earth's origin can be placed about 4,600 million years ago, but the part that really interests palaeontologists begins about 542 million years ago. It was at this point that animals first evolved hard shells and from then onwards their fossils have been easy to find. Since that time the geography of the Earth has changed, with the continents constantly moving to new positions.

4100mya
3300mya
2500mya
1700mya
900mya

4500mya
3700mya
2900mya
2100mya
1300mya

Devonian   Permian

Silurian   Carboniferous   Triassic

Precambrian | Cambrian | Ordovician | Jurassic | Cretaceous | Tertiary

Present

590mya   505mya   438mya 408mya 360mya   286mya 248mya 213mya   144mya   65mya

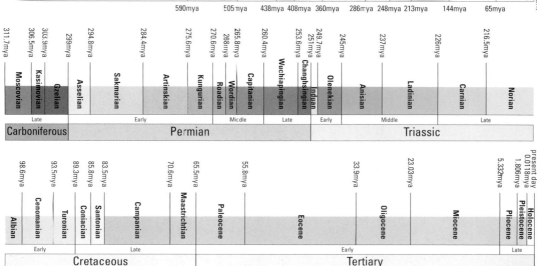

311.7mya	305.5mya	303.9mya	299mya	294.8mya	284.4mya	275.6mya	270.6mya	268mya	265.8mya	260.4mya	253.6mya	251mya	249.7mya	245mya	237mya	226mya	216.5mya
Moscovian	Kasimovian	Gzelian	Asselian	Sakmarian	Artinskian	Kungurian	Roadian	Wordian	Capitanian	Wuchiapingian	Changhsingian	Induan	Olenekian	Anisian	Ladinian	Carnian	Norian
Late				Early			Middle			Late		Early		Middle		Late	
Carboniferous			Permian									Triassic					

98.6mya	93.5mya	89.3mya	85.8mya	83.5mya	70.6mya	65.5mya	55.8mya	33.9mya	23.03mya	5.332mya	1.806mya	present day 0.0118mya
Albian	Cenomanian	Turonian	Coniacian	Santonian	Campanian	Maastrichtian	Paleocene	Eocene	Oligocene	Miocene	Pliocene	Pleistocene / Holocene
Early					Late		Early					Late
Cretaceous							Tertiary					

EARLY EVOLUTION

Where did life come from? We are not quite sure, but it seems that living things of one kind or another have been around since the Earth was cool enough to have liquid water on its surface. The process of evolution has meant that there has been an uninterrupted stream of living creatures ever since then.

What is life? There are several definitions, but each agrees that a living thing absorbs materials and energy, grows and reproduces. The tiniest bacteria and single cells conform to this definition, and these are the living things that existed way back when the Earth had just begun to cool.

By far the greatest part of Earth's history is encompassed by Precambrian time, but there is little direct evidence about what living things were like then. Bacteria and single-celled organisms do not leave much in the way of fossils. However, we have indirect evidence that things lived then, and gradually evolved into soft-bodied, multi-cellular creatures during that period. This vast span of time is called the Cryptozoic, meaning "the time of hidden life". The end of the Precambrian period (542 million years ago) and the beginning of the fossil record proper is usually marked by an event called the "Cambrian explosion".

At this time, the beginning of the Cambrian period, evolution perfected the hard shell. Organisms had the ability to absorb the mineral calcite from the seawater and lay it down as a living shell, or from organic compounds they built up a kind of natural plastic called chitin – the material from which our fingernails are made. This had two results. First it meant that there was suddenly a kind of evolutionary arms race. Animals had always been hunting and eating one another. Now some animals could defend themselves, and consequently the hunters evolved new structures and techniques to get the upper hand.

Evolution in the ocean

Suddenly the oceans (for all life was in the oceans at this time) were full of all kinds of creatures that had not existed before. And what strange beasts they were! There were animals with many legs or none, with shelled heads, with shelled tails, with spikes, and with burrowing tools – it was as if nature was trying out anything just to see what worked. By the end of the Cambrian period this vast array of strange beings had whittled itself down to a dozen or so well-tried evolutionary lines that have

continued until the present day. The second result of evolution producing hard shells was based on the fact that hard-shelled animals leave good fossils. The history of life from that point forward is well documented, which is why the time span from the Cambrian to the present day is called the Phanerozoic (meaning "obvious life").

One of the surviving strands of life consisted of worm-like animals with a nervous system running down their length, supported by a jointed framework. The brain was at the front, protected by a box. The mouth and sensory organs were also at the front. From simple animals like this evolved the first vertebrates, the first animals with backbones.

The first vertebrates

Fish were the first vertebrates that we would recognize, and they came to prominence in Ordovician and Silurian times (488–416 million years ago).

The first fish are known as the "jawless fish". Rather like the modern

Below: Evidence of life in the Precambrian period (more than 542 million years ago) is vague. However, what we do know is that all the major evolutionary lines of living things had evolved by the dawn of the Cambrian

period (542 million years ago) and were leaving their imprint on the Earth, as well-preserved fossils.
The vast majority of animals at this time, both living and fossil, are invertebrates.

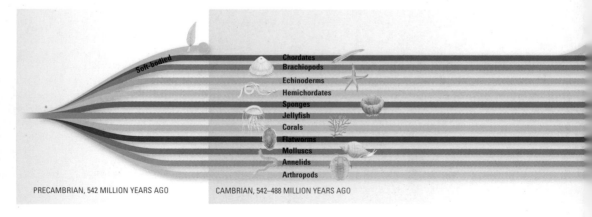

Soft-bodied

Chordates
Brachiopods
Echinoderms
Hemichordates
Sponges
Jellyfish
Corals
Flatworms
Molluscs
Annelids
Arthropods

PRECAMBRIAN, 542 MILLION YEARS AGO   CAMBRIAN, 542–488 MILLION YEARS AGO

lamprey, they had a sucker instead of jaws, and they probably lived by sucking up nutritious debris from the seafloor as they swam along. A fin along the underside of the tail ensured that they swam head down. Proper jaws and a more organized skeleton then developed. The first skeletons were not made of bone but of cartilage. The cartilaginous fish are represented today by the sharks and rays. They appeared in Silurian times.

The next stage was the evolution of bone around the cartilage framework. Bone formed the skeleton and also the armour plates for protection. Then came the kinds of scales we would recognize from the fish we see today. These bony, armoured fish, and the scaly fish, appeared in the Devonian period. By this time there were so many different fish that the Devonian period (416–359 million years ago) is often termed the "Age of Fishes".

A changing environment

Meanwhile, changes were taking place out of the water. In early Precambrian times the atmosphere was a bitter mix of noxious gases, which is why all early life evolved in the sea. Gradually the by-products of early living systems were seeping into the atmosphere and changing it. Oxygen, a product of the photosynthesis process, which keeps plants alive,

started to build up in the atmosphere and make land habitable. The first green patches of land appeared between tides probably during Silurian times. When plants pioneered life on land, animals followed behind. One kind of fish developed lungs to enable it to breathe the oxygen of the atmosphere. It also developed paired muscular fins that would allow it to crawl on a solid surface as well as swim in the water. The vertebrates were poised to take a step on to the land. As soon as the continents became habitable, life spread there from the oceans, and a vast array of animals have been present ever since.

Life on land

Creatures had been venturing out on to land for hundreds of millions of years. Tracks of arthropods (the first land-living animals) are known from beach sediments formed in Ordovician times. There are mysterious marks from dry land deposits that look like motor-cycle tracks back in the Cambrian period. They seem to have been tentative explorations, but it does not appear that animal life out of the water was permanent until plants had gained a foothold. Insects and spiders infested the primitive early plants that clothed the sides of streams in the Silurian period. The first fish ventured out in the subsequent Devonian time.

It is not clear why fish first appeared on land. Some scientists say that the newly evolved arthropod fauna that had established itself among the plants was too tempting a food source to be ignored. Others suggest that land-living was an emergency measure – if a fish became trapped in a drying puddle of water it would need to be able to survive and travel over land to find more water in which it could live. There is also a theory that the waters became too dangerous due to predatory animals; there were clawed arthropods as big as alligators at the time, and some fish found it safer to take up a land-living existence.

Tiktaalik was typical of the kind of fish that was able to spend time on land. The major adaptation was the lung. Fish normally breathe through gills – feathery structures that can filter dissolved oxygen from the water. Now lungs enabled oxygen to be extracted directly from the air. Then there was the manner of

Below: By Ordovician (488–444 million years ago) and Silurian (444–416 million years ago) times the backboned animals had evolved, in the form of the most primitive fish. The backbone supported the whole body, the limbs were arranged in pairs at the side, and the brain was encased in a box of bone. The next stage came when these swimming animals evolved to be able to breathe air.

Jawless fish

Cartilagenous fish

ORDOVICIAN, 488–444 MILLION YEARS AGO

SILURIAN, 444–416 MILLION YEARS AGO

locomotion. A typical fish's fin consists of a ray of supports with a web between, spreading from a muscular stump. In *Tiktaolik* and its relatives the fins consisted of muscular lobes, supported by a network of bones, with the fin material forming mere fringes along the edge. Two pairs were arranged on the underside of the body, and they could be used both for swimming and for heaving the animal across open land.

The first amphibians

By the end of the Devonian period, the next stage in the evolution of land vertebrates had been accomplished, and the first amphibian-like animals appeared. These animals were much more complex in their variety and relationships than the single term "amphibian" implies. *Ichthyostega* was one of the earliest of these animals. The difference between *Ichthyostega* and the lobe-finned fish was in the limbs. Now the fins were clearly jointed, with leg and toe bones. It seems likely that they evolved for pulling the animal along through weeds in shallow water, but they were ideal for clambering on land as well. The *Ichthyostega* foot was odd by

Below: Many of the evolutionary lines that existed at the start of the Precambrian continued to evolve. Some evolutionary lines, such as the armoured fish, ceased to exist. Other lines split with new creatures evolving and beginning new evolutionary lines.

modern standards because there were eight toes. The standard arrangement of a maximum of five toes for a land-living vertebrate had yet to be established. For all its land-living abilities, *Ichthyostega* and its relatives still had the head and tail of a fish, and had to return to the water to breed.

The next great advance in evolution was the ability of animals to breed on land. This was achieved by the first of the amniotes, named after the amnion – the membrane that contained the developing young within the egg. A hard-shelled egg evolved, that nourished the young in what was essentially a self-contained watery pond, that could be laid away from the water. At last the link with the ancestral seas was severed. Early examples include *Westlothiana* from Scotland and *Hylonomus* from Nova Scotia, both dating from the early Carboniferous period.

The first reptiles

The true reptiles then established themselves along a number of evolutionary branches. In the simplest form of classification they can be classified by the number and arrangement of holes in the skull behind the eye socket. The anapsids had no such holes because the skull was a solid roof of bone behind the

eye. The anapsids were prominent in the Permian period in a group called the pareiasaurs. Modern relatives of pareiasaurs include tortoises and turtles.

The synapsids, however, had a single hole in the skull at each side. They became the mammal-like reptiles, the major group of the Permian period. When they died away in Triassic times they lived on as the humble mammals, and did not really come to prominence again until the Tertiary period.

The diapsids were different because they had two holes behind the eye. Modern diapsids include snakes, crocodiles and lizards. However, one group of Mesozoic (the combined Triassic, Jurassic and Cretaceous periods) diapsids was much more important than the others. They were the dinosaurs.

Dinosaur evolution

The dinosaurs evolved from the diapsid line that we call the archosaurs, meaning the ruling reptiles. Other archosaurs were the pterosaurs, and the crocodiles and alligators that we have today. A typical Triassic archosaur was a swift, two-footed, running meat-eater, usually no bigger than a wolf and usually much smaller. In fact an advanced archosaur would have looked very much like a typical,

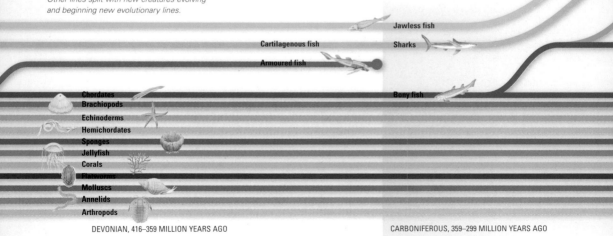

Jawless fish

Cartilagenous fish

Sharks

Armoured fish

Chordates
Brachiopods
Echinoderms
Hemichordates
Sponges
Jellyfish
Corals
Flatworms
Molluscs
Annelids
Arthropods

Bony fish

DEVONIAN, 416–359 MILLION YEARS AGO

CARBONIFEROUS, 359–299 MILLION YEARS AGO

small, meat-eating dinosaur. What made the dinosaur different from its archosaurian ancestor lay mostly in the structure of the leg and hips.

Most reptiles have legs that stick out at the side, with the weight of the animal slung between them. This gives the animal a sprawling gait. To enable it to run quickly it must throw its body into S-shaped curves to give the sideways-pointing limbs the reach that is needed. In contrast, a dinosaur's leg was straight and vertical. It was plugged into the side of the hip where it was held in place by a shelf of bone.

This meant that a dinosaur's weight was at the top of the leg, and transmitted straight downwards. Vertical limbs can support a greater weight than sprawling limbs. This is the arrangement that we see in a typical mammal, and this upright stance is seen in all dinosaurs, whether two- or four-legged.

Saurischians and ornithischians

So, the first dinosaur was probably like an archosaur, and good at running. From there dinosaur evolution

diverged into two main lines – the saurischia and the ornithischia. The difference between the two lines lies in the structure of the hips.

The saurischia had hip bones arranged like those of a lizard, a structure that radiated from the leg bone socket, with a pubis bone that pointed down and forward. This line is further divided into two groups; the first group developed along the evolutionary line pioneered by the earlier archosaurs, the two-footed hunters. They were termed theropods

Below: As the fish developed into more complex forms, some became land dwellers, with jointed limbs and lungs able to breathe the air. These became the first amphibians. From them, evolved animals able to live on land all the time, without resorting to water at any stage in their growth. The reptiles, with their hard-shelled eggs, represented this stage, and they diversified into all kinds of land-living types.

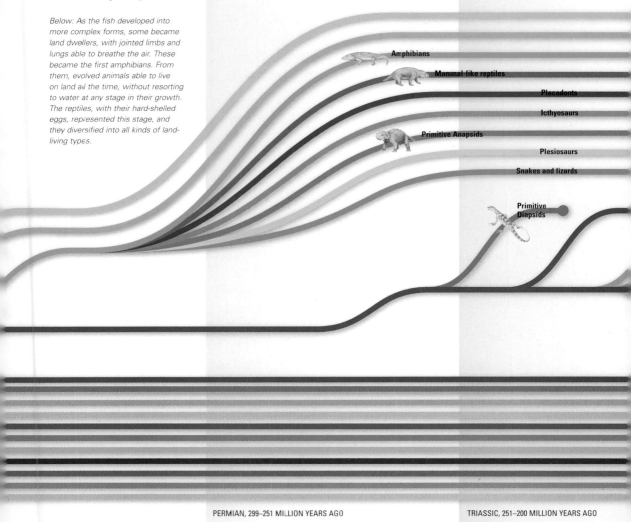

Amphibians

Mammal-like reptiles

Placadonts

Icthyosaurs

Primitive Anapsids

Plesiosaurs

Snakes and lizards

Primitive Diapsids

PERMIAN, 299–251 MILLION YEARS AGO

TRIASSIC, 251–200 MILLION YEARS AGO

or "beast-footed", by the Victorian scientists who detected a similarity between their foot bones and the bones of mammals. All the meat-eating dinosaurs were theropods, from small chicken-size scampering insect-eaters, to massive 15m- (50ft-) long beasts.

The other saurischian group are the sauropods, meaning "lizard-footed", and so called because of the similarity in the structure of the foot to that of a modern lizard. They were the huge, long-necked plant-eaters of the dinosaur world, and their body shape evolved in response to a changing vegetarian diet. The shape of the saurischian hip, with its forward-pointing pubis bone, meant that the big digestive system of a plant-eater had to be carried in front of the hind legs. The result is an animal that would be unable to walk solely on its hind legs, and in response the smaller front legs became stronger to take the weight. This development reduced the mobility of the animal, and so a long neck developed to enable it to reach enough food. As a group of dinosaurs sauropods encompass the biggest land animals that ever existed.

Below: The dinosaurs, once they evolved, soon developed into a number of distinctive groups. Some were meat-eaters, others were plant-eaters. Some moved on four legs and others on two. They were the most significant land animals of the time – between the late Triassic and the end of the Cretaceous. However, at the end of the Cretaceous they, and many other animal groups, became extinct.

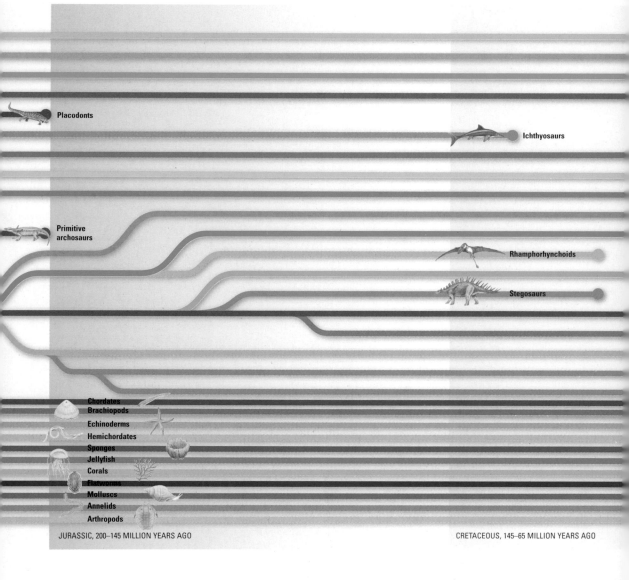

Placodonts

Ichthyosaurs

Primitive archosaurs

Rhamphorhynchoids

Stegosaurs

Chordates
Brachiopods
Echinoderms
Hemichordates
Sponges
Jellyfish
Corals
Flatworms
Molluscs
Annelids
Arthropods

JURASSIC, 200–145 MILLION YEARS AGO

CRETACEOUS, 145–65 MILLION YEARS AGO

The second line of dinosaurs was the ornithischians. They were plant-eaters but had a different arrangement of hip bones. The pubis bone was swept back and lay along the backward-facing ischium bone. This meant the typical ornithischian could carry the weight of its body beneath the hips, and so it could still walk on its hind legs balanced by the heavy tail. A typical two-footed ornithischian was the ornithopod, the bird-footed dinosaur with three splayed toes.

The sauropods and ornithopods also had different eating methods. The sauropods could not chew their food; they had to eat so much that they would not have had time to. Their teeth showed that they raked leaves and twigs from the branches and swallowed what they took without processing it. In contrast, ornithopods had teeth that could chew food thoroughly before swallowing it.

Other developments from the basic ornithopod involved the development

of armour. The stegosaurs had plates, the ankylosaurs had armour, and the ceratopsians had horns.

Mammals

After 160 million years, the dinosaurs became extinct, but not before the theropods gave rise to the birds. The way was open for the mammals. Since the end of the Cretaceous, the mammals have expanded and occupied all ecological niches once occupied by the dinosaurs.

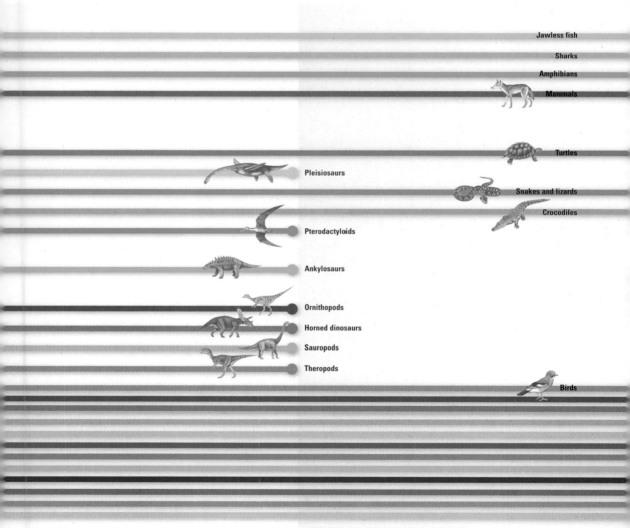

Jawless fish
Sharks
Amphibians
Mammals
Turtles
Pleisiosaurs
Snakes and lizards
Crocodiles
Pterodactyloids
Ankylosaurs
Ornithopods
Horned dinosaurs
Sauropods
Theropods
Birds

TERTIARY, 65 MILLION YEARS AGO–PRESENT DAY

REPTILE CLASSIFICATION

The reptiles appear to have evolved along three major lines, distinguished from one another by the arrangement of holes in the skulls. Scientists have found it difficult to classify the early amphibians – their fossils are too rare and scattered to give a coherent picture of their evolution.

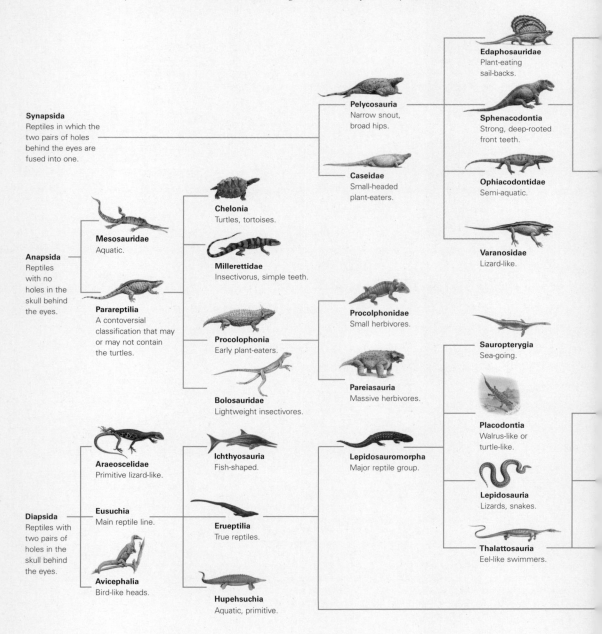

Synapsida
Reptiles in which the two pairs of holes behind the eyes are fused into one.

Pelycosauria
Narrow snout, broad hips.

Edaphosauridae
Plant-eating sail-backs.

Sphenacodontia
Strong, deep-rooted front teeth.

Caseidae
Small-headed plant-eaters.

Ophiacodontidae
Semi-aquatic.

Varanosidae
Lizard-like.

Anapsida
Reptiles with no holes in the skull behind the eyes.

Mesosauridae
Aquatic.

Parareptilia
A contoversial classification that may or may not contain the turtles.

Chelonia
Turtles, tortoises.

Millerettidae
Insectivorus, simple teeth.

Procolophonia
Early plant-eaters.

Bolosauridae
Lightweight insectivores.

Procolophonidae
Small herbivores.

Pareiasauria
Massive herbivores.

Diapsida
Reptiles with two pairs of holes in the skull behind the eyes.

Araeoscelidae
Primitive lizard-like.

Eusuchia
Main reptile line.

Avicephalia
Bird-like heads.

Ichthyosauria
Fish-shaped.

Erueptilia
True reptiles.

Hupehsuchia
Aquatic, primitive.

Lepidosauromorpha
Major reptile group.

Sauropterygia
Sea-going.

Placodontia
Walrus-like or turtle-like.

Lepidosauria
Lizards, snakes.

Thalattosauria
Eel-like swimmers.

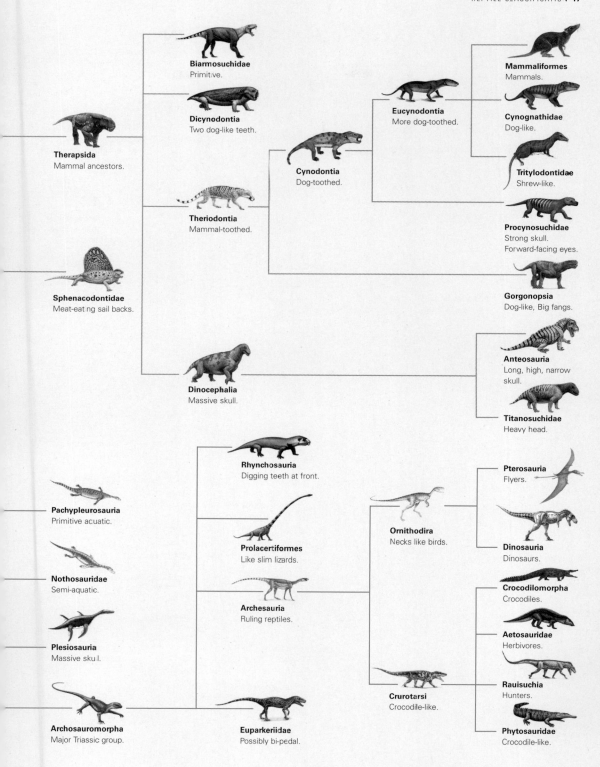

Biarmosuchidae
Primitive.

Dicynodontia
Two dog-like teeth.

Therapsida
Mammal ancestors.

Theriodontia
Mammal-toothed.

Mammaliformes
Mammals.

Eucynodontia
More dog-toothed.

Cynognathidae
Dog-like.

Cynodontia
Dog-toothed.

Tritylodontidae
Shrew-like.

Procynosuchidae
Strong skull.
Forward-facing eyes.

Sphenacodontidae
Meat-eating sail backs.

Gorgonopsia
Dog-like, Big fangs.

Anteosauria
Long, high, narrow
skull.

Dinocephalia
Massive skull.

Titanosuchidae
Heavy head.

Rhynchosauria
Digging teeth at front.

Pachypleurosauria
Primitive acuatic.

Prolacertiformes
Like slim lizards.

Pterosauria
Flyers.

Ornithodira
Necks like birds.

Dinosauria
Dinosaurs.

Nothosauridae
Semi-aquatic.

Archesauria
Ruling reptiles.

Crocodilomorpha
Crocodiles.

Plesiosauria
Massive skull.

Aetosauridae
Herbivores.

Rauisuchia
Hunters.

Crurotarsi
Crocodile-like.

Archosauromorpha
Major Triassic group.

Euparkeriidae
Possibly bi-pedal.

Phytosauridae
Crocodile-like.

DINOSAUR CLASSIFICATION

The various dinosaurs evolved from common ancestors – in technical terms they were "monophyletic".
Early in their evolution they split into two major evolutionary lines, and these in turn split into a number
of different families, each with its own character and specialization.

The dinosaurs fall into two major groups –
the saurischians and the ornithischians. The
saurischians are divided into the plant-eating
sauropodomorphs and the meat-eating
theropods, while the latter are divided into a
number of different plant-eating types. Note
that the formal classifications (e.g. Theropoda)
are used interchangeably with the less formal
(e.g. theropods) throughout the book. This is
customary in palaeontology.

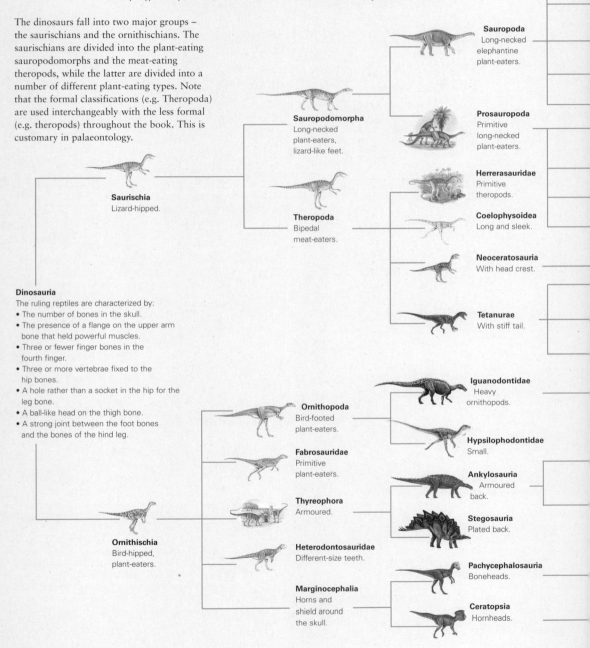

Sauropoda
Long-necked
elephantine
plant-eaters.

Sauropodomorpha
Long-necked
plant-eaters,
lizard-like feet.

Prosauropoda
Primitive
long-necked
plant-eaters.

Saurischia
Lizard-hipped.

Herrerasauridae
Primitive
theropods.

Theropoda
Bipedal
meat-eaters.

Coelophysoidea
Long and sleek.

Neoceratosauria
With head crest.

Dinosauria
The ruling reptiles are characterized by:
• The number of bones in the skull.
• The presence of a flange on the upper arm
  bone that held powerful muscles.
• Three or fewer finger bones in the
  fourth finger.
• Three or more vertebrae fixed to the
  hip bones.
• A hole rather than a socket in the hip for the
  leg bone.
• A ball-like head on the thigh bone.
• A strong joint between the foot bones
  and the bones of the hind leg.

Tetanurae
With stiff tail.

Iguanodontidae
Heavy
ornithopods.

Ornithopoda
Bird-footed
plant-eaters.

Hypsilophodontidae
Small.

Fabrosauridae
Primitive
plant-eaters.

Ankylosauria
Armoured
back.

Thyreophora
Armoured.

Stegosauria
Plated back.

Ornithischia
Bird-hipped,
plant-eaters.

Heterodontosauridae
Different-size teeth.

Pachycephalosauria
Boneheads.

Marginocephalia
Horns and
shield around
the skull.

Ceratopsia
Hornheads.

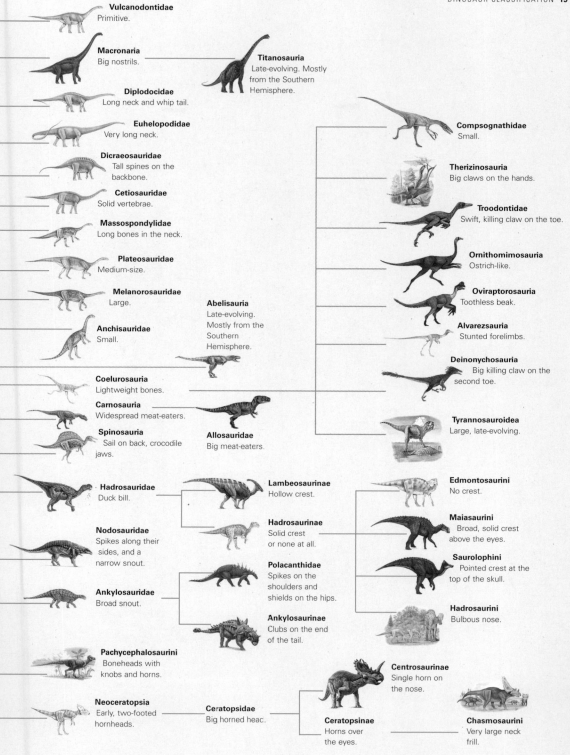

Vulcanodontidae
Primitive.

Macronaria
Big nostrils.

Titanosauria
Late-evolving. Mostly
from the Southern
Hemisphere.

Diplodocidae
Long neck and whip tail.

Euhelopodidae
Very long neck.

Compsognathidae
Small.

Dicraeosauridae
Tall spines on the
backbone.

Therizinosauria
Big claws on the hands.

Cetiosauridae
Solid vertebrae.

Troodontidae
Swift, killing claw on the toe.

Massospondylidae
Long bones in the neck.

Ornithomimosauria
Ostrich-like.

Plateosauridae
Medium-size.

Oviraptorosauria
Toothless beak.

Melanorosauridae
Large.

Abelisauria
Late-evolving.
Mostly from the
Southern
Hemisphere.

Alvarezsauria
Stunted forelimbs.

Anchisauridae
Small.

Deinonychosauria
Big killing claw on the
second toe.

Coelurosauria
Lightweight bones.

Carnosauria
Widespread meat-eaters.

Tyrannosauroidea
Large, late-evolving.

Spinosauria
Sail on back, crocodile
jaws.

Allosauridae
Big meat-eaters.

Hadrosauridae
Duck bill.

Lambeosaurinae
Hollow crest.

Edmontosaurini
No crest.

Nodosauridae
Spikes along their
sides, and a
narrow snout.

Hadrosaurinae
Solid crest
or none at all.

Maiasaurini
Broad, solid crest
above the eyes.

Saurolophini
Pointed crest at the
top of the skull.

Polacanthidae
Spikes on the
shoulders and
shields on the hips.

Ankylosauridae
Broad snout.

Ankylosaurinae
Clubs on the end
of the tail.

Hadrosaurini
Bulbous nose.

Pachycephalosaurini
Boneheads with
knobs and horns.

Centrosaurinae
Single horn on
the nose.

Neoceratopsia
Early, two-footed
hornheads.

Ceratopsidae
Big horned head.

Ceratopsinae
Horns over
the eyes.

Chasmosaurini
Very large neck
frill.

BIRD AND MAMMAL CLASSIFICATION

In Tertiary and Quaternary times it has been the birds and the mammals that have represented the main groups of tetrapods. There are several opinions about the classification of the mammals, and the diagram that is shown here represents a workable compromise.

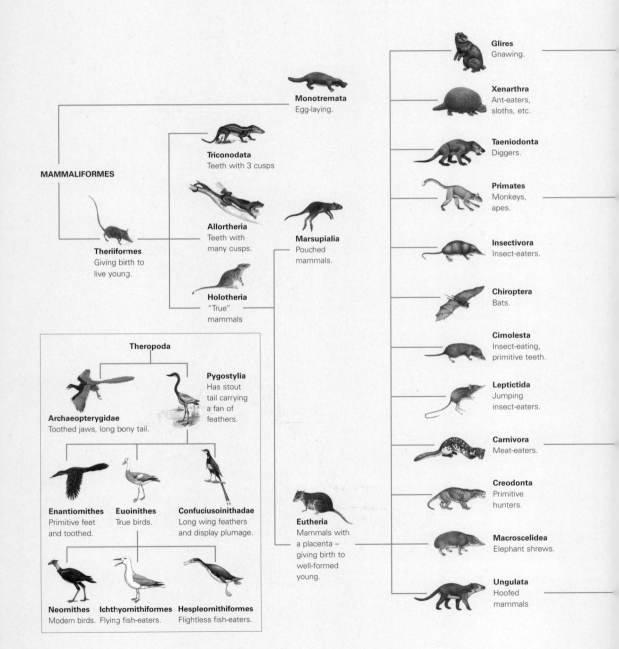

Glires
Gnawing.

Xenarthra
Ant-eaters, sloths, etc.

Taeniodonta
Diggers.

Primates
Monkeys, apes.

Insectivora
Insect-eaters.

Chiroptera
Bats.

Cimolesta
Insect-eating, primitive teeth.

Leptictida
Jumping insect-eaters.

Carnivora
Meat-eaters.

Creodonta
Primitive hunters.

Macroscelidea
Elephant shrews.

Ungulata
Hoofed mammals

Monotremata
Egg-laying.

Triconodata
Teeth with 3 cusps

MAMMALIFORMES

Allortheria
Teeth with many cusps.

Marsupialia
Pouched mammals.

Theriiformes
Giving birth to live young.

Holotheria
"True" mammals

Eutheria
Mammals with a placenta – giving birth to well-formed young.

Theropoda

Pygostylia
Has stout tail carrying a fan of feathers.

Archaeopterygidae
Toothed jaws, long bony tail.

Enantiornithes
Primitive feet and toothed.

Euoinithes
True birds.

Confuciusoinithadae
Long wing feathers and display plumage.

Neornithes
Modern birds.

Ichthyornithiformes
Flying fish-eaters.

Hespleornithiformes
Flightless fish-eaters.

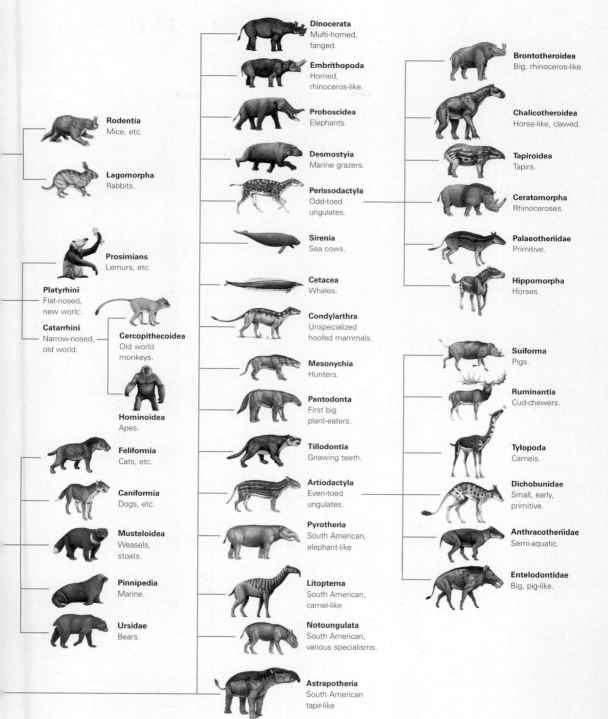

Rodentia
Mice, etc.

Lagomorpha
Rabbits.

Prosimians
Lemurs, etc.

Platyrhini
Flat-nosed,
new world.

Catarrhini
Narrow-nosed,
old world.

Cercopithecoidea
Old world
monkeys.

Hominoidea
Apes.

Feliformia
Cats, etc.

Caniformia
Dogs, etc.

Musteloidea
Weasels,
stoats.

Pinnipedia
Marine.

Ursidae
Bears.

Dinocerata
Multi-horned,
fanged.

Embrithopoda
Horned,
rhinoceros-like.

Proboscidea
Elephants.

Desmostyia
Marine grazers.

Perissodactyla
Odd-toed
ungulates.

Sirenia
Sea cows.

Cetacea
Whales.

Condylarthra
Unspecialized
hoofed mammals.

Mesonychia
Hunters.

Pantodonta
First big
plant-eaters.

Tillodontia
Gnawing teeth.

Artiodactyla
Even-toed
ungulates.

Pyrotheria
South American,
elephant-like

Litopterna
South American,
camel-like

Notoungulata
South American,
various specialisms.

Astrapotheria
South American
tapir-like

Brontotheroidea
Big, rhinoceros-like.

Chalicotheroidea
Horse-like, clawed.

Tapiroidea
Tapirs.

Ceratomorpha
Rhinoceroses.

Palaeotheriidae
Primitive.

Hippomorpha
Horses.

Suiforma
Pigs.

Ruminantia
Cud-chewers.

Tylopoda
Camels.

Dichobunidae
Small, early,
primitive.

Anthracotheriidae
Semi-aquatic.

Entelodontidae
Big, pig-like.

THE ANATOMY OF A DINOSAUR

Of all fossil animals, it is the dinosaurs that are the most popular and attract the greatest public attention.
Here we use the dissection of a typical dinosaur as an introduction to the anatomy of tetrapods in general.
The most basic dinosaur was the theropod meat-eater.

The theropods appeared at the beginning of the Age of Dinosaurs, in the latter part of the Triassic period. In appearance the early theropods would have resembled their thecodont ancestors. The thecodonts are the reptile group that had teeth in sockets, rather than in grooves as lizards had. The main differences would have been in the stance of the legs and the structure of the skull. The thecodonts had been active hunters, and the early theropods carried on this tradition.

A theropod is an ideal shape for a hunter. The head, jaws and teeth are carried well forward, and are the first part of the animal to make contact with its prey. The arms and the claws are also well forward. The body is quite small, as befits a meat-eating animal. The legs are powerful, with strong muscles working against the bones of the lizard-like hips. The tail is big and heavy, used for balancing and keeping the upper body well forward.

We know that at least some of the theropods were warm-blooded. The term warm-blooded does not necessarily refer to the temperature of the blood, but instead implies that a mechanism exists that keeps the animal's body at the same temperature regardless of the temperature of its surroundings. Nowadays warm-bloodedness is found in mammals and birds, the latter being direct descendants of the theropods.

The dinosaur group

The theropods were not the only dinosaurs. Their close relatives were the sauropods – the long-necked plant-eaters. These originated as animals

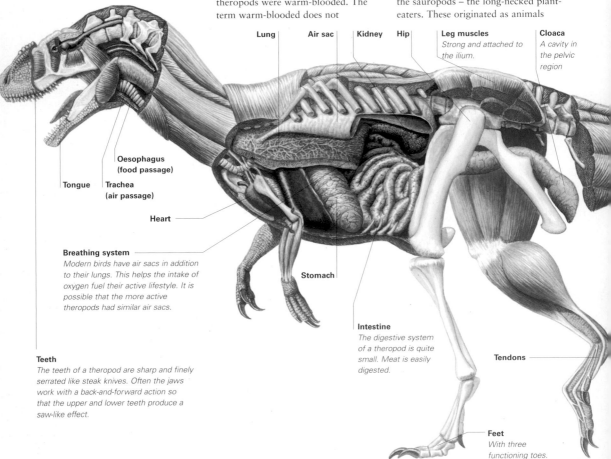

Lung

Air sac

Kidney

Hip

Leg muscles
Strong and attached to the ilium.

Cloaca
A cavity in the pelvic region

Oesophagus
(food passage)

Tongue

Trachea
(air passage)

Heart

Breathing system
Modern birds have air sacs in addition to their lungs. This helps the intake of oxygen fuel their active lifestyle. It is possible that the more active theropods had similar air sacs.

Stomach

Teeth
The teeth of a theropod are sharp and finely serrated like steak knives. Often the jaws work with a back-and-forward action so that the upper and lower teeth produce a saw-like effect.

Intestine
The digestive system of a theropod is quite small. Meat is easily digested.

Tendons

Feet
With three functioning toes.

very similar to the theropods but with much more sophisticated digestive systems, as would be required by a diet of plants. This much larger digestive system had to be carried well forward of the hips and so the animal was too unbalanced to walk on its hind legs, and resorted to a four-footed stance. The long neck then evolved allowing the animal to reach enough food to fuel the big slow-moving body.

The theropods and sauropods together make up the saurischians, or lizard-hipped dinosaurs – so-called because of the hip bones.

The other major dinosaur group are the ornithischians, or bird-hipped dinosaurs. In these the pubis bone – the hip bone that sticks down and forward in the theropod – now sweeps back along the ischium. This makes for a big space beneath the hips – a space that can be occupied by a big plant-eating gut. The ornithischians were all plant-eaters, and many were two-footed, being able to balance on their hind legs with their heavy gut beneath their hips instead of in front. Other ornithopods were covered in armour of

one sort or another, or had heavy horn-covered heads, and these types reverted to the four-footed stance to support the extra weight.

Ornithischians

The second major group of dinosaurs were those with the bird-like hip structure. These were the ornithischians, and they were all plant-eaters. The smallest were two-footed, but the bigger ones had armour and went about on all fours.

Above: Sphaerotholus *shows a number of ornithopod features – an armoured head and the ability to walk on hind legs.*

Sauropods

The sauropods were closely related to the theropods, in that they both had the lizard-shaped hip bone structure. However they were plant-eaters. They went on all fours and had heavy bodies, long necks and tiny heads.

Right: Brachiosaurus *was a typical sauropod, and one of the biggest. The strong legs held directly beneath the body helped to support the great weight.*

Skin covering *Usually scaly, but sometimes feathered in small types of dinosaur.*

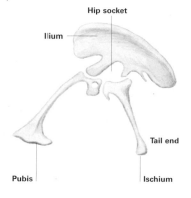

Hip socket

Ilium

Tail end

Pubis

Ischium

Hip *(left) A theropod meat-eater had the saurischian hip – with the arrangement of bones like those of a lizard. At each side there was a top bone called the ilium, which held the leg muscles. Pointing forward and down from the hip socket was the pubis, which took the weight of the animal as it lay down. Sweeping down and back was the ischium, to which the tail muscles and the inside leg muscles were attached.*

Skull *(above) A dinosaur's skull is a lightly-built framework of struts of bone. The teeth constantly grow and drop out, each one being replaced by another that is growing beneath it.*

COPROLITES AND DIET

We have plenty of information that shows what extinct animals ate and how they ate it. The evidence
includes close examination of their fossilized jaws and teeth, fossilized stomach contents and also
of fossil dung, known as coprolites.

We know that the carnivorous mammals have a particular kind of tooth and jaw apparatus for shearing meat. The teeth of plant-eaters are designed for raking leaves from trees, chewing, chopping and munching, and processing plant food before swallowing it. Teeth and jaws represent physical evidence of the type of diet that extinct animals had. But there are other lines of investigation.

Palaeontologists describe fossil dung as "coprolite". A coprolite can be a very valuable tool to work out the diet of something that is dead and fossilized. Coprolites of land-living animals are rare. Most coprolites we find come from marine animals, such as fish. As with all other fossils, marine conditions provide a far better preservation medium than any land habitat. A piece of dung deposited on land will be trampled on, decomposed or eaten by bacteria, fungi or other organisms. In fact, one of the ways of identifying a structure as a coprolite of a land-living animal is by the presence of dung-beetle burrows through it. Dung is a remarkably nutritious substance to certain creatures. Yet for all that, there have been a fair number of dinosaur and mammal coprolites

Above: A Tyrannosaurus *upper jaw.*

discovered, and they give a good insight into their diets.

Theropod coprolites seem to be more common than those of plant-eating dinosaurs. This is probably because they contain a high proportion of bone material that makes them more robust than the stuff produced by a plant-eater. Studies have been done to try to determine what that bone material might be. The chemistry of a coprolite reflects the chemistry of the meal, and chemical studies on coprolites from a *Tyrannosaurus* or one of its relatives suggest a diet predominantly of duckbilled ornithopods.

The coprolites of herbivorous dinosaurs are more problematic. They are very difficult to identify and, when they are found, it is almost impossible to determine what dinosaur produced them. Coprolites in Jurassic rocks in Yorkshire, England, are identified as dinosaur droppings simply because of their size – no other plant-eating animals about at that time could produce such a volume of dung. These coprolites consist of a mass of pellets like deer droppings, and contain the partly digested remains of cycad-like

plants. Coprolites have been found close to duckbill nesting sites in Montana, and they contain shredded conifer stem material. Duckbills had powerful enough teeth to allow them to chew woody twigs to extract the nourishment. Grass structures in sauropod coprolites from India show that grass existed in late Cretaceous times, much earlier than first thought.

Cololites are similar to coprolites.

Below: Some dinosaurs swallowed stomach stones to help grind up food once in the stomach. These polished stones would have been regurgitated and swapped for stones with a sharper edge.

Below: Fossilized dung, shaped like huge droppings, can provide a wealth of information about what extinct animals ate.

These are fossilized stomach contents that we are sometimes lucky enough to find in the body cavity of a fully preserved dinosaur skeleton. The best examples of stomach contents are those that consist of a meal that has not been digested. The excellent skeleton of *Compsognathus* found in the lagoonal limestones of Solnhöfen, in southern Germany, contains the bones of a little lizard in its stomach area, its last meal. The skeleton of one of the *Coelophysis* that died of hunger as a water hole dried up contains the bones of a small crocodile.

Most cololites are in less good condition. A cololite found in the skeleton of the Australian ankylosaur *Minmi* contains the seeds of flowering plants and the spores of ferns. The concentration of seeds suggests that the ankylosaur went for those in particular, and may have been important in spreading the seeds around through its faeces. The chopped up state of the material indicates that it had been well chewed before being swallowed, more proof that these animals

Right: In extreme times dinosaurs are known to have eaten their own young in an attempt to survive, as observed in Majungotholus.

had cheeks. The skeleton of badger-sized Cretaceous mammal *Repenomamus* from China has the bones of young dinosaurs in its stomach region.

Some stomach contents consist of the stomach stones swallowed by sauropods. In the flood plain deposits of the North American Jurassic Morrison Formation there are little heaps of polished stones.

They have been identified as stomach stones that became too worn and smooth to be of use. The animal vomited them up and replaced them with sharper stones.

More indirect evidence of diet comes from tooth marks. *Allosaurus* teeth marks have been found on the bones of Jurassic sauropods. Broken and shed *Allosaurus* teeth have also even been found near such remains. Marks that match *Tyrannosaurus* teeth have been found on the bones of *Triceratops*. And *Tyrannosaurus* teeth have even been found wedged in the bones of duckbilled *Hypacrosaurus*. There can hardly be better indications of feeding behaviour.

Below: Coelophysis was once thought to have been a cannibal, as two well-preserved adult skeletons seemed to have the bones of youngsters in their stomachs. However, recent studies show that in one specimen the bones are those of a small crocodile, and in the other the baby is lying beneath the adult's rib cage, not within it.

FOSSIL FOOTPRINTS

Footprints are the best tools we have to tell how an animal lived. The study of footprints is known as ichnology, and the study of fossil footprints is known as palaeoichnology. The science involves an understanding of how sediments and animals behave.

A dead animal may have left a skeleton, but in all likelihood there will only be a few bones left for us to study once the rest of it has been eaten or eroded. It is even more likely that there will be no surviving physical remains. At best, there can only be one skeleton. However, that single animal will have made millions of footprints during its lifetime, and these footprints, if preserved, can tell us all sorts of things about the animal's lifestyle.

The specialist study of fossil footprints is known as palaeo-ichnology. It has a long pedigree. In the early nineteenth century, three-toed fossil footprints were found in the Triassic sandstones of Connecticut, USA. They were studied by local naturalist Edward Hitchcock, who thought they were the footprints of giant birds (there was no concept of dinosaurs at that time). Even today it is almost impossible to match a footprint to the animal that made it. Palaeoichnologists overcome this problem by attributing "ichnospecies" names to them. This allows each footprint or footprint track to be catalogued and studied without

Above and right: Fossil footprints can tell us whether an animal travelled alone or in groups, whether there were different sizes of animal passing by and the speed at which the animal travelled. Here are several Apatosaurus *trackways from the Morrison Formation, USA.*

referring to the animal that may have made it. The name of an ichnospecies often ends with the suffix *opus*. For example, *Anomoepus* is a footprint that may or may not have been made by a small ornithopod, *Tetrasauropus* is a footprint that was probably made by a prosauropod, and *Brontopodus* is almost certainly the footprint of a big sauropod, but which one? Most footprints are so vague that we can only tell that an animal of some kind passed this way.

If we find a series of footprints (or a trackway), made as the animal was moving along, we can tell if it had been travelling alone, in a pair, or in a larger group. Sometimes we find different sizes of the same kind of footprint, indicating that a family group of old and young had passed. Some trackways can be traced over tens or hundreds of kilometres, usually in separate outcrops. They are known as megatracksites.

Below: A Cheirotherium *footprint in Triassic rocks is 29cm/12in long.*

Below: A true print in the top layer usually has the mud squeezed up around it, and may preserve the skin texture of the foot.

Dinosaur trackways

The dimensions measured when studying fossil tracks are:
- Foot length (a).
- Foot width (b).
- Step length (the distance between successive prints) (c).
- Stride length (the distance between successive prints made by the same foot) (d).
- Pace angulation (the angle between three successive prints) (e).
- Angle of rotation (the angle that the direction of the foot makes with the direction of travel).
- Various equations using these parameters can be used to estimate a dinosaur's speed.

Below and left: Palaeoichnology can be a highly mathematical science. Using the values measured here, and the dimensions of a dinosaur's leg (if known), we can calculate its walking speed.

However, gauging the size of the footprint can be tricky. The print that fossilized in the rock may not be the actual print impressed by the dinosaur's foot. When an animal leaves a footprint in the top layer of sediment, another less distinct print is impressed into the layer beneath. Often the top layer was washed off and only the impression in the lower layer (the underprint) is preserved. It is essential to recognize underprints so that we do not make erroneous measurements; they tend to be smaller than true footprints and, if confused, would make calculations of speed unreliable. Fortunately, a true print can usually be recognized by its detail. If we find a print that actually preserves the skin

Below: Many prints are preserved as casts. A footprint may be filled by later sediments, and when these sediments become rock they preserve a three-dimensional impression of the print that stands proud of the bedding plane.

texture of the underside of the foot, then we can be confident that we are looking at a true print.

Trackways

Sometimes, tracks can be so spectacular that it is easy to jump to unwarranted interpretations in the excitement of discovery. A track that was clearly made by a herd of sauropods may have a theropod track running parallel to it. This would appear to show a sauropod herd being stalked by a theropod that is waiting to attack. Or the theropod may have followed the same route days later – it is difficult to tell. And a track of sauropod footprints that only show the tips of the front toes and nothing else, may be interpreted as the marks made by the animal swimming and poling

Below: The shapes of footprints can be misleading. The very long print (top), was made in soft mud, where the toes dragged a blob of mud along (bottom).

Left: Footprint in top layer of earth.

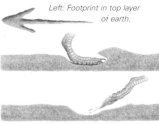

itself through the water using its front feet. More likely it is a set of underprints, with the narrower front feet penetrating deeper into the sediment and forming underprints, while the broader hind feet spread the weight and remain on the surface.

Below top: A dinosaur makes a footprint, with the true print impressed into the top layer of sediment and an underprint pressed into the layer below.

Below middle: The top layer of sediment is washed away.

Below bottom: Only the underprint is preserved.

FOSSIL EGGS

We obtain a unique insight to the life of an extinct animal when we find the fossils of its eggs. This, of course, only applies to egg-laying animals like reptiles and birds. Over the past few decades there has been a great interest in the study of fossil eggs, particularly those of dinosaurs.

It has always been assumed that dinosaurs laid eggs, as other large reptiles do. However, it was in the 1920s that the first dinosaur eggs were actually found. A series of expeditions into the Gobi Desert, led by Roy Chapman Andrews of the American Museum of Natural History, USA, uncovered several nests containing eggs and a large number of skeletons of the ceratopsian *Protoceratops*. For 70 years these "*Protoceratops* eggs" were the most famous and important dinosaur eggs found. At the site, the structure of the nest, the arrangement of eggs within it, and the proximity of the nests to one another were all clearly visible. It was suggested that there was evidence that the nests had been attacked. A small theropod named *Oviraptor*, or egg-stealer, was found close to one of the nests, having been overwhelmed by a sandstorm while trying to rob it. Then, in the 1980s, more of the same nests were found, again by an expedition from the American Museum of Natural History, but this time one of the nests had the skeleton of an *Oviraptor* actually sitting on it. The nests and eggs are now believed to be those of the *Oviraptor*.

At another site known as Egg Mountain in Montana, USA, nests with eggs and remains of the ornithopod *Orodromeus* were found, with remains of the theropod *Troodon*. The *Troodon* was first believed to be at the site intending to rob the nests, but the nests are now believed to be those of the *Troodon*. The *Orodromeus* remains are thought to be from corpses that the parent theropods

Above: Sauropod eggs are almost spherical in shape.

had brought back to feed their young. The provenance of the *Troodon* eggs was confirmed when some of the eggs were dissected and baby *Troodon* were found inside. However it is unusual to be able to dissect a dinosaur egg. A fossilized egg is an egg that died before it had a chance to hatch. When the egg dies the embryo inside is usually destroyed. Beetles burrow through the shell and lay eggs, and when the larvae

Left: There are skeletons of Oviraptor *parents, in Mongolia, actually lying over their eggs in the nest, wings spread out over them to keep them warm while they incubated.*

Left and right: Skeletons of baby Troodon *have been found inside eggs. As in other animals, the babies have proportionally larger heads and feet than the adults.*

hatch they eat the dinosaur egg's contents. Most fossilized eggs contain only scraps of bone that are so jumbled that they are unidentifiable.

Shell fragments are often found associated with nesting sites, where the eggs came to maturity and hatched. Modern microscopy techniques have been used on shell fragments so that dinosaur egg shells can be compared with those of modern animals. It was once assumed that dinosaur eggs had soft or flexible shells, like modern reptiles. However, study of the crystalline structure of fossil egg shells shows that the majority of them were hard and rigid, like those of birds.

Egg identification

As with footprints, it is difficult to assign a fossil egg to a dinosaur species, and it can only be successfully accomplished in rare occurrences when an identifiable embryo can be found inside the egg, or if a nest has the skeleton of a parent nearby.

The south German Solnhöfen skeleton of *Compsognathus* has a number of spherical objects associated with it. They are thought to be unlaid eggs that burst out of the body cavity after the animal died. Such an association of eggs and skeleton is very rare, though.

As with footprints, palaeontologists give names to particular fossil eggs – names that do not imply

Above: Maiasaura nests show all stages of nesting behaviour – complete eggs, broken eggshell, bones of hatchlings and skeletons of well-grown youngsters that have not yet left the nest.

identification of the egg layer. Such a classification is an "oospecies".

Most dinosaur eggs are found in nests. Occasionally they look as if they have been laid in holes in the ground without much preparation – a spiral of eggs found in France is thought to have been laid by a stegosaur that just deposited them and left. Most of the nests that we know about are quite complex structures, and are found together in rookeries or nesting sites. The nests laid by the duckbill ornithopod *Maiasaura* in Montana are the best studied. A typical nest consists of a mound of mud or soil, with a depression in the top. That depression is filled with ferns

Above: Ornamentation on the fossil shell surface as well as the crystal structure of the eggshell itself help palaeontologists to identify oospecies.

and twigs, providing insulation for the eggs that are laid inside it. Each nest is positioned about an adult dinosaur length from the next, so that they do not disturb one another.

Right: Egg nests have been found in Montana, USA, at a site known as Egg Mountain.

DINOSAUR BABIES AND FAMILY LIFE

Fossils found at nesting sites can tell us how quickly dinosaurs grew, for example, by looking at the structure of the individual bones and analysing their growth lines. We can also tell that growing dinosaurs were subject to diseases and injuries throughout life.

The *Maiasaura* nests in Montana, USA, have been the subject of the most studies on dinosaur family life. The Montana nesting site was by a lake in the uplands. Besides nesting behaviour, this site also provided all sorts of information about how the animals lived and grew. Once out of the egg, a baby *Maiasaura* was a small, vulnerable animal, about 45cm (18in) long. It remained in the nest and was nurtured by its parents, who brought food in the form of leaves and fruits to the nest. Nests were protected against raids by large lizards and theropod dinosaurs, such as *Troodon*. Seasonal climates meant that once in a while the *Maiasaura* herd had to leave the upland nesting site and migrate to the coastal plains, where food was still plentiful. The nestling grew to twice its birth length in about five months. At this juvenile stage the young dinosaur

Below: The quite recent find of a parent Psittacosaurus *with 34 young demonstrates that some dinosaurs did nurture their young in the nest.*

Above: Evidence suggests that some genera of dinosaurs looked after their young, nurturing the hatchlings and bringing back food until they were five months old.

would have left the nest, and followed the parents, learning how to find food for itself. Juveniles continued to grow, and after about a year reached a length of about 3.5m (11½ft).

The microscopic structure of the bones show us that the full adult length of about 7m (23ft) was reached in about six years, after which time the growth rate became very much slower. The conclusion is that dinosaurs, or at least duckbill dinosaurs, grew extremely quickly in the early part of their lives, and that the growth rate then slowed. This is the same growth pattern that we see in modern mammals and birds, but not in reptiles. The *Maiasaura* herds returned to their nesting area when it was time to breed, a round trip of around 300km (185 miles).

Sauropods

Similar evidence for the family life of sauropod dinosaurs comes from a site called Auca Mahuevo, in Argentina. This late Cretaceous site has a vast array of nests from a titanosaurid sauropod, probably *Saltasaurus*, made on a flood plain. During the breeding season herds of pregnant females descended on the area – a region that was just out of reach of most floodwaters from nearby streams. Their nests were simple compared with the nest structures of *Maiasaura* – they merely scooped out a shallow basin, about 1m (3ft) in diameter, with their clawed front feet, and laid their eggs in this. Each female laid 15–40 eggs in nests positioned 1.5–5m (5–16½ft) apart. Having laid their eggs, the females abandoned them, without any intention of looking after the nestlings once they hatched. It is possible that they remained in the area to discourage the big meat-eating dinosaurs, such as *Aucasaurus*, that roamed the region.

Below: Young Maiasaura *left their nests after about five months of growing and being tended by their parents.*

The young sauropods, on hatching, would have been about 30cm (12in) long – little animals compared with the adults who were 30 times their length – but they were equipped to find their own food immediately. Youngsters able to look after themselves immediately, are known as "precocious hatchlings" by biologists, and this is true of most reptiles today. Once out of the nest, the perils of the outside world awaited the young dinosaurs.

Dinosaur remains have been found showing signs of sickness and injury that would have prevented many from reaching old age. Palaeopathology, the study of ancient diseases, has shown that a range of traumas and diseases afflicted the dinosaurs.

Many ceratopsian skeletons have healed fractures to their rib-cages, suggesting injuries due to fighting. One ceratopsian skull has been found with a hole in it that matches the dimensions of a ceratopsian horn,

Above: As Maiasaura *youngsters grew, they joined the herd in their annual migrations to the productive feeding grounds.*

suggesting a head-on fight between the two animals. Tyrannosaurs suffered from gout. There are at least two examples of this disease that would have been brought on by a surfeit of red meat in the diet. Arthritis has also been identified in quite a few dinosaurs, including *Iguanodon*, where foot bones have been fused together. Additionally, a number of the bigger ornithopods have fractures to the upward-projecting spines from the tail vertebrae close to the hips. It has been suggested that these breakages were caused by excessive violence during mating.

DEATH AND TAPHONOMY

The study of what happens to a body between its death and its fossilization is known as taphonomy. The taphonomy of an individual fossil organism is extremely important to the scientist who wishes to find out about that animal, and the conditions under which it lived and died.

Fossil-bearing rocks are usually full of bivalves, or sea urchins, or corals, or even fish. Fossils of sea-living things are common because when they die, creatures that live in the sea are likely to sink to the bottom and be covered by sand, mud and other marine sediments. As these sediments solidify, the dead creatures become fossils. An animal that lives and dies on land, on the other hand, may have been killed by a hunting animal and may be eaten on the spot. Once the hunter has eaten its fill, scavenging animals will take their pick of the prey. The bones are broken up and carried away, and what is left is eaten by insects and broken down by fungi and bacteria. After a few weeks there is nothing left of the dead animal but a smear on the ground – certainly nothing left to fossilize.

The vast majority of tetrapods have left nothing to tell us of their existence. However if an animal's body falls into a river and is buried in flood deposits, or if it dies in a toxic lake where no scavengers can live, or if it is engulfed and buried in a sandstorm, there is a chance that the remains will survive in fossil form.

Fossil finds

Tetrapod fossils are found in a number of forms. The most spectacular is the "articulated skeleton". This is the ultimate prize because it has all the bones still joined together as when the dinosaur died. However, the temptation is to leave it as found, and so information is lost if the skeleton is not dissected and studied minutely.

The fate of a *Stegosaurus*

A tetrapod skeleton is rarely found complete. Many things can have happened to the dead body before it became buried and fossilized.

1 Above: A *Stegosaurus* is killed by a hunting *Allosaurus*. The big meat-eater eats its fill. Inevitably much of the body is left uneaten.

2 Above: Smaller meat-eaters, such as *Ceratosaurus*, wait until the *Allosaurus* has fed and departed and then scavenge most of the flesh left behind.

3 Above: The scraps that remain are eaten by even smaller scavengers, and the rest is broken down by insects and fungi. By this time the skeleton is scattered.

4 Above: Eventually the scattered bones are buried by river deposits during the wet season as the plain floods. Now the process of fossilization can begin.

Above: The skin impression of a hadrosaur was created as mud solidified around a corpse before it decayed – a rare occurence.

The next best thing is an "associated skeleton". This is a jumble of bones that obviously comes from the same animal, but which have been broken up and scattered. Usually something is missing, carried away by the forces that pulled it apart. More common is an "isolated bone". Sometimes its origin is obvious, but not always, and scientific errors have been perpetrated by the misidentification of isolated bones. *Dravidosaurus* was thought to be a stegosaur, living in India where no

Below: A partly excavated isolated femur (thigh bone) that belonged to a sauropod dinosaur, preserved in Upper Jurassic redbed mudstones of the Upper Shaximiao Formation, in Sichuan, China.

other stegosaurs had been found, and existing tens of millions of years after the stegosaurs were supposed to have died out. The isolated bones on which the identification was based were actually parts of a plesiosaur, a marine reptile.

Bits of bone that have no scientific value whatsoever are termed "float". Some dinosaur excavations map and catalogue every bit of float found, in the hope that they might one day yield some information. Most excavations just ignore them.

Above: The associated skeleton of a sauropod in Jurassic sediment in the Dinosaur National Monument, Utah, USA.

Finally there are trace fossils, footprints, droppings, eggs and other lines of evidence that an animal existed, but without any physical remains. Paradoxically it may be that trace fossils tell us more about the animal's life than the body fossils.

Below: An articulated skeleton of Lambeosaurus *where the components have been kept together as in life.*

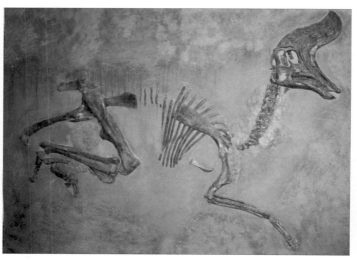

COAL FORESTS

Perhaps the earliest environment populated by tetrapods was the delta swamps of late Devonian and Carboniferous times. Primitive plants grew on sandbanks at river mouths and formed forests. These were ideal habitats for the amphibians, the earliest reptiles and the invertebrates that fed them.

During Devonian and Carboniferous times, the constantly but slowly moving continents were beginning to move closer to one another ultimately forming the single supercontinent of Pangaea. As the continents converged and collided they threw up massive mountain ranges along their edges. Vigorous weathering cut into these mountains and huge amounts of eroded debris, such as sand and silt, was washed into the surrounding shallow seas, building up vast areas of deltas and swamps.

This geographical activity coincided with the first widespread development of plant life, and soon these deltas and swamps were clothed in early plantlife, such as forests of ferns, giant horse-tails and huge trees that were the relatives of the present day's insignificant club mosses. In the shallow weed-choked waters and moist undergrowth land-living amphibians and the first egg-laying reptiles thrived.

The unstable nature of the deltas – their frequent inundation by the sea and their subsequent re-building – led to a cyclical sequence of sediments, in which marine sediments were followed by river sediments, followed by sediments that formed dry land, followed by marine sediments again. The prolific growth of plants at these times produced thick beds of peat that eventually became seams of coal when the whole sequence solidified into rocks.

The first animals to inhabit this swampy environment were land-living arthropods such as insects. Over time vertebrates left the water and evolved into land-living animals, using this insect fauna as a food resource.

The abundant animal life of the time is found as fossils in various parts of the rock sequence.

Rock sequences such as these are today found throughout the traditional coal-mining areas of Europe and North America.

Below: The sequence from the bottom upwards is mud (forming shale), sand with current bedding (forming sandstone), seat-earth (leached soil with plant roots) and peat (forming coal).

1 Giant club mosses on river banks.
2 Giant horsetails in shallow water.
3 Undergrowth of ferns.
4 Back swamps choked with vegetation.
5 Early reptile, *Hylonomus*, in tree stump.
6 Alligator-like amphibian, *Eogyrinus*, hunting fish in the waterways.

7 Giant invertebrates such as dragonflies, millipedes and scorpions.
8 Iron in river water collects around organic remains.
9 Iron nodules in shale, containing fossils of fish, amphibians and reptiles.

ARID UPLANDS

Arid habitats such as limestone plateaux rarely preserve fossils. However, occasionally animals venture into the cave systems beneath, or are washed into fissures by flash floods, or stumble into swallow holes. There they may be buried and their remains preserved as jumbled bones mixed up with stony debris.

During the later Triassic southern England consisted of an arid, limestone plateau where fossils have been found. Early prosauropods, such as *Thecodontosaurus*, eked out a living here on the scrappy vegetation. The plateau was riddled with gullies and caves, and moist underground air seeped to the surface producing a slightly lusher vegetation at the cave mouths. Plant-eaters were tempted to these areas, and occasionally lost their footing and fell to their deaths. There they were devoured by cave-dwelling

Below: An arid limestone plateau habitat is the setting for a swallowhole into which lizards and a Thecodontosaurus *have fallen. Eventually the swallowhole is filled with debris and the bones of the dead animals fossilize.*

animals, or were covered by cave debris and later fossilized. This shows how Triassic dinosaur remains are found in much older Carboniferous rocks. The Carboniferous limestones formed the uplands and the Triassic animals were fossilized in them.

The Naracoort Caves in South Australia acted as a similar trap, and have preserved the remains of Pleistocene mammals such as *Diprotodon* and *Thylacoleo*. We can often tell if a dinosaur died in an arid environment. As a dead body dries out in the heat, the tendons that link the bones shrink. When these tendons dry and contract, they pull the tail up and the neck back, drawing the skull over the shoulders and back.

Above: A sandstorm could mean death for a large animal like a dinosaur. It could also mean that its skeleton was preserved entirely and articulated. An animal engulfed in tonnes of sand suffocated quickly, but the sheer weight of the sediment kept it in place and once the soft tissues decayed the bones remained undisturbed.

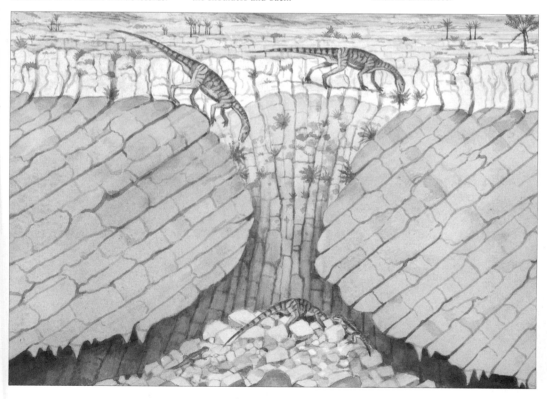

OASES AND DESERT STREAMS

*In an arid landscape, the presence of water is a great attraction to life. Where desert depressions reach
down to the moist rocks of an aquifer and water bubbles to the surface in cool pools, or where streams
flow down from mountains to parched lowlands, vegetation grows and animals live.*

Rivers and streams flowing away from rapidly eroding uplands can carry vast quantities of broken rock fragments in the form of grains of sand and silt. They are spread across the lowlands as the current slackens. The result may be accumulating beds of fertile soil in which plants grow. So, in desert areas, the river supports the riverside vegetation and the seasonal floods spread layers of silt over the landscape, fertilizing it. The River Nile, with its annual floods, keeps the Egyptian desert alive and was responsible for

producing one of the earliest prolific farming civilizations.

In Triassic times the vast supercontinent of Pangaea was beginning to break up. The interior was still arid, being so far from the sea, and life could only really flourish in the flood plains of rivers. It was the time of the prosauropods – the first big plant-eating dinosaurs – the early relatives of the huge long-necked sauropods. They fed on the cycad-like plants that grew close to the ground, and also reached up to browse the

coniferous trees that lined the water courses. This heavy browsing led to the evolution of trees and foliage that had strong, sharp, sword-like leaves as a defence, such as the monkey puzzle tree that still exists in the mountains of South America and in ornamental gardens across Europe. Among the predators, the various crocodile relatives, such as the phytosaurs, were giving way to the first of the theropods. When the plant-eaters were attracted to the water and became trapped in quicksands, their great

weight preventing them from escaping, the meat-eaters would take advantage of their helplessness. Remains of all these animals are found in the rocks formed from the river sandstones and the silty flood deposits of the time. Occasionally their fossils tell the story of trapped animals devoured by predators, or even of animals killed by the changing seasonal conditions.

Below: A prosauropod, attracted to the water, is trapped in quicksand. Small theropods and crocodile-like predators close in for the kill.

Petrified Forest National Park
The most famous fossil occurrence of a Triassic desert waterside habitat is Petrified Forest National Park in Arizona, USA. Close by, at Ghost Ranch, a pack of meat-eating theropods, *Coelophysis*, was discovered as articulated skeletons. The evidence shows that they died from dehydration around a drying water hole. The best evidence for prosauropods being caught in quicksand and killed by predators comes from the Frick Brick Quarry, Switzerland. Almost identical circumstances are known from the Molento and Lower Elliot Formations of South Africa, showing how widespread these habitats were on Pangaea.

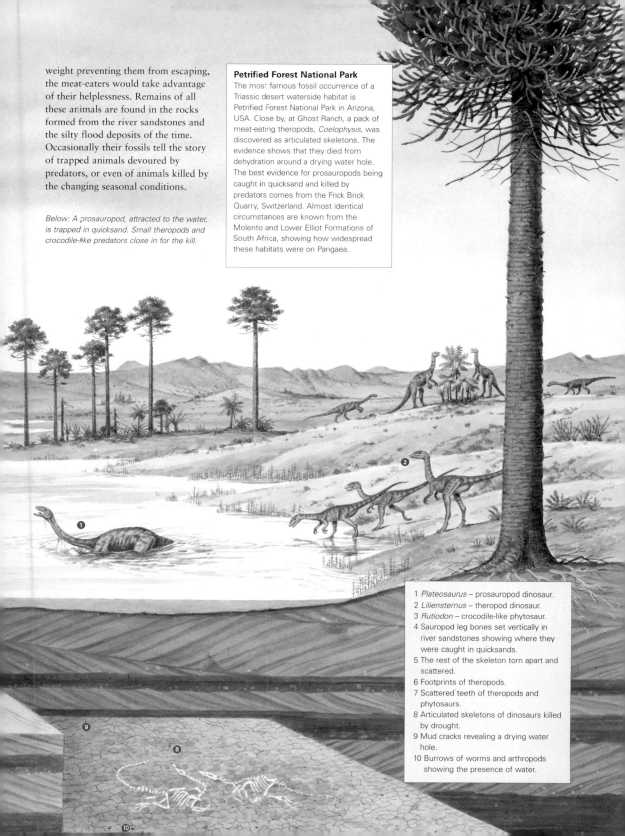

1 *Plateosaurus* – prosauropod dinosaur.
2 *Liliensternus* – theropod dinosaur.
3 *Rutiodon* – crocodile-like phytosaur.
4 Sauropod leg bones set vertically in river sandstones showing where they were caught in quicksands.
5 The rest of the skeleton torn apart and scattered.
6 Footprints of theropods.
7 Scattered teeth of theropods and phytosaurs.
8 Articulated skeletons of dinosaurs killed by drought.
9 Mud cracks revealing a drying water hole.
10 Burrows of worms and arthropods showing the presence of water.

CALM LAGOONS

Seawater trapped behind reefs forms shallow lagoons that evaporate slowly under a hot, tropical sun.
The salt concentration in the water increases and the water becomes toxic, killing anything that enters it.
Detailed evidence of life in dinosaur times comes from deposits formed under these conditions.

In late Jurassic times the break up of Pangaea was well under way. A huge embayment, known as the Tethys Sea, separated the continent of Europe and Asia from the landmasses of the south; those that became Africa, South America and Australia.

Along the northern flanks of the Tethys a deep-water reef developed, formed from sponges. The remains of this reef can be found today in rocks from Spain to Romania. As the reef grew, and as continental movements raised the sea floor, the reef approached the surface where the deep-water sponges died. The reef growth was continued by corals building on the sponge-formed

structures. The reefs reached the surface where they cut off lagoons between the deep Tethys waters and the shoreline of the continent to the north. Debris from the reefs filled the lagoons and the water became shallow. The heat of the sun evaporated the water from these shallows, and salt and other minerals settled on the lagoon floor. The water was constantly replenished from the ocean beyond the

Solnhöfen

There is a concentration of lagoon deposits in Solnhöfen, southern Germany. These so-called lithographic limestone quarries have yielded famous fossils of flying animals and land-living creatures. There are fossils of ammonites that have plunged into the limy mud, horseshoe crabs that have dropped dead at the end of their tracks, and floating animals, such as brittle stars and jellyfish, that drifted into the poisonous waters.

Right: This pterosaur fossil is from Solnhöfen, Germany. The details of its anatomy are clearly defined.

Below: The lagoon environment is best known from Solnhöfen, Germany, where some of the finest fossils have been uncovered, but lagoons of this kind probably extended across the whole of southern Europe.

reefs, but the lower layer of the lagoon water became poisonous with the concentration of minerals. Any fish that swam in it died and sank to the bottom. Any arthropod that crawled in was poisoned and died. The bodies lay undisturbed as the water was poisonous for scavengers too.

Islands were scattered across the lagoons which, with the arid shoreline, were formed from the stumps of the sponge reefs that had emerged from the water as the land rose. The animals of these dry lands consisted of pterosaurs, the first-known bird *Archaeopteryx*, little lizards and small dinosaurs, such as the chicken-sized theropod *Compsognathus*. All of these animals have been found as articulated skeletons in the deposits formed in the lagoons.

1 *Rhamphorhynchus* – rhamphorhynchoid pterosaur.
2 *Pterodactylus*.
3 *Archaeopteryx* – primitive bird.
4 *Compsognathus* – theropod dinosaur.
5 *Bavarisaurus* – lizard.
6 Articulated skeletons preserved in thinly-bedded limestone.

RIPARIAN FORESTS

Many dinosaur fossils from North America come from the late Jurassic Morrison Formation that stretches across much of the Midwest. It consists of river- and flood-deposits. Under seasonally dry conditions, vegetation was mostly restricted to river banks and only flourished in the wet season.

When the pioneers headed west across North America in the late nineteenth century, they found the bones of strange animals. The finds drew the attention of scientists who, within 30 years, discovered more than 100 previously unknown dinosaur types. Most came from a sequence of rocks called the Morrison Formation, which was formed from river sediments in late Jurassic times. The landscape was once that of a riparian forest. In the late Jurassic a shallow seaway stretched north-south along the length of the North American continent. To the west the ancestral Rocky

Mountains arose along the edge of the ocean. Between the two stretched a plain built up of sediment washed down from the mountains. Rivers meandered across the plain, flooding frequently and depositing sediment. In times of flood, the sediment built up on the riverbanks, forming levees, until the surface of the river became higher than the elevation of the surrounding plain. The water frequently broke through the levees spreading sediment in a fan-shaped deposit, and river water often seeped through the levees as springs filled freshwater ponds and lakes across the plain. Other lakes,

formed at times of flood, dried out and became poisonous with alkaline minerals leached from the soil.

Below: During the late Jurassic, in North America, in the wet seasons the rains were plentiful and floods spread across the plains. Flood sediment was deposited over the landscape causing prolific plant growth on which animals browsed.

The original flood deposits were disturbed by the trampling feet of the animals – called "bioturbation". The alkaline lakes formed beds of limestones. Ribbons of river sediments run through the whole sequence. Dinosaur fossils are found as isolated bones on the flood deposits or sometimes as associated skeletons in the river sediments.

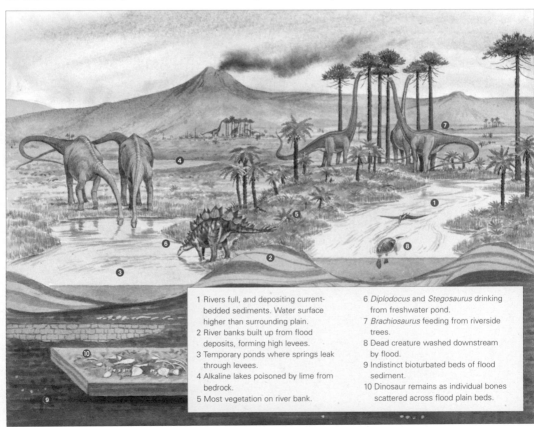

1 Rivers full, and depositing current-bedded sediments. Water surface higher than surrounding plain.
2 River banks built up from flood deposits, forming high levees.
3 Temporary ponds where springs leak through levees.
4 Alkaline lakes poisoned by lime from bedrock.
5 Most vegetation on river bank.

6 *Diplodocus* and *Stegosaurus* drinking from freshwater pond.
7 *Brachiosaurus* feeding from riverside trees.
8 Dead creature washed downstream by flood.
9 Indistinct bioturbated beds of flood sediment.
10 Dinosaur remains as individual bones scattered across flood plain beds.

Morrison Formation

The Morrison Formation, named after the Colorado town of Morrison, USA, consists of 30–275m (100–900ft) of shale, siltstone and sandstone, and stretches from Montana to New Mexico. Its landscape was once that of a riparian forest, or forest growing on riverbanks. The most famous outcrops of the Morrison Formation are in Dinosaur National Monument in Utah, and the Fruita Palaeontological Area and Dry Mesa Quarry, both in Colorado. At Tendaguru in Tanzania, a similar environment existed at exactly that time and almost the same range of animals lived there.

Above: The Morrison Formation is seen most obviously as a ridge along the foothills of the modern Rocky Mountains.

On this topographic surface the plants were confined largely to the river banks and around the freshwater pools. The plants formed forests and isolated stands of primitive conifers and ginkgos, with a lower-storey of cycad-like plants and tree ferns, and an undergrowth of ferns and beds of horsetails close to the water. The open plain had a scrappy growth of ferns. The alkaline pools were barren. The climate was seasonal with dry periods interrupted by times of abundant rain.

Herds of sauropods migrated across this landscape. *Diplodocus,* *Apatosaurus, Camarasaurus, Brachiosaurus* and many others, moved from thicket to thicket wherever there was food. The plated *Stegosaurus* fed from the trees and undergrowth. The main ornithopod was *Camptosaurus*. There were theropods aplenty, ranging from the enormous *Allosaurus*, through the medium-size *Ceratosaurus* to smaller animals like *Ornitholestes* and *Coelurus*. The abundant fossil remains of these dinosaurs are found mostly as isolated bones in flood deposits, but there are also associated skeletons and articulated skeletons in river channel deposits.

Below: In the dry season rivers dried to a trickle, and ponds became hard-packed mud. Water evaporation brought up the mineral calcite and left it as lumpy beds of limestone beneath the soil surface. Animals migrated into areas where there was still food to be had.

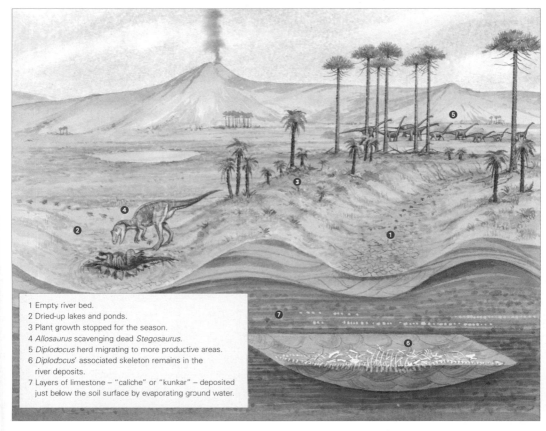

1 Empty river bed.
2 Dried-up lakes and ponds.
3 Plant growth stopped for the season.
4 *Allosaurus* scavenging dead *Stegosaurus*.
5 *Diplodocus* herd migrating to more productive areas.
6 *Diplodocus*' associated skeleton remains in the river deposits.
7 Layers of limestone – "caliche" or "kunkar" – deposited just below the soil surface by evaporating ground water.

LAKE ENVIRONMENTS

Lakes tend to be rather ephemeral. Water-filled hollows left by glaciers or landslide-blocked valleys soon fill with sediment and disappear. However, lakes formed in rift valleys, such as those in modern East Africa may be long-lasting landscape features. Good fossil remains are found in rift valley lake deposits.

At the end of the Jurassic period and the beginning of the Cretaceous, the eastern part of the northern continent, in the area where China now lies, was beginning to break up. The continental surface was split by fault lines, running north-east to south-west, which produced rift valleys. These rift valleys contained lakes. The water of these lakes was clear, and the only sediment was from very fine particles. In quiet times the valleys were filled with lush forests of long-needled conifers and ginkgoes. Horse-tails, ferns and mosses flourished around the lake edges. This environment supported an astonishing assortment of animals. The biggest were the dinosaurs. A similarly

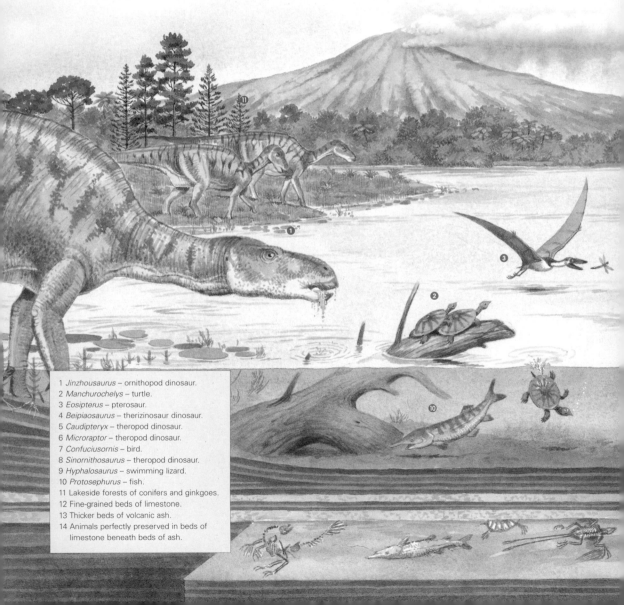

1 *Jinzhousaurus* – ornithopod dinosaur.
2 *Manchurochelys* – turtle.
3 *Eosipterus* – pterosaur.
4 *Beipiaosaurus* – therizinosaur dinosaur.
5 *Caudipteryx* – theropod dinosaur.
6 *Microraptor* – theropod dinosaur.
7 *Confuciusornis* – bird.
8 *Sinornithosaurus* – theropod dinosaur.
9 *Hyphalosaurus* – swimming lizard.
10 *Protosephurus* – fish.
11 Lakeside forests of conifers and ginkgoes.
12 Fine-grained beds of limestone.
13 Thicker beds of volcanic ash.
14 Animals perfectly preserved in beds of limestone beneath beds of ash.

important fossil lake site is known from Messel, near Frankfurt, in Germany. The Eocene lake lay in a depression formed by the earth movements that produced the nearby Rhine valley. Volcanic gases periodically poisoned the water and the air round about. The lake-deposited shales contain a remarkable collection of early Tertiary mammals, birds, fish and insects. The sterile

Below: Lake environments existed throughout the changing landscape of the Age of Dinosaurs. The lake deposits found in Liaoning, China, are beds of fine limestone interspersed with beds of volcanic ash that formed in early Cretaceous times.

waters resulted in the fine preservation of hair and feathers, and even the coloration of insect wings. These specimens were gradually uncovered as the shales were exploited for the oil they contained. There were also true birds, some of which still had primitive claws on the wings. These smaller creatures hunted the swarms of insects that lived close to the water.

We know of the existence of these animals because the mode of preservation is so fine that even the finest structures are visible. Every now and again the volcanoes that formed along the fault lines would erupt, engulfing the valleys in poisonous gas and showering the lakes with a thick

deposit of ash. Gassed by volcanic fumes and buried in fine volcanic ash, the smallest details of skin and feathers have been preserved around the almost perfectly articulated skeletons. We also have the fossils of their insect prey, and all sorts of other animals, such as turtles and other swimming reptiles, and also early mammals, that lived in the area.

WETLANDS

Wetlands tend to be most extensive in moist, tropical climates, close to upland areas. Heavy erosion from exposed hills brings masses of sediment down to low-lying regions. The sediment spreads out to form mud flats through which streams meander.

The most extensive, modern, tropical wetlands are found in the papyrus-choked Nile drainage area in Sudan and Uganda. Wetlands also appear further south in the Okovango flood plain at the edge of the Kalahari Desert. The mud flats support vegetation peculiarly adapted to the conditions. In these places the dominant plants are water-loving grasses and reeds, which developed in the Tertiary period. In the Mesozoic,

the water courses and mud banks supported thick beds of horsetails. Like grasses, horsetails spread by means of underground stems which held the sediments together, and provided a firm surface on which more permanent vegetation could grow. Like reeds, the horsetails could grow in shallow water. On stabilized sandbanks, thickets of ferns such as *Weichselia* grew. Stable ground developed and on it thrived thickets of conifers and cycads.

During the Cretaceous period, herds of *Iguanodon* roamed the wetlands of northern Europe, grazing on horsetail beds. Smaller herbivores, such as the fleet-footed *Hypsilophodon*, also scampered there. In the more permanent thickets, armoured dinosaurs such as *Polacanthus* grazed. Where plant-eating animals flourished,

Below: Dinosaur bones may be quite abundant and diverse in wetland deposits, but they tend to consist of isolated bones or, at best, associated skeletons.

European wetlands

The dinosaur wetlands of northern Europe are best represented by the Wealden and Wessex formations, laid down during the early Cretaceous. They outcrop along southern England and the Isle of Wight, and across the Paris Basin. At the time, mountains of limestone and metamorphic rock to the north supplied the sediment that washed down to the lowlands. The result is a thick sequence of sandstone, mudstone and clay, with frequent beds of plant-rich material. Dinosaur remains mostly occur as isolated bones, some showing signs of gnawing. They are often found with the teeth of freshwater fish, and are sometimes encrusted with the eggs of freshwater snails, showing that they were deposited in stream beds or in the backwaters of rivers.

meat-eaters that preyed on them followed, and in the horsetail swamps of northern Europe they included *Megalosaurus* and *Neovenator*. Wading in the waters were fish-eating theropods, such as *Baryonyx*.

Dinosaurs were not the only animals that existed here. In the water were crocodiles, including the dwarf form, *Bernissartia*, no bigger than a domestic cat, and turtles such as *Chitracephalus*, about the size of a dinner plate. In the skies there wheeled pterosaurs, such as the condor-sized *Istiodactylus*. Disarticulated bones of all these animals have been found in the swamp deposits laid down at that time. The isolated fossil bones in these deposits may be worn and polished,

showing that they were washed about for a while before settling. They are sometimes encrusted with the eggs of water snails, suggesting that they lay on the stream bed for some time before becoming buried.

1 *Oviraptor* – theropod dinosaur.
2 *Eotyrannus* – theropod dinosaur.
3 *Neovenator* – theropod dinosaur.
4 *Bernissartia* – crocodile.
5 *Pelorosaurus* – sauropod dinosaur.
6 *Polacanthus* – armoured dinosaur.
7 *Baryonyx* – theropod dinosaur.
8 *Iguanodon* – ornithopod dinosaur.
9 *Hypsilophodon* – ornithopod dinosaur.
10 River and flood deposits.
11 Lens-shaped stream deposits.
12 Fossils appear as isolated bones in river deposits.

SWAMP FORESTS

By the end of the Cretaceous period the vegetation of the world was taking on an appearance familiar to humans. There were still no grasses, but flowering plants had appeared in the undergrowth, and deciduous trees were beginning to take over from conifers as the main woodland flora.

It is easy to imagine dinosaurs as being the inhabitants of deep, dark jungles. Where there were deep forests and woodlands, there were dinosaurs well adapted to living there. As the advanced ornithopods, especially the duckbills, evolved and diversified during the Cretaceous period, plant-life evolved with them. The broad mouths and low-slung necks of these ornithopods showed that they fed close to the ground. The evolutionary response would have been for plants, having been grazed, to develop survival strategies that would allow them to rebuild populations quickly. Flowers and enclosed seeds do this, allowing the main part of the germination process to take place after the parent has been destroyed. Flowering plants in deciduous woodlands evolved with the low-feeding habits of the later plant-eaters such as the broad-mouthed duckbills. Many of these dinosaurs had extravagant head structures, linked to their nasal passages. They made grunts and trumpet noises, producing sounds that would penetrate dense forest undergrowth so that the animals could communicate with one another.

The deciduous woodlands spread across the broad plains and deltas, between the newly arisen Rocky Mountains and the spreading inland sea, that now reached from the Arctic Ocean down across the middle of North America. The deposits formed the Hell Creek Formation. On higher ground the forests consisted of vegetation such as tall stands of primitive conifers with an understorey of cycads and ferns. These forests were inhabited by the last of the long-necked sauropods that had been the most important plant-eaters since the beginning of the Jurassic period.

Fossil survival

Duckbilled dinosaurs are sometimes found as "mummies", with the skin still fossilized around them. This occurred when the animal died and was stranded in the open, possibly on a sandbank in a delta. One flank of the dead animal would be pressed down into the mud, impressing the skin texture into the mud. The insides would have shrivelled away and the skin would have dried to leather, shrinking around the bones of the leg and the rib cage. The exposed part of the skeleton would have deteriorated quickly, the bones carried off by scavengers or washed away in the river. The next flood would have filled its insides with sand swirled along by the floodwaters. Fish have been found fossilized inside the rib cages of such dinosaurs. The skin would not have survived the subsequent fossilization process but by then the impression in the surrounding sediment would have solidified. The result is an articulated skeleton that is surrounded by the impression of the skin, giving a valuable insight into what the outer covering of a dinosaur was like.

Below: All manner of creatures from the tiniest insects to the largest dinosaurs have been found fossilized in areas that were once river beds and swamps.

1 *Kritosaurus* – ornithopod dinosaur.
2 *Troodon* – theropod dinosaur.
3 *Tyrannosaurus* – theropod dinosaur.
4 *Ceramornis* – modern-type bird.
5 Skeleton of *Kritosaurus* drying out, with underside buried in mud.
6 Deciduous trees.
7 Flowering herbaceous undergrowth.
8 Swamp deposits.
9 Channel deposits.
10 Dinosaur remains appear as articulated or associated skeletons.

OPEN PLAINS

Herd-living animals have always lived on wide open plains. This was also true in earlier times. The late Cretaceous rocks formed on the plains of North America have ample evidence of herding behaviour in the horned dinosaurs – including the fossilization of entire herds in deposits known as "bone beds".

Modern open landscapes are dominated by herds of plant-eating animals. Look at the high Serengeti plain of Tanzania today, with its herds of wildebeest, impala, elands, gazelles, zebras and so on. So it was with the open plains of the Cretaceous period.

The low vegetation that clothed the open landscapes in the Cretaceous period consisted largely of ferns but no grass. This was where herds of ceratopsians roamed. The Cretaceous herds consisted of a number of different types of ceratopsian. They differed from one another by the arrangement of horns and frills, and other head ornamentation. The youngsters all looked the same – when they were young they were sheltered by the rest of the herd and had little contact with other ceratopsians. As adults, however, the range of horn and frill types was very marked. This suggests that the ceratopsians lived in tightly organized herds, moving from place to place as a unit, and keeping away from herds of other types, as happens in modern grassland animals. Additionally, the ceratopsians migrated with the seasons to where

food was most abundant.

We can see evidence of herd behaviour in the fossil occurrences known as bone beds. They consist of a mass of bones, usually of one species of ceratopsian. The bones lie on the bottom of stream channel deposits and tend to be of the same size, suggesting that they have been sorted out by flowing water. They may consist of the remains of more than 1,000 individuals of the same species. It is easy to visualize the scenario. A herd of ceratopsians was overcome by water while crossing a river during

Below: Herds of various species of ceratopsian roamed the plains of North America in the late Cretaceous, harassed by the big meat-eaters of the time.

migration, and the bodies were washed downstream. The bodies washed ashore on a river beach, where they decayed and were scavenged by other dinosaurs and pterosaurs. Then the remains were picked up by the flooding river and deposited on the channel bottom, along with shed teeth from the scavengers.

It is not just ceratopsians that have been found as bone bed fossil deposits. A bone bed of the brontothere *Mesatirhinus* has been found in late Eocene rocks in Wyoming. The condition of this bed indicates that it formed in exactly the same way as the ceratopsian bone beds of earlier times.

Dinosaur Provincial Park
The best-studied ceratopsian bone beds are in the late Cretaceous Dinosaur Park Formation, in Dinosaur Provincial Park, in Alberta, Canada, where there are at least eight occurrences, one stretching for almost 10km (6 miles). Others occur in Montana, USA, and as far north as the North Slope of Alaska.

Below: Dinosaur Provincial Park, Alberta.

1 *Chasmosaurus* in defensive circle.
2 *Albertosaurus*.
3 *Quetzalcoatlus* – pterosaur.
4 *Centrosaurus* in migrating herd.
5 *Styracosaurus* displaying to other ceratopsians.
6 Plain formed of flood sediments.
7 Lens-shaped channel deposits.
8 Bone beds at the bases of channel deposits.

Formation of a bone bed
A Herd of ceratopsians crosses a river.
B Herd panics and individuals drown.
C Bodies are washed up on a sandbar.
D Bodies are scavenged by meat-eaters.
E Remains are washed into the river and settle on the bed.

SHORELINES AND ISLANDS

The edge of the sea is a popular habitat for modern birds. It was probably the same for their ancestors, the dinosaurs, though evidence is sparse. Shoreline deposits tend to be restricted and rather ephemeral since they are constantly disrupted by waves and tides.

In Cretaceous times the seaway that spread down the length of the North American continent stretched from Alaska to the Gulf of Mexico, splitting the dry land in two. In some areas the hinterland was thickly forested, and migrating dinosaurs travelled along the beaches where the walking was easier. North-south trackways have been found in Oklahoma, Colorado and New Mexico, USA. These discoveries have led to the concept of the "dinosaur freeway" – a beach migration route indicated by the consistent direction of Cretaceous dinosaur footprints. There are also others.

Island inhabitants

Crocodiles inhabited the rivers that opened out into the continental sea, and their toe marks have been found where they clawed their way across the river sediments. Birds found plenty to eat on the tidemarks, and their three-toed footprints abound. They can be distinguished from the footprints of small dinosaurs by their more splayed toes.

Out at sea, islands, as always, can support a variety of specialized life forms. Animals reach newly formed

Below: The sequence of early Cretaceous rocks known as the Dakota Group was formed as seashore deposits along the interior sea of contemporary North America. It is famous for its footprints and trackways.

Dinosaur Ridge

The most publicized exposure of the dinosaur freeway is Dinosaur Ridge, in the Rocky Mountain foothills, just a few kilometres west of Denver, Colorado.

the mainland's animal life. In either scenario, if the island becomes a permanent geographical feature the animal life will adapt and evolve to survive there. One such adaptation is the development of dwarf forms. Small animals need less food to survive, and islands have limited natural resources. In the Ice Age there were elephants the size of pigs on the Mediterranean islands, and giant ground sloths on the Caribbean islands that were much smaller than their mainland relatives. A modern example is the Shetland pony, a breed well adapted to the bleak habitat of the Scottish islands. The northern edge of the Tethys Sea in Mesozoic Europe had a whole archipelago of islands, thrown

up by earth movements as the northern and southern continents approached one another. Bones of dinosaurs unearthed in Romania show that duckbills, only one-third of the size of their relatives, and ankylosaurs the size of sheep, existed there.

islands by being rafted there on logs or other land debris. Also, an island that formed when an area of land was cut off by the sea usually retains some of

1 *Planicoxa* – ornithopod dinosaur.
2 *Acrocanthosaurus* – theropod dinosaur.
3 Pterodactyloid pterosaur.
4 Birds.
5 Alternating beds of beach sands and shallow sea sediments.
6 Sand ripples.
7 Ornithopod trackways following the shoreline as herds migrated.
8 Theropod trackways at right angles to the shoreline, as individuals came out of the forest to scavenge or to attack the migrating herds.
9 Bird trackways in random feeding pattern.
10 Individual dinosaur bones in sea sediments – badly worn and encrusted with sea life.

DESERTS

We often associate vertebrate remains with water, and animals living near it are more likely to be buried in sediment once they have died than those that lived far away from it. However, there are a number of occurrences where animals of very dry habitats have been found fossilized.

The Gobi Desert was a desert even in Cretaceous times. Ephemeral lakes surrounded by plains of scrubby vegetation were separated by vast areas of shifting sand dunes. Here roamed herds of *Protoceratops*, in huge numbers. Occasionally, the harsh, dry winds of the desert overwhelmed these herds in a sandstorm or a dust storm, and sometimes they were engulfed by sliding sand as an unstable dune collapsed. Either occurrence had the effect of killing the animals where they stood, and preserving them in a tomb of dry sand that eventually solidified into desert sandstone.

A specimen that shows just how quickly such an event occurred was found in the 1970s. The skeletons

of a *Protoceratops* and a fierce little *Velociraptor* where found wrapped around one another. The meat-eater's arm was seized in the ceratopsian's beak, while the long claws of the former were clutched tightly to the latter's head shield. The two had been buried and died while in the middle of their struggle. Elsewhere in the desert, dinosaur nests and eggs have been preserved in sandstone. One nest, belonging to the theropod *Oviraptor*, even had the mother dinosaur sitting across it, in a vain attempt to incubate her eggs as the sandstorm engulfed her.

In the Triassic, at the beginning of the Age of Dinosaurs, northern Scotland consisted of several arid basins. Here the remains of desert

animals buried in sand rotted away completely, leaving bone-shaped holes in the subsequent sandstone. Skeletons of animals, such as the aetosaur *Stagonolepis*, have been reconstructed from latex poured into these natural moulds.

The oldest evidence for sand dunes in the Sahara Desert dates from 7 million ears ago – the Miocene. These deposits contain the bones of the early hominid *Ardipethicus*, suggesting that our ancestors were versatile enough to leave the woodlands and forests.

Below: The shifting sands and airborne dust clouds of the desert were constant threats to the dinosaurs of the late Cretaceous Gobi Desert. Often they were buried in hollows or on the flanks of collapsing sand dunes.

TAR PITS

Fossils formed of the hard parts of organisms unaltered are rare and most valuable. The tar pits of Rancho La Brea in Los Angeles, USA, provide an environment in which such fossils are found. The natural pools of tar are visible today, preserved at the Page Museum in Hancock Park.

During the last few million years, as the coast ranges of western North America were pushed up, natural oils collected in the twisted rock structures below ground and formed reservoirs of petroleum. In some places, as the rocks twisted and cracked, the oil seeped to the surface and spread out forming sticky pools. In the open air, the volatile components of the crude oil evaporated and left behind a layer of viscous asphalt or tar. This is how the tar pits of Los Angeles, USA originated.

Rains came, and formed pools of water on top, and from a distance these pools looked, to animal life, like refreshing watering holes. Animals came to drink, and once they had waded into the pools they became enmired in the tar beneath and could not escape. Predatory animals in the area converged on these spots to feed on the captive beasts, and often became trapped themselves. In this way the situation resembled that found in desert oases and streams, except instead of river quicksand trapping the animals, the danger was formed by natural deposits of tar. In both cases the result was a "predator trap" in which animals that came to feed on trapped animals became trapped themselves.

Even in dry seasons when water was not present as an attraction, the tar pits were a danger. They were disguised by wind-blown dust and leaves, which would stick to the surface. The warmer weather made the tar sticky, and unsuspecting animals wandering across it became quickly bogged down.

The bones that sank into the tar were well preserved, although stained dark brown by the asphalt chemicals. Those left on the surface were weathered by heat, cold, wind and rain. They were also damaged by the gnawing of rodents that somehow escaped entrapment themselves. Similar tar pits are found in Peru and Iran.

Below: The tar pits of Los Angeles are a cluster of pits that contain an incredibly rich and diverse collection of fossils of animals, insects, and plant life that existed 40,000 years ago.

1 Mammoth trapped by tar beneath its feet.
2 Sabre-toothed cat, attracted by the struggle of a ground sloth, becomes trapped too.
3 Dire wolves do not venture on to the tar, and so their remains are rarer.
4 Vultures can take carrion and can avoid the tar.
5 Bones buried quickly are preserved.

TUNDRA

The most recent Ice Age, known as the Pleistocene Ice Age, is the period of time in which conditions were noticeably different from those that exist nowadays. Cold climate animals inhabited the chilly wastes that surrounded the greatly expanded glaciers and ice caps of the time.

During the Ice Age the temperature dropped worldwide. Ice caps expanded outwards from the poles, and the glaciers crept downwards from the mountains. The Earth had the modern range of vegetation, stretching from the cold tundra, through coniferous forests, temperate forests, tropical deserts and grasslands, to equatorial rain forests, but these zones were all compressed towards the equator with large areas of cold climate environments on the northern continents. Tundra conditions, or the point at which tree growth is hindered, prevailed here.

The factor that controlled the landscape conditions in the tundra was the permafrost. This was a layer of permanently frozen subsoil with, above it, soil that thawed and then froze as the seasons changed. During the summer the soil thawed, but the water could not drain away through the permafrost below, and resulted in boggy landscapes of lakes and swamps, clothed in hardy low-growing plants, such as grasses, sedges, mosses, lichens and heathers, as well as dwarf species. The soils themselves consisted of sand and rubble washed out of glaciers and spread by rivers of water from the melting ice.

Cold climate animals, such as mammoths, woolly rhinoceros and Irish elk migrated with the seasons across this bleak landscape, feeding on the summer vegetation, in the same way that modern-day reindeer and musk-ox cross the contemporary equivalent in far northern Canada and Siberia.

Occasionally the animals became trapped in muds and streams, and their bodies were preserved by freezing as the cold winter conditions came around.

This was the landscape inhabited by our human ancestors in the Ice Age. These were the animals that they hunted for food.

Below and above: As the ice caps expanded and contracted with the seasons, the rivers pouring from them deposited rubble and sand. These soils supported cold-climate plants that were grazed by a number of migrating cold-adapted animals.

1 Rivers of meltwater flowing from the glaciers, choked by sediment.
2 Glacier debris forming a landscape of drumlins and eskers – mounds of sand and rubble.
3 Wet landscape of poorly drained soil.
4 Soil saturated with water below the water table.
5 Permafrost – permanently frozen soil.
6 Dead mammoth in pond mud, frozen in the permafrost.
7 Mammoth buried under collapsing riverbank.
8 Herds of mammoth.
9 Irish elk.
10 Woolly rhinoceros.

EXTINCTION

Mass extinctions are a fact of geological history. Palaeontologists have identified at least five. That which caused the demise of the dinosaurs 65 million years ago, although not the greatest, was certainly the one that has received most attention.

There have been many theories regarding the reasons behind the extinction of the dinosaurs. Mostly the theorists have been divided into those who favour gradual extinction as an explanation and those who favour sudden catastrophe. "Sudden", in geological terms, can actually cover half a million years, but looks instantaneous in the geological record.

The gradualists say that for most of the late Cretaceous period, conditions had been stable, with not much variation in climate for tens of millions of years. It may be that the specialized animal life of the time had become too specialized and too adapted to these conditions. A slight change in the environment, such as the raising of the temperature or the cooling of the atmosphere, may have put intolerable stresses on the dinosaurs so that they could not cope. Another theory suggests that the gradual movement of

continents brought landmasses into close proximity to one another. As a result animals would have been able to migrate and carry diseases with them to which the endemic population would have been vulnerable.

Above: One theory suggests that climatic conditions changed at the end of the Cretaceous, killing the dinosaurs.

However, some view the demise of the dinosaurs as a sudden occurrence. In the 1970s, a discovery was made that changed this discussion for ever. It was found that the end of the Cretaceous and the beginning of the Tertiary (the K/T boundary) was marked by a bed that was rich in the element iridium. Iridium is rare at

Below and left: Another theory backed by scientific understanding is that a meteor hit the Earth's surface, resulting in huge changes to the Earth's atmosphere and the extinction of the dinosaurs. The meteor is said to have caused the Chicuxlub Crater, at the edge of the Gulf of Mexico.

Above: Intense volcanic activity, with its release of gases and vast quantities of dust, could be a factor in mass extinctions.

Above: The eruptions that caused vast lava flows in Cretaceous India would have altered the climate considerably.

the surface of the Earth but quite common in meteorites. So a new theory was put forward that a massive meteorite had struck the Earth 65 million years ago. The immediate result would have been shock waves and fires. The explosion would have sent masses of dust and steam into the atmosphere, blanketing the Earth and causing temperatures to fall as the sunlight was blocked out. Plants would have died. Plant-eating dinosaurs would have starved. Then meat-eating dinosaurs, denied their plant-eating prey, would have perished. After months or years the skies would have cleared and plants would have started to grow. By that time all the dinosaurs would be dead.

A huge buried structure, on the coast of Yucatan in Mexico, was identified as a meteorite crater dating from the end of the Cretaceous period. Signs of shattered rocks and sea wave damage were identified around the Caribbean. Corroborative evidence was being amassed.

Other theories were put forward too. Iridium is also found beneath the Earth's crust, and could be brought out by volcanic activity. There was indeed a great deal of volcanic activity at that time – half of the continent of India is made up of lava flows that erupted just

then. Volcanic activity as intense as that could have had exactly the same effect as a meteorite strike, blanketing the Earth with smoke and fumes. India and Yucatan were on exactly opposite sides of the globe 65 million years ago. A massive meteorite impact in one could have sent shock waves through the crust and instigated volcanic activity in the other. Or the meteorite could have split into two, one part falling in India, the other in Yucatan. In any case, the theories are popular.

However, there were other subtleties. Statistical analysis of fossil finds suggested that the dinosaurs had been dying out before the end of the Cretaceous. Perhaps they were already on their way when an impact finished them off. Whatever the cause, the reign of the dinosaurs came to an end. The land was swept clear, to be colonized by mammals – small and insignificant while the dinosaurs were alive, but adaptable enough to take over once they had gone. They evolved and spread into every niche that had been previously occupied by the dinosaurs.

The first major mass extinctions identified in the fossil record are those at the Cambrian-Ordovician boundary and during the Silurian, both of which reduced the numbers of species of marine invertebrates. Then came the

extinction at the Devonian-Carboniferous boundary that may have eliminated 70 per cent of all species. The first that affected the tetrapod fauna was at the Permian-Triassic boundary, and this ended the dominance of the mammal-like reptiles and allowed the dinosaurs to gain a foothold. At the Triassic-Jurassic transition the big amphibians and most of the dinosaurs' relatives were wiped out leaving the dinosaurs themselves with a clear run. It is possible that we are currently living through a sixth mass extinction.

Below: Volcanic eruptions occurred throughout the Age of Dinosaurs producing the Triassic ash beds seen here. However, none was so great as that which occurred at the end.

THE FOSSILIZATION PROCESS

After an animal dies and its body is subjected to the different destructive forces of taphonomy, what is left may be fossilized. This process involves the burial and preservation of the remains. The sediments in which the remains are buried become sedimentary rock, and the original remains are turned into mineral.

Taphonomy is the study of what happens to an organism after its recent death and before it becomes a fossil. Diagenesis is what happens thereafter to turn what remains of the body into something that will withstand the passage of time.

The fossils that we pick up from the rocks, or that we see in museums, have changed a great deal in substance and appearance since they were parts of living organisms.

Fossil formation

The fossil is formed at the same time as the sediment in which the organism lies is turned to sedimentary rock. This diagenesis usually involves two processes – compression and cementation. Compression is produced by the sheer weight of the overlying sediments, and acts to compact the grains of the sediment into a more coherent mass. In cementation, the ground water percolating through the rock deposits mineral, often calcite (the same as builders' cement), between the grains gluing them together as a solid mass. Further compaction and the input of heat from the Earth's interior may alter the mineral content of the rock. If this

Below: A dinosaur's body is washed up on a river sandbank. It has been dead a few days, and its stomach is bloated by decomposition. Beneath its body are the river sands.

happens it will take the rock from the realms of a sedimentary rock to a metamorphic rock, a new type in which usually all the fossils are destroyed.

In the rare occurrences in which an embedded organism becomes a fossil in a sedimentary rock, one of a number of different processes may be involved. An organism, or part of an organism, may remain unaltered. This is a very rare form of fossilization, but we sometimes see it in an insect preserved in a piece of amber. The procedure is simple. An insect settles on a glob of resin seeping from a tree and becomes trapped. It is engulfed and preserved before it has a chance to decay. Later the resin becomes buried and turned into amber through the normal processes that turn sediment to rock, and the insect is preserved within. Sometimes, though, this perfect preservation is illusory as the insects innards may have decayed through the action of its own bacteria. There are no dinosaur fossils that have been

Below: In a dry season the following year, the vegetation has not survived, the river has dried up and mud cracks appear in the mudbanks, and the dinosaur's body is now a skeleton. The neck and tail are pulled back, as the tendons have dried and shrunk. Other scavenging dinosaurs take the last bits of nutrition from the skeleton.

preserved like this, and no dinosaur blood in preserved mosquito stomachs.

The hard parts of an organism may, however, remain unaltered. We find this in the teeth of sharks from the last few tens of millions of years. The teeth are so much tougher than the rest of the skeleton that they survive for a long period. The bones of Ice Age mammals trapped in the tar pools of Los Angeles are another example. Again, no dinosaurs are preserved like this. Sometimes only the original carbon of the organic substance remains. This is seen in the black shapes of leaves that are sometimes seen in plant fossils. Taken to an extreme, this process gives us coal.

Over long periods of time, the cellular structure of the original may be replaced by a totally different substance. Silica in ground-water

Below: Over subsequent years mud and sand are deposited by flooding. The movement of the sediment means that the skeleton begins to break up and be dispersed. The skeleton is still white, fresh bone.

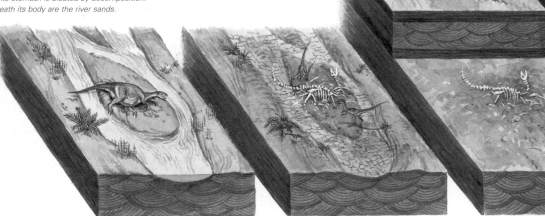

passing through rocks may replace the original carbon, molecule by molecule, and give a fossil that shows the original microscopic structure, but made of silica. Petrified wood is a good example. It is also seen in some Australian plesiosaurs, in which the

Below: Eventually the skeleton is buried so deep below the surface that the sediments become rock and the skeleton becomes a fossil. The previously white skeleton turns black as the bone becomes mineralized. Above the dinosaur may be river sands and muds, beds of conglomerate (fossilized shingle), and marine limestone.

bone has been replaced by opal.

Groundwater percolating through the rock may also dissolve away all traces of the original organism. The result is a hole in the rock called a "mould", in exactly the same shape as the original. Permian reptiles from the desert sandstones of Elgin, in Scotland, occur in this way, and pouring latex into the moulds produces casts of the original bones. This casting is sometimes done naturally, with

Below: The mountain is rising. The sedimentary rocks are uplifted and distorted. The skeleton is distorted too.

dissolved groundwater minerals filling the moulds. Sea urchins in chalk are sometimes found replaced by flint. Then there are fossils which contain no part or impression of the original animal at all. They are called trace fossils and encompass footprints and trackways, coprolites and eggs.

All these processes mean nothing to us unless the fossil is returned to the surface. This only happens when the rocks containing the fossil are uplifted through earth movements, usually mountain building associated with the movement of the tectonic plates. Then erosion has to wear away all the rocks above so that the fossil is exposed. If this erosion is too vigorous, the exposed fossil will not last for long as it will be eroded away as well. All in all, even if a dinosaur does become fossilized, the odds are very much against our finding and excavating it. There are good reasons why dinosaur fossils are rare.

Below: Tens of millions of years later the rocks erode, revealing the fossilized skeleton.

EXCAVATION IN THE FIELD

Once the fossil animal is discovered, the job of excavation can take on military-style logistics. Not only does the skeleton have to be extracted without damage, but the setting and the surrounding rocks must be analysed as well to give as full a picture as possible.

Usually a fossil skeleton is found by chance, by somebody out walking, or by a quarry worker turning over a rock. A *Stegosaurus* in Colorado was found recently when a palaeontologist working on another dig threw his hammer at random into a cliff. Another, in Montana, was found by a farmer digging a hole for a fence post. Once found, the skeleton is reported to a museum, a university or another institution that has the means to excavate it. Planning can take months or even years, and much of this involves finding the money to do the work – because practical palaeontology can be a very expensive business.

On site, the first thing is to find out how much of the skeleton there is. The overburden – that is the rock directly over the skeleton – has to be removed. This can be done with earth movers. Then, when the rock is down to a few centimetres of the bed that contains

Below: A palaeontologist marks out a grid over the bones, providing a reference to all finds in preparation for drawing a site map.

the skeleton, the last of the overburden is removed carefully by hand, usually with fine tools and brushes.

Once the skeleton has been exposed, the next phase is to catalogue what is there. A site map is drawn. This is a plan of the site showing where each

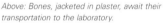

Above: Bones, jacketed in plaster, await their transportation to the laboratory.

bone lies, and the presence of anything else that may be of interest. Only when all this has been done can the excavation begin. Fossils that are newly exposed to the air may be very fragile. They may be unstable, and chemicals in them may react with the atmosphere and cause the fossil to decay quickly. Exposed bones are treated as quickly as possible with a chemical or varnish to seal them, and stop the air causing deterioration. The exact nature of the chemical or varnish has to be recorded for the technicians at the laboratory or the museum.

Fossil bones can be so fragile that they crumble away if they are lifted. To avoid this, they are encased in a jacket. It consists of a layer of moist paper, then a layer of plaster bandages – just like the plaster cast used in medicine for repairing a broken limb. (In the days of the "bone wars", this technique was pioneered by a fossil hunter, using rice, the staple food of the expedition.) The exposed part of the bone is covered in this plaster jacket. Then the rock surrounding and underlying it is dug away, the bone turned over, and the rest of the jacket applied to the other side. How many jackets, and how big a part is jacketed at one time, will depend on the state of the skeleton. In an associated skeleton nearly every individual bone needs to be jacketed. If the skeleton is articulated and there is access to heavy lifting equipment, the whole skeleton can be jacketed in one go. The complete Colorado *Stegosaurus* mentioned above was airlifted from the site by helicopter, but it is rare for such expensive resources to be put at the disposal of a dinosaur excavation.

Once the skeleton has been removed the adjacent area is sifted for other specimens. Palaeontologists will look for the teeth of animals that may have scavenged the skeleton, the seeds of the plants that lived at the same time, and all sorts of other information that can be used to build up a picture of the animal's life.

Back in the laboratory, preparators, the technicians skilled in handling fossil material, prepare the fossil so

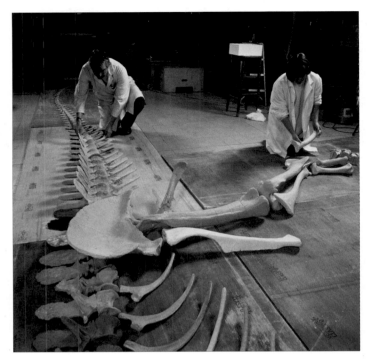

Below: The mounted skeleton of Triceratops *is exhibited in life-like pose. It is made from a cast of the fossils.*

Above: A mounted skeleton of Baryonyx *is being prepared for exhibition. The casts, made of glass-reinforced plastic, are a lightweight material better suited to exhibition.*

that it can be studied by the palaeontologist. Jackets are removed from the skeleton, and any stabilizing chemical is removed or replaced with one that is more appropriate for laboratory study.

In the past, the ultimate result was a mounted skeleton, with the fossil bones clamped to a welded steel frame erected in the public display area of a museum. Today a display skeleton is formed using casts of the skeleton rather than the actual fossils. Technicians make moulds from the fossils and cast reproductions of the bones in a lightweight material that is easier to handle, and which can be mounted more efficiently. This allows for the actual fossils to be stored and kept available for study. Missing bones are replaced by casts from other skeletons, or sculpted by artists to complete the mount.

THE PALAEONTOLOGISTS

There are many hundreds of palaeontologists, the scientists who have discovered and studied the fossils,
who should receive a mention in this list, all having pushed forward the frontiers of palaeontology or
still actively doing so. The following is merely a selection of those who have contributed to the science.

Pyrotherium *discovered by Ameghino.*

Florentino Ameghino
(1857–1911)
More famous for his work on the
unique fossil mammals of South
America, Ameghino pioneered the
study and excavation of dinosaurs and
other extinct vertebrates in Argentina
in the late nineteenth century. Most of
his discoveries are now housed in the
La Plata Museum, Argentina.

Roy Chapman Andrews
(1884–1960)
Andrews led several expeditions from
the American Museum of Natural
History into the Gobi desert in the
1920s. The intention was to find the
earliest remains of human ancestors.
Instead he found a vast array of new
dinosaurs in Cretaceous rocks.
Perhaps the most important find was
the first example of a dinosaur nest
with its eggs. He pioneered the use
of motorized transport to reach
fossil sites.

Robert T. Bakker (1945–)
The concept of warm-bloodedness in
dinosaurs is associated more with the
charismatic Bakker than anyone else.
Since the 1960s he has maintained that
dinosaurs were lively active animals,

using many lines of evidence, including
comparative anatomy and fossil
population studies. He has named
several new dinosaur genera.

Rinchen Barsbold (1935–)
Barsbold is a Mongolian
palaeontologist who, since the 1980s
has worked with the Palaeontological
Centre, Mongolian Academy of
Sciences, Ulan Baatar, and has done a
great deal to uncover and name the
central Asian dinosaurs.

José F. Bonaparte (1928–)
The most famous contemporary
Argentinean palaeontologist,
Bonaparte has added hugely to the
understanding of dinosaurs in South
America. His work includes studies of
the late Cretaceous armoured
titanosaurs and an investigation of
South American pterosaurs that led to
a renaissance of the subject.

Robert Broom (1866–1951)
Broom was born in Scotland and
moved as a doctor to South Africa in
1903. He was the principal discoverer
of the Permian and Triassic mammal-
like reptiles of southern Africa. He also
did important work on early hominids.

William Buckland (1784–1856)
The first professor of geology at
Oxford University, Buckland was the
first to publish a scientific description
of a dinosaur – *Megalosaurus* in 1824.
Dinosaur bones had been noted before
but never studied seriously. He
recognized that the fossil jawbone and
teeth that had been brought to him
had come from a giant reptile of some
kind. This was before the concept of a
dinosaur had been established.

Edwin Colbert (1905–2001)
This curator of the American Museum
of Natural History and later the

Camarasaurus *discovered by Cope.*

Museum of Northern Arizona is
famous for his discovery, in the 1960s,
of the fossil of a mammal-like reptile,
Lystrosaurus, in Antarctica. This
discovery helped to confirm the
understanding of the movements of the
continents caused by plate tectonics.
He famously studied the bonebed of
Coelophysis found in New Mexico.

Edward Drinker Cope (1840–97)
Cope was one of the two figures in the
nineteenth century "bone wars" – a
long-lasting rivalry with Othniel
Charles Marsh to find and describe as
many dinosaur specimens from the
newly-opened Midwest of North
America as possible. He worked from
Philadelphia, USA.

Eberhard Fraas (1862–1915)
Most dinosaur discoveries had been
made in Europe and in North America.
Fraas, a German authority on
dinosaurs sent an expedition, led by
Werner Janensch, to German East
Africa (now Tanzania) in 1907 and
uncovered the Jurassic dinosaur
deposits at Tendaguru, including the

Moschops *discovered by Broom.*

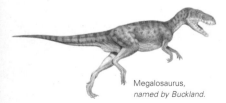

Megalosaurus,
named by Buckland.

huge *Brachiosaurus* (now renamed *Giraffatitan*) that for years stood in the Humboldt Museum in Berlin as the biggest mounted skeleton in the world.

John R. (Jack) Horner (1946–)

As state geologist for Montana, USA, Horner was the principal investigator of the dinosaur nesting sites of Egg Mountain and Egg Island, and named *Maiasaura*. The result of his work has been a new understanding of dinosaur family and social life.

Louis S. B. Leakey (1903–72)

Born in Kenya, Leakey is noted for his discoveries of early hominids in Africa. His belief that early humans were to be found in Africa rather than Asia was borne out by his finds, most famously the fossils in the Olduvai Gorge in Tanzania, which he excavated with his wife Mary.

Joseph Leidy (1823–91)

The first dinosaur to be studied in North America, was found in New Jersey, and named by Leidy. *Hadrosaurus* is now regarded as a *nomen dubium* as there was no skull to identify it. Leidy was based in Philadelphia and went on to name more dinosaurs. He is best known for his work on fossil Tertiary mammals.

Gideon Mantell (1790–1856)

As a country doctor in south-east England, Mantell collected fossils in his spare time. He and his wife Mary Ann found the bones of *Iguanodon* and named it in 1825. As time went on he spent less time as a doctor and devoted his time to amassing fossil collections, which he established in Brighton. This collection was opened to the public as a museum of the local geology and fossils, and it was eventually sold to the British Museum in 1838.

Maiasaura *was named by John R Horner.*

Othniel Charles Marsh (1831–99)

Marsh was Cope's opponent in the "bone wars", working from Yale University. They sent rival teams to sites to try to outdo one another. Before this time there had been only six genera of dinosaurs described. By the time the frenzy of the bone wars was over there were more than 130.

Herman von Meyer (1801–1896)

The first German palaeontologist, von Meyer described and named the first bird *Archaeopteryx*, as well as some of the pterosaurs from the same area of southern Germany. He also discovered *Plateosaurus*, and pioneered the study of dinosaurs as well as invertebrate fossils in Germany and northern Europe.

John H. Ostrom (1928–2005)

With his discovery and description of *Deinonychus* in 1969, American palaeontologist Ostrom established the evolutionary connection between birds and dinosaurs. This also gave rise to the theory that the dinosaurs, at least the theropods, were warm-blooded like birds. The theory was eventually vindicated by the discoveries of the feathered dinosaurs and dinosaur-like birds in China.

Richard Owen (1804–92)

It was Sir Richard Owen, a British anatomist working at the British Museum (Natural History) – now the Natural History Museum – who is credited with creating the concept of dinosaurs by announcing at a meeting of the British Association for the Advancement of Science, in 1841, a new group of animals named the Dinosauria – based on *Megalosaurus*, *Iguanodon* and the ankylosaur *Hylaeosaurus*, which was also found by Mantell.

Ernst Stromer von Reichenbach (1870–1952)

This German palaeontologist was the first to excavate the dinosaur sites of Egypt. He discovered such dinosaurs as *Aegyptosaurus*, *Carcharodontosaurus* and *Spinosaurus*. His specimens were lost during World War II when the museum in Munich was bombed. His excavation sites were forgotten too, until rediscovered by a team from Washington University in 2000.

Harry Govier Seeley (1839–1909)

A British palaeontologist, his main contribution was to divide the dinosaurs into the orders Saurischia and Ornithischia based on their hip structures. He named a number of dinosaurs, but most were fragmentary and have been renamed or declared *nomen dubia*. He also published an early pioneering account of pterosaurs.

Paul Sereno (1957–)

Perhaps the most prolific dinosaur hunter today, Sereno, based in Chicago, USA, has discovered new dinosaurs in North Africa and in central Asia. In 1986 he advanced the understanding of dinosaurs by reclassifying the ornithischians in a system that is still used today.

Dong Zhiming (1937–)

The most famous modern Chinese palaeontologist, Dong opened up the vast dinosaur beds of Sichuan and in north-west China. Working with the Institute of Vertebrate Palaeontology and Palaeoanthropology in Beijing, China, he named about 20 new dinosaur genera. Among his wealth of discoveries he established the homalocephalid family of bonehead dinosaurs.

Coelphysis, *named by Cope but extensively studied by Colbert.*

Anserimimus *discovered by Barsbold.*

THE DEVONIAN, CARBONIFEROUS AND PERMIAN PERIODS

The Devonian, Carboniferous and Permian periods are collectively termed the upper Palaeozoic era. At this time in Earth's history, the scattered landmasses of the world were moving towards each other. As they collided, mountain ranges arose along the leading edges. These mountain ranges rapidly eroded, with the sand and rubble spreading out in the surrounding shallow deltas and swamps. Moist lowlands became colonized by plants. This is where the first vertebrates set foot on land.

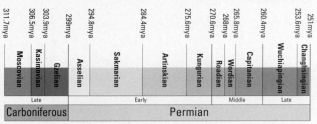

Above: The timeline shows the different chronological stages of the Devonian, Carboniferous and Permian periods of Earth's history. (mya = million years ago)

1 *Edaphosaurus*
2 *Dimetrodon*
3 *Diadectes*
4 *Eryops*
5 *Diplocaulus*
6 *Eudibamus*

THE EARLIEST TETRAPODS

The term "tetrapod" refers to all vertebrates excluding what we would call the fish. The word literally means "four-footed". However, the classification also covers things like birds that only have two feet, and whales and snakes that have none. The main point is that they are all descended from four-footed ancestors, and the remains of these are to be found in Devonian and Carboniferous rocks.

Ichthyostega

This is the classic animal that shows transitional features between those of a fish and an amphibian. The skull is very fish-like and the tail carries a swimming fin. However, the legs are those of a tetrapod, but seemingly evolved for swimming and pushing through aquatic plants in shallow water.

Right: The front feet of Ichthyostega *have never been found, so it is impossible to say if they had five or more toes.*

Features: The skull is very much like that of a fish. The backbone is stronger than that of the most closely related fish, and the rib-cage is a strong structure of overlapping ribs – all weight-bearing features. The legs are built very much like those of every other tetrapod, with a single upper bone (a humerus in the front legs and a femur in the rear) and two lower bones (the radius and ulna, and the tibia and fibula). The hind limbs are smaller than the forelimbs and acted as paddles. The hind foot has seven toes. The ears appear to have been adapted to work primarily underwater.

Distribution: East Greenland.
Classification: Tetrapod.
Meaning of name: Fish roof.
Named by: Säve-Söderbergh, 1932.
Time: Famennian stage of the Late Devonian.
Size: 1.5m (about 5ft).
Lifestyle: Predator.
Species: *I. stensioei, I. watsoni, I. eigili, I. kochi.*

Acanthostega

This was primarily an aquatic animal, shown by the limb joints that were set for swimming, the paddle tail and the lateral line sensory apparatus, like that of a fish, which was used for finding its way about underwater. Although the limbs were similar to those of a land-living animal, it is thought that it only occasionally clambered out of the water. The backbone was too weak to support the weight of the animal on land, but the presence of heavy scales on the underside shows that it lived close to the ground – presumably in flooded wetlands and stream margins. It seems likely that *Acanthostega* spent its time in shallow rivers.

Features: The jaws and fang-like teeth are very fish-like. Both front and back feet consist of at least eight digits, indicating that the standard pattern of five digits had not yet evolved. The bony structure of the shoulder girdle is like that of a fish, allowing the flow of water past it, and this seems to indicate the presence of gills. The backbone and ribs are not stoutly built, suggesting that *Acanthostega* did not spend much time out of the water.

Distribution: East Greenland.
Classification: Tetrapod.
Meaning of name: Spiky roof.
Named by: Jarvik, 1952.
Time: Famennian stage of the late Devonian.
Size: 60cm (2ft).
Lifestyle: Predator.
Species: *A. gunneri.*

Right: The braincase of Acanthostega *seems to fall somewhere between that of a lobe-finned fish and one of the later true amphibians.*

Pederpes

Distribution: Central Scotland.
Classification: Amphibian.
Meaning of name: Peter's foot.
Named by: Clack, 2002.
Time: Tournasian stage of the early Carboniferous.
Size: 1m (just over 3ft).
Lifestyle: Predator.
Species: P. finneyae.

Pederpes was found in 1971 and lay misclassified in the Hunterian Museum in Glasgow. It was first thought to have been the skeleton of a lobe-finned fish. Later study by Jenny Clack of Cambridge showed that it was a tetrapod, and important because it was one of the few fossils from part of the fossil record that seems to be missing – a stretch of time between the late Devonian and early Carboniferous in which we would expect to find the remains of animals that show the transition between completely aquatic tetrapods and those adapted to land life. This is the so-called "Romer's gap".

Features: The feet seem to have five toes, although this is not very clear on the only specimen that has been found. However, the feet are turned forward, with the middle toe pointing forward in the direction of travel. This puts the feet in the ideal position to work on land making *Pederpes* the earliest-known purely land-living animal. The skull is narrow and deep, suggesting a land-living animal, but the structure of the ear indicates that it spent much of its time in water.

Left: The narrow skull suggests that Pederpes *did not breathe by pumping a throat pouch, but by drawing air into the lungs by muscular action of the body.*

ROMER'S GAP

There are several gaps in the fossil record – periods of time when nothing is known of a particular group and scientists have to wonder about evolutionary developments in that period. One classic example is "Romer's Gap", named after the influential American palaeontologist Alfred Sherwood Romer, whose special study was the evolution of primitive terrestrial vertebrates from the ancestral fish. This is a 30 million year period between the late Devonian and into the early Carboniferous, during which time the clumsy early swimming tetrapods had developed into a wide range of amphibians and amphibian-like animals, many of which pursued a land-living existence. Very few fossils exist from this period of time, possibly because soil or climate conditions did not allow any to form. The recent discovery of *Pederpes* in Scotland, along with the slightly later *Whatcheeria* in America and another called *Ossinodus* in Queensland, Australia – all three from Romer's Gap – is beginning to shed light into this dark area of vertebrate evolution.

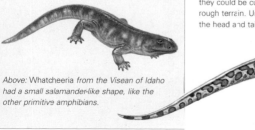

Above: Whatcheeria *from the Visean of Idaho had a small salamander-like shape, like the other primitive amphibians.*

Casineria

Until the discovery of *Pederpes*, *Casineria* was the earliest-known animal with a terrestrially adapted skeleton. It is so advanced that it is regarded as being on the line that led to the amniotes – the animals that laid eggs on land. Its small size – most of which is tail – suggests that the land-living animals did not evolve from the big aquatic Devonian tetrapods, but from much smaller relatives. It probably lived like a modern salamander, scuttling about in moist undergrowth, feeding on the insect life that abounded in the early land vegetation.

Features: The most obvious feature of the single skeleton of *Casineria* that has been found is the clear adaptation to land-living. The backbone is much stronger than in any of the water-dwelling Devonian amphibians, and the humerus is weight-bearing like that of a reptile. The five fingers and toes have notches that must have held tendons, indicating that they could be curled – an adaptation to walking over rough terrain. Unfortunately the skeleton lacks the head and tail.

Distribution: Central Scotland.
Classification: Tetrapod.
Meaning of name: Cheese Bay.
Named by: Smithson and Clack, 1999.
Time: Visean epoch of the early Carboniferous.
Size: 15cm (6in).
Lifestyle: Insectivore.
Species: C. kiddi.

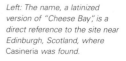

Left: The name, a latinized version of "Cheese Bay," is a direct reference to the site near Edinburgh, Scotland, where Casineria *was found.*

THE AGE OF AMPHIBIANS

By the time the Carboniferous period was under way, the group that we loosely call the amphibians had developed into a wide range of animals. Many had adapted well to their life on land, and indeed the thick deltaic forests of the time supported a wealth of invertebrate life that acted as a food source, but strangely some abandoned this lifestyle and returned to the water.

Eucritta

By early in Carboniferous times the tetrapods had diverged – one route leading to the temnospondyls, the ancestors of modern amphibians, and another leading to the amniotes, those animals that laid hard-shelled eggs. The fact that *Eucritta* is known from several specimens means that we can see how the animal grew throughout life. In particular the shoulder girdle became more sturdy as it grew, probably an adaptation to land-locomotion.

Features: At the time *Eucritta* existed the divergence between temnospondyls and amniotes had not really established itself. This animal, known from about five specimens, seems to show features of both. It seems to be very close to the most primitive of the temnospondyl amphibians in the characters of the skull. However, there is a palate on the roof of the mouth that amphibians tend to lack. The legs are spread wide and the feet have five toes.

Left: The full name of Eucritta melanolimnetes *translates as "the true creature from the Black Lagoon."*

Distribution: Central Scotland.
Classification: Amphibian, Baphetidae.
Meaning of name: True creature.
Named by: Clack, 1998.
Time: Visean epoch of the early Carboniferous.
Size: 20cm (8in).
Lifestyle: Insectivore.
Species: *E. melanolimnetes.*

Crassigyrinus

This was a totally aquatic hunter, swimming with a sinuous motion through the weed-choked ponds of early Carboniferous central Scotland, snapping up fish and smaller amphibians. Its big eyes would have been well adapted to hunting in murky waters, and the powerful jaws and strong spiky teeth would have formed an effective fish trap. It would have hunted through the muddy swamps with powerful sideways movements of its body and tail, stabilizing and steering itself with its tiny limbs.

Features: The body is long and sinuous, and the limbs, particularly the forelimbs, are very tiny and of no use for locomotion on land. The skull is known from three complete and several incomplete specimens, but they do not give enough information for the animal to be precisely classified. The jaw bones have anchor points for strong muscles and the roof of the mouth is very much like that of a lungfish but carries huge teeth that are unique to this genus.

Above: Fossil amphibian expert Alec Panchen visualized the lifestyle of Crassigyrinus *as similar to that of a Moray eel.*

Distribution: Scotland.
Classification: Amphibian, Anthracosaurs.
Meaning of name: Thick frog.
Named by: Watson, 1929.
Time: Visean epoch of the early Carboniferous
Size: 2m (6½ft).
Lifestyle: Fish-hunter.
Species: *C. scoticus.*

Greererpeton

Distribution: West Virginia, Illinois, USA.
Classification: Amphibian, Colostidae.
Meaning of name: Amphibian from Greer.
Named by: Romer, 1969.
Time: Visean epoch of the early Carboniferous
Size: 1.5m (5ft).
Lifestyle: Fish-hunter.
Species: *G. burckemorani.*

Features: *Greererpeton* has a broad flat head, a short neck and a long body and tail. The head is about a quarter of the length of the body and the back has about 40 vertebrae – twice as many as is found in contemporary amphibians. There appears to be no ear structure, but instead the skull has open grooves that may indicate the presence of a lateral line sensory system like that used by fish for sensing prey underwater.

Left: Greererpeton *evidently reverted to an aquatic way of life soon after its ancestors left the water and came out on to land.*

Several eel-like tetrapods swam about in the early Carboniferous swamps and deltas, making little use of their newly evolved four legs. They would have swum by lateral undulations of the body, stabilized by the tiny limbs. *Greererpeton* was one, discovered in the vicinity of Greer, in West Virginia, and more recently found as jaw fragments in Illinois, USA. Several specimens are known, showing different sizes and ages of individuals, the youngest being about a quarter of the size of the adults.

FROM FISH TO TETRAPOD

The development of ancestral tetrapods from the fish involves physical changes.
1. The loss of several bones from the skull. This isolated the head from the shoulders and produced a recognizable neck.
2. The loss of bones that support internal gills. The gills themselves may have been retained in some genera.
3. The development of a backbone of strong vertebrae as the main organ of support, rather than a notochord – a rigid rod of gristle – and the subsequently reduced notochord not reaching into the skull.
4. The development of muscular limbs with separate fingers and toes.
5. The formation of a bony connection between the backbone and the hips (the sacrum). This is a weight-bearing adaptation.
6. The loss of the fin rays – structures that would get in the way on land.
7. The first feet did not have the standard five-toed pattern.

Ophiderpeton

Ophiderpeton was an aistopod – one of a range of specialized limbless amphibians that existed from early Carboniferous times to the earliest Permian. They diverged from the main amphibian group quite early and probably pursued a burrowing lifestyle in the peaty soil of the coal forests where worms, centipedes, snails and all sorts of other invertebrates thrived. The large eyes were positioned to point forward – an adaptation to a hunting way of life.

Features: The body is extremely long, containing something like 200 vertebrae, while the tail is very short. There is no sign of limbs or of limb girdles. The skull is reduced, with some bones missing and others fused together into fewer structures, resulting in just a few bones supporting the braincase and the jaws. *Ophiderpeton* has a large number of species, and there are many similar genera, such as *Phlegethonia* and *Sillerpeton*. There is also one, *Lethiscus*, known from Devonian rocks.

Distribution: Widespread from Ohio, USA, to the Czech Republic.
Classification: Amphibian, Aistopoda.
Meaning of name: Snake amphibian.
Named by: Huxley, 1865.
Time: From the early Carboniferous to the early Permian.
Size: 70cm (27½in).
Lifestyle: Burrowing insectivore.
Species: *O. granulosum, O. vicinum, O. pectinatum, O. zieglerianum.*

Right: Ophiderpeton *must have looked rather like a snake, and probably pursued the same kind of lifestyle as modern legless lizards or amphibians like a caecilian.*

AMPHIBIANS OF THE COAL SWAMPS

The Carboniferous was the time of the coal swamps. The closing of the continents, which brought Europe and Africa into contact with North America, threw up a huge mountain range which immediately began to erode. The quantities of sediment that were produced spread across the seas and formed deltas and swamps, which supported a thick tropical vegetation. These swamps were full of swimming amphibians.

Proterogyrinus

Right: Proterogyrinus *was one of those amphibians that seemed at home in the tangled undergrowth of the forests as well as in the murky waters of the swamps.*

This was a large amphibian well adapted for life both on land and in water. The hind part, including the tail, was fish-like, while the forequarters were more adapted to a terrestrial way of life – leading fossil amphibian expert Jenny Clack to surmise that in the transition from water-living to land-living animals, the changes started at the tail and worked forward. The skull is very amphibian-like, but its hips, limbs and vertebrae are almost reptilian. The toes, in particular the number of toe bones, seem to be intermediate between the amphibians and the reptiles.

Features: The well-developed limbs show *Proterogyrinus* to have been able to walk on land, and the flattened tail is obviously a swimming aid. The neck is short and the body has only 32 vertebrae in front of the hips – quite a small number compared to more aquatic amphibians. The skull is lightly built and the jaw could flex during opening, like that of ancestral fish. The eardrum seems adapted to work underwater rather than on land.

Distribution: Scotland; West Virginia, USA.
Classification: Amphibian, Embolomeri.
Meaning of name: Early twister.
Named by: Romer, 1970.
Time: Serpukhovian stage of the early Carboniferous.
Size: 1m (about 3ft).
Lifestyle: Fish-hunter.
Species: *P. scheeleri*, *P. pancheni*.

Pholidogaster

Pholidogaster is known from two specimens found in the early Carboniferous rocks of the Midland Valley of Scotland. Historically it was one of the first to show science the evolutionary link between fish and amphibians. Although the early Carboniferous was before the time of the coal measures, the swamps and forests were already beginning to spread, providing the ideal environment for large, amphibious animals.

Below: The species name, Pholidogaster pisciformes, *reflects the similarity of this creature's body shape to a fish.*

Features: This animal has a very long and slender body, with small and feeble limbs. The shoulder structure is further back than is usual. Belly scales are present, suggesting that in life it did not just swim, but crawled over hard surfaces as well. The structure of the jaw is not clear, since the jaw bones on both specimens are not well preserved. However, there are large fangs in the front of the mouth, presumably used in hunting.

Distribution: Scotland.
Classification: Amphibian, Embolomeri.
Meaning of name: Scaly stomach.
Named by: Huxley, 1862.
Time: Visean epoch of the early Carboniferous.
Size: 1m (about 3ft).
Lifestyle: Fish-hunter.
Species: *P. pisciformis*.

Eogyrinus

Distribution: Northern England.
Classification: Amphibian, Embolomeri.
Meaning of name: Dawn twister.
Named by: Watson, 1926.
Time: Moscovian stage of the late Carboniferous.
Size: 4.5m (15ft).
Lifestyle: Fish-hunter.
Species:
E. attheyi.

This monster must have lived like a crocodile, cruising the murky waters of the coal swamps, lying in wait among the spreading roots of the giant club moss trees and in the reed beds of the great horsetails, waiting for unsuspecting prey to swim by.

Below: Recent studies by Jenny Clack suggest that an amphibian Pholiderpeton described by Huxley in 1869 is the same animal as Eogyrinus. If this is so then Pholiderpeton's name takes priority.

Features: The body is long and sinuous and the tail muscular and flattened from side to side. This must have been the main organ of propulsion, assisted by the dorsal fin along the tail. It is one of the largest-known inhabitants of the coal swamps. It is completely covered by scales, more like a reptile than any modern amphibian. This feature seems to be left over from the ancestral fish, and suggests that these ancient amphibians did not breathe partly through their skins as modern amphibians do.

SHALLOW WATER FEEDERS

Many of these Carboniferous amphibians became highly adapted to living and feeding underwater in the shallow pools and streams of the tropical swamps. Their limbs became weaker and their bodies flatter. One problem, though, was the articulation of the jaw. In primitive forms the articulation is at the back of the skull, behind the point of attachment of the vertebra column. This means that the mouth could only be opened by dropping the jaw downwards – the stiff neck being unable to lift the skull itself. This is no problem in open water, but on land and in shallow water there would not have been room for the mouth to open properly. Eventually the articulation changed, so that the jaw joint moved directly beneath the condyle – the joint between the backbone and the skull – and this enabled the skull to tilt upwards to open the mouth while the jaw was still on the ground.

Below: If the backbone and the jaw are in the same plane, the mouth opens easily.

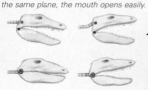

Cochleosaurus

Judging by the number of remains found *Cochleosaurus* seems to have been one of the main medium-sized predators of the coal swamps of Eastern Europe. All of the big water-dwelling amphibians of this time were carnivorous. Many different amphibian types are known from here. It was quite a widespread animal, another species being known from Canada. At that time North America and Europe were part of the one landmass, and there were continuous swamps along the northern edge. The swamps were full of lush vegetation and supported a variety of animal life that acted as prey for the big amphibians.

Below: The Canadian specimen is one of those found in the fossilized tree stumps in Nova Scotia.

Distribution: Nova Scotia; Eastern Europe.
Classification: Amphibian, Temnospodili.
Meaning of name: Ear lizard.
Named by: Fritsch, 1885.
Time: Moscovian stage of the late Carboniferous.
Size: 1.6m (5ft).
Lifestyle: Hunter.
Species: C. bohemicus, C. florensis.

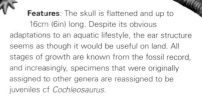

Features: The skull is flattened and up to 16cm (6in) long. Despite its obvious adaptations to an aquatic lifestyle, the ear structure seems as though it would be useful on land. All stages of growth are known from the fossil record, and increasingly, specimens that were originally assigned to other genera are reassigned to be juveniles of *Cochleosaurus*.

PRIMITIVE AMNIOTES AND REPTILES

Amphibians lay eggs in water and their young pass the first part of their life in an aquatic environment, like the contemporary tadpole that transforms into a land-living adult frog. The amniotes developed the ability to live without the water-living part of this cycle by laying leathery-shelled eggs lined by an amniotic membrane that kept water in but allowed oxygen to pass in and carbon dioxide out.

Westlothiana

Known from two almost complete skeletons *Westlothiana* was found in early Carboniferous oil shale deposits near Edinburgh, in Scotland. It was a land-living animal that scampered through the undergrowth around a tropical freshwater lake, along with huge terrestrial invertebrates such as scorpions and millipedes that were larger in size than it was. Perhaps forest fires drove these animals into the lake where they drowned and fossilized. *Westlothiana* was at first hailed as the earliest reptile, but later researchers are not so sure.

Features: The original classification of *Westlothiana* as a reptile is based on the skull, the humerus and the vertebrae. However, there is not enough information to fit it into any known reptile group. It is now regarded as an amniote but something more primitive than any reptile. There are heavy lizard-like scales all round the body. The arrangement of bones in the wrist, the primitive articulation of the back-bones and the structure of the roof of the mouth are like those of an amphibian or an early amphibian-like vertebrate. In appearance and lifestyle *Westlothiana* must have looked very much like a modern lizard.

Below: When first found by professional fossil hunter Stan Wood, this creature was nicknamed "Lizzy the Lizard". This eventually found its way into the species name as Westlothiana lizziae.

Distribution: Central Scotland.
Classification: Uncertain, possible amniote or pre-amniote.
Meaning of name: From the county of Westlothian.
Named by: Smithson, 1989.
Time: Visean stage of the early Carboniferous.
Size: 30cm (1ft).
Lifestyle: Insectivore.
Species. *W. lizziae.*

Hylonomus

The mode of preservation of the best-preserved fossil of *Hylonomus* is interesting. It was found inside the hollow stump of a dead lycopod tree that had grown within the coal forests. Such a stump would be colonized by insects, and insect-eating animals would seek them out. The skeleton of *Hylonomus* was found curled up in the fossil of such a stump. It may be that the *Hylonomus* individual climbed into the stumps to feed on the insects, or it is possible that it fell in and could not escape. Nevertheless its presence in the stump is taken as an indication that these were land-living animals.

Features: Long regarded as the earliest reptile known, *Hylonomus* has the appearance, and probably had the lifestyle, of a modern lizard. There are five long toes on each foot and the tail is long and the neck short. Its insectivorous diet is shown by the short conical teeth, although the teeth at the front are a little longer than the rest. Tooth specialization such as this is usually only found on more advanced reptiles. The legs are held out to the side in the typical sprawling posture of lizards.

Below: A similar-looking animal, Palaeothyris, is also known from the Joggins area of Nova Scotia. Another, Petrolacosaurus, comes from the late Carboniferous of Texas, USA.

Distribution: Nova Scotia, Canada.
Classification: Reptile, parareptilia.
Meaning of name: Forest mouse.
Named by: Dawson, 1860.
Time: Bashkirian stage of the late Carboniferous.
Size: 20cm (8in).
Lifestyle: Insectivore.
Species: *H. lyelli.*

Solenodonsaurus

Distribution: Czech Republic.
Classification: Uncertain, probably an early amniote or an amphibian close to the diadectomorphs.
Meaning of name: Single tooth lizard.
Named by: Broili, 1924.
Time: Moscovian stage of the late Carboniferous
Size: 80cm (2½ft).
Lifestyle: Insectivore.
Species: *S. janenschi.*

Solenodonsaurus was an enigmatic lizard-like animal that has been classed with the earliest ancestors of the reptiles, even though its date makes it too late to be any kind of an ancestor. Latest research seems to put it just on the amphibian side of the line although various studies by different scientists are constantly revising its position in relation to other primitive tetrapods. It is one of the ongoing mysteries of palaeontology. It is known from incomplete skeletons, lacking the hind limbs and tail.

Features: Another small animal of lizard-like appearance, it has sharp slightly curved teeth, suggesting that it had an insectivorous diet. The skull is flat and the ear structure is simple, indicating that its hearing did not have much sensitivity to high-pitched airborne sounds. The back is covered in circular bony scales and there are long scales on its belly, presumably as protection while crawling over the ground. The backbone is more like that of contemporary land-living amphibians than a reptile. However, the teeth do not have the deeply folded enamel structure that we see in contemporary amphibians – the structure that gives rise to the term "labyrinthodont" to describe them.

Below: The rear part of this restoration is speculative, and based on the appearance of other tetrapods of the time.

THE AMNIOTE

The amphibian lifestyle involves spending the early part of life in a watery environment, as an aquatic organism, and then emerging on to dry land as a fully formed animal. The amniote does this, but the water phase takes place inside an egg – the individual's own private pond. The shell of the egg is made of mineral or leathery substance and allows oxygen in and carbon dioxide out. It is able to retain the water inside the egg, and is tough enough to protect the contents.

Three important membranes line the inside –
• The amnion that encloses the embryo with a water fluid.
• The chorion that encloses a bag of food – the yoke.
• The allantosis that encloses a cavity for solid waste.

The embryo must be fertilized before these shells and membranes develop, and so fertilization takes place within the female's body. This is the origin of intimate sexual intercourse – something absent from the lifestyle of fish and amphibians.

Above: A cross-section of an egg.

Diadectes

Diadectes was a big fully terrestrial animal, classed as an amphibian. It is important in being the earliest-known vegetarian amphibian. It was one of the largest animals of its time. Jaw material of a similar animal has recently been found in early Carboniferous rocks of Tennessee.

Features: The skeleton, built for terrestrial living, is very much like that of a reptile, but the skull shows that it is not a member of this group. The teeth are short and adapted for grinding tough food, and set in powerful jaws. The teeth at the front are peg-like and were used for nipping off mouthfuls of vegetation, and those at the back are flat like molars, used for grinding. The limbs stick out sideways, and the bones of the foot are simplified to minimize the stresses involved in moving its great weight with the resulting sprawling posture.

Below: The rather pig-like appearance of Diadectes *is the result of the broad body needed to house the digestive system that would cope with a vegetarian diet.*

Distribution: New Mexico, Oklahoma, Texas, USA; and Germany.
Classification: Amphibian, order uncertain.
Meaning of name: Crossways biter
Named by: Watson, 1917.
Time: Late Carboniferous to early Permian.
Size: 3m (10ft).
Lifestyle: Low browser.
Species: *D. absitus, D. sideropelicus.*

AQUATIC AMPHIBIANS

The dawn of the Permian period brought the Age of Reptiles, but despite the changing landscape, the amphibians continue to be important. The Permian was largely a time of deserts and dry climates, but the desert streams and the ponds of the oases were full of water-dwelling animals, and here the amphibians were top of the food pyramid.

Diplocaulus

The spectacular head of this amphibian may have had two functions. It may have been a hydrofoil device, allowing the animal to remain on a stream bed while facing into a strong current, or it may have made it difficult for any predator to swallow. It is known from a large number of specimens.

Below: The wide tips of the head may have been linked to the body by skin flaps, as shown here, or they may have been kept free of such flaps.

Features: The obvious feature of this amphibian is the boomerang-shaped head that seems out of proportion to the rest of the body. Apart from that the body is like that of any other aquatic amphibian – broad and flat, long-tailed, with four toes on the front feet and five on the back. We know the whole growth sequence of *Diplocaulus*, showing that the head grew from a normal salamander-like shape in the young form to the full boomerang shape in the adult.

Distribution: Texas, USA.
Classification: Amphibia, Nectridia.
Meaning of name: Double stalk.
Named by: Cope, 1882.
Time: Early to late Permian.
Size: 1m (about 3ft).
Lifestyle: Fish-hunter.
Species: *D. marginocolis.*

Microbrachis

The microsaurs were a group of about 30 tiny amphibians, with small legs and short tails, that were thought to have been restricted to the late Carboniferous and Permian beds. However, the discovery of early–late Carboniferous microsaur *Utahaerpeton* in 1991 and an unnamed early Carboniferous form in 1995 pushed back their history considerably. *Microbrachis* was a typical member of the group.

Right: Microbrachis would have looked and behaved something like a modern salamander or an axolotl.

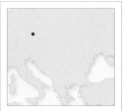

Features: The teeth of this creature are typical of microsaurs in being flattened from side to side. The small-scale structure of the front teeth suggests that the group may be ancestral to modern amphibians. It has the long body of an aquatic animal, with a flexible backbone of many vertebrae and tiny legs. The presence of external gills throughout life, like the modern axolotl (aquatic salamander), is a phenomenon known as "paedomorphosis" – the retention of juvenile features in an adult stage.

Distribution: Czech Republic.
Classification: Amphibia, Microsauria.
Meaning of name: Little branch.
Named by: Fritsch, 1876.
Time: Early Permian.
Size: 15cm (6in).
Lifestyle: Insectivore.
Species: *M. pelikani, M. obtusatum.*

Eryops

Distribution: Texas, USA; and Europe.
Classification: Amphibia, Temnospondyli.
Meaning of name: Big eye.
Named by: Miner, 1925.
Time: Permian.
Size: 1.5m (5ft).
Lifestyle: Fish-hunter.
Species: *E. megacephalus*.

When we think of the giant amphibians of the late Palaeozoic, it is usually *Eryops* that comes to mind. At home in the water as on the land, this was one of the biggest vertebrates of its time, preying on fish, small reptiles and amphibians which it probably hunted in the water. Teeth on the palate suggest that once it caught its prey it would swallow it by throwing its head upwards and backwards, just as crocodiles and alligators do today.

Features: *Eryops* has a stout body, supported by a strong spine with very wide ribs, four short and strong legs and a short tail. It has a big head with broad jaws armed with strong teeth, showing the folded enamel pattern of the typical labyrinthodont. The eyes are on the top of the skull, suggesting that it could lie in wait in the water with only its eyes showing, much like a modern alligator. It would have breathed by expanding the area inside its mouth and pumping air to the lungs using its throat muscles.

Left: Eryops *is known from several complete skeletons, but the head bones are so robust that they are often found as lone fossils.*

LABYRINTHODONTS

The early amphibians are known as labyrinthodonts. This is on account of the structure of their teeth. The teeth were conical, but the enamel, the tough outer covering, was convoluted and infolded into a maze-like pattern along with the bony dentine that underlay it. In some examples there was no pulp, the interior of the tooth consisted of nothing but folded dentine. This is similar to the pattern in the lobe-finned fishes, indicating that the earliest amphibians evolved from this group.

Below: The cross-section of a tooth.

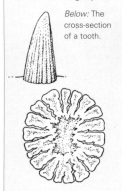

Bageherpeton

The crocodile-like long-bodied swimming hunters, characterized by *Archegosaurus* of the Carboniferous, continued into the Permian. *Bageherpeton* was typical, living in lakes and desert streams and feeding on fish, and other small vertebrates, as well as the many invertebrates that were present. The archegosaurs were first studied in Europe, but the discovery of *Bageherpeton* in South America shows that they were much more widespread.

Features: The notable features of this animal are the long narrow jaws. Long jaws and sharp teeth like this evolved several times in different amphibian lineages, and were evidently an adaptation for catching fish. The lower jaw is strengthened on the inside and this is interpreted as a reinforcement for the powerful jaw action that would crush a fish as soon as it was caught. The presence of fish scales in the same rock as the skull of *Bageherpeton* indicates that it lived among fish in shallow water.

Distribution: Rio Grande do Sul State, Brazil.
Classification: Amphibia, Temnospondyli.
Meaning of name: Crawling animal from Bagé city.
Named by: Dias and Barberena, 2001
Time: Late Permian.
Size: Over 2m (6ft).
Lifestyle: Fish-hunter.
Species: *B. longignathus*.

Left: The long jaws and sharp teeth of Bageherpeton *were similar to those of the modern-day gavial – a fish-eating crocodile of the rivers of South-east Asia.*

LAND-LIVING AMPHIBIANS

The amphibians were primarily water-based animals, needing the water as a place in which to breed. The adult forms of many of the Permian amphibians, however, were well adapted to life on land, despite the fact that the reptiles were also establishing themselves as the major group of land dwellers. These terrestrial amphibians began to show a wide range of adaptation to the dry environments.

Seymouria

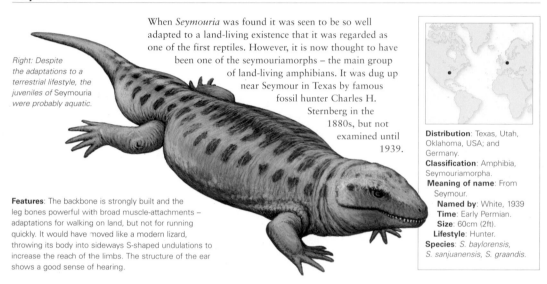

Right: Despite the adaptations to a terrestrial lifestyle, the juveniles of Seymouria were probably aquatic.

When *Seymouria* was found it was seen to be so well adapted to a land-living existence that it was regarded as one of the first reptiles. However, it is now thought to have been one of the seymouriamorphs – the main group of land-living amphibians. It was dug up near Seymour in Texas by famous fossil hunter Charles H. Sternberg in the 1880s, but not examined until 1939.

Features: The backbone is strongly built and the leg bones powerful with broad muscle-attachments – adaptations for walking on land, but not for running quickly. It would have moved like a modern lizard, throwing its body into sideways S-shaped undulations to increase the reach of the limbs. The structure of the ear shows a good sense of hearing.

Distribution: Texas, Utah, Oklahoma, USA; and Germany.
Classification: Amphibia, Seymouriamorpha.
Meaning of name: From Seymour.
Named by: White, 1939
Time: Early Permian.
Size: 60cm (2ft).
Lifestyle: Hunter.
Species: *S. baylorensis*, *S. sanjuanensis*, *S. graandis*.

Platyhystrix

A recurring feature of the land-living animals found in the Texas Red Beds is the presence of a sail on the back. It is found, most notably, among the reptiles in the pelycosaur group but there was also at least one amphibian that showed a structure like this. *Platyhystrix* may have used its sail for temperature regulation or for signalling.

Features: The spines grow upwards from the vertebrae, derived from armour plates that ran along the backs of earlier relatives (see the arrangement on *Cacops*). The spines are flat and expanded at the ends, and covered with coarse pits that suggest that in life they carried a sail of skin or even of horn. However, it seems that the skin sail did not continue to the tops of the spines – only about half way up.

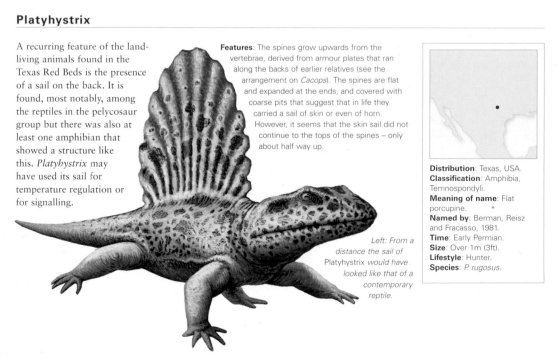

Left: From a distance the sail of Platyhystrix would have looked like that of a contemporary reptile.

Distribution: Texas, USA.
Classification: Amphibia, Temnospondyli.
Meaning of name: Flat porcupine.
Named by: Berman, Reisz and Fracasso, 1981.
Time: Early Permian.
Size: Over 1m (3ft).
Lifestyle: Hunter.
Species: *P. rugosus*.

Peltobatrachus

Distribution: Tanzania.
Classification: Amphibia, Temnospondyli.
Meaning of name: Armoured amphibian.
Named by: Panchen, 1959.
Time: Late Permian
Size: 70cm (2½ft).
Lifestyle: Insectivore.
Species: *P. pustulatus.*

An obvious adaptation to life out of the water would be the evolution of armour, especially in the presence of such fierce terrestrial reptiles as existed in southern Africa in the late Permian. *Peltobatrachus* had an arrangement of armour over its upper surface that was reminiscent of the modern armadillo. With such an armour this animal could have been only a slow mover. However well it was adapted to land life, *Peltobatrachus* probably had to return to the water to lay its eggs.

Features: Many other primitive amphibians show a covering of armour plates, but none is quite as thorough as *Peltobatrachus*. The armour of *Peltobatrachus* consists of bony plates arranged over the shoulders and hips, and in transverse bands in between, in the same pattern as the armour of an armadillo. In life these bony plates would have been covered in horn. The teeth are unknown, but it seems likely that such a slow-moving animal would have eaten insects and other invertebrates. Comparison with closely related amphibians indicates that the jaws probably worked with a snapping action, similar to that of today's angler fish.

Left: In life Peltobatrachus must have looked rather like an armadillo.

NEW ENVIRONMENTS – NEW LIFE

As the Carboniferoust slipped into the Permian, a number of changes took place in the world's geography. For one thing, an ice age gripped the southern continents. The evidence for this can be found in South Africa, India, Australia and South America as well as Antarctica. This reached its most intense in the early Permian and then the glaciers died away gradually.

In addition the continents of the globe were moving together and beginning to assemble into the supercontinent of Pangaea, crushing up mountain ranges in between.

These events meant that the environments were changing. The loss of shallow seas meant that the coal swamps were drying up, and deserts spread everywhere across the vast landmass, producing such dry-land sediments as the Texas Red Beds. The only places that were comfortable for life were the coastal regions.

New plants evolved. One of the most common was a kind of fern that reproduced by a seed. This was called *Glossopteris*, and became the most abundant plant on the southern continent. It became the main food for the land-living insects and invertebrates, and these, in turn, became prey for the land-living amphibians and newly evolved reptiles.

Left: Fossil of the typical oval Glossopteris leaf.

Cacops

Perhaps the amphibians that were most well adapted to a land-living existence were the cacopids. These evolved quite late in the Permian and did not die out until the succeeding Triassic period. *Cacops* was the most typical genus of this family. There have even been trackways found that have been referred to *Cacops*. It may have been a nocturnal animal like most modern frogs.

Features: Bony plates cover the body and there is a row of thick armour plates down the backbone. The legs are well adapted for walking and are almost like those of reptiles. A notch at the rear of the skull would have supported an eardrum in life, showing a sensitivity to airborne sounds – as opposed to the other main hearing mechanism of amphibians involving detecting ground-borne sounds through the jaw bone on the ground.

Distribution: Oklahoma and Texas, USA.
Classification: Amphibia, Temnospondyli.
Meaning of name: Blind-looking.
Named by: Williston, 1910.
Time: Early Permian.
Size: 40cm (16in).
Lifestyle: Hunter or insectivore.
Species: *C. aspideporus.*

Below: The bony plates on the body may have been for protection, against other big land-living amphibians, or to prevent water loss in the dryness and heat of the environment.

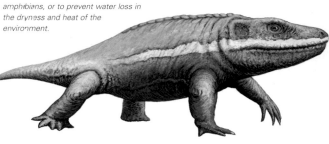

PARAREPTILES AND CAPTORHINIDS

The traditional classification of the reptiles is by the arrangement of holes in the side of the skull behind the eyes. The most primitive in this classification are the anapsids – those that lacked any hole. Nowadays this group is known as the parareptiles because certain features differ from those in the other reptilian lineages. Some members such as Milleretta, *however, did confusingly have holes in the skull.*

Acleistorhinus

When this creature was discovered in the 1960s it was thought to have been an early member of a fairly advanced group, but later studies by Debraga and Reisz in 1996 show that it is actually the earliest known of the parareptiles. Comparison with other vertebrates suggests that the parareptiles evolved from the main reptile stem as far back as the late Carboniferous.

Features: *Acleistorhinus* is a small lizard-like animal. Its closest relatives are *Lanthanosuchus* and *Macroleter* from later deposits in Russia, showing that the parareptiles were very widespread during the early part of their history. The head is quite solid and the teeth unspecialized for anything in particular – they are all about the same size, conical and only slightly curved. The jaw is short and very powerful. *Acleistorhinus* probably fed on the insects that abounded at the time.

Distribution: Oklahoma, USA.
Classification: Reptile, Parareptilia.
  Meaning of name:
    Anchor nose (a reference to a bone in the shoulder rather than the nose).
Named by: Daly, 1969.
Time: Kungurian stage of the early Permian.
Size: 30cm (1ft).
Lifestyle: Insectivore.
Species: *A. pteroticus.*

Right: Apart from the head, Acleistorhinus *would have been difficult to tell apart from any member of the other primitive reptile groups of the time.*

Milleretta

The anapsids were a group of reptiles that had no openings in their heads, aside from the eye sockets and the nostrils. Nevertheless one group, the millerettids, did have a pair of openings behind the eyes, although the rest of the skull was distinctly anapsid. These openings evolved independently. The group represents a side branch of the anapsid reptile line.

Features: *Milleretta* has a pair of openings in the skull behind the eyes. This led to speculation that it represented a stage in the evolution of one of the other reptile lines, but this is not now thought to be the case. A concave area in the back of the skull suggests the presence of an eardrum and complex hearing system. The rest of the animal is built like other primitive reptiles – an agile little insect-eater, just like a modern lizard.

Distribution: South Africa.
Classification: Reptile, Parareptilia, Millerettidae.
Meaning of name: Miller's little one.
Named by: Watson, 1957.
Time: Late Permian.
Size: 60cm (2ft).
Lifestyle: Insectivore.
Species: *M. rubidgei.*

Left: Milleretta *must have lived like a modern lizard, scuttling about dry areas and hunting insects.*

THE HIND FOOT OF PRIMITIVE REPTILES

On the hind foot of a typical modern lizard the outside toe is quite short, but the next one is extremely long, the next only a little shorter, and the other two shorter still. We can see the sense in this if we look at the hind leg of the lizard in its usual position, sprawled out at the side. The foot is essentially making contact with the ground along its leading edge, and this arrangement of toes means that it can grip the ground with four claws while in this position. This arrangement of toes in a primitive fossil reptile gives a clue that this animal must have moved like a modern lizard with its limbs sprawling out at the side.

Sometimes, however, the feet have a symmetrical shape. The middle toe is long and those to the side become progressively smaller. This is the foot of an animal that walks with its lower leg held vertically, rather than out to the side, like a crocodile when it is walking with its belly off the ground. In an animal such as this the foot is turned to point forwards in the direction of travel.

We can tell how ancient animals walked by looking at their foot bones.

Far left: A reptile foot with a vertical lower leg.
Left: The foot of a sprawling reptile.

Captorhinus

Another group of early amniotes were the captorhinids. These were among the most primitive of the eureptilia (the "true reptiles" as opposed to the parareptilia, the "almost reptiles") *Captorhinus* itself is known from several species that differ from one another by the arrangement of teeth.

Features: The main species, *C. aguti*, has multiple tooth rows giving a broad insect-grinding surface, as had its relative *Labidosaurus*, and extra cheek teeth, while *C. laticeps* has the cheek teeth but only a single tooth row. The articulation of the hind limb, especially in the bone structure of the ankle, shows that *Captorhinus* must have moved like a modern lizard. The structure of the forelimb is quite loose, allowing a great deal of movement.

Below: The leg articulation was similar to those of most modern reptiles, except the turtles to whom, ironically, the captorhinids were most closely related.

Distribution: Oklahoma, USA.
Classification: Reptile, Anapsida
Meaning of name: Stem nose.
Named by: Cope, 1918.
Time: Leonardian stage of the early Permian.
Size: 25cm (10in).
Lifestyle: Insectivore.
Species: *C. aguti*, *C. magnus*, *C. laticeps*.

Mesosaurus

Distribution: Brazil; South Africa.
Classification: Reptile, Mesosauridae.
Meaning of name: Middle lizard.
Named by: Gervais, 1864.
Time: Early Permian.
Size: 45cm (1½ft).
Lifestyle: Filter feeder.
Species: *M. brasiliensis*, *M. tenuidens*, *M. timidum*.

Another animal that was sometimes classed as a parareptile, but is now thought to have been a separate animal close to the ancestry of the reptiles, is *Mesosaurus*. It was a freshwater swimming animal, and the presence of its fossils in both Brazil and South Africa is taken as an early proof of continental drift. It was a freshwater animal and could not have crossed the Atlantic Ocean, and so it must have existed at a time when South America and Africa were part of the same continent.

Below: Mesosaurus and its close relatives Brasileosaurus and Stereosternum were freshwater filter feeders, close to the beginning of the reptile line.

Features: For such an early reptile, *Mesosaurus* is very specialized. Its flattened tail, webbed hind feet, and thickened ribs to counteract buoyancy, show it to have been an aquatic animal. Its eyes and nostrils are high on the skull, for easy breathing on the water surface. The teeth are many and fine, arranged like those of a comb. It seems likely that they were used, not for seizing prey, but for filtering shrimp-like invertebrates from the waters of the lakes and rivers in which it lived.

PAREIASAURS

The heyday of the parareptiles came in the Permian and early Triassic periods, with the evolution of the pareiasaurs. These were the major plant-eaters, and were big, slow-moving hippopotamus-like animals. They evolved quickly to become huge though they tended to become smaller and increasingly armoured as they evolved. They are represented in the modern fauna by turtles and tortoises.

Pareiasaurus

"Ancient ugliness" and "the reptile with the Queen-Anne legs" are two descriptions given to this animal. It is typical of the pareiasaur family – massive, squat and with a head armoured with bony plates. It would have grazed the meadows of seedferns that represented the main plant-life of the time – especially along stream banks. Most of their remains are found in river deposits.

Features: The skull is heavy and studied with lumps and spikes. The teeth are small and leaf-shaped, serrated like vegetable graters for dealing with tough vegetable matter. Grinding teeth are present on the palate to help to pulp the food. The jaws are broad, and hinged to produce a strong bite rather than a wide gape. The body is huge and barrel-shaped, evidently to hold the enormous digestive system needed to allow the animal to break down the plant material.

Left: Pareiasaurus was the main plant-eater of the time, grazing the boggy sides of the desert streams and rivers.

Distribution: South Africa; Eastern Africa; Eastern Europe.
Classification: Reptile, Procolophonia, Pareiasauria.
Meaning of name: Helmet-cheeked lizard.
Named by: Owen, 1876.
Time: Wuchiapingian stage of the late Permian.
Size: 2.5m (8ft).
Lifestyle: Low browser.
Species: *P. americanus, P. baini, P. omoeratus, P. nasicornis, P. peringuegi, P. serridens*.

Scutosaurus

As the pareiasaurs evolved they developed more armour, which became more and more compact. *Scutosaurus* was one of the most heavily armoured, although its later relative, the tiny *Anthodon* (only about half a metre long), was even more densely armoured. The armour is a suggestion of the evolution of a turtle shell.

Features: *Scutosaurus* is similar to *Pareiasaurus* but much taller, due in part to the fact that the legs are held more directly beneath the body in an upright stance. The back is heavily armoured and the head covered in spikes and horns, with a prominent horn on the nose and a pair on the angles of the jaw. The eardrum is quite well developed, indicating that *Scutosaurus* would have been able to hear high frequency airborne sounds.

Above: Scutosaurus was the biggest plant-eater of the day.

Distribution: Russia.
Classification: Reptile, Procolophonia, Pareiasauria.
Meaning of name: Shielded lizard.
Named by: Amalitzky, 1922.
Time: Wuchiapingian stage of the late Permian.
Size: 2.5m (8ft).
Lifestyle: Low browser.
Species: *S. karpinskii*.

Elginia

Distribution: Scotland; possibly Russia.
Classification: Reptile, Procolophonia, Pareiasauria.
Meaning of name: From Elgin.
Named by: Newton, 1892.
Time: Changhsingian stage of the late Permian.
Size: 60cm (2ft).
Lifestyle: Low browser.
Species: *E. mirabilis*.

One of the last of the pareiasaurs to thrive was *Elginia*, and it was also one of the smallest. It is known from only one skull. Towards the end of the Permian, conditions were drying out and the loss of habitat involved probably drove the pareiasaurs to extinction. The last were small specialized forms such as this.

Features: On the skull of *Elginia* the customary pareiasaur bumps and knobs are expanded into a whole array of spectacular horns, one pair in particular being very long, sweeping out sideways and curving backwards. The horns were probably used for display rather than defence. This creature existed perhaps two million years after the big pareiasaurs died out.

Below: Elginia is known from only a single skull – although a specimen from Russia may be the tail of a juvenile – and so the restoration of the rest of the animal is rather speculative.

OTHER PAREIASAURS

The pareiasaurs were quite conservative in build. They all possessed the big mouths and tubby bodies to process vast quantities of low-nutritional vegetation. The broad area of the laterally expanded skull and especially the cheek region may have supported a complex system for producing saliva, to help in the breakdown of near-indigestible material.

Despite this general similarity, they represented quite a widespread group, more so than the four genera illustrated here would suggest.

Above: Bradysaurus – a primitive pareiasaur from the middle Permian of South Africa.

Bradysaurus. South Africa. 2.5m (8ft) long. An early form that was not as heavily armoured as the others.
Deltavjatia. Northern Russia. 3m (10ft) long. Had skull features that were distinctly turtle-like.
Shihtienfenia. Shansi Province, China. 2m (6ft) long. The first to have been found so far east.

Eunotosaurus

Eunotosaurus was once regarded as the direct ancestor of the turtles and tortoises, because of the expanded ribs that seemed to indicate an evolution towards a rigid body that would support a solid shell. There was also a thin armour of bony scutes across the broad back, which suggested an evolutionary stage in a solid shell's formation. However, this is not now regarded as so, and the *Eunotosauridae* is now thought to have been a separate aberrant family.

Features: The remarkable feature of this animal is the rib-cage. It consists of only eight pairs of ribs and these are expanded and leaf-shaped, almost touching each other, and there are only ten vertebrae in the back. However the shoulder blade is outside the rib cage as is usual among vertebrates and not on the inside as it is in turtles and tortoises. The skull is quite unlike that of turtles and tortoises, suggesting that *Eunotosaurus* is not ancestral to them.

Distribution: South Africa.
Classification: Reptile, Procolophonia.
Meaning of name: True southern lizard.
Named by: Seely, 1892.
Time: Capitanian stage of the middle Permian.
Size: 20cm (8in).
Lifestyle: Low browser.
Species: *E. africanus*

Left: The broad ribcage of Eunotosaurus must have made it look like a very squat lizard, probably like the horned lizard of today.

PRIMITIVE DIAPSIDS

The reptile group that was to dominate the Mesozoic era had its origins in the Carboniferous period and went on to develop into a number of specialist forms in the Permian. These were the diapsids, those with two holes for muscle attachment in the skull behind the eyes. However, at the time, the group was insignificant compared with the other large creatures that were around.

Araeoscelis

The oldest known diapsid was *Petrolacosaurus* from the late Carboniferous of Kansas, USA. Its close relative, *Araeoscelis*, from the early Permian of Texas, USA, was a similar animal but had a more highly adapted dentition. They were both small lizard-like animals, as were the early examples of all the reptile groups.

Features: The front and back legs are quite long but are the same length – an unspecialized condition. The teeth are massive, blunt and conical, probably adapted for crushing the hard carapaces of insects. One of the pairs of holes in the skull is closed over with bone, presumably to anchor stronger jaw muscles.

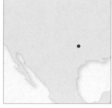

Distribution: Texas, USA.
Classification: Reptile, Araeoscelidae.
Meaning of name: Thin legs.
Named by: Williston, 1913.
Time: Artinskian stage of the early Permian.
Size: 60cm (2ft).
Lifestyle: Insectivore.
Species: *A. casei, A. gracilis.*

Above: The lizard-like feet of Araeoscelis, or something closely related, are thought to have been the makers of a trail of fossil footprints in Nova Scotia dating from this time.

Coelurosauravus

The diapsids quickly adapted to all sorts of lifestyles in all sorts of environments. *Coelurosauravus* was the earliest gliding vertebrate known. The wing arrangement was so unusual that a researcher in the 1920s removed the wing bones from an early specimen thinking that they were parts of a fish skeleton that had become mixed up with it. It became extinct at the very end of the Permian during the great mass-extinction of that time.

Distribution: Madagascar; Germany; England.
Classification: Reptile, Diapsida order uncertain.
Meaning of name: Flying hollow lizard.
Named by: Piveteau, 1926.
Time: Changhsingian stage of the late Permian.
Size: 30cm (1ft).
Lifestyle: Gliding insectivore.
Species: *C. jaekeli.*

Features: *Coelurosauravus* has wings formed from extra struts of hollow bone, about two dozen to each side, that grew from the skin of the sides, and these supported a web of skin (called a patagium). This is similar to that of later reptiles such as *Icarosaurus* and *Khuneosaurus* and even the modern *Draco*, the flying dragon of Southeast Asia. Half of its length was made up of a long tail that probably stabilized it in flight. The smooth head with the frill around the rear edge of the skull may have helped to streamline *Coelurosauravus* while it glided from tree to tree or from cliff-face to cliff-face.

Above: The wing could be folded out when Coelurosauravus took to the air, and allowed it to glide from tree to tree.

Thadeosaurus

Distribution: Madagascar.
Classification: Reptile, Eosuchia.
Meaning of name: Thadeo's lizard.
Named by: Piveteau, 1926.
Time: Chaghsingian stage of late Permian to Induan stage of early Triassic.
Size: 60cm (2ft).
Lifestyle: Amphibious hunter.
Species *T.colcanapi*.

Like the modern monitor lizard, *Thadeosaurus* was an amphibious hunter. When swimming, it used only its tail, keeping its legs for manoeuvring and for walking on land. This animal represented the start of an evolutionary trend that would eventually lead to the almost fully aquatic nothosaurs and the totally marine plesiosaurs of the Mesozoic era.

Above: Thadeosaurus hunted on land in the shallow waters of ponds and streams.

Features: The tail of *Thadeosaurus* is about twice the length of the body to help in swimming. The five clawed toes of the hind foot are arranged with the longest on the outside, an adaptation to efficient movement on land. The breastbone is well developed, also as a response to an increase in the efficiency of terrestrial motion. This was an animal at home in the water as well as on land.

RULERS IN WAITING

*Top left: The anapsid condition with no holes behind the eyes.
Top right: The synapsid condition with a single hole.
Right: The diapsid condition with two holes.*

The diapsids as a group are characterized by two holes in the side of the skull behind the eyes. These held the muscles that worked the jaws. At first they were small lizard-like animals, but soon they evolved into all sorts of specialized shapes – fliers, swimmers and burrowers. The main ecological niches for big animals, were at this time taken by the parareptiles and the mammal-like reptiles, and so the diapsids remained small and insignificant.

It is almost the same situation as that of the mammals during the Age of Dinosaurs – quite diverse but small and insignificant. It took the extinction of the dinosaurs to allow the mammals to develop into what they now are.

Likewise, it was the mass-extinction of the parareptiles and the bulk of the mammal-like reptiles at the end of the Permian that allowed the diapsids to develop. They would become the crocodiles, lizards, snakes, and even the great swimming reptiles of the Mesozoic as well as the dinosaurs and birds.

Hovasaurus

Distribution: Madagascar.
Classification: Reptile, Eosuchia.
Meaning of name: Hova lizard.
Named by: Piveteau, 1926.
Time: Changhsingian stage of late Permian to Induan stage o early Triassic.
Size: 50cm (1½ft).
Lifestyle: Aquatic hunter.
Species: *H. boulei*.

Early diapsid aquatic life advanced with *Hovasaurus*. The tail was paddle-like and twice the length of the rest of the body. Skin impressions have been found, and these show lizard-like scales. Close relatives *Tangasaurus* and *Barasaurus* are almost identical, differing only in the height of the spines on the tail. *Hovasaurus* is known from several specimens, all of which had stomach stones that must have been swallowed to counteract excessive buoyancy while swimming. It seems possible that *Hovasaurus* and its relatives survived the end-Permian extinction event.

Features: This creature is very similar to *Thadeosaurus*, to which it is closely related, but it shows much greater adaptations to an aquatic way of life. The spines on the tail are long, supporting a deep paddle-like tail, and those on the back are also long, suggesting the presence of a fin-like stabilizing structure. Stones are present in the stomach area of some specimens, and these would have been swallowed as a buoyancy control.

Above: Hovasaurus swam with powerful side-to-side sweeps of its broad paddle-like tail.

THE PELYCOSAURS

The major group of reptiles in the Permian period were those that we call the mammal-like reptiles. This was a group of synapsid reptiles that evolved more and more mammal-type features as they developed, and eventually evolved into the mammals themselves. The earliest forms were not particularly mammal-like but showed some of the early features that define the group.

Ophiacodon

Ophiacodon was quite a large animal and may have spent much of its time in or near the water, where it would have hunted amphibians or fish, as well as on land where it would have preyed upon more terrestrial reptiles. There were several species, appearing at slightly different times. The teeth in the long jaws resemble those of the big fish-eating amphibians more than those of the later reptiles. This is one of the reasons for the suggestion that *Ophiacodon* was a semi-aquatic fish-hunter.

Features: The skull is quite deep and narrow, giving a good attachment area for strong jaw muscles, suggesting a powerful bite. Unlike the earlier pelycosaurs that are sprawling and lizard-like, the hind legs of *Ophiacodon* were held more directly beneath the body, indicating an ability to run down its prey. The forequarters are strong, presumably to hold up the big head. The teeth are quite small and beginning to show a differentiation in size.

Below: Some scientists disagree with the proposed aquatic lifestyle of Ophiacodon, *thinking the tall narrow skull would make this impractical.*

Distribution: Texas and Ohio, USA.
Classification: Reptile, Pelycosauria, Ophiacodontia.
Meaning of name: Snake tooth.
Named by: Marsh, 1878
Time: Late Carboniferous to early Permian
Size: 3.5m (12ft).
Lifestyle: Fish hunter.
Species: *O. uniformis, O. retroversus, O. major, O. navajovicus, O. mirus, O. tranguala, O. watsoni.*

Varanops

Most of the fossils that have been found of *Varanops* are of juveniles, and so it is not easy to see what the adult was like. The skulls and jaws were very long and fragile, and their reconstruction into a complete set of head bones is very difficult. *Varanops* was one of the latest-surviving of the pelycosaurs. It became extinct in the earliest part of the late Permian, like the rest of its type, when more advanced mammal-like reptiles became the most important meat-eaters.

Features: The teeth of *Varanops*, and of its close relative *Varanosaurus*, which may in fact turn out to be the same genus, are flattened and strongly curved. These are evidently the teeth of a flesh-eating hunter. It appears that the head is quite primitive, like that of an ophiacodont, while the rest of the body is more like a more advanced member of the pelycosaurs, a sphenacodont.

Distribution: Texas and Oklahoma, USA.
Classification: Reptile, Pelycosauria, Varanopseidae.
Meaning of name: Like a monitor lizard.

Left: Varanops *probably lived like its modern namesake the monitor lizard of Africa and Asia.*

Named by: Williston, 1914.
Time: Early Permian.
Size: 1.5m (5ft).
Lifestyle: Hunter.
Species: *V. brevirostris.*

DEVELOPMENT OF EARLY PELYCOSAURS

The synapsid group of reptiles are classified by the fact that the holes in the skull behind the eyes were fused into a single opening. This big opening provided a very strong anchoring area for the jaw muscles and led to powerful and widely-opening jaws in some of the examples. The main evolutionary line of these became the mammal-like reptiles and ultimately the mammals.

The earliest representatives were the pelycosaurs. The name "pelycosaur" literally means "basin lizards" and is a reference to the arrangement of the bones in the hip, which produced a basin-like structure.

The oldest pelycosaur was *Protoclepsydrops* from the Carboniferous. It was also the oldest-known synapsid, from the tree-stump locality of Joggins, Nova Scotia. The earliest pelycosaurs were small, lizard-like animals , but they evolved into respectably-sized animals, some with spectacular sails on their backs.

We can see the transition from the sail-less *Sphenacodon* to a big-sailed form such as *Dimetrodon* through an intermediate type, *Ctenospondylus*, which was identical to these two in body arrangement, but had a sail that was only half the size of that of *Dimetrodon*. Another form, *Secodontosaurus*, was similar to *Dimetrodon* but had a much flatter head.

The pelycosaurs did not survive the middle Permian.

Sphenacodon

This fierce hunter was built with a huge head and killing teeth backed up by a body that was adapted for a strong frontal attack. It was the biggest of the sphenacodont group. Others included the 1.2m (4ft) *Haptodus*, which looked just like a miniature *Sphenacodon* and must have hunted smaller prey.

Features: In *Sphenacodon* we find the beginnings of the elongation of the vertebrae, which would eventually evolve into the back sails that are so distinctive of the most spectacular of pelycosaurs. Here, though, they did not carry the sail, but probably acted as anchor points for strong back muscles that may have allowed *Sphenacodon* to lunge at its prey. As in the rest of the group the teeth are specialized into dagger-like front teeth and meat-shearing teeth at the back.

Right: Two species of Sphenacodon *are known. The lightweight species is* S. ferox *(fierce) and the heavier is* S. ferocior *(fiercer).*

Distribution: Texas, USA.
Classification: Reptile, Pelycosauria, Sphenacodontidae.
Meaning of name: Wedge tooth.
Named by: Marsh, 1878.
Time: Sakmarian stage of the early Permian.
Size: 2.5m (8ft).
Lifestyle: Hunter.
Species: *S. ferox, S. ferocior.*

Dimetrodon

Distribution: Texas, USA.
Classification: Reptile, Pelycosauria, Sphenodontidae.
Meaning of name: Two shapes of teeth.
Named by: Cope, 1878
Time: Kungurian stage of the early Permian.
Size: 3m (10ft) – the biggest was *D. grandis.* Earlier species were smaller.
Lifestyle: Hunter.
Species: *D. angelensis, D. grandis, D. limbatus.*

Below: The sail was probably an adaptation to harsh climate conditions, or it may have been brightly coloured and used for signalling.

Probably the most familiar of the pelycosaurs, notably because of the spectacular sail, *Dimetrodon* predated the dinosaurs by 50 million years and belonged to a totally different line of reptile evolution. It was the biggest and fiercest terrestrial predator of the time.

Features: The significant feature, the sail, is formed from elongated spines that grow upwards from the vertebrae of the back. In life this would have been covered in skin. Turned to the sun in the early morning, the sail would have absorbed heat into the blood stream and made the animal active. If the temperatures became too hot in the day the sail would cool the blood in the wind, like a car radiator. Apart from that, *Dimetrodon* is very similar to *Sphenacodon*, from which it probably evolved.

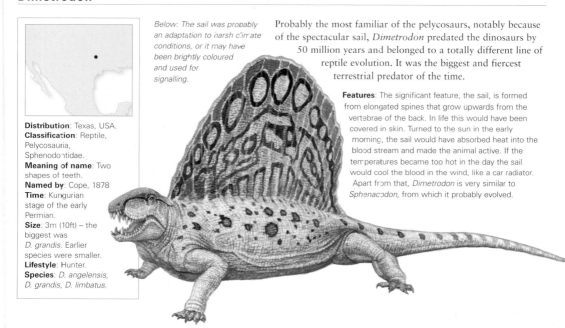

PELYCOSAURS AND OTHER EARLY SYNAPSIDS

The pelycosaurs were not all fierce hunters. Some developed into herbivores, and as such became the first of the plant-eating reptiles. At the same time the early synapsids branched out into other specialized forms, but eventually settled into the evolutionary line that became the advanced mammal-like reptiles.

Edaphosaurus

This is the best-known plant-eating pelycosaur. The vegetarian habit is inferred from the small head and the crushing teeth that formed a broad pavement across the roof of the mouth. It lived in the same place and at the same time as the other famous sail-backed pelycosaur, *Dimetrodon*, and may have been part of its prey. It evolved from much smaller insectivorous pelycosaurs, such as *Ianthasaurus* from the late Carboniferous. This looked like *Edaphosaurus* but had the jaws and teeth of an insectivorous animal and was only about 60cm (2ft) long.

Right: The teeth of Edaphosaurus *were arranged into a broad grinding surface on the palate.*

Features: The characteristic sail is supported by spines growing up from the vertebrae of the back. Those at the front are curved forward, those at the back are curved backwards, and those in the middle are straight. The spines have cross-pieces, like the spars on the masts of square-rigged sailing ships. These increased the turbulence of air passing by, and improved the heat exchange process, enabling the animal to have a smaller sail than its relatives.

Distribution: Texas, USA; Czech Republic.
Classification: Reptile, Pelycosauria, Edaphosauridae.
Meaning of name: Pavement lizard.
Named by: Cope, 1882.
Time: Late Carboniferous to early Permian.
Size: 3m (10ft).
Lifestyle: Low browser.
Species: *E. bonaerges, E. pogonias, E. cruciger.*

Casea

The caseids represented an abundant group of herbivorous reptiles related to the pelycosaurs, from middle and late Permian times. The largest were 3m (10ft)-long *Caseoides*, and *Angelosaurus* weighing up to 300kg (661lb), but there were much lighter examples, with *Caseopsis* being about the same length but built as an agile animal. For a long-lasting group they were very conservative in build. *Casea* was one of the last and most abundant, living at the end of the Permian period. With its big body and tiny head *Casea* must have resembled *Edaphosaurus* without the sail.

Features: *Casea* has the massive pig-like body that is to be expected of herbivorous reptiles, accommodating the complex digestive system needed for breaking down plant material. The neck vertebrae are small and the ribs are big and curved, giving the barrel-shape to the body. The teeth in the upper jaw are thick and peg-like, while those in the lower jaw are totally absent. The holes in the skull are enormous, and the upper jaw overhangs the lower quite considerably.

Below: The tiny head and the overhang of the upper jaw of Casea *give the animal a goofy appearance.*

Distribution: Texas, USA; and France.
Classification: Reptile, Caseidae.
Meaning of name: Cheesy.
Named by: Williston, 1910.
Time: Late Permian.
Size: 1.2m (4ft).
Lifestyle: Low browser.
Species: *C. broilli, C. rutena.*

Cotylorhynchus

Distribution: Texas, USA.
Classification: Reptile,
Caseidae.
Meaning of name:
Hollow snout.
Named by: Stovall, 1937
Time: Arinskian stage of the
late Permian.
Size: 6m (20ft).
Lifestyle: Low browser.
Species: C. hancocki,
C. romeri.

This, as far as we know, was the biggest of the caseids.
It was the biggest land animal of its time, relying on its
sheer size rather than armour or swiftness of foot to
protect it from the carnivores of the time and place.
These enormous caseids probably had the same lifestyle
as the huge pareiasaurs that preceded them. With the
extinction of the caseids there seems to have been a gap
of several million years before big plant-eaters evolved
once more, in the form of the herbivorous dinocephalians
of the latest Permian.

Features: The skull is tiny, almost
ridiculously so, and perched on a
longish neck in front of truly
massive shoulders. The sprawling
limbs must have had massive
muscles, like those of the
modern giant tortoise, to support
and move the 2-tonne weight of
this enormous animal. The truly
giant species, C. hancocki, was
probably a descendant of the
earlier and much smaller
C. romeri.

Left: Cotylorhynchus *was the
biggest of several large caseids
from the end of the Permian.*

EARLY RUNNER

Among all these big reptiles it is possible to lose
sight of the fact that many of them were quite
small animals. One in particular, *Eudibamus*, was
about 26cm (10in) long, and found in Germany in
1993, but not scientifically described until 2000. It
looked rather like a little lizard, but its front
legs were small and weak and the hind legs
more than one and a half times their size. This,
and the alignment of the hip, knee and ankle
joints, led researcher Robert Reisz to speculate
that *Eudibamus* ran on its hind legs, as many of
the later reptiles did. This would make it the
earliest reptile to do so. As *Eudibamus* was
unrelated to any of the later bipedal dinosaurs it
shows that bipedality is one of those
features that evolved independently
several times.

*Right: Sprightly
Eudibamus – a
two-legged runner.*

Tetraceratops

With a name like that, *Tetraceratops* would
be expected to appear among the
ceratopsian dinosaurs of the late Cretaceous.
However, it was a big synapsid reptile that
carried four horns on its head, so hence its
name. Its position in the pelycosaur family
tree is unclear, but current thinking is that it
was a sphenacodontid related to
Dimetrodon. It probably lived in upland
areas far from the swamps where the
pelycosaurs lived. This would explain why
so few fossils have been found.

Features: All we know of *Tetraceratops* is a single
skull. This has features that show it to have been a
meat-eating synapsid, but the nature of its
preservation, in a particularly hard piece of ironstone,
makes it difficult to examine. At first it was thought
to have been an early gorgonopsid, one of the
advanced mammal-like reptiles, and even now
scientists are unsure of its actual affinities.

Distribution: Texas, USA.
Classification: Reptile, order
uncertain, but probably
Pelycosauria
Meaning of name: Four-
horned head.
Named by: Matthew, 1908.
Time: Sakmarian stage of the
early Permian.
Size: 1m (about 3ft).
Lifestyle: Hunter.
Species: T. insignis.

*Right: The horns would have been used for display
and may have been brightly-coloured. It is quite
possible that only the males of* Tetraceratops
*possessed them, the females being much
duller and unspectacular
in appearance.*

CARNIVOROUS THERAPSIDS

The therapsids were the main line of mammal-like reptiles that dominated the Permian world and evolved to become the mammals. They probably evolved from the sphenacodontid line of the pelycosaurs. The eotitanosuchians were the first of the therapsids to appear, followed by the titanosuchians. They were fearsome crocodile-like animals, possibly with leanings towards vegetarianism.

Biarmosuchus

The biarmosuchians were intermediate between the sphenacodonts and the more mammal-like therapsids. Their remains are mostly found to the west of the Ural Mountains in Russia, but there are close relatives in South Africa.

Features: *Biarmosuchus* and the other biarmosuchians are moderately sized carnivores, a little more lightly built and athletic than their ancestral sphenacodonts. The skull is very much like that of a sphenacodont, but has a single pair of long canine teeth in both the upper and lower jaws. The rear of the skull is large, holding powerful muscles that would have given a strong bite. The feet are quite symmetrical, showing that they point forward and do not sprawl out at the side.

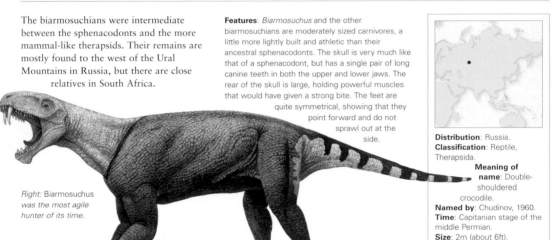

Right: Biarmosuchus *was the most agile hunter of its time.*

Distribution: Russia.
Classification: Reptile, Therapsida.
Meaning of name: Double-shouldered crocodile.
Named by: Chudinov, 1960.
Time: Capitanian stage of the middle Permian.
Size: 2m (about 6ft).
Lifestyle: Hunter.
Species: *B. tener.*
B. antecessor, B. tchudinovi.

Titanosuchus

The name, meaning "gigantic crocodile", was given by Sir Richard Owen to an incomplete set of teeth. The rest of the animal appears to have been very crocodile-like as well, with short sprawling limbs, a long, low body and a long tail.

Features: The dentition, consisting of strong incisors adapted for both stabbing and crushing, having both points and flat grinding surfaces, the big canines and the many serrated teeth at the back, along with the general crocodile-like build of the body, has always suggested that *Titanosuchus* was a meat-eater. However, in recent times some scientists have come to doubt this, seeing the dentition as just as well adapted for plant food as for meat – the serrated rear teeth could have worked like vegetable graters.

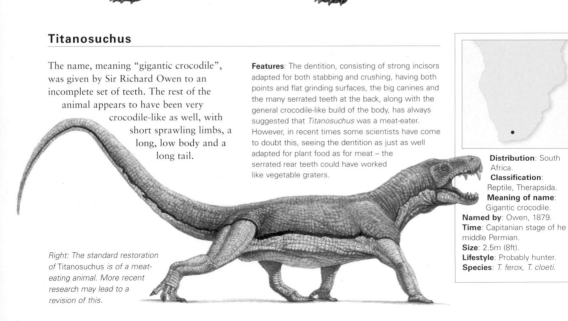

Right: The standard restoration of Titanosuchus *is of a meat-eating animal. More recent research may lead to a revision of this.*

Distribution: South Africa.
Classification: Reptile, Therapsida.
Meaning of name: Gigantic crocodile.
Named by: Owen, 1879.
Time: Capitanian stage of he middle Permian.
Size: 2.5m (8ft).
Lifestyle: Probably hunter.
Species: *T. ferox, T. cloeti.*

Jonkeria

Distribution: South Africa.
Classification: Reptile,
Therapsida.
Meaning of name: From
Jonkers.
Named by: Van Hoepen,
1916.
Time: Middle Permian.
Size: 5m (16ft).
Lifestyle: Low browser.
Species: *J. truculenta,
J. ingens, J. vanderbyli,
J. haughtoni, J. koupensis,
J. parva. J. rossowi,
J. boonstrai.*

Most of the titanosuchids resembled the earlier eotitanosuchids in their general crocodile-like forms, but some became very much larger and slower-moving – seemingly a stage towards the development of plant-eating types. They must have preyed on the biggest and slowest-moving of the contemporary herbivores, the anapsid pareiasaurs or their own relatives the dinocephalians.

Features: *Jonkeria* has a long snout, expanded at the end, and large incisors and canines. The teeth show it to have been a meat-eater, but it has a very large and heavy build for an animal with such a lifestyle, and the limbs are very stout. It is quite possible that it was omnivorous, eating mostly plants but also taking carrion and live animals, like some of today's nominal carnivores such as bears.

Below: Jonkeria *may have been the equivalent of the modern bear, eating anything that it could find, and dangerous to anything that it approached.*

Anteosaurus

The anteosaurids – a group distinguished by the long front teeth and the upturned front to the jaw – are known from both Russia and South Africa, two widely-separated regions even in late Permian times, the one located about 50 degrees north and the other about 60 degrees south. The wide distribution of these and other therapsids led the way to their spectacular spread and dominance in the succeeding early Triassic.

Features: *Anteosaurus* is known from more than 30 skulls and parts of the body skeleton. Sixteen skulls are well preserved and were originally used to establish about ten different species. Nowadays they are all regarded as different growth stages of the one species, *A. magnificus*. It has the skull of a powerful predator, with big cutting teeth at the front and reduced teeth at the back making room for huge biting muscles. The hips are like those of a modern crocodile.

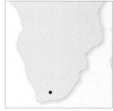

Distribution: South Africa.
Classification: Reptile,
Therapsida.
Meaning of name: Before
lizards.
Named by: Watson,
1921.
Time: Capitanian
epoch of middle
Permian.
Size: 6m (20ft).
Lifestyle: Hunter.
Species:
A. magnificus.

Left: The crocodile-like hips suggest that Anteosaurus *was an ambush hunter, with a hunting style very much like that of the modern crocodile, when on land.*

HERBIVOROUS THERAPSIDS

As the therapsids blossomed they developed a number of off-shoots, one of which consisted of the massive dinocephalians. The dinocephalians represent an early branch of the therapsids, and show many primitive features. They were mostly rhinoceros-size animals with heads formed from greatly thickened head bones. They had a reduced number of foot bones, suggesting that they were not very agile.

Moschops

The enormously thickened skull of *Moschops* (up to 10cm (4in) thick) suggests that they must have head-butted one another for dominance or display. However, such contests were unlikely to have been the energetic head-butting that we see in today's mountain goats, more a sedate pushing action, in keeping with an animal the size of a rhinoceros and feet like a giant tortoise. It was a herd-living animal.

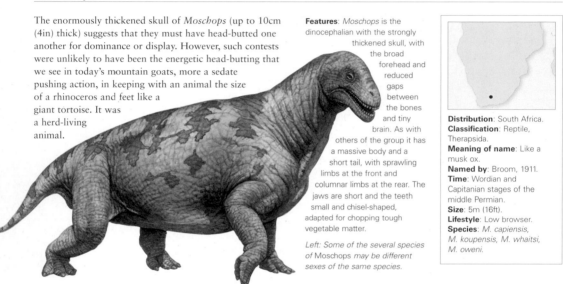

Features: *Moschops* is the dinocephalian with the strongly thickened skull, with the broad forehead and reduced gaps between the bones and tiny brain. As with others of the group it has a massive body and a short tail, with sprawling limbs at the front and columnar limbs at the rear. The jaws are short and the teeth small and chisel-shaped, adapted for chopping tough vegetable matter.

Left: Some of the several species of Moschops *may be different sexes of the same species.*

Distribution: South Africa.
Classification: Reptile, Therapsida.
Meaning of name: Like a musk ox.
Named by: Broom, 1911.
Time: Wordian and Capitanian stages of the middle Permian.
Size: 5m (16ft).
Lifestyle: Low browser.
Species: *M. capiensis, M. koupensis, M. whaitsi, M. oweni.*

Estemmenosuchus

When the caseid pelycosaurs declined as the main plant-eaters of the middle and late Permian, it was the estemmenosuchids that took their places. They were massive bull-sized animals with sprawling gaits and spectacular ornamentation on their heads. Their huge bodies indicate that they were vegetarians, despite the big canine teeth.

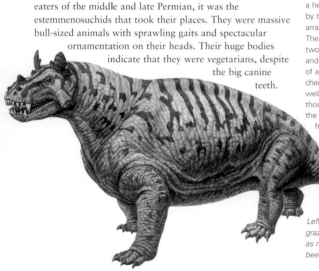

Features: *Estemmenosuchus* has a heavy skull that is ornamented by the most remarkable arrangement of bumps and horns. There were two by the nostrils, two in the middle of the snout and two that are like the antlers of a moose over the eyes. The cheek bones flare outwards as well. The hind legs are erect, like those of a mammal, for bearing the weight of the animal, but the front legs sprawl out to the side, probably for ease of steering and manoeuvring, allowing the fantastic head to be presented to any rival.

Left: Estemmenosuchus *probably grazed in low-lying marshy areas, as most of their remains have been found in flood deposits.*

Distribution: Ocher District, Russia.
Classification: Reptile, Therapsida.
Meaning of name: Crowned crocodile.
Named by: Tchudinov, 1960.
Time: Capitanian stage of the middle Permian.
Size: 3m (10ft).
Lifestyle: Low browser.
Species: *E. uralensis, E. mirabilis.*

Ulemosaurus

Distribution: Isheevo, Tatarstan.
Classification: Reptile, Therapsida.
Meaning of name: Lizard from Ulema River.
Named by: Rjabinin, 1938.
Time: Capitanian stage of the middle Permian.
Size: 2m (About 6ft).
Lifestyle: Low browser.
Species: *U. svijagensis*.

Ulemosaurus was rather like *Estemmenosuchus* without the head ornamentation. It was originally thought to have been a specimen of *Moschops* but it is considerably more primitive in its build than this. As with its relatives it has a big spherical body, short neck and tail, stout hind legs and sprawling front legs that were used for steering.

Features: The skull has a narrow tapering snout, and is wide and very tall behind the eyes. *Ulemosaurus* also has a more specialized tooth pattern. The small side teeth fully interlock as the jaws close, giving them a tight grip on the tough vegetation. The rest of the mouth structure, especially the lack of a palate, shows that *Ulemosaurus* could not have chewed its food, and that most of the food processes took place in a complex stomach contained in the huge body.

Left: The very heavy jaws have led some scientists to suggest that Ulemosaurus *was actually a meat-eater, and used the powerful jaw muscles to tear into big animals with its strong front teeth.*

DINOCEPHALIAN TEETH

The front teeth of the dinocephalians tended to differ from genus to genus, but the back teeth were usually of the same pattern. The back teeth were much smaller than those of the front. Each tooth was the same shape as the next in each jaw, and the same in the opposing jaw. It consisted of a curved crown with a rounded blunt tip, and a ledge near the root. When the jaws closed the curved crowns interlaced, and the tips fitted into sockets on the ledges of the opposite set. As a result the teeth meshed perfectly, allowing no gaps for the escape of food and making the chopping and crushing action very efficient.

Features: The distinction of *Styracocephalus* lies in the ornamentation of the head and the size and arrangement of the teeth. The front teeth and the canines are conical and quite small, and there is a large number of back teeth, as well as a set of crushing teeth on the palate. This is a very primitive arrangement – later forms had prominent interlocking incisor teeth at the front and reduced rear and palate teeth.

Styracocephalus

The head ornament of *Styracocephalus* meant that it could be recognized from a distance. The crest stuck upwards and backwards, but there is some variation in its shape, and this suggests that it changed throughout life and that it was different between the male and the female. *Styracocephalus* was a large herbivore that may have been fully terrestrial or partly aquatic like the modern hippopotamus, which resembles it in shape and size. It may have evolved from the estemmenosuchid group. Its remains are known from South Africa but it probably had a wider distribution.

Distribution: South Africa.
Classification: Reptile, Therapsida.
Meaning of name: Spike head.
Named by: Houghton, 1929.
Time: Capitanian stage of the middle Permian.
Size: 4.5m (15ft).
Lifestyle: Browser.
Species: *S. platyrhynchus*.

Left: Styracocephalus *probably spent much of its time resting while digesting its meal of tough plant material.*

GORGONOPSIANS AND OTHER MEAT-EATING THERAPSIDS

A rough description of the appearance of a gorgonopsian would be a cross between a crocodile and a sabre toothed tiger. They were hunters by nature, and had light agile skeletons, with long legs well adapted for running. They were typical of the fauna of late Permian South Africa.

Lycaenops

The long canine teeth of *Lycaenops* were evidently adapted for stabbing, and would have inflicted deep bleeding wounds in the flanks of larger prey. Rather than going for the quick kill it would have waited for its prey to bleed to death. It is possible that *Lycaenops* hunted in packs, although no evidence of this has been found.

Features: The skull is long, low and slender. The canine teeth are very long, in both jaws. The blades are serrated, and the roots are deep, running well into the bone, accounting for the deep skull. There are few teeth in the back of the mouth, and those show little wear, suggesting that *Lycaenops* bolted down food without chewing. There is a zygomatic arch – a curved structure in the

cheek bone under which the jaw muscle passed – a feature that mammals have. Not only did *Lycaenops* have mammal-like teeth but also mammal-like jaw muscles. The leg articulation shows that it could have both sprawled and walked with a high stance.

Right: The sabre teeth of Lycaenops would have been used for stabbing and tearing flesh – for both killing and eating.

Distribution: South Africa.
Classification: Reptile, Therapsida.
Meaning of name: Like a dog.
Named by: Colbert, 1948.
Time: Late Permian.
Size: 1m (about 3ft).
Lifestyle: Hunter.
Species: *L. ornatus, L. angusticeps.*

Hipposaurus

This creature belongs to a family called the ictidorhinids – the "weasel-noses". These have always been regarded as gorgonopsids although they may be more primitive than that – harking back to the most primitive therapsids. *Hipposaurus* appears to be quite primitive. Even though it was quite a large animal it seems to have preyed upon quite small creatures.

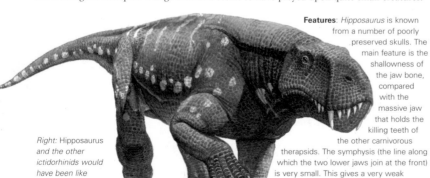

Right: Hipposaurus and the other ictidorhinids would have been like modern stoats and civets – hunting the smaller animals of the time and place.

Features: *Hipposaurus* is known from a number of poorly preserved skulls. The main feature is the shallowness of the jaw bone, compared with the massive jaw that holds the killing teeth of the other carnivorous therapsids. The symphysis (the line along which the two lower jaws join at the front) is very small. This gives a very weak mouth apparatus, which is unable to withstand much twisting, in contrast to the tearing and shearing mechanisms of its contemporaries.

Distribution: South Africa.
Classification: Reptile, Therapsida.
Meaning of name: Horse lizard.
Named by: Broom, 1940.
Time: Wuchiapingian stage of the late Permian.
Size: 1.2m (4ft).
Lifestyle: Hunter of small animals.
Species: *H. seelyi, H. boonstrai.*

THE GORGONOPSIAN NOSE

Most meat-eating vertebrates rely on their eyesight to find their prey. The eyes of a cat or an owl are positioned so that they point forward, focussing on their prey and using their binocular vision to judge their distance to it. Alternatively they use acute hearing to track their prey, and this is reflected in the complex structure of a mammal's ears.

The gorgonopsians did not benefit from either of these features. Their eyes were small and mounted on the sides of their heads, unable to produce a stereoscopic image. Their hearing mechanism was relatively unsophisticated, the bones that form a mammal's ear structure were at this point in evolution still part of the jaw apparatus. Instead, it seems that they relied on a sense of smell.

The nasal cavity of a gorgonopsian was divided into two chambers. The larger one had complex folds of thin bone and cartilage that, in life, would have been covered with moist membranes that must have been sensitive to smells. These structures are called turbinals and are found in modern mammals. Their presence suggests that they had a very well-developed sense of smell, and used it for hunting. It is possible that the turbinals were used for heat regulation as well, indicating that the gorgonopsians were to some extent warm blooded.

Phthinosuchus

This animal is known only from the rear part of the skull, although the front part has been described and has since been lost. It is another of the hunting therapsids that does not quite fit into any accepted classification. As such it has been assigned to its own suborder Phthinosuchia.

Features: *Phthinosuchus* is something of a wastebasket taxon. All sorts of bits of bone from this period and area have been attributed to it, but the only thing that is really known is the partial skull. This is high and narrow and the snout seems to have been short. The lower jaw is quite slender and obviously does not support the big fangs of the other hunting therapsids. All in all the skull seems to be similar to that of the sphenacodonts, but has a much bigger synapsid opening behind the eyes.

Distribution: Russia.
Classification: Reptile, Therapsida.
Meaning of name: Withered crocodile.
Named by: Efremov, 1954.
Time: Wordian stage of the middle and late Permian.
Size: 1.5m (5ft).
Lifestyle: Hunter.
Species: *P. discors*, *P. horissiaki*.

Left: Because of the paucity of material, any restoration of Phthinosuchus such as this must be regarded as highly speculative.

Theriognathus

Distribution: South Africa.
Classification: Reptile, Therapsida.
Meaning of name: Mammal jaw.
Named by: Broom, 1920.
Time: Wuchiapingian stage of the late Permian.
Size: 1.1m (3½ft).
Lifestyle: Hunter.
Species: *T. microps*.

The animals in the therocephalian family, such as *Theriognathus*, resembled the gorgonopsians and had a similar lifestyle. They flourished at the end of the Permian and continued on into the Triassic. It was once thought that the mammals themselves evolved from the therocephalians, but this idea is not now generally held. However, they had developed several mammal-type features.

Right: Covered in fur, Theriognathus would have looked like a mammal. Naked, it would have looked like a gorgonopsid.

Features: *Theriognathus* has a flat jaw with a wide muzzle. In life it must have resembled a modern rodent or small carnivore, and it is quite possible that the mammal-like features of a high metabolic rate backed up by an insulating layer of fur first appeared. Turbinals (coiled bones in the nose that help temperature regulation) were present, as were a palate and a complex ear structure, all mammalian features. However, it still retains the reptilian jaw structure made up of several bones, rather than the single bone we find in the lower jaw of mammals.

DICYNODONTS AND CYNODONTS

Probably the most abundant of the mammal-like reptiles, in both Permian and Triassic periods, were the dicynodonts. The name "two dog-like teeth" derives from the pair of long canines that were such a prominent feature of these mammals. The front of the mouth became toothless and armed with a beak. The whole of the jaw structure seems to have been adapted to making these features function.

Cistecephalus

The fossils of *Cistecephalus* are so common that it has given its name to a zone of stratification in the Permian rocks of South Africa. It was also one of the first of the dicynodonts to be described, back in the 1870s. It was a burrowing animal (possibly explaining the abundance of fossils), living underground like a modern mole, devouring insects, worms and probably roots and rhizomes as well. It is the oldest-known burrowing synapsid.

Features: The skeleton of *Cistecephalus* is compact and streamlined, and the skull is wedge-shaped and flattened, with broad muscle attachments at the rear for strong neck muscles. These are just the features for a burrowing mode of life. The humerus is short and the front legs very muscular as shown by the big muscle scars. The front feet have broad toes. These would have been used for digging.

Below: Cistecephalus *had the body shape of other burrowing animals, a compact streamlined teardrop shape with stout little digging limbs.*

Distribution: South Africa; India.
Classification: Reptile, Therapsida.
Meaning of name: Box head.
Named by: Owen, 1876.
Time: Wichiapingian stage of the late Permian.
Size: 30cm (12in).
Lifestyle: Burrowing insectivore.
Species: *C. microrhinus*.

Diictodon

Like *Cistecephalus*, *Diictodon* was a burrowing dicynodont. Recent work done by Corwin Sullivan and his associates at Harvard suggests that the tusks were only present in the males, indicating that they may have had a display function. Full skeletons have been found curled up in their burrows where they died.

Features: Although *Diictodon* was one of the first of the dicynodonts to evolve, it already has all the specializations of the group. The tusks are there and also the beak. Also present is a notch in front of the tusks that seems to have been used for holding and snipping stems, presumably of the ubiquitous seed fern *Glossopteris* that formed the main part of the southern flora at the time. Bone structure shows that *Diictodon* was a fast-growing animal in its youth. There were five sharp claws on each forefoot, used for digging. *Diictodon* seems to have had a good sense of smell.

Below: The stout barrel-shaped body of Diictodon *and the thick stubby limbs show it to have been a burrowing animal.*

Distribution: South Africa.
Classification: Reptile, Therapsida.
Meaning of name: Two weasel teeth.
Named by: Owen, 1876.
Time: Changhsingian stage of the late Permian.
Size: 45cm (1½ft).
Lifestyle: Burrowing omnivore.
Species: *D. feliceps*.

Dicynodon

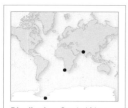

Distribution: South Africa; Tanzania; India; Antarctica.
Classification: Reptile, Therapsida.
Meaning of name: Two dog teeth.
Named by: Owen, 1845.
Time: Changhsingian stage of the late Permian.
Size: 1.4m (4ft).
Lifestyle: Omnivore.
Species: *D. lacerticeps, D. trautscholdi, D. turpior, D. trigonocephalus, D. whaitsi, D. grimbeeki, D. bolorhinus, D. leontops, D. lissops, D. osborni, D. plateceps.*

Dicynodon, from which the whole suborder to which it belongs takes its name, was one of the most common and widespread of the dicynodonts. Its wide range was once used as proof that continental drift had taken place. It is regarded as the rabbit of Permian times because of its lifestyle and its abundance.

Left: The two fangs could have been used for digging up roots or for display.

Features: The skull is toothless, except for the prominent tusks on the upper jaw. The turtle-like beak was the main food-gathering device. It is thought to have been an undergrowth dweller, feeding in the thick vegetation of the coastline of southern Pangaea. The end of the Permian period is marked by the sudden disappearance of fossils of *Dicynodon* in the South African beds. *Dicynodon* may well have been the ancestor of the big herbivorous dicynodonts in the later Triassic period. Two species of *Dicynodon* have been described from Russia, but recent researches suggest that these belong to a different genus altogether.

THE PERMIAN EXTINCTION

The Permian period began with an ice age, but it closed with a mass extinction. The sea beds had been full of a diverse fauna of sea-lilies (plant-like members of the starfish family), brachiopods (like modern bivalves but totally unrelated) and moss-like bryozoans. Over the surface crawled the last trilobites (joint-legged animals that had been an important part of the Palaeozoic fauna). In the waters swam shelled cephalopods (relatives of the squid and octopus). Then something happened and all these animals disappeared or were severely reduced. It has been estimated that 96 per cent of all species died out at this time.

We do not know the reason for all this destruction. It may not have been an abrupt event. There is evidence that animals were declining ten million years before then. The accumulation of continents to form Pangaea and the consequent fall in sea level may have been a factor, as may have been the presence of intensive volcanic activity in what is now Siberia.

Plants were also affected. The *Glossopteris* seed fern that was such an important part of the flora disappeared. Forests that clothed the coastal mountain ranges of the northern continents were wiped out. This must have affected the plant-eaters of the time.

Among the land-living animals, the most important group, the mammal-like reptiles, almost disappeared. A few of these survived and continued into the Triassic where they were to evolve, and eventually develop into the modern mammals.

Procynosuchus

The cynodonts are the success story of the mammal-like reptiles. Late to evolve, they survived the end of the Permian mass extinction and developed to become the most important of the Triassic types. *Procynosuchus* was an early cynodont from the Permian. A related animal *Parathrinaxodon* from Tanzania may be another species of *Procynosuchus*.

Features: For all it was an early type, *Procynosuchus* shows some specialist adaptations. The tail vertebrae are flattened and the tail and hindquarters very flexible. The hind limbs are paddle-like. These are adaptations to a swimming lifestyle. The chewing mechanism is much more sophisticated than any of its contemporaries, with greater areas for muscle attachment. The pattern of points on the teeth is very mammal-like.

Below: Procynosuchus *must have looked and behaved like an otter of modern times, chasing fish in shallow ponds and rivers.*

Distribution: South Africa; Zambia; Germany.
Classification: Reptile, Therapsida.
Meaning of name: Before dog crocodile.
Named by: Broom, 1931.
Time: Wuchiapingian stage of the late Permian.
Size: 65cm (2ft).
Lifestyle: Fish-hunter.
Species: *P. delaharpeae, P. rubidgei.*

THE
TRIASSIC
PERIOD

The Triassic period is regarded as the beginning of the Age of
Dinosaurs, although the dinosaurs did not appear until its end.
At that time the landmass of the world was joined together in a
supercontinent known as Pangaea. Its centre was so far from the sea
that desert conditions prevailed, and the only habitable areas were
around the coastal rim. Here the dinosaurs developed, alongside the
mammals that were later destined to replace them. The dinosaurs
evolved along with the vegetation that was to support them (from seed
ferns in the early Triassic to conifers in the late Triassic), and later they
evolved to prey upon other creatures which ate the plantlife.

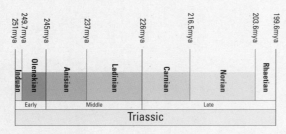

*Above: The timeline shows the different chronological stages of the Triassic
period of Earth's history. (mya = million years ago)*

1 *Eudimorphodon.*
2 *Plateosaurus.*
3 *Placerias.*
4 *Liliensternus.*
5 *Proganochelys.*

PRIMITIVE AMPHIBIANS

Mass-extinction at the end of the Permian cleared the way for all sorts of new animals to evolve.
However, some of the old groups did survive and continued to flourish until the end of the Triassic. One
such group consisted of the temnospondyls – an order of amphibians named for a complex structure of
the backbone – some of which lasted until the Cretaceous period.

Metoposaurus

The metoposaurs – the big-headed amphibian family to which *Metoposaurus* belonged – were amazingly successful, considering that the true age of amphibians had passed. Their remains are found almost worldwide, from North America, through Europe to Morocco and India. There are several genera, all similar to one another, and they may eventually all be classed as *Metoposaurus*. Other genera include the giant *Buettneria* and the tiny *Apachesaurus*, both from North America.

Features: *Metoposaurus* is a large aquatic amphibian. It has a broad flat triangular head with the eyes, crocodile-like, on the top, but further forward than on other amphibian skulls. Its limbs are very small, and would have been used for poling the animal across the bed of the rivers and lakes in which it lived. There is evidence for a lateral line system that picks up vibrations and changes in pressure in the water and allows the animal to sense what is going on around it – the sensory system that fish possess – along the side of the skull. This and the shape of the animal show it to have been almost completely aquatic.

Distribution: Germany; Poland; Texas, USA.
Classification: Amphibian, Temnospondyli.
Meaning of name: Front lizard.
Named by: Lydekker, 1890.
Time: Carnian stage of the late Triassic.
Size: 2.5m (8ft).
Lifestyle: Fish-hunter.
Species: *M. diagnosticus, M. krasiejowensis, M. stuttgartiensis, M. heimi, M. santaecrucis, M. bakeri.*

Left: Metoposaurus would either have been an ambusher, lunging up from the lake bed to catch fish, or a mid-water hunter, swimming actively after its prey.

Mastodonsaurus

The name of this gigantic amphibian was once thought to have referred to its size – as big as the elephantine mammal Mastodon. In fact, it was named after the nipple-shaped tooth that was first discovered by Jaeger in the 1820s. This was actually a damaged tooth, and its true shape would have been very similar to that of the typical labyrinthodont amphibian – a conical structure with the dentine and enamel folded in the characteristic maze-like pattern.

Features: The tooth that Jaeger found is actually the biggest amphibian tooth known. These fang-like teeth are found on the lower jaw and must have been used to catch fish. A feature shared with other amphibians of the group is a third 'eye' on top of the skull. This consists of a lens, a retina and an optic nerve, but no iris. It was probably used as a heat-sensitive organ that could register the intensity of the light, and was part of the system that regulated the temperature of the body.

Below: The two tusk-like teeth at the front of the lower jaw protruded through slots on the skull above.

Distribution: Europe.
Classification: Amphibian, Temnospondyli.
Meaning of name: Nipple-toothed lizard.
Named by: Jaeger, 1828.
Time: Middle Triassic.
Size: 6m (20ft).
Lifestyle: Fish-hunter.
Species: *M. torvus, M. cappelensis, M. giganteus.*

Gerrothorax

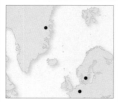

Distribution: Germany; Sweden; Greenland.
Classification: Amphibian, Temnospondyli.
Meaning of name: Plated chest.
Named by: Nilsson, 1937.
Time: Norian stage of the late Triassic.
Size: About 1m (3ft).
Lifestyle: Fish- and invertebrate-hunter.
Species: G. pulcherrimus, G. rhaeticus.

Lake-bottom-dwelling amphibians tend to be very flat, and in *Gerrothorax* this flattening is taken to an extreme. It would have spent all of its time in the water, as indicated by the presence of external gills, and may have hunted like the modern angler fish – lying still on the pond bed and attracting prey into its broad mouth.

Features: The skull is wide and flat, with two small closely set eyes on the top. The limbs are small and webbed and the tail short. A remarkable feature of *Gerrothorax* is the presence of three pairs of external gills, and the fact that these gills are supported by bony structures. External gills are usually only found in the larval stages of amphibians, but in *Gerrothorax* they existed throughout the animal's life. There is evidence of armour plating both above and below.

Below: Gerrothorax would have had no need to leave the water.

TEMNOSPONDYLS

The land-living amphibians declined in the face of the development of the reptiles in Permian and Triassic times. Only the purely aquatic types were able to survive. The temnospondyls, which had not developed the robust legs needed for prolonged action on land, and had retained external gills into adult-hood, continued to thrive in the water. Some groups survived the extinction at the end of the Triassic and continued on into the Jurassic and Cretaceous, especially on the southern continents.

The latest temnospondyl known is *Koolasuchus* from the early Cretaceous of Australia. It thrived in rift valley lakes very close to the contemporary south pole. It is assumed that crocodiles, their prospective competitors, were unable to tolerate the cold conditions.

Modern amphibians belong to a group called the lissamphibia, and they evolved in the Triassic. They have the following characteristics:
• Skin that is moist and permeable. This allows breathing through the skin. Glands in the skin help keep it moist.
• The hearing mechanism is sensitive to low sound frequencies.
• Eye sensitivity. The retina has so-called "green rods" that are unique to amphibians.
• Tooth structure. The tooth has two layers separated by a layer of fibrous tissue
• Eye muscles. These allow the eyeball to move, helping to push food down the throat.
These are all secondary characteristics, evolved after the amphibians as a group developed. None of these features is seen in any of the early fossil groups.

Pelorocephalus

The chigutisaurs were a family of temnospondyls that have been found only in the southern continents, where they would have occupied the southern part of Pangaea, or Gondwana, in Triassic times. They are known from Argentina and India. The chigutisaurs were one of the few amphibian groups to survive the mass-extinction at the end of the Triassic and continue into the Jurassic and Cretaceous periods.

Features: *Pelorocephalus* is a typical chigutisaur, with its broad head and body, its stubby limbs that are, nonetheless, adapted for swimming, and its finned tail. It would have hunted like a crocodile, and in fact its Jurassic and Cretaceous relatives grew to great lengths and took the place of crocodiles, flourishing in areas where those reptiles were not present.

Distribution: Argentina.
Classification: Amphibian, Temnospondyli.
Meaning of name: Monstrous head.
Named by: Cabera, 1944.
Time: Carnian stage of the late Triassic.
Size: About 1m (3ft).
Lifestyle: Fish-hunter.
Species: P. cacheutensis, P. ischigualastensis, P. mendozensis, P. tenax.

Above: Pelorocephalus was a fearsome hunter in the lakes and rivers of the southern supercontinent.

TRIASSIC PARAREPTILES

The most primitive group of reptiles used to be known as the anapsids – those with no holes in the skull behind the eye. Recent research, however, shows that this is not a true grouping, and that several separate primitive groups of reptiles developed this feature independently. The classification has now been replaced with that of the parareptiles – the "almost reptiles".

Hypsognathus

The procolophonids were a family of parareptiles that ranged worldwide in Permian and Triassic times. They were mostly insectivorous but towards the end of their time range many of them became adapted for plant-eating. *Hypsognathus* was one of the latter, a slow-moving vegetarian, feeding on the sparse desert plants of the time.

Features: *Hypsognathus* has a wide, squat body, suggesting that it was not a very agile animal. Its broad cheek teeth are ideal for grinding up fibrous material found in tough desert plants. Small nipping teeth at the front suggest that it was selective about what it ate. The array of spikes around its head would have been a defensive measure – a protection against the carnivorous dinosaurs such as *Coelophysis* that were coming to the fore at that time.

Distribution: New Jersey, USA.
Classification: Reptile, Captorhinida.
Meaning of name: High jaw.
Named by: Gilmore, 1928.
Time: Norian to Rhaetian stages of the late Triassic.
Size: 33cm (13in).
Lifestyle: Low browser.
Species: *H. fenneri.*

Left: Hypsognathus *lived in rather barren terrain that left few fossils of other plants or animals for comparison, and so its actual age in the late Triassic is not clear.*

Proganochelys

The turtles and the tortoises represent a group that evolved a very successful shape very early, and have had no need to make any changes since. Structurally they seem to be very similar to the pareiasaurs of the Permian, particularly in the arrangement of skull bones and the fact that the pareiasaurs were beginning to develop an armoured covering, and are possibly descended from them. The shell is made up of bones fused to the ribcage, covered in horn.

Features: The main differences between *Proganochelys* and modern turtles are in the neck and tail. The neck could not be pulled back into the shell, and so it is armoured by plates and spikes. The tail is also long and armoured with spikes, with a club at the end. The shell, however, is similar to that of modern types. The shoulder girdle is inside the ribcage. There is a very modern-looking ear opening and there are teeth only on the palate. It is possible that the anapsid skull evolved independently, and that the turtles are actually derived from diapsid reptiles.

Distribution: Europe.
Classification: Reptile, Chelonia.
Meaning of name: Early shell.
Named by: Baur, 1887.
Time: Norian stage of the late Triassic.
Size: About 1m (3ft).
Lifestyle: Plant-eater.
Species: *P. quenstedi.*

Left: Although Proganochelys *was found in Europe, there are related forms from the Triassic of North America and South-east Asia.*

Procolophon

Distribution: South Africa; Antarctica; Brazil.
Classification: Reptile, Procolophonidae.
Meaning of name: Before the end.
Named by: Owen, 1876.
Time: Induan stage of the early Triassic.
Size: 30cm (12in).
Lifestyle: Burrowing plant-eater.
Species: P. trigoniceps, P. braziliensis.

A recent find in Brazil, *Procolophon*, from which the Procolohonidae derive their name, is known only from the early Triassic but also from South Africa and Antarctica. These regions were close to one another at that time. It has a close relative in *Tichvinskia* from Russia. Recent studies show that it was a burrowing animal, probably feeding on roots and underground tubers, the kinds of structures used by desert plants to survive long periods of drought.

Features: The body is stout and lizard-like with a triangular-shaped skull and strong chisel-like teeth. The three middle fingers of the hand are broad, and this would give the hand a spade shape indicating an adaptation for a burrowing lifestyle. Bones of *Procolophon* have been found in a fossil burrow in a South African Triassic riverbank deposit, supporting this interpretation.

Left: Procolophon burrowed away from the harshness of the Triassic desert day.

THE SKULL WITH NO HOLES BEHIND THE EYES

Some scientists prefer the term "parareptiles" (not-quite reptiles) to "anapsids", but define the term in different ways. In one definition parareptiles includes all that we would regard as anapsid, except for the mesosaurs. Another includes the mesosaurs but not the turtles. Yet another excludes both mesosaurs and turtles. The term "parareptile" implies that every other reptile group must be regarded as "eureptiles" (true reptiles) and this will include birds. More work is needed in the field before the situation can be clarified.

As a final confusion, certain procolophonids, have recently been found that have holes in the skull behind the eye socket, flying in the face of the traditional definition of anapsids.

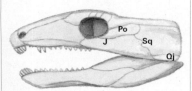

Above: The main bones of a reptile's skull behind the eye are the jugal (J), the postorbital (Po), the squamosal (Sq) and the quadratojugal (Qj). The positions of holes between these bones, holding the jaw muscles, define the anapsid, the diapsid and the synapsid conditions.

Colubrifer

The burrowing habit adopted by certain anapsids, such as *Procolophon*, was taken to an extreme in *Colubrifer*. When it was first studied it was thought to have been an early burrowing lizard, similar to many of the legless skinks of today. However, independent work by Sue Evans from University College, London, and Robert Reisz from the University of Toronto, pursuing quite different lines of research, in 2000, showed that it was a specialized anapsid. There is a possibility that it is actually a Triassic species of the Permian *Owenetta*. The structure of the skull suggests that *Owenetta* may be close to the ancestry of the turtles. If this is so it is a fine example of a great range of shapes of animals within a single group.

Features: The single skeleton of *Colubrifer* that has been found shows an elongated skink-like animal with tiny limbs. It presumably lived like a modern sand-dwelling skink, the similarity in shape being a result of convergent evolution (the development of similar shapes in response to similar environmental conditions and lifestyles).

Distribution: South Africa.
Classification: Reptile, Parareptilia.
Meaning of name: Snake-like.
Named by: Carroll, 1982
Time: Induan stage of the early Triassic.
Size: 30cm (12in).
Lifestyle: Burrowing insectivore.
Species: C. campi.

Below: A long spindly body shape, such as that possessed by Colubrifer, is good for burrowing in the ground.

PLACODONTS

The placodonts were a strange group of swimming reptiles that appeared in the middle Triassic and became extinct at the Triassic/Jurassic boundary. They were heavy-bodied swimmers that mostly fed on shellfish: their teeth were specialized for picking shells from rocks and crushing them between their jaws. In the Triassic seas they would have been the equivalent of modern walruses.

Paraplacodus

The placodonts consisted of several families, including the unarmoured placodontoids, which must have resembled gigantic newts, and the armoured cyamodontids, which evolved into turtle shapes. *Paraplacodus* was typical of the former. The species lived in shallow seas and lagoons at the edge of the Tethys Ocean and fed from the banks of shellfish that grew there.

Features: The jaws of *Paraplacodus* are uniquely adapted to picking up shellfish, with three pairs of protruding teeth in the top and two in the bottom. The teeth project from the front of the mouth. They have a series of rounded crushing teeth in the upper and lower jaws. The thick ribs produce a distinctly box-like body with an almost square cross-section, a strong set of belly ribs forming the flat floor of the body – a heavy design that kept it close to the seabed.

Distribution: Northern Italy.
Classification: Placodontia, Placodontoidea.
Meaning of name: Almost *Placodus* (see below).
Named by: Peyer, 1931.
Time: Anisian to Ladinian stages of the middle Triassic.
Size: 1.5m (6ft).
Lifestyle: Shellfish-eater.
Species: *P. broilii.*

Left: A specimen of a similar placodont called Saurosphargis *was destroyed during World War II; this may have been a close relative.*

Placodus

Aquatic animals adopt several methods of dealing with buoyancy. The placodonts used what is known as pachystasis: the development of particularly thick and heavy bones to keep the body submerged. Voluminous lungs allowed them to adjust their buoyancy. This technique is often used by animals that feed while walking along the sea bed, such as modern dugongs and sea-otters.

Features: Like *Paraplacodus, Placodus* has protruding teeth at the front. However they are shorter, thicker and more spoon-shaped. The crushing teeth are not confined to the edges of the mouth but form a broad pavement across the palate, and the skull is particularly strong to withstand the stresses of crushing seashells. A row of bony scutes forms a jagged ridge along the back. A gap at the top of the skull may have held a light-sensitive organ.

Distribution: Germany, Italy.
Classification: Placodontia, Placodontoidea.
Meaning of name: Flat toothed.
Named by: Agassiz, 1833.
Time: Anisian to Ladinian stages of the Triassic.
Size: 3m (10ft).
Lifestyle: Shellfish-eater.
Species: *P. gigas.*

Left: When the teeth were first found in 1830 the ichthyologist Louis Agassiz misidentified them as fish teeth, and gave the name Placodus. *Richard Owen, in 1858, recognized them as reptile teeth.*

Cyamodus

Although some of the placodontids had armour plates embedded in the skin, it was the cyamodontids that took this to an extreme. The bodies became broad and flat, almost turtle-like, and they probably spent much of their time close to the shallow sea bed, pulling themselves over the sand and searching for shellfish, very much as modern rays do.

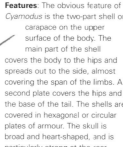

Features: The obvious feature of *Cyamodus* is the two-part shell or carapace on the upper surface of the body. The main part of the shell covers the body to the hips and spreads out to the side, almost covering the span of the limbs. A second plate covers the hips and the base of the tail. The shells are covered in hexagonal or circular plates of armour. The skull is broad and heart-shaped, and is particularly strong at the rear.

Left: Juvenile specimens of Cyamodus *appear to have an extra tooth on the roof of the mouth, compared with adults. It seems possible that they reduced the number of teeth as they grew to maturity.*

Distribution: Germany, Italy.
Classification: Placodontia, Cyamodontidae.
Named by: Meyer, 1863.
Time: Anisian to Ladinian stages of the middle Triassic.
Size: 1.3m (4ft).
Lifestyle: Shellfish-eater.
Species: *C. rostratus, C. hildegardis, C. kuhnschneyderi.*

OTHER PLACODONTS

We do not know the ancestors or the origin of the placodonts. The latest understanding is that they may be diapsids (a group of reptiles with two pairs of holes in the skull behind the eyes) and may be related to the ancestors of the plesiosaurs, although their limbs were not so highly adapted for swimming.

Helveticosaurus from the Anisian stage of the Triassic in Switzerland was once thought to have been a primitive, possibly ancestral, placodont. It resembles *Placodus* in appearance. However it now seems to be totally unrelated.

Psephoderma from several localities in Europe resembled *Cyamodus* with its two-part carapace. Its shell has three distinct ridges running longitudinally along it.

Placochelys from the Landinian stage of the Triassic in Hungary has jaws that were pointed at the front, and no front teeth. It is possible that it possessed a horny beak to peck the shellfish from the reef.

Protenodontosaurus appears to be intermediate between *Cyamodus* and the more advanced turtle-like placochelyids.

Right: The skulls of Paraplacodus (top) and Placodus (bottom). Though the two creatures have a similar appearance they are quite different.

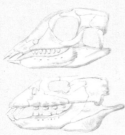

Henodus

We can imagine *Henodus* like some kind of reptilian ray, paddling its broad flat body across the bed of a shallow lagoon, foraging in the rippled sand with its broad mouth. Its plate-like body would have made it better adapted to searching along flat sea beds than to the shellfish-encrusted reefs frequented by its relatives. Its weak limbs suggest that it did not spend much time on land.

Features: This is the placodont that most resembles the turtles. There is a carapace over the whole of the body that stretches out well beyond the span of the limbs, and as in the other cyamodonts this is matched by a plastron, a lower shell, which covers the undersurface. Both carapace and plastron are made up of a geometric array of individual plates. The head is squared off at the front and is shortened in front of the eyes.

Distribution: Southern Germany.
Classification: Placodontia, Cyamodontidae.
Meaning of name: Armoured.
Named by: Huene, 1936.
Time: Carnian stage of the late Triassic.
Size: 1m (3ft).
Lifestyle: Sea floor-forager.
Species: *H. chelyops.*

Left: Henodus is the only placodont so far found in non-marine deposits. It seems to have lived in brackish or freshwater lagoons. Its broad mouth was adapted for lifting shellfish from soft sandy beds.

NOTHOSAURS

If the ichthyosaurs are the whales and dolphins of the Mesozoic world, then the plesiosaurs can be regarded as the seals and sea lions. Before the plesiosaurs were fully established, a more primitive offshoot from their line, the nothosaurs, were the most abundant fish-catchers of the time. The nothosaurs had the same long necks as their plesiosaur cousins but were not so well adapted to water.

Ceresiosaurus

With a long body and powerful tail *Ceresiosaurus* has features usually indicative of an animal that swims by undulations of its body. However the bone structure, especially in the thick tail and strong hips, suggests that it was a pursuit diver and hunted underwater like a penguin, manoeuvring with its strong paddles. Analysis of its stomach contents has shown it to be a hunter of marine reptiles.

Features: *Ceresiosaurus* has much longer toes than other nothosaurs. The additional length was gained by an increase in the number of bones in each toe (polyphalangy). The toes are fused into a swimming paddle. The tail bones are thick and evolved to support powerful muscles. The skull is short, much shorter than that of any other nothosaur, and looks very much like that of a plesiosaur with the nostrils placed well forward. The front paddles are bigger than the hind, suggesting that they played a bigger part in the animal's locomotion.

Left: The preserved stomach contents of Ceresiosaurus *contain the remains of* pachypleurosaurs. Ceresiosaurus *must have been a fast hunter to catch such agile prey.*

Distribution: Europe.
Classification: Sauropterygia, Nothosauridae.
Meaning of name: Lizard of Ceres.
Named by: Peyer, 1931.
Time: Anisian stage of the middle Triassic.
Size: 4m (13ft).
Lifestyle: Swimming hunter.
Species: *C. calcagnii*, *C. russelli*.

Nothosaurus

The main fish-eaters on the northern shores of the Tethys Ocean in Triassic times were the nothosaurs, and *Nothosaurus* itself was typical. It swam in the surf and shallow inshore waters but rested and bred on the beaches and rock caves of the shoreline. It may have laid eggs in the sand as modern turtles do, or it may have given birth at sea.

Features: The skull is long and flat, and the long jaws are equipped with sharp interlocking teeth, some in the form of paired fangs, making a formidable fish trap. Each foot is webbed and has five long toes, and can be used for walking on land and for swimming. The body, neck and tail are long and flexible. A muscular tail helped in swimming.

Distribution: Europe; North Africa; Russia; and China.
Classification: Sauropterygia, Nothosauridae.
Meaning of name: False lizard.
Named by: Munster, 1834.
Time: Anisian and Ladinian stages of the middle Triassic.
Size: 3m (10ft), although a newly found species, *N. giganteus*, reached 6m (20ft).
Lifestyle: Swimming hunter.
Species: *N. mirabilis*, *N. giganteus, N. procerus*.

Pachypleurosaurus

The pachypleurosaurs were not true nothosaurs but were closely related. They were quite small and probably lived close to shore or in lagoons, rather like the modern marine iguana. They may have originated in China and migrated to Europe along the northern shores of the Tethys Ocean. The small head suggests that they hunted small fish or shellfish.

Features: The tail is deep and obviously used as a swimming organ. The hips and shoulders, although adapted for swimming, are still strong enough to support the animal on land. The head is very small and the structure of the ear suggests that it was sensitive to sounds above the sea surface rather than underwater. Some of the bones are thick, an adaptation associated with buoyancy control in some aquatic animals.

Distribution: Italy; Romania; Switzerland.
Classification: Sauropterygia, Pachypleurosauria.
Meaning of name: Thick rib lizard.
Named by: Cornalia, 1854.
Time: Anisian to Ladinian stages of the middle Triassic.
Size: 1m (3ft).
Lifestyle: Swimming shellfish-eater.
Species: *P. edwardsi.*

Right: The pachypleurosaurs appear to be the link between the placodonts and the group comprising the nothosaurs and plesiosaurs. However, the group has such extreme skeletal specializations that it is difficult to be sure.

SEA-LIVING CREATURES

It is a strange fact that as soon as vertebrates evolved to live on land, away from the water, casting off the need to return to the water to breed as amphibians do, there was an evolutionary movement that took a certain number back to the sea. Adaptations for living in the sea – such as streamlined body shape, limbs that worked as paddles, a body density that allowed floating or sinking – which had all been lost began to evolve once more.

In some groups, such as that of the ichthyosaurs, this re-adaptation became almost total. In others, such as the nothosaurs, it was partial. It is as if the nothosaurs adopted a kind of a half-way stage so that they could exploit both the marine environment and that on the land. Such adaptations included legs that were webbed for swimming but still retained the bone structure and musculature for moving about on land, and the specialist teeth that would catch slippery prey like fish.

In common with all other secondarily marine vertebrates, even those as highly adapted as the ichthyosaurs, they still needed to breathe air. Gills that would have allowed underwater breathing never re-evolved once they had been lost by their ancestors.

Below: Nothosaurus breathing out of the water.

Lariosaurus

It appears that at least this genus of nothosaur was viviparous (able to bear live young). Several skeletons have been found associated with embryos indicating that they carried their young to maturity in their bodies. In one specimen two juvenile placodonts of the genus *Cyamodus* have been found in the stomach area, a clue to the diet of this nothosaur. Much of our modern knowledge of nothosaurs comes from the work of Dr Olivier Rieppel of the Field Museum in Chicago, the present day specialist in these uniquely Triassic marine reptiles.

Features: The primitive features of this small nothosaur include the short neck and toes. The back legs are five-toed with claws, and slightly webbed. The front legs are adapted into paddles. Both pairs of legs are quite short and do not give the impression of powerful swimming structures. The front legs are stronger than the hind, suggesting that they were the main swimming organs, unlike in the pachypleurosaurs. Fangs at the front of the broad head interlock as the jaws are closed, forming a vicious trap for catching fish and other aquatic animals.

Distribution: Spain; France; Italy; Germany; Switzerland; and China.
Classification: Sauropterygia, Nothosauridae.
Meaning of name: Lizard from Lake Lario.
Named by: Curioni, 1847.
Time: Anisian to Ladinian stages of the middle Triassic.
Size: 60cm (2ft).
Lifestyle: Fish- and crustacean-eater.
Species: *L. valceresii, L. balsamii, L. curioni, L. xingyiensis.*

Left: The front legs are stronger than the hind suggesting that they were the main swimming organs, unlike in the pachypleurosaurs.

NOTHOSAURS AND THE PLESIOSAUR CONNECTION

The plesiosaurs were far better adapted for aquatic life than their relatives the nothosaurs. Their bodies were flat and rigid, and they were powered by paddles that worked with a flying action, rather like modern turtles. In Triassic times there were many that seem to be transitional between the two groups.

Pistosaurus

The pistosaurs seem to have been intermediate between the nothosaurs and the true plesiosaurs. Their remains are known from both sides of the Atlantic, *Pistosaurus* was found in Germany, while its relative *Augustasaurus* was found in America. It is not known whether they swam like crocodiles as nothosaurs did, or like turtles like plesiosaurs.

Below: The waters of the Tethys Ocean were full of shoals of fish, and many fish-eaters, like Pistosaurus, *evolved to exploit these.*

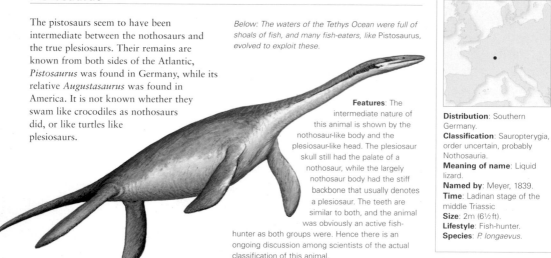

Features: The intermediate nature of this animal is shown by the nothosaur-like body and the plesiosaur-like head. The plesiosaur skull still had the palate of a nothosaur, while the largely nothosaur body had the stiff backbone that usually denotes a plesiosaur. The teeth are similar to both, and the animal was obviously an active fish-hunter as both groups were. Hence there is an ongoing discussion among scientists of the actual classification of this animal.

Distribution: Southern Germany.
Classification: Sauropterygia, order uncertain, probably Nothosauria.
Meaning of name: Liquid lizard.
Named by: Meyer, 1839.
Time: Ladinan stage of the middle Triassic.
Size: 2m (6½ft).
Lifestyle: Fish-hunter.
Species: *P. longaevus.*

Dactylosaurus

Another problematic group of early marine reptiles are the pachypleurosaurs – the lizards with thick ribs. They were once thought to have been a part of the nothosaur group but are currently believed to be separate from them, and more primitive. The heads were short and the necks were long. *Dactylosaurus* was the smallest of this group of small reptiles.

Right: Dactylosaurus *was the smallest of the marine reptiles of the time.*

Features: The distinctive feature of *Dactylosaurus* is the shape of the humerus, indicating that the muscle attachment would mean a different swimming action from its contemporaries. There is a reduced number of bones in the flippers, showing a specialization to a swimming mode of life. However the primitive nature of the rest of the skeleton indicates that it was at best semi-aquatic. Like the other pachypleurosaurs, the ribs are thick (hence the name of the group), to help to control the buoyancy while submerged.

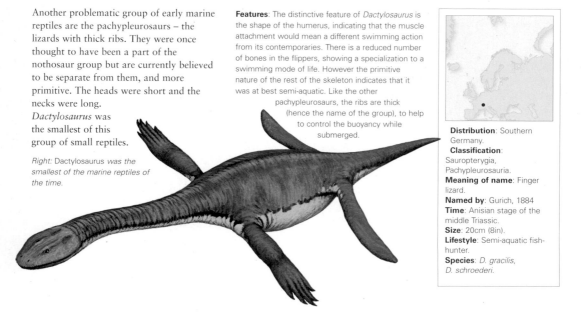

Distribution: Southern Germany.
Classification: Sauropterygia, Pachypleurosauria.
Meaning of name: Finger lizard.
Named by: Gurich, 1884
Time: Anisian stage of the middle Triassic.
Size: 20cm (8in).
Lifestyle: Semi-aquatic fish-hunter.
Species: *D. gracilis, D. schroederi.*

Corosaurus

Distribution: Wyoming, USA.
Classification: Sauropterygia, Nothosauria.
Meaning of name: North-west wind lizard.
Named by: Case, 1936.
Time: Early to late Triassic.
Size: About 1m (3ft).
Lifestyle: Fish-hunter.
Species: C. alcovensis

Originally thought to have been a pachypleurosaur *Corosaurus* is now regarded as the most advanced nothosaur so far known. It lived in quiet waters, as shown by the presence of stromatolites (mineral mounds built up by algae and bacteria) in the same shallow marine strata. It is possible that these mounds may have made suitable nesting areas for these reptiles.

Features: The shoulder and hip girdles of *Corosaurus* are very similar to those of the early plesiosaurs. The paddles are also highly adapted, although not quite in the plesiosaur manner. They would have been ideal for paddling through quiet waters, but less suited for clambering about on land. Nevertheless, *Corosaurus* probably spent some time on land, basking or laying eggs.

Above: Corosaurus is the only nothosaur known from North America.

PANGAEA

The fact that all the continents of the world had united in Triassic times to form the single supercontinent of Pangaea meant that all the ocean areas had also united to form the single superocean of Panthalassa. With the convergence of landmasses, the total coastline was shorter than at any other time in the Earth's history.

A deep embayment, the Tethys Ocean, almost split the continent apart, and the shallow waters of the edges of this produced thick limestone beds in the region of central and southern Europe. It is from these deposits that we have most knowledge of the marine reptiles of the time. However, with the discoveries of remains as far afield as the Mid-West of North America and Central China, it seems that similar sea animals existed all around the coast of Pangaea in Triassic times.

Above: The long coastline of the single continent of Pangaea had many shallow shelves and embayments – ideal habitats for sea reptiles.

Thalassiodracon

Although only named in 1996 this animal has been known since 1838, when Owen gave it the name *Plesiosaurus hawkinsii*. It was part of the extensive marine fauna found by Thomas Hawkins, the noted Victorian collector of fossil sea animals, in Somerset, southwest England. There is some disagreement between palaeontologists as to whether this is a primitive plesiosauroid or a pliosauroid.

Features: *Thalassiodracon* is a small primitive plesiosaur with a small head, about one tenth of the length of the whole animal, but this is large compared with later plesiosaurs. The snout is short and pointed and the skull lightly built. Unlike later plesiosaurs the hind limbs are bigger than those at the front. It has the long neck of the plesiosauroids, but the lower jaw, the hip bones and the larger hind limbs of the pliosauroids.

Distribution: England.
Classification: Sauropterygia, Plesiosauria.
Meaning of name: Ocean dragon.
Named by: Storrs and Taylor, 1996
Time: Rhaetian to Hettangian stages of the late Triassic and early Jurassic
Size: 2m (6½ft).
Lifestyle: Fish-hunter.
Species: T. hawkinsi.

Above: Thalassiodracon hunted fish in the shallow waters of northern Europe at a time when the Triassic period gave way to the Jurassic.

EARLY ICHTHYOSAURS

Perhaps the best known of fossil sea reptiles are the fish-lizards, the ichthyosaurs. The best are known from fossils in Jurassic rocks, but their history reaches back into Triassic times. There are several distinct early shapes of ichthyosaurs, before the classic dolphin-like appearance was standardized. Some are eel-like and others are enormous and whale-shaped.

Cymbospondylus

Although *Cymbospondylus* is formally regarded as part of the shastasaurid group of primitive ichthyosaurs, recent studies suggest that it may be more primitive than originally thought – maybe too primitive to be regarded as an ichthyosaur. Indeed, in appearance it does not seem to have the physical features, such as the dorsal fin and fish-like tail, which are so distinctive of the later members of the group. *Cymbospondylus* is the state fossil of Nevada.

Features: *Cymbospondylus* is more eel-shaped than the other ichthyosaurs, with a narrow body and a long flexible tail that takes up about half the length of the animal. The legs have already evolved into paddles but these were probably used more for stabilization, while the swimming action was produced by undulations of the body. The head is quite small in relation to the body but is quite typical of the ichthyosaurs, with the long snout and the small, sharp fish-catching teeth already present.

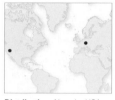

Distribution: Nevada, USA; and Germany.
Classification: Ichthyosauria, Shastasauridae.
Named by: Leidy, 1868.
Time: Middle Triassic.
Size: 6m (18ft).
Lifestyle: Fish-hunter.
Species: C. natans, C germanicus, C. nevadanus, C. parvus, C. piscosus, C. grandis, C. petrinus, C. buchseri.

Mixosaurus

In appearance *Mixosaurus* seems intermediary between the eel-like forms typified by *Cymbospondylus*, and the more familiar dolphin-shaped forms such as the later *Ichthyosaurus*, hence the name. The long tail with its low fin suggests that it was a slow swimmer, moving forward by undulation of the tail.

Features: The body and tail are long, with the tail having a low fin that is not as well-developed as the shark-like fin of the later types, and there is a stabilizing dorsal fin on the back. The paddle-like limbs are still made up of five toes each, unlike the multi-toed structure found in later ichthyosaurs. However, each toe has more individual bones than is usual (polyphalangy) and the front limbs are longer than the rear.

Distribution: China; Timor; Indonesia; Italy; Alaska, Nevada, USA; and Spitsbergen, Svalbard.
Classification: Ichthyosauria, Mixosauridae.
Meaning of name: Mixed lizard.
Named by: Baur, 1887.
Time: Middle Triassic.
Size: 1m (3ft).
Lifestyle: Fish-hunter.
Species: M. atavus, M. kuhnschnyderi, M. cornalianus.

Left: Mixosaurus *lived at the same time as* Cymbospondylus.

Shonisaurus

In the 1920s miners found a deposit of huge bones in Nevada. When they were excavated 30 years later they were found to be the remains of 37 individual enormous ichthyosaurs, and were named *Shonisaurus*. They may have been a shoal that was beached, or cut off from the sea. More recently, the scientific view is that they died and sank to the sea bed.

Features: This ichthyosaur has a whale-like shape and long narrow paddles. It has teeth only at the front of the jaws. *S. popularis* held the record as the biggest of the ichthyosaurus, but the skeleton of an even bigger species, *S. sikanniensis*, was found in the 1990s in British Columbia. With an estimated length of 21m (70ft), this was so big it could only have been seen in its entirety from the air. Studies of this new species suggest that *Shonisaurus* may not have been as deep-bodied as originally thought.

Distribution: Nevada, USA; to British Columbia, Canada.
Classification: Ichthyosauria, Shastasauridae.
Meaning of name: Lizard from the Shoshone Mountains.
Named by: Camp, 1976.
Time: Norian stage of the late Triassic.
Size: 15m (50ft).
Lifestyle: Ocean hunter.
Species: *S. popularis*, *S. sikanniensis*.

Right: Of the skulls found, only the smaller ones had teeth. Shonisaurus may have become toothless as it grew.

ICHTHYOSAUR CLASSIFICATION

The exact classification of the ichthyosaurs has always been a mystery. For most of the history of their study it was thought that they belonged to an evolutionary line that was totally distinct from that of any other reptile. Then there was the puzzle of the origin of their appearance and lifestyle. How did such fish-like animals evolve from land-living creatures? Indeed, were their ancestors ever land-living? For a while there was a theory that sometime in the late Palaeozoic they evolved directly from swimming amphibians without going through a land-living phase at all.

This intriguing suggestion was laid to rest in 1998 with the re-examination of the original *Utatsusaurus* skeleton by Ryosuke Motani and Nachio Minoura from Berkeley and Hokkaido Universities. By using computer imagery, they were able to reverse the distortion of the original specimen and study it carefully. They found that *Utatsusaurus*, despite its fishy shape, was actually quite closely related to the lizard-like diapsid reptiles such as *Petrolacosaurus*, making it distantly related to the ancestors of the lizards and snakes, and even the dinosaurs. They were more closely related to these than to the other main line of sea-going reptiles, the turtles.

Utatsusaurus

This, the earliest-known ichthyosaur, was found in 1982 in Japan, but it was not until a study of the skeleton was published in 1998 that its significance as the most primitive of the group was noted. Like most primitive ichthyosaurs it swam by undulations of its body rather than by sweeps of the tail, and it probably inhabited the shallow waters of the continental shelf.

Features: The skull is rather broad and tapers gradually towards the snout, unlike the narrow jaws of most other ichthyosaurs. The teeth are small for the size of skull and arranged in a groove – a primitive feature. Its paddles are small and unusually, the hindlimb is bigger than the forelimb. The paddle is made up of four fingers, unlike the usual five found in other ichthyosaurs. Its undulating swimming action is shown by the large number of very thin narrow vertebrae, making for a very flexible body.

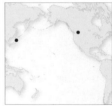

Distribution: Japan; Canada.
Classification: Ichthyosauria, Utatsusauridae.
Meaning of name: Lizard from Utatsugyoryu in Japan.
Named by: Shikama, Kamei and Murata, 1998.
Time: Olenekian stage of the early Triassic.
Size: 3m (10ft).
Lifestyle: Swimming hunter.
Species: *U. hataii*.

Left: The side-to-side swimming action of Utatsusaurus would have been fairly inefficient and would have restricted it to life in shallow seas on continental shelves.

MORE ICHTHYOSAURS

Early in their evolution the ichthyosaurs spread across the world-embracing ocean of Panthalassa, hunting in the shelf seas around the supercontinent of Pangaea. At this time they tended to be quite eel-like with straight tails, and had not evolved the streamlined fish shape and shark-like fins that was to become so distinctive in later times.

Barracudasaurus

Remains of this animal found in 1965 were named *Mixosaurus maotaiensis*. More complete remains were found in 2005 and these show it to have been a different genus altogether. The discovery suggests that the mixosaur family may have originated in the eastern waters of the Tethys Ocean and spread west rather than the other way around as had originally been thought.

Features: The new specimens consist of two skulls and a shoulder girdle with attached front paddles. These show *Barracudasaurus* to be sufficiently different from *Mixosaurus* as to represent a completely new genus. The differences are in the teeth. There is only one row of teeth in each lower jaw, and those at the back are wide, blunt and rounded, presumably indicating a different diet from other mixosaurids. The skull is longer behind the eye socket than that of other mixosaurids.

Distribution: Ginzhou Province, China.
Classification: Ichthyosauria, Mixosauridae.
Meaning of name: Barracuda lizard.
Named by: Jiang, Maisch, Hao, Matzke and Sun, 2005.
Time: Middle Triassic.
Size: About 1m (3ft).
Lifestyle: Fish-hunter.
Species: *B. maotaiensis*.

Right: Barracudasaurus swam the shallow seas that existed over China in middle Triassic times.

Besanosaurus

This ichthyosaur would have been an eel-like swimmer, cruising at moderate speed but with rapid acceleration and versatile manoeuvring, using undulations of its body. It lived in the lagoons that lay along the north coast of the Tethys Ocean. The Besano Outcrop in which it was found is one of the most fossil-rich marine beds of the Triassic.

Features: *Besanosaurus* is similar to the other shastasaurids (the family of large ichthyosaurs that evolved early in the group's history and were restricted to Triassic times) in having a thin snout, small teeth, no dorsal fin and a long straight tail without the fish-like fin on the end. The one skeleton found took six years to prepare from the limestone matrix and was found to have the remains of unborn young within its body cavity, showing that it was a female, and that *Besanosaurus* bore its young alive, as is known from the later ichthyosaurs.

Below: Besanosaurus and the other ichthyosaurs bore their young alive, and so did not have to come ashore to lay eggs.

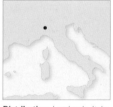

Distribution: Lombardy, Italy.
Classification: Ichthyosauria, Shastasauridae.
Meaning of name: Lizard from Besano.
Named by: Dal Sasso and Pinna, 1996
Time: Middle Triassic.
Size: 6m (20ft).
Lifestyle: Fish- or cephalopod-hunter.
Species: *B. leptorhynchus*.

Omphalosaurus

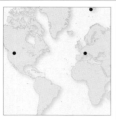

Distribution: Western North America, Spitzbergen; Austria.
Classification: Order uncertain, probably Ichthyosauria.
Meaning of name: Button lizard.
Named by: Merriam, 1906.
Time: Ladinian stage of the middle Triassic.
Size: 2m (6½ft).
Lifestyle: Cephalopod-hunter.
Species: *O. nettarhynchus*, *O. nevadanus*, *O. wolfi*.

The original specimen, a scrappy skull, was so poorly preserved that the discoverer, Merriam, did not think it was an ichthyosaur at all, but something related to the placodonts or rhynchosaurs, because of the broad shell-grinding teeth. Even now there is a great deal of controversy over whether or not it really is a member of the ichthyosaur order.

Features: The teeth are strong and blunt, adapted for crushing the shells of molluscs. "Durophagous" is the technical term for this, and the speciality is found in many groups of marine reptiles. In fact the name "button lizard" refers to the button-shaped teeth. The ichthyosaur features are the shortness of the vertebrae and the articulation of the ribs, but many scientists say that these are insufficient to identify the animal as an ichthyosaur.

Grippia

Distribution: Spitsbergen, Spalberg.
Classification: Ichthyosauria.
Meaning of name: Anchor.
Named by: Wiman, 1929.
Time: Anisian stage of the middle Triassic.
Size: About 1m (3ft).
Lifestyle: Generalist feeder.
Species: *G. longirostris*.

Grippia was always thought of as the classic durophagous ichthyosaur (like *Omphalosaurus*) – one with strong blunt teeth for crushing shellfish. However, recent studies show that the dentition was not quite right for this. It is now thought to have been a more generalist feeder, eating a wide range of foodstuffs. It is difficult to tell, as the original material on which the genus is based was destroyed in a bombing raid during World War II, along with other important fossils.

Features: The back teeth of *Grippia* are broad and flat, and are arranged in two rows in each jaw. They would have made a broad crushing platform for dealing with tough shellfish. However, studies by Ryosuke Motani, from the University of Oregon, in 1997 showed that the inner row is not high enough to make contact, and

instead consisted of replacement teeth for when the outer ones wore out. The teeth are smaller and do not have the specialist strength-giving structures found in true shellfish-eaters.

THE SHAPE OF EARLY ICHTHYOSAURS

The earliest ichthyosaurs are a mystery. Many are known only from partial remains and so it is difficult to tell what the whole animal looked like. Those that are well known seem to lack the distinctive fish shape that characterizes the later types. They had long eel-like bodies and probably swam with a sinuous action of body and tail together.

It is not known if they had the fins on the back and tail as the later ones had. Some of them show a very slight downturn on the tail as though a fin of some sort were supported above, but it would not be the spectacular shark-like fin of later ichthyosaurs.

As a result, when an early or middle Triassic ichthyosaur is known only from a scrap of bone or a few teeth, it is customary to restore the whole animal as a long slender beast, not well endowed with swimming fins.

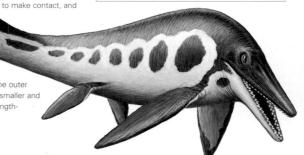

Right: Rather than being a specialist shellfish-feeder, Grippia probably ate anything that came its way.

MONKEY LIZARDS AND OTHER SMALL DIAPSIDS

The diapsids, the reptile subclass that was destined to be the most important, started off as a small group of specialized climbing and gliding animals. Some appear to have been very agile tree-living animals, which has led to their classification as the order Simiosauria, the "monkey lizards".

Cosesaurus

When discovered, *Cosesaurus* caused a scientific stir. The skeleton seemed to be surrounded by plumes of some sort which have been interpreted as feathers. Several scientists see this as indicating that *Cosesaurus* was ancestral to the birds. However, the consensus is now that these plumes were marks in the rock formed as the rock was split.

Features: In build, *Cosesaurus* is lizard-like. The main difference between it and a lizard is in the structure of the hips, which are broader than usual and accommodate more than the usual number of vertebrae. This suggests an ability to walk or run on its hind legs. The muscle scars on the arm bones and the breastbone show that it could pull its arms together strongly. This has been interpreted as an adaptation to climbing.

Distribution: Catalonia, Spain.
Classification: Archosauromorpha, Prolacertiformes.
Meaning of name: Cose lizard (from the local indigenous people).
Named by: Ellenberger and de Villalta, 1974.
Time: Late Triassic.
Size: 15cm (6in).

Left: Cosesaurus must have been an active little hunter, running on its hind legs, or climbing trees, or both.

Lifestyle: Climbing or running insectivore.
Species: *C. aviceps*.

Megalancosaurus

This primitive diapsid had features that were undoubtedly adaptations to a tree-climbing lifestyle. The bird-like nature of its forelimbs and the superficial resemblance of the big-eyed, pointed-jaw skull to that of a bird, as well as the bird-like way the skull is articulated with the neck, have led some to suggest that this may be the ancestors of the birds, a view not widely accepted. It may have evolved its bird-like features independently, as adaptations to life in the branches of trees.

Features: The body is very lightweight and bird-like. The feet have opposable toes, the front feet have the first three toes opposed to the other two, like a chameleon, and the rear feet have the first toe opposed to the rest, like a perching bird. The tail is strong and prehensile, with a hooked claw at the end.

Right: Megalancosaurus probably had a similar lifestyle to a modern chameleon, climbing in bushes and catching insects.

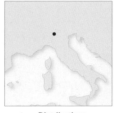

Distribution: Northern Italy.
Classification: Avicephala, Simiosauria.
Meaning of name: Big forelimb lizard.
Named by: Calzavara, Muscio and Wild, 1981.
Time: Norian stage of the late Triassic.
Size: 18cm (7in).
Lifestyle: Arboreal insectivore.
Species: *M. preonensis*.

Macrocnemus

Distribution: Switzerland; Italy.
Classification: Archosauromorpha, Prolacertiformes.
Meaning of name: Big tibia.
Named by: Nopcsa, 1930.
Time: Anisian to Ladinian stages of the middle Triassic.
Size: 1m (about 3ft).
Lifestyle: Insectivore.
Species: *M. bassanii*.

The area of northern Italy and Switzerland lay at the northern shores of the Tethys Ocean in Triassic times. Calm lagoons deposited limestone under low oxygen conditions, meaning that dead animals were slow to decay and were easily fossilized. Among the sea-living animals preserved here are the remains of reptiles from the surrounding shoreline, including lizard-like *Macrocnemus*. Footprints that were formed on tidal mud flats at the time have been found in central Europe, and have been given the name *Rhynchosauroides* (an ichnospecies – a scientific name based only on footprints). They were probably made by *Macrocnemus*, as the size is right and they show the presence of a very small outer toe.

Features: *Macrocnemus* is a medium-sized lizard-like animal, with a long neck and very long hind legs. The head is quite large and the jaws pointed. The teeth are small and sharp, indicating an insectivorous diet. Remains of the skin have been found on the tail of a juvenile specimen, and this shows a pattern of scaling like that of a lizard.

Right: Macrocnemus *may have hunted invertebrates along the mud flats of the shoreline of the Tethys Ocean.*

Drepanosaurus

Drepanosaurus is closely related to *Megalancosaurus* and shared some of its arboreal specializations. It lived in the same area at the same time. The fact that two such delicate tree-living animals have been preserved in these fine sediments suggests that perhaps tree-living reptiles were common everywhere but not usually fossilized.

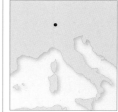

Distribution: Northern Italy.
Classification: Avicephala, Simiosauria.
Meaning of name: Sickle lizard.
Named by: Pinna, 1979.
Time: Norian stage of the late Triassic.
Size: 50cm (20in).
Lifestyle: Arboreal insectivore.
Species: *D. unguicaudatus*.

Right: Drepanosaurus *may have used its huge claw, like the modern anteater, to rip bark off trees in its hunt for insects.*

Features: The only skeleton known lacks the head and neck, but the prehensile tail has the same hook on the end as *Megalancosaurus*. The forelimbs, however, are quite different. The second finger has an enormous claw, as big as the rest of the hand, and the bones of the arm are adapted to working this. A hump on the shoulders may show the attachment of strong arm muscles. A similar animal, *Dolabrosaurus*, is known from the late Triassic deposits of Arizona, USA.

FLYING DIAPSIDS

If you stood in the riverside forests that bordered the seasonal streams draining from the arid interior of the Triassic supercontinent, you would probably be surrounded by colourful little flying animals. These would have resembled butterflies or little birds. However, neither of these existed at that time. These flying creatures would have been gliding diapsid reptiles.

Longisquama

This animal, with its long scales along the back, may have represented an early attempt at flight. Sharov, the discoverer, suggested that spread sideways these structures could have formed an elementary gliding wing. Another interpretation, by Robert Bakker, is that they were brightly coloured and used for display or heat regulation. They may have performed all these functions. Fine striations on the surface of the scales, and their central vein, suggest the structure of a feather. This has led to the suggestion that *Longisquama* lies on the evolutionary line that led to birds, but few scientists accept this idea.

Features: The most remarkable feature of this little lizard-like animal is the double row of long scale-like structures that run along the back, forming six to eight pairs. There is one pair for each pair of ribs on the body. They have a central hollow vane, like the feathers of a bird, but unlike feathers they appear to be formed of flat sheets rather than plumes. Only the front part of the body is preserved in the only known specimen and so the hindquarters here are conjectural.

Distribution: Kyrgystan.
Classification: Diapsid, order uncertain.
Meaning of name: Long scales.
Named by: Sharov, 1970.
Time: Early Triassic.
Size: 15cm (6in).
Lifestyle: Insectivore.
Species: *L. insignis.*

Above left: If Longisquama were a flying animal, it would have glided between Triassic trees like a big tropical butterfly.

Hypuronector

When first found, *Hypuronector* was regarded as a swimming reptile, hence the name. This was a reasonable assumption for the deep-finned tail could easily have been a swimming organ. However, the current interpretation is that it was a gliding animal, possibly with gliding membranes between its legs, and using the tail as an air rudder.

Features: Dozens of specimens of *Hypuronector* have been found in lake deposits of New Jersey, USA. The build of the skeleton is similar to that of the contemporary tree-living reptiles of Italy. The deep tail is strengthened by spines reaching downwards from the vertebrae. The legs are long suggesting that they may have supported a gliding membrane. The feet have not been found, and so it is not known if it has the same perching toes as its Italian cousins.

Distribution: New Jersey, USA.
Classification: Avicephala, Simiosauria.
Meaning of name: Deep-tailed swimmer.
Named by: Colbert and Olsen, 2001.
Time: Carnian stage of the late Triassic.
Size: 15cm (6in).
Lifestyle: Gliding insectivore.
Species: *H. limnaios.*

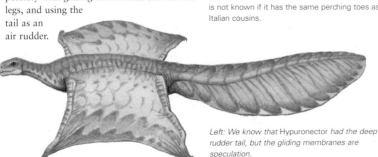

Left: We know that Hypuronector *had the deep rudder tail, but the gliding membranes are speculation.*

Kuehneosaurus

Distribution: Southern England.
Classification: Lepidosauria, Squamata.
Meaning of name: Kuehne's lizard.
Named: Robinson, 1967.
Time: Late Triassic.
Size: 65cm (26in).
Lifestyle: Gliding insectivore.
Species: *K. latus*.

Since the 1850s scientists have been discovering the remains of Triassic reptiles in Carboniferous limestones in Somerset, in southern England. In Triassic times these limestones formed a desert ridge full of caves and gullies. Contemporary animals became trapped in these gullies and were fossilized in rocks that were several million years older than they were. Among these fossils was the gliding reptile *Kuehneosaurus*.

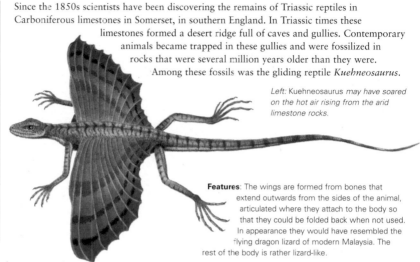

Left: Kuehneosaurus may have soared on the hot air rising from the arid limestone rocks.

Features: The wings are formed from bones that extend outwards from the sides of the animal, articulated where they attach to the body so that they could be folded back when not used. In appearance they would have resembled the flying dragon lizard of modern Malaysia. The rest of the body is rather lizard-like.

ICAROSAURUS LOST AND FOUND

The gliding reptile *Icarosaurus* was found in 1961 in a quarry in North Bergen, New Jersey, USA, by three teenagers who were collecting fossils. One of them, Alfred Siefker, took it to the American Museum of Natural History in New York, USA, so that it could be studied properly.

Then in 1989 Siefker realized that under US law the fossil was rightfully his, and so he claimed it back from the museum. It then dropped out of scientific sight until Siefker put it up for auction in 2000 to raise money for his medical expenses. Scientists feared that this priceless specimen would disappear forever into a private collection and be lost to science. It was bought for $167,000, only about half of what was expected, by Californian retired businessman Dick Spight who donated it back to the American Museum of Natural History.

The incident raises many questions about ownership of scientifically valuable fossils.

Icarosaurus

The only specimen of this gliding reptile, consisting of the front part of the body and the bones of one wing, was discovered in 1961 and was subsequently extensively studied by the American Museum of Natural History in New York, USA. It appears to be closely related to *Kuehneosaurus* from England and has a similar wing structure.

Distribution: New Jersey, USA.
Classification: Lepidosauria, Squamata.
Meaning of name: Icarus' lizard.
Named by: Colbert, 1961.
Time: Carnian stage of the late Triassic.
Size: 18cm (7in).
Lifestyle: Flying insectivore.
Species: *I. siefkeri*.

Left: The Triassic woodlands were home to brightly coloured flying reptiles.

Features: Named after the son of Deadalus in Greek mythology, who tried to fly using manufactured wings this is another of the Triassic diapsids specialized for flight. Like its relative *Kuehneosaurus* its wings are supported by extensions of the ribs that held a gliding membrane. In 1991 yet another flying diapsid, this one with a longer neck, was found in the same age of rocks in Virginia. It has yet to be named.

THALATTOSAURS

The thalattosaurs (ocean lizards) were a widespread group of marine diapsid reptiles. They are known from California, USA, across Canada, Europe and China. They are fairly small animals and all tended to have the same eel-like body shape and long swimming tails. Apart from the fact that they were diapsids, their exact relationship to the rest of the reptile class is uncertain.

Anshunsaurus

With its long neck and its paddle-like limbs, *Anshunsaurus* was thought to be some sort of ichthyosaur-like reptile when its fossils were first found. However, further study of the fragmentary remains showed that it was a thalattosaur. This came as a surprise as previously animals from this order were known only from Europe and North America. Scientists currently place it within the Askeptosauridae family, though it is ackowledged that the relationships between the three families within the thalattosaur order are not entirely clear. Only one specimen of *Anshunsaurus* has been found, so scientists have as yet described only one species. Like other animals that returned to a life in the ocean, *Anshunsaurus* is adapted to its marine environment but retains features that are similar in appearance to its land-living ancestors.

Features: As in the other thalattosaurs, *Anshunsaurus* has a neck and body that are long and slim. The tail is extremely long and ribbon-like, propelling the animal along with a sinuous side-to-side action. The broad webbed feet would have been used for steering and manoeuvring. These reptiles must have come on to land to lay eggs, but their movements would have been clumsy there.

Below: Agile in the water, awkward on land, Anshunsaurus was a hunter of the shallow waters.

Distribution: Guizhou, China.
Classification:
Lepidosauromorpha,
Askeptosauridae,
Thalattosauria.
Meaning of name: Lizard from Anshun.
Named: Liu, 1999.
Time: Middle Triassic.
Size: 1m (about 3ft).

Lifestyle: Fish-hunter.
Species:
A. huangguoshouensis.

Askeptosaurus

Askeptosaurus is typical of the thalattosaurs. It has a long slim neck and body with an extremely long ribbon-like tail. The whole structure is built to articulate from side to side giving an eel-like swimming motion, stabilized by the limbs, rather like that of the marine iguana of the Galapagos Islands. The fact that the toes were well-webbed means that the limbs could have been used to propel the animal through the water when moving at low speeds, as well as for steering when moving at higher speeds propelled by the tail. *Askeptosaurus* has given its name to the family Askeptosauridae to which it belongs. This is one of three families within the thalattosaur order, the others being the Endennasauridae and the eponymous Thalattosauridae.

Features: All the swimming adaptation seems to be in the shape of the body and tail. The limbs are not greatly specialized for swimming and were probably used only for steering and stabilization. The nostrils are far back on the skull, close to the eyes, as in many air-breathing aquatic animals. Its large eye sockets contain sclerotic rings, reinforcing rings of bone, which suggest that it hunted fish by diving into deep waters.

Below: The jaws of Askeptosaurus are long and armed with many sharp teeth.

Distribution: Switzerland.
Classification:
Lepidosauromorpha,
Askeptosauridae,
Thalattosauria.
Meaning of name: Lizard without a stick.
Named by: Nopcsa.
Time: Middle Triassic.
Size: 2m (6½ft).
Lifestyle: Fish-hunter.
Species:
A. italicus.

Xinpusaurus

Distribution: Guizhou, China.
Classification:
Lepidosauromorpha,
Thalattosauria,
Thalattosauridae.
Meaning of name: Lizard
from Xinpu.
Named by: Yin, 2000.
Time: Carnian stage of
the late Triassic.
Size: 1.5m (about 5ft).
Lifestyle: Fish-hunter.
Species: *X. suni, X. kohi.*

Xinpusaurus is quite a large thalattosaur, discovered in China, in 2000. The two species of *Xinpusaurus* and one of *Anshunsaurus* brings to three the total of thalattosaurs found in China to date. The limbs of this thalattosaur are clearly adapted as fins, rather than as the webbed feet that are more ususal among the thalattosaurs. This would have made this genus even more clumsy on land than its relatives. It has been placed within the Thalattosauridae family by scientists who have studied its fossils.

Features: The obvious distinguishing feature of *Xinpusaurus* is the snout. The long toothless upper jaw projects well beyond the lower, giving the appearance of a swordfish – a feature that it shares with the European *Endennasaurus*. The purpose of this is unclear, but it could have been used for probing in the sand and mud of the sea bed for prey. This could indicate that the different genera of thalattosaurs were adapted for different lifestyles in the shelf seas around Pangaea.

Below: The Xinpusaurus *is instantly recognizable by its toothless upper jaw and peddle-like limbs.*

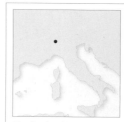

Endennasaurus

This is known from two well-preserved specimens. The general structure seems to indicate that it was more at home on land than the other thalattosaurs were, despite its obvious adaptations to a swimming lifestyle. It has been suggested that this genus, of which only one species has been found, was more of a shoreline creature than the other thalattosaurs. It may have spent its life walking the beaches and splashing about in the shallows in search of its prey. Certainly it would have had no difficulty moving about on dry land when the need arose, such as when laying eggs.

Below The purpose of the upper beak is puzzling, but the Endennasaurus *probably fed on soft-bodied shoreline animals.*

Features: The body of *Endennasaurus* is much stouter than that of the other thalattosaurs. The ribs are heavy, as a restriction to the natural buoyancy of the animal, and there is a massive set of belly ribs. The tail is very long and compressed from side to side. The forearm is stout and broad, suggesting use as a paddle. The long jaws have no teeth but there is a long upper beak.

Distribution: Northern Italy.
Classification:
Lepidosauromorpha,
Thalattosauria,
Endennasauridae.

Meaning of name: Lizard
from Endenna.
Named by: Renesto, 1989.
Time: Norian stage of the
late Triassic
Size: 70cm (28in).
Lifestyle: Aquatic hunter of
soft-bodied animals.
Species: *E. acutirostris.*

DIAPSIDS – THE RHYNCHOSAURS

The Permian extinction saw the end of the seed ferns such as Glossopteris *as the dominant form of vegetation. The animals, such as dicynodonts, that fed on them also died out. The Triassic seed ferns, like* Dicroidium, *were smaller and less abundant, and the rhynchosaurs evolved to feed on them. These flourished until* Dicroidium *was replaced by the conifers at the end of the Triassic.*

Mesosuchus

Like the rest of the rhynchosaurs, *Mesosuchus* had a beak at the front of the snout and a single nostril hole on the mid-line of the skull: Either it was a very early member of the group that came to be the most significant plant-eater for a short time, or it represents nothing more than a primitive relative. The body was like that of a small lizard and showed little specialization to any particular way of life.

Below: Mesosuchus was one of the first plant-eaters to exploit the sudden expansion of the low-growing ferns.

Features: As the most primitive of the rhynchosaurs, *Mesosuchus* has many unspecialized features. It has conventional teeth in the front of the mouth and a single row at each side, which is quite different from the usual rhynchosaurian pattern of no teeth at the front and multiple rows at the side. In fact, it is so different that *Mesosuchus* is often regarded as being outside the main group of rhynchosaurs.

Distribution: South Africa.
Classification: Archosauromorpha, Rhynchosauria.
Meaning of name: Middle crocodile.
Named by: Watson, 1912.
Time: Early to middle Triassic.
Size: 30cm (1ft).
Lifestyle: Low browser.
Species: *M. browni.*

Hyperodapedon

This rhynchosaur was one of the most common herbivores of the late Triassic. Its bones constitute the principal terrestrial fossils of the period in Brazil. Many other genera, including *Scaphonyx* and *Paradapedon*, are so similar that they are now regarded as being the same genus. The name alludes to the structure of the jaws. The hind limbs were adapted for scratching in the soil and so it is probable that *Hyperodapedon* fed on the roots and underground shoots of the seed ferns.

Features: This is the typical rhynchosaur, with its barrel-shaped body and its broad head. The jaws narrow to a beak. There are two broad plates of teeth on the upper jaw consisting of several tooth rows. There is a groove down the middle of each. This groove accommodates the tooth row of the lower jaw, producing a powerful chopping and crushing action. The front of the mouth bears a pointed beak.

Distribution: Scotland; Brazil; India.
Classification: Archosauromorpha, Rhynchosauria.
Meaning of name: Best pestle surface.
Named by: Huxley, 1859.
Time: Carnian stage of the late Triassic.
Size: 1.3m (4ft).
Lifestyle: Low rooter or browser.
Species: *H. gordoni, H. huxleyi, H. huenei.*

Left: A widespread plant-eater, Hyperodapedon *would have been a common sight all over the supercontinent.*

Rhynchosaurus

Distribution: England.
Classification:
Archosauromorpha,
Rhynchosauria.
Meaning of name: Beaked
lizard.
Named by: Owen, 1841.
Time: Middle Triassic.
Size: 60cm (2ft).
Lifestyle: Rooter or low
browser.
Species: R. articeps,
R. brodiei, R. spenceri.

This rhynchosaur was much smaller than the others of the group. It lived in groups in semi-arid areas on the banks of rivers. Some remains show that it had been caught in a flash-flood and transported some distance before being buried. Fossilized footprints that match the feet of *Rhynchosaurus*, and given the name *Rhynchosauroides*, have been found in the southeastern Alps.

Features: The skull of *Rhynchosaurus* is broad and flat. The lower jaw is deep. The body is more slimly built than in other rhynchosaurs. The hind limbs are semi-erect, like a crocodile on land, and the hind foot seems to be adapted for digging and scratching at the ground. The teeth of all rhynchosaurs are different from those of other reptiles, being living organs rather than dead tissue. The structure of the skull indicates that *Rhynchosaurus* had a good sense of smell but poor hearing.

Below: Its digging specializations suggest that Rhynchosaurus fed on the underground rhizomes of the ferns as well as the fronds.

PESTLE-AND-MORTAR TEETH

The rhynchosaurs had very distinctive teeth. They worked with a pure up and down action, pulping the food like a pestle and mortar rather than grinding it with a sideways motion.

The teeth grew more like those of a mammal than those of a reptile. In most reptiles the teeth wear out and are replaced by new ones growing up beneath them from within the jaw. The working teeth are essentially dead tissue, easily pushed out by new ones growing up from the bone when they are worn out. In the rhynchosaurs they were replaced from the rear of the jaw as the animal grew. The teeth were hollow and had root canals that were filled with nerves and blood vessels. This meant that the teeth were living organs and could continue to grow from the inside. This is more like the tooth arrangement of a mammal than a reptile.

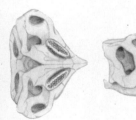

Above: The skull of Hyperodapedon was particularly broad and the rows of massive teeth narrowed sharply towards the snout.

Trilophosaurus

This is one of those animals that is difficult to classify. In some respects, particularly the strange tooth structure, *Trilophosaurus* is similar to the rhynchosaurs but is much more primitive. Judging by the number of individuals found in desert flood deposits in Texas, USA, this is where it was thought to have lived, although some scientists regarded it as being a tree-climber.

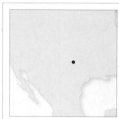

Distribution: Texas, USA.
Classification:
Archosauromorpha,
Trilophosauridae.
Meaning of name: Three-
crested lizard.
Named by: Gregory, 1945.
Time: Carnian stage of the
late Triassic.
Size: 2.5m (8ft).
Lifestyle: Low- or possibly
tree-living browser.
Species: T. buettneri,
T. jacobsi.

Features: As in the rhynchosaurs, the jaws of *Trilophosaurus* are articulated so that they work in a simple up and down action. There are no teeth at the front but those at the back have very broad surfaces. The snout is very narrow and it appears to have carried a beak. The eyes are high up on the small head. The legs, with their long lizard-like toes show it to have been a four-footed animal. The tail is long and heavy.

Above: Despite its greater length, Trilophosaurus appears to have been more lightly built then the main run of rhynchosaurs.

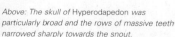

A DIAPSID VARIETY

The Triassic diapsids were a very varied group. Some developed long necks for various feeding purposes. These tended to exist along the shoreline, feeding on the variety of foods to be found there. Others remained more generalized, with a familiar lizard-like build, and evolved into the group that survives today in the form of the lone genus of tuatara.

Tanystropheus

The neck bones, when first found, were thought to have been limb bones, possibly the wing bones of flying reptiles. The first remains were discovered in the 1850s and until the 1920s it was reconstructed as a gliding creature rather like a flying squirrel. Study of the animal recommenced in the 1970s, led by Rupert Wild of Stuttgart, and it is on his findings that modern restorations are based. There is no agreement about how the neck was held or was used. What does seem to be evident is that *Tanystropheus* lived along the shoreline. The head and neck make it a water animal, but the rest of the body is distinctly terrestrial.

Left: The front legs of Tanystropheus are short with small feet, while the hind legs are long with very long feet.

Features: The most obvious feature of *Tanystropheus* is the neck consisting of about two thirds of the length of the animal, made up of 12 extremely long vertebrae, making it stiff but not totally inflexible. The skull is large and in the adult carries teeth like those of a fish-eater, but in the juvenile more like those of an insectivore. The adult teeth are long and interlocking at the front, like those of a plesiosaur, and strong and sharp at the back. The body is slender and the tail moderately long with a thick base.

Distribution: Northern Italy; southern Germany; Israel.
Classification: Archosauromorpha, Prolacertiformes.
Meaning of name: Long strap (a reference to the neck).
Named by: Meyer, 1852.
Time: Middle to late Triassic.
Size: 6m (20ft).
Lifestyle: Possibly an ambush hunter.
Species: *T. longobardicus*, *T. conspicuous*, *T. meridensis*, *T. fossai*.

Dinocephalosaurus

The structure of the neck of this aquatic animal has led scientists to speculate that it had an unusual hunting technique. It is possible that it thrust its head forward, opened its jaws and expanded its neck using its neck ribs. It would then engulf the volume of water before it, along with any fish it contained. Most modern hunting fish catch their prey by thrusting out their mouthparts, causing an influx of water that brings the food into the mouth. It is also seen in modern snapping turtles that have the ability to depress the tongue suddenly to the floor of the mouth and produce a suction effect.

Features: This is the longest-necked prolacertiform ever found. The neck consists of 25 vertebrae, each with a pair of ribs, the articulation of which seems to suggest that they were used to expand the volume of the neck. These cervical ribs are thin and flexible and each covers several vertebrae when folded back, making the neck relatively inflexible. The limbs are short and broad and contain few bones, indicating that *Dinocephalosaurus* spent most of its life at sea.

Distribution: Guizhou, China.
Classification: Archosauromorpha, Prolacertiformes.
Meaning of name: Terrible headed lizard.
Named by: Li, 2002.
Time: Triassic.
Size: 3m (10ft).
Lifestyle: Fish or squid hunter.
Species: *D. orientalis*.

Right: The first part of the skeleton to be found was the toothy skull, and its name celebrates this rather than the remarkable neck.

Brachyrhinodon

Distribution: Scotland.
Classification: Lepidosauria, Sphenodontida.
Meaning of name: Short nose tooth.
Named by: Huene, 1910.
Time: Carnian stage of the late Triassic.
Size: 15cm (6in).
Lifestyle: Insect-hunter.
Species: *B. taylori.*

Brachyrhinodon is one of the oldest of the sphenodonts, a group whose only modern representative is the tuatara of New Zealand. In Triassic times these took the place of lizards as the main small insectivores and were very common. Similar animals such as *Planocephalosaurus*, *Polysphenodon* and the larger *Cleveosaurus* are found elsewhere, and abounded until the lizards took over in Jurassic times.

Features: In structure *Brachyrhinodon* is almost identical to the modern tuatara, but has a very short snout – probably an adaptation to some kind of feeding specialization. In general the front teeth are conical or wedge-shaped, and there are small teeth covering the palate that are adaptations for crushing. The bones that edge the jaws are rigid, and this may be to resist twisting motions – an adaptation to coping with struggling prey.

Above: Brachyrhinodon *would have hunted hard-shelled insects, crushing their carapaces with the broad expanse of teeth on the palate.*

TUATARA

The modern tuatara is found only on a few islands off the north coast of New Zealand. The name comes from a Maori word meaning "spiky back". It was always thought to have been a lizard, but in 1867 Albert Gunther of the British Museum, England, realized that it was a member of a group that was quite abundant in Triassic times.

In 1989 Charles Dougherty at Victoria University in Wellington, Australia, identified two species, *Sphenodon punctatus* and *S. guntheri*. They grow to of 60cm (2ft) long and weigh 1kg (2lb).

They differ from lizards in having specialist teeth: a single row on the lower jaw fits between a double row in the upper. They have a simple third eye on the roof of the head and lack a lizard's visible ear openings.

Above: The modern tuatara is similar to Triassic sphenodontids.

Diphydontosaurus

The sphenodonts were all largely similar to one another but differed in the shape of the head and jaws, indicating that they had different diets. The presence of two kinds of teeth referred to in the name of this animal suggests a specialist diet, possibly hunting different insects from its contemporaries. A very similar animal, *Planocephalosaurus*, is known from England.

Features: From a distance *Diphydontosaurus* would have been indistinguishable from all the other tuatara-like sphenodontids that existed at the time. The number of teeth and the shapes and structures of them are all that separate them from others of the group. Presumably they would have been differently coloured as well and may have borne different ornamentation, just like the range in colours and shapes we see among modern genera of lizards.

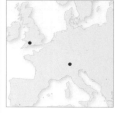

Distribution: Southern England; Northern Italy.
Classification: Lepidosauria, Sphenodontida.
Meaning of name: Double-toothed lizard.
Named by: Whiteside, 1986.
Time: Norian stage of the late Triassic.
Size: 15cm (6in).
Lifestyle: Insectivore.
Species: *D. avonis.*

Above: There must have been a wide variety and a large number of desert-living insects and arthropods to support the diverse types of insect-eating reptiles of the time.

MORE MISCELLANEOUS DIAPSIDS

The diapsids were rapidly becoming the most important group around in Triassic times, and were adapted to a wide range of habitats, lifestyles and food sources. They were also found all over the world and inhabited the arid wastes of central Pangaea as well as the moister shorelines and coastal plains where the climates were more equable and most of the vegetation was to be found.

Nanchangosaurus

Nanchangosaurus and its relatives are so unusual that it is not quite clear where they fit into the classification of the diapsids. One idea is that they were related to the ichthyosaurs. Their aquatic lifestyle suggests this, as does their streamlined shape, the expanded paddle-like hands, and the long thin jaws. However, these features could well be the result of convergent evolution – the phenomenon whereby different animals evolve similar features as adaptations to a similar lifestyle. A gap in the skull in front of the eye sockets suggests that they are more related to the archosaurs.

Below: Nanchangosaurus swam in shallow inland waters like a little crocodile.

Features: This small aquatic reptile has a cylindrical body, tapered at both ends ("fusiform" in anatomical parlance), paddle-like limbs, the larger in the front, and a long flattened tail. There is a high ridge down the back, protected by a row of bony plates. The neck is quite long and furnished with neck ribs. The jaws are long, flattened and toothless, possibly adapted for probing the lake bed for invertebrates.

Distribution: Nanzhang, China.
Classification: Eosuchia, Hupehsuchia.
Meaning of name: Lizard from Nanzhang.
Named by: Wang, 1959.
Time: Anisian stage of the middle Triassic.
Size: 1m (about 3ft).
Lifestyle: Fish- or crustacean-hunter.
Species: *N. suni.*

Hupehsuchus

Hupehsuchus was very similar to *Nanchangosaurus* but had heavier armour and the back spines were more finely divided. The main difference is in the structure of the feet, which were adapted to paddles enabling the animal to swim swiftly. Despite these evident differences there is a distinct possibility that *Hupehsuchus* and *Nanchangosaurus* are actually the same genus. The state of fossilization makes it unclear if the specimens representing the latter have the distinctive extra digits. If this is the case, the name *Nanchangosaurus* would be used for both.

Features: The most surprising feature of *Hupehsuchus*, noticed in a newly discovered specimen in 2004, is that the feet have more than the usual number of toes, seven in the front feet and six in the hind – a phenomenon known as "polydactyly". This is part of the development of a swimming paddle, and seems to indicate a close relationship with the ichthyosaurs. However, the structure of these extra digits, encompassing the hand bones as well as just the finger bones, is quite different from that of the extra digits found in some ichthyosaurs, which consists of finger bones only.

Below: The extra toes would have produced a paddle-like structure to help Hupehsuchus to swim or possibly to push itself through the weeds.

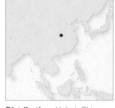

Distribution: Hubei, China.
Classification: Eosuchia, Hupehsuchia.
Meaning of name: Crocodile from Hubei.
Named by: Young, 1972.
Time: Olenekian stage of the early Triassic.
Size: 1m (about 3ft).
Lifestyle: Fish- or invertebrate-hunter.
Species: *H. nanchangensis.*

Tanytrachylos

Distribution: Virginia, USA.
Classification:
Archosauromorph,
Prolacertiformes.
Meaning of name: Long
neck.
Named by: Olsen, 1979.
Time: Late Triassic.
Size: 50cm (20in).
Lifestyle: Aquatic insectivore.
Species: *T. ahynis*.

With its long neck, *Tanytrachylos* is possibly related to *Tanystropheus* from Europe. Unlike *Tanystropheus*, it was a freshwater animal. The remains of more than 90 individuals have been found in a single lake deposit in Virginia, USA, but none is complete. However, there are enough pieces to allow scientists to put together an accurate picture of the beast – a hunter of the myriad insects that must have infested the vegetation that grew by the lakeside.

Features: The body is long and slender, and the hind legs are much longer than the front. The long toes of the hind foot appear to be webbed. The jaws are slender and armed with pointed, conical teeth. These appear to be well-adapted to hunting insects, suggesting that it pursued aquatic invertebrates in water and probably insects on land. Fracture marks on the tail vertebrae suggest that this animal could shed its tail like modern lizards, but this theory has yet to be proved. About 40 per cent of the specimens from the single site have a slightly different pelvis from the rest. As there is no other difference between the two types it is likely that this represents a difference between the sexes.

Right: Tanytrachylos *probably hunted small prey both in the water and on land.*

BIPEDAL RUNNING

When a small diapsid like *Langobardisaurus* takes to its hind legs to run, the motion is not like the two-footed stance of a dinosaur. In a dinosaur the legs are held beneath the body so that the weight bears down on them like a roof supported by pillars.

Non-dinosaur reptiles had their legs sticking out at the side and so the gait was quite different. Usually a sprawling reptile's foot, such as that of a lizard, has toes that become longer from the first to the fourth, so that the toe tips contact the ground as the foot or hand is held out to the side.

In *Langobardisaurus* there was not much difference in the length of toes, and this suggests that the foot could have been held so that the toes pointed forward, even though the legs were sprawled out to the side.

Above: When escaping from predators Langobardisaurus *had a running gait that was quite different from anything else.*

Langobardisaurus

The longer hind legs of this animal suggest that it may have spent much of its time running on its rear feet. The big head on the long neck would have been held aloft and back to keep the balance – as we see in some modern lizards, like the frilled lizard of Australia – when it runs.

Features: This lizard-like animal has a big-eyed, short-snouted skull, and a very long neck. The dentition is strange, consisting of thin pointed teeth at the front, followed by large three-crowned teeth, and a big grinding tooth at the back. The toes of the feet are not as asymmetrical as in modern lizards, suggesting that it walked with its feet pointing forward, rather than sprawled out at the side, as it ran away from bigger predators.

Distribution: Northern Italy.
Classification:
Archosauromorph,
Prolacertiformes.
Meaning of name: Lizard
from Lombardy.
Named: Renesto, 1994.
Time: Norian stage of the
late Triassic.
Size: 50cm (20in).
Lifestyle: Insect-hunter.
Species: *L. pandolfii*,
L. tonelloi.

Left: The triangular, deep snout and varied teeth suggest that Langobardisaurus *hunted and ate a variety of foods including tough-shelled items, such as crustaceans.*

THECODONTS – TOWARDS THE DINOSAURS

The most widespread and successful group of the diapsid reptiles were the thecodonts, the creatures that had their teeth in individual sockets in the jaws. They are grouped into a superorder known as the archosauria, or ruling reptiles. In early Triassic times they radiated into a number of groups, some of which were short-lived, but others became the crocodiles, the pterosaurs, and ultimately the dinosaurs.

Marasuchus

This thecodont often appears under the name *Lagosuchus* (rabbit-crocodile). The skeleton of *Lagosuchus* was very scrappy, and was later found to be identical to skeletons which had already been named *Marasuchus*. *Lagosuchus* was then declared a *nomen dubium* (a doubtful name) and *Marasuchus* was adopted instead. Confusingly, the name Lagosuchidae was retained for the family.

Below: Hopping or running, Marasuchus would have been a fierce little hunter of small animals on the South American plains.

Features: There has been much discussion as to whether or not this rabbit-crocodile actually jumped like a rabbit. The anatomy is ambiguous. The length of the hind legs and the fact that the weight of the animal was concentrated towards the rear, as in a kangaroo, suggest that it may have been an efficient jumper, but there is no evidence of the strong tendon that usually powers a leaper's hind foot.

Distribution: Argentina.
Classification: Ornithodira, Archosauria.
Meaning of name: Mara (a South American rabbit-like rodent) crocodile.
Named by: Sereno and Arcucci, 1994 (formerly *Lagosuchus*, Romer, 1971).
Time: Ladinian stage of the middle Triassic.
Size: 50cm (20in).
Lifestyle: Hunter.
Species: *M. illioensis*.

Left: The largest of today's anolis lizards is about the size of Marasuchus.

Euparkeria

Euperkeria, or an animal very similar to it, is regarded as the ancestor that led to the radiation of the thecodont group. Covered in scales, it had the appearance of a crocodile. It walked on all fours, but ran and hunted on its hind legs and had the lifestyle of a dinosaur. This proved to be a successful pattern that survived.

Features: *Euparkeria* has the sharp, deeply rooted teeth that we see in all thecodont groups. The hip bones allow the legs to come under the body, resulting in an erect posture that is much more efficient for moving a heavy body swiftly. The hind legs are bigger than the front, but not as markedly so as in its descendants. The ankle joint is still quite primitive for a two-footed animal, and so it must have spent much of its time on all fours. A row of armour plates protects the spine.

Distribution: South Africa.
Classification: Archosauromorpha, Euparkeriidae.
Meaning of name: Parker's true one.
Named by: Broom, 1913.
Time: Anisian stage of the middle Triassic.
Size: 60cm (2ft).
Lifestyle: Hunter.
Species: *E. capensis*.

Left: With its fierce teeth, its erect hind legs and its armoured scales, Euparkeria *anticipates many of the thecodont group to come.*

Erythrosuchus

Distribution: South Africa.
Classification: Reptile, Archosauriformes.
Meaning of name: Red crocodile.
Named by: Broom, 1905.
Time: Early Triassic.
Size: 5m (17ft).
Lifestyle: Ambush hunter.
Species: *E. africanus*.

If *Euparkeria* illustrates the development of the lightweight thecodonts that were so successful, then *Erythrosuchus* is typical of the heavier types that thrived for only a short time. These erythrosuchids were big carnivores, equivalent to the dinocephalians of the Permian, and the rauisuchids of the late Triassic. They were the biggest land animals of the time. Other examples included *Vjushkovia* and *Shansisuchus* from China.

Features: *Erythrosuchus* is typical of the family to which it gives its name, in having a massive head that is almost half the length of the body. It is armed with huge teeth and jaws powered by a thickly muscled neck. The sheer weight of the animal and the relatively weak nature of its feet indicate that it must have been an ambush predator, hiding in riverside vegetation and leaping out on unsuspecting large herbivores on which it fed.

Left: Erythrosuchus *was the most feared predator of its day, probably hunting the big cow-sized dicynodonts like Kannemeyeria.*

Silesaurus

This animal is about as close as evolution can get to being a dinosaur, without being one. Its mouth is very much like that of a plant-eating ornithischian, with its beak and its leaf-shaped teeth, but other features of the skeleton show that it is more primitive. It seems as if the features that define a dinosaur did not evolve all at once, but piecemeal. It was discovered as a collection of bones from several individuals fossilized in a lake deposit. The most complete skeletons were all pointing in the same direction, as though the heavy part of the body – the hips – dragged along slowly while the lighter forequarters were washed along by water current.

Distribution: Southern Poland.
Classification: Reptile, Archosauria, Ornithodira.
Meaning of name: Lizard from Silesia.
Named by: Dzik, 2003.
Time: Carnian stage of the late Triassic.
Size: 2m (6½ft).
Lifestyle: Low browser.
Species: *S. opolensis*.

PSEUDOSUCHIANS

A classification often seen in old books is the pseudosuchians (the "fake crocodiles"). This was a term once used to describe the archosaurs that were the common ancestors of the pterosaurs and the dinosaurs. As the true complexity of the dinosaur ancestry gradually came to light, we could see that this was an artificial classification and it is no longer used.

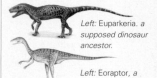

Left: Euparkeria. *a supposed dinosaur ancestor.*

Left: Eoraptor, *a true dinosaur.*

Features: The narrow hips with the long pubic bones are very dinosaur-like, as are the jaws with the beak at the front of the mouth, and the low tooth count. However, the articulation of the neck bones and the fact that the hind limb is attached to the hip in a socket rather than a hole show that it is not a dinosaur. The non-dinosaur-like long front legs are not thought by palaeontologists to be important in the classification – they could have evolved separately.

Left: A dinosaur or an advanced thecodont? Silesaurus *shows features of both.*

RAUISUCHIANS

The dinosaurs had not yet evolved, but by middle Triassic times there was an ecological niche that would be filled with large, ferocious, long-jawed, sharp-toothed terrestrial hunters. The time of the meat-eating dinosaurs seemed to have come, but before they evolved their niche was occupied by a family of giant land-living crocodiles – the rauisuchians.

Lotosaurus

With no teeth *Lotosaurus* is an exception to the rauisuchian group. Although this fact has been used to show it had a vegetarian diet, it is quite possible that it survived on a diet of shellfish, as it lived in an area that was periodically flooded by shallow seas. Another sail-backed form *Ctenosauriscus* is known, but not well studied.

Features: The skeleton is entirely that of a typical rauisuchian, except for the mouth and the tall fin down its back. The jaws are toothless and quite smooth and the snout is turned down like that of a tortoise. The jaws are powerful and would have anchored strong muscles, suggesting a diet of hard-shelled things like molluscs. Spines on the back bone would have supported a sail of skin, possibly used for signalling or for temperature regulation.

Distribution: Hunan Province, China.
Classification: Archosauria, Rauisuchia.
Meaning of name: Lotus lizard.
Named by: Zhang, 1975.
Time: Middle Triassic.
Size: 2.5m (8ft).
Lifestyle: Herbivore or shellfish-eater.
Species: *L. adentus.*

Right: With the sail on its back, and its toothless jaws, Lotosaurus *was the most unusual of the rauisuchians.*

Ticinosuchus

A moderate-sized land-living predator, *Ticinosuchus* was more lightly built than other rauisuchians, that roamed the arid hinterland of the northern coastline of the Tethys Ocean. It probably hunted many of the semi-aquatic vertebrates of the Tethys beaches. The distribution of the *Cheirotherium* footprints attributed to it, indicate that it and its closest relatives were quite widespread.

Features: Enough of this animal is known to show that it has a heel structure in its foot and held its legs directly under its body, like a dinosaur, not sprawled out to the side like a lizard, or something in between like a crocodile. The heel is a lever structure, which when working with a strong tendon produces a powerful run. An outer toe is well developed that reinforces this action – it is the toe that gives the "thumb" print in the *Cheirotherium* trackways. See box.

Distribution: Switzerland; Italy.
Classification: Archosauria, Rauisuchia.
Meaning of name: Crocodile from the Tessin River.
Named by: Krebs, 1965.
Time: Anisian and Ladinian stages of the middle Triassic.
Size: 2.5m (8ft).
Lifestyle: Hunter.
Species: *T. ferox.*

Right: Ticinosuchus *was a lightly-armoured hunter with a double row of bony plates down its back.*

Arizonasaurus

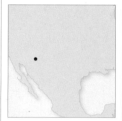

Distribution: Arizona, USA.
Classification: Archosauria, Rauisuchia.
Meaning of name: Lizard from Arizona.
Named by: Welles, 1947.
Time: Anisian stage of the middle Triassic.
Size: 6m (20ft).
Lifestyle: Hunter.
Species: *A. babbitti*.

Palaeontology, like history, repeats itself. The first part of *Arizonasaurus* to be found in 1947 consisted only of a jawbone and there was nothing to indicate that it was much different from the other rauisuchids. The discovery of a 50 per cent complete skeleton in 2000 and published in 2003 showed just how different it actually was.

Features: The new specimen of *Arizonasaurus* consists of the skull, the backbone as far as the hips and the limb girdles. It shows that it has a sail on its back, not unlike that of the Permian *Dimetrodon* and of the dinosaur *Spinosaurus* that was to come. Presumably this was used for the same purposes as the sail in these other two top predators of their times, as a heat regulation device or as a signal and display structure.

Left: The sail of Arizonasaurus would have enabled it to signal to others over wide areas.

MYSTERIOUS "HAND BEAST"

In 1839, in the Storton sandstone quarries near Liverpool, England, quarrymen extracting the Triassic sandstones for building stones came across a sequence of fossils. They looked as if they had been formed by human hands, with four fingers pointing forward and a thumb out at the side. They had evidently been formed by a four-footed animal, as the trackway incorporated these large "hand prints" and smaller ones in front of them. They were given the name *Cheirotherium*, or "hand beast".

The oddest thing about these prints was that the "thumb" print was on the outside of the tracks, rather than the inside, where we would have expected it to be. Similar footprints were found in Germany and in various places in North America. No bones of the animal had ever been found, and there was a great deal of speculation as to what kind of beast it was, and why its thumb was on the outside. The main suspects were a kangaroo-shaped animal, a crocodile-shaped animal and some kind of amphibian. This last was championed by the most famous naturalist of the day, Sir Richard Owen. Sir Charles Lyell came up with the novel explanation that they were formed by an animal that walked cross-legged – with its left foot forming the prints on the right-hand-side, and vice versa.

It was not until the discovery of *Ticinosuchus* in 1965 that the likely maker was revealed. The unique rauisuchian foot structure, with the long fifth toe and the lever heel arrangement, fitted the footprints perfectly.

Saurosuchus

This is probably the biggest and the best-known of the rauisuchians. It was the top predator of the time and probably fed on the big plant-eating dicynodonts and rhynchosaurs. It lived in the same time and place as the first dinosaurs, but was much larger. Its long jaws contained teeth of different sizes. As they became worn they fell out and were replaced by new ones growing from within the jaws. Old individuals may have died of starvation after their teeth stopped regenerating.

Features: This has the appearance of the typical rauisuchian, with the huge head, the short neck, the four short legs with the dinosaur-like erect gait and the long tail. The erect gait evolved separately from that of dinosaurs. In the latter it is produced by the joint on the femur being off to the side, while in the rauisuchians it is brought about by the angling of the hip joint downwards. This is the main argument against the rauisuchians being the ancestors of dinosaurs, as some have suggested.

Distribution: San Juan Province, Argentina.
Classification: Archosauria, Rauisuchia.
Meaning of name: Lizard crocodile.
Named: Reig, 1959.
Time: Carnian stage of the late Triassic.
Size: 7m (23ft).
Lifestyle: Hunter.
Species: *S. galilei*.

Left: Saurosuchus was the ultimate predator of the time, and was at the apex of the local food pyramid.

LITTLE CROCODILES

The crocodiles of Triassic times were not the sluggish aquatic ambush predators of today. Instead they were small and active hunters that stalked the arid landscapes searching for small vertebrates. They were all fast runners, some of them rearing up on their hind legs like the dinosaurs to come. Those that do not fit into the modern order of Crocodylia are categorized as Crocodylomorpha – "crocodile-shaped".

Terrestrisuchus

Not all members of the Triassic crocodile clan were big. Some, like *Terrestrisuchus*, were little lizard-like animals and scampered around on long legs hunting insects and small vertebrates. It is one of the animals found in the Triassic fissure deposits in Carboniferous limestone of the Welsh borderlands.

Features: The body is delicate with very long limbs and feet. The tail is longer than the body and head combined. It may have spent time running on its hind legs. Its slim build has led to its description as the greyhound of the Triassic. Recent finds suggest that it may actually be the young of another two-legged crocodile relative *Saltoposuchus*.

Distribution: Wales; southern England.
Classification: Archosauria, Crocodylomorpha.
Meaning of name: Ground-living crocodile.
Named by: Crush, 1984.
Time: Norian stage of the late Triassic.
Size: 50cm (18in).
Lifestyle: Hunter of insects or small vertebrates.
Species: *T. gracilis*.

Above: The short body and long legs of Terrestrisuchus show it to have been a fast runner either on two legs or on four.

Gracilisuchus

This crocodile was so uncrocodile-like with its big boxy head, its long running hind limbs, its grasping hands that it was classed as an ancestor of the dinosaurs until the 1980s. This and the ambiguities over *Terrestrisuchus* show how similar these groups of crocodile relatives were to one another. They were all small, swift hunters, preying on the smallest vertebrates of the time.

Features: The lightweight body has a disproportionately large head, balanced by a long tail. Held straight, this would have allowed it to run, dinosaur-like, on its hind legs. However, its skull, vertebrae and ankle joints are distinctly crocodilian. A double row of armour plates runs down the backbone. The head is shorter than expected in crocodiles, and there is a palate separating the mouth from the nasal cavity – allowing it to breathe and eat at the same time.

Distribution: Argentina.
Classification: Archosauria, Crocodylomorpha.
Meaning of name: Graceful crocodile.
Named: Romer, 1972.
Time: Ladinian stage of the middle Triassic.
Size: 30cm (12in).
Lifestyle: Hunter.
Species: *G. stipanicocorum*.

Right: Gracilisuchus would have been agile enough to hunt the fastest lizards over the arid landscape of the time.

Erpetosuchus

Distribution: Scotland;
Connecticut, USA.
Classification: Archosauria,
Crocodylia.
Meaning of name: Crawling
crocodile.
Named: Newton, 1894.
Time: Late Triassic.
Size: 30cm (12in).
Lifestyle: Insect-hunter.
Species: E. granti.

This primitive crocodile was one of the vertebrates found in the Lossiemouth desert sandstones of northern Scotland, and preserved only as moulds in the rock. The remains were able to be studied by pouring liquid rubber into these hollows in the sandstone and the resulting casts pulled out when still flexible producing a three dimensional representation of the original.

Features: This seems to be the most closely related to the line of the modern crocodiles of all animals known from this period. It has big eyes and only a few teeth in the front of the mouth. It probably ran on its hind legs, judging from the structure of the forelimb, which is not well adapted for running. It seems that it was a swift animal that may have fed on insects.

*Above: Although we know only the forequarters of
Erpetosuchus, we can assume that the rest of the
body was similar to that of the other land-living
crocodiles of the time.*

WARM-BLOODED TO COLD

The array of lightweight running crocodiles from the Triassic period shows that this group was quite different from the sluggish water-bound representatives of today. It is tempting to think that the ancestors were once active and warm-blooded and have since reverted to a low metabolic, cold-blooded lifestyle. The anatomy of modern crocodiles supports this idea. A warm-blooded metabolism implies high blood flow rates and so a specialist heart is necessary to handle this. Modern crocodiles do, indeed, have a complex four-chambered heart although they have a cold-blooded metabolism. They use this to adjust their blood flow while diving. It seems very likely that this four-chambered heart actually evolved among the crocodiles' land-living ancestors as part of an increased metabolism, and became part of the diving physiology much later as they reverted to a cold-blooded lifestyle, possibly adopting an aquatic role in response to increased competition from the dinosaurs as the main land-living predators.

Water-living crocodile-relatives did exist at this time, such as *Chasmatosuchus*, which looked more like the modern forms, but these were the exceptions.

*Left: The four-chambered heart
of a modern crocodile.*

Protosuchus

The protosuchia were a widespread group of medium-sized terrestrial crocodiles. In contrast to the other terrestrial crocodiles they had skulls that were much more similar to those of modern crocodiles, with a broad rear that anchored strong jaw muscles. Other protosuchids include *Erythrochampsia*, *Nothochampsa* and *Orthosuchus* from South Africa.

Features: Its skeleton is very much like that of other crocodiles except for the articulation of the backbone, and the presence of a strong set of belly ribs that probably aided breathing while it was active. A double row of armour plates guards the back. The lower jaw has a pair of canine teeth that fit into a slot in the upper jaw when closed, very much like the arrangement in modern crocodiles.

*Below: Protosuchus lived alongside the earliest
dinosaurs. They probably
competed for the
same prey.*

Distribution: Nova Scotia,
Canada; Arizona, USA.
Classification: Archosauria,
Crocodylia.
Meaning of name: First
crocodile.
Named by: Brown, 1934.
Time: Late Triassic to early
Jurassic.
Size: 1m (about 3ft)
Lifestyle: Terrestrial hunter.
Species: P. richardsoni,
P. micmac, P. haughtoni.

PHYTOSAURS – ANTICIPATING THE MODERN CROCODILE

The crocodiles had become the top predators on land. However, a specialized offshoot of the thecodont or archosaur line anticipated the aquatic lifestyle and took to the rivers and lakes of Triassic Pangaea. This was the phytosaur family, which had habits and appearances similar to those of modern crocodiles.

Rutiodon

Rutiodon is what we think of as the typical phytosaur. In fact it is something of a wastebasket taxon, with other phytosaurs from later deposits in Europe attributed to it. These will probably be classified as different genera eventually, with names such as *Machaeoroprosopus* and *Pseudopalatus* already in use.

Features: The crocodile shape, with the long body, the swimming tail and the sprawling limbs are the features we find in all phytosaurs. What distinguishes them from the crocodile line is the position of the nostrils on the top of the head, between the eyes, rather than at the tip of the snout. *Rutiodon* has long narrow jaws and the teeth are easily identified by their strong vertical grooving when found as isolated fossils: As in all phytosaurs the armour is thicker and stronger than in modern crocodiles.

Below: Rutiodon *had the same lifestyle as a modern crocodile, and so convergent evolution dictates that it had a very similar appearance.*

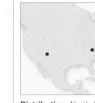

Distribution: North America.
Classification: Archosauria, Phytosauridae.
Meaning of name: Wrinkled tooth.
Named by: Emmons, 1856.
Time: Carnian stage of the late Triassic.
Size: 2.5m (8ft).
Lifestyle: Fish-hunter.
Species: *R. carolinensis, R. megalodon, R. gregorii, R. crosbiensis, R. lepturus, R. grandis, R. manhattanensis.*

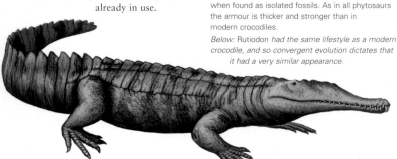

Nicrosaurus

Well-preserved remains of prosauropod dinosaurs such as *Plateosaurus* are known from Triassic river sandstones in Germany and Switzerland. The evidence points to their being trapped in quicksand and pulled apart by carnivorous dinosaurs and phytosaurs. The presence of the teeth at these sites indicates that *Nicrosaurus* was the phytosaur responsible. A second species has been identified in New Mexico, USA.

Features: The obvious difference between *Nicrosaurus* and the other phytosaurs is the tall crest that runs down the middle of the skull. This would have been used for display and communication purposes. Otherwise the snout is very crocodile-like, being turned down and with an expanded tip. There are differently sized teeth along the jaws, with long fangs at the tip of the snout and shorter conical teeth, D-shaped in cross section for strength, further back. The back of the animal is covered with bony plates.

Distribution: Germany; New Mexico, USA.
Classification: Archosauria, Phytosauridae.
Meaning of name: Lizard from the Neckar River.
Named by: Fraas, 1866.
Time: Norian stage of the late Triassic.
Size: 2.5m (8ft).
Lifestyle: Hunter or scavenger.
Species: *N. kapffi, N. buceros, N. meyeri.*

Right: Nicrosaurus *hunted dinosaurs.*

A VARIETY OF PHYTOSAURS

The obvious difference between a phytosaur and a crocodile is the position of the nostril on the top of the head. This was an adaptation to breathing while the animal was submerged. The crocodile's nostrils at the tip of the snout show a different way of doing the same job. More subtle differences include the structure of the ankle, which is more primitive in phytosaurs. Also modern crocodiles have a bony palate, separating the nasal passages from the mouth area, allowing them to breathe even though the mouth is full of water. Phytosaurs did not have this. However, they may have had a fleshy palate that served the same function, but there would be no fossil evidence of such a thing.

Three general types are recognized.
• Dolichorostral, or long-snouted, such as *Rutiodon* aand *Mystriosuchus*, and also *Paleorhinus* from Europe and India, and *Arganarhinus* from Morocco. These probably hunted fish.
• Brachyrostral, or short-snouted, such as *Nicrosaurus* and *Smilosuchus*, and also *Brachysuchus* and *Redondosaurus* from New Mexico. These were hunters, probably ambushing terrestrial animals that came to the streams and lakes to drink.
• Altirostral, or high-snouted, such as *Angistorhinus* and *Plseudopalatus* which were intermediate between the two. They were probably generalist feeders.

Mystriosuchus

The recent discovery of an almost complete skeleton of *Mystriosuchus* in shallow sea limestones in northern Italy suggests that it lived in river mouths, in brackish water or even in the sea itself. It is more lightly built and seems to have had greater aquatic adaptations than the others. It was once thought that the slimly built phytosaurs like *Mystriosuchus* were the female forms of the much heavier types like *Nicrosaurus*. However, they do not occur at the same times and in the same places.

Features: Its jaws are long, narrow and cylindrical with an expanded spoon-shaped extension at the tip. The snout is quite distinct from the high cranium, which has upward-pointing eye sockets and a raised nostril. The teeth are all the same size and narrow and fluted. The head is so similar to the modern fish-eating gavial crocodile that it seems evident that it had the same diet and lifestyle. Four rows of plates run down the back and the throat is protected by a mass of smaller plates.

Distribution: Northern Italy; southwest Germany.
Classification: Archosauria, Phytosauridae.
Meaning of name: Spoon crocodile.
Named by: Von Meyer, 1863.
Time: Norian stage of the late Triassic.
Size: 4m (13ft).
Lifestyle: Fish-hunter.
Species: *M. planirostris*, *M. westphali*.

Below: Mystriosuchus was an active hunter of fish.

Smilosuchus

This is the biggest phytosaur known, with a length of up to 12m (40ft) rivalling some of the biggest crocodiles that ever existed. It was probably the largest land animal of the late Triassic. Usually the largest land animals are plant-eaters. In Triassic times it seems they were meat-eaters – see also early Triassic *Erythrosuchus*.

Features: *Smilosuchus* is one of the group of heavily built phytosaurs with the distinct sharp-edged crest on the head. It used to be known as *Machaeroprosopus* (knife face) because of this crest. The jaws are short (for a phytosaur) broad and massive, and there are different-sized teeth at the front from those at the back. The three at the front are particularly big, and must have been used for seizing struggling prey and subduing it. The crest may have strengthened the jaws for this purpose.

Distribution: Arizona, USA.
Classification: Archosauria, Phytosauridae.
Meaning of name: Sabre crocodile.
Named: Long and Murray, 1995.
Time: Carnian stage of the late Triassic.
Size: 12m (40ft).
Lifestyle: Hunter.
Species: *S. gregorii*.

Left: The largest meat-eater of the time must have preyed on the largest of the contemporary plant-eaters.

AETOSAURS OR STAGONOLEPIDS – VEGETARIAN CROCODILES

That the Triassic was the heyday of the crocodiles and their relatives is illustrated by the fact that one branch of the family tree produced a group that were purely plant-eaters. Armoured, heavy-bodied and pig-snouted, the aetosaurs (or stagonolepids) ranged across Pangaea from Scotland to Argentina.

Doswellia

It is not clear where *Doswellia* fits in to the crocodile dynasty. It may not belong here and might be more closely related to the rauisuchians. In fact the only resemblance it had to the aetosaurs is the presence of highly ornamented armour plates that run in rows along its back.

Features: *Doswellia* has a very elongated and flattened crocodile-like body, rather square-shaped in cross section. Armour is present on the tops and sides, forming about ten rows, but not underneath. The snout is long and narrow, but the teeth are missing from the only specimen known. The legs show that it was not a fast runner. This fact and the squareness of the body suggest a herbivorous lifestyle, but the lower jaws are those that we usually associate with a meat-eater. It is a puzzle.

Distribution: Maryland, USA.
Classification: Archosauria, Order uncertain.
Meaning of name: From Doswell.
Named by: Weems, 1980.
Time: Carnian stage of the late Triassic.
Size: 2m (about 6ft).
Lifestyle: Unknown.
Species: *D. kaltenbachi*.

Right: We can reconstruct the appearance of Doswellia – *the square, box-like body, the heavy limbs, the long narrow snout – with some confidence, but we cannot infer much about its lifestyle.*

Stagonolepis

Stagonolepis seems to have been the most common of the aetosaurs at the end of the Carnian stage of the late Triassic. The first species, *S. robertsoni*, was found in Scotland – one of the animals preserved as hollow moulds in the Losseimouth sandstone – while others have been found in the Americas, *S. wellesi* in Arizona and *S. scagliai* in Argentina.

Features: The body is long and narrow, but with a deep stomach area to hold the big digestive system needed for processing plant material and covered in armour consisting of small plates. The plates are highly sculptured, often with a radiating pattern, articulated on the forward edge, flat on the back and curved on the tail. These were originally thought to have belonged to a fish. The weak teeth are leaf- and chisel-shaped, and the snout is turned up, pig-like, evidently for rooting about on the ground for plant food.

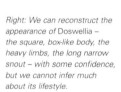

Distribution: Scotland, USA; South America.
Classification: Archosauria, Aetosauridae.
Meaning of name: Ornamented scale.
Named by: Agassiz, 1844.
Time: Carnian stage of the late Triassic.

Size: 2.7m (9ft).
Lifestyle: Low browser.
Species: *S. robertsoni, S. wellesi, S. scagliai.*

Left: Stagonolepis would have foraged on the ground, snuffling about with its pig-like snout, looking for roots and shoots to eat.

Desmatosuchus

Distribution: Arizona, Texas, USA.
Classification: Reptile, Archosauria.
Meaning of name: Link crocodile.
Named by: Case, 1920.
Time: Carnian and Norian stages of the late Triassic.
Size: 5m (16ft).
Lifestyle: Low browser.
Species: *D. haploceros.*

The first specimen found was thought to have been part of a phytosaur, but the armour plates of the two groups are different – the ornamentation on a phytosaur's plates consists of pits, while that on a stagonolepid's scales consists of ridges. Nests found in Petrified Forest National Park, USA, have been attributed to stagonolepids like

Features: *Desmatosuchus* is the largest stagonolepid so far known. Apart from the usual stagonolepid pattern of scales, several parallel rows down the back and sides, the armour consists of a number of spines that stick out sideways from the neck, shoulders and sides, with the longest above the front legs, anticipating those of the ankylosaur dinosaurs to come. This was probably for protection against the earliest of the meat-eating dinosaurs or the last of the rauisuchians.

Desmatosuchus.

Left: With its spiky armour and its size, Desmatosuchus *would have been a daunting adversary.*

A HOST OF STAGONOLEPIDS

Many other stagonolepids are known:
Calyptosuchus – a delicate and slender type, 3–4m (10–13ft) long from the Carnian and Norian strata of Petrified Forest National Park in Arizona.
Coahomasuchus – 1m (about 3ft) long, known from a complete skeleton from the late Carnian deposits of Texas, USA.
Acaenasuchus – bearing thorn-like spikes along the armour of the back, and known from the early Triassic of North America.
Paratypothorax – similar to *Longosuchus* but found in Germany.
Redondasuchus – a 2m (6½ft)-long type with distinctive keeled scutes from the Rhaetian stage of the late Triassic of New Mexico, USA.
Aetosaurus – the animal from which the group derives its name. A small type only 1.5m (5ft) long from Norian rocks of New Jersey and Connecticut, USA, Greenland, Italy and Germany. Scutes have even been found in Madagascar.
Aetosauroides – an early Triassic form from South Africa.

Neoaetosauroides (above) – a typical early Triassic aetosaur.
Argentinosuchus – from Argentina.
Lucasuchus – named after Spencer Lucas who initiated the study of stagonolepids.

Longosuchus

The skeleton of *Longosuchus* was misidentified by Cope in 1875 as a phytosaur. Its true nature as a stagonolepid was not realized until the 1980s. These spined stagonolepids seem to have been restricted to the western part of Pangaea. It is used as an index fossil. This means that it has been dated so precisely that whenever a fragment of its armour is found in a rock the date of that rock is known.

Features: The skull of this large stagonolepid is quite small, with small eyes. The armour consists of rows of square plates, and a series of spines along the sides. However, it does not have the very long spines on the shoulders as has its larger relative *Desmatosuchus*. The limbs are quite small and slim for such a heavy animal. Presumably it relied on armour like an armadillo rather than flight for defence.

Below: Like an armadillo, Longosuchus *would have presented its well-armoured back and sides to an enemy when it was attacked.*

Distribution: Arizona, USA.
Classification: Reptile, Archosauria.
Meaning of name: Long crocodile (not because of its length but after the palaeontologist Robert A. Long).
Named by: Hunt and Lucas, 1990.
Time: Carnian stage of the late Triassic.
Size: 4m (13ft).
Lifestyle: Low browser.
Species: *L. meadei.*

PTEROSAURS

As soon as reptiles evolved there were a certain number that could exploit the possibilities of a life of flight. During the Permian and Triassic periods there were several reptiles that could glide on flaps of skin. However, it was only with the evolution of the pterosaurs in the late Triassic that true flight developed among the reptiles.

Sharovipteryx

Although not a pterosaur, *Sharovipteryx* shows what might be regarded as a primitive equivalent of the pterosaur flying membrane. Its long hind legs suggest that it jumped like a grasshopper, and once airborne the gliding

Below: Sharov originally named this pterosaur Podopteryx *in 1971, but that name was found to have been taken, or pre-occupied, by a beetle.*

membrane would then catch the air and allow the animal to glide. The membrane was possibly also used as a display device.

Features: The legs of *Sharovipteryx* are extremely long and support a panel of skin – a patagium. This stretches from the hind legs to the tail and from the hind legs to the little front limbs, undoubtedly forming a gliding surface. The leg bones are too lightly built for powered flight. The forelimbs are tiny and possibly only used for adjusting the trim of the patagium while in flight.

Distribution: Madygen, Kirgyzstan.
Classification: Archosauromorpha, possibly Ornithodira.
Meaning of name: Sharov's wing.
Named by: Cowen, 1981.
Time: Early Triassic.
Size: 30cm (1ft) long.
Lifestyle: Probably a gliding insectivore.
Species: *S. mirabilis*.

Eudimorphodon

Distribution: Northern Italy.
Classification: Pterosauria, Rhamphorhynchoidea.
Meaning of name: The true two-formed tooth.
Named by: Sambelli, 1973.
Time: Norian stage of the late Triassic.
Size: 1m (3ft) wingspan.
Lifestyle: Fish-eater.
Species: *E. ranzii*.

This is one of the earliest pterosaurs known, and it was a fully formed flying animal. It is known from several articulated skeletons, including those of juveniles. Its claws show that it lived on steep surfaces such as cliffs or tree trunks, from which it could launch its fish-hunting flights over water. Scales of fish have been found in the stomach regions.

Features: The teeth mark this animal out as a fish-eater, since they are mostly small and densely packed in the jaw with more than 100 in the whole mouth. Some have several points. Some are larger than others, forming big fangs near the front of the mouth. This is an ideal arrangement for dealing with slippery prey. No later pterosaur had as elaborate an arrangement of teeth as this. The hind legs are quite sturdy for the size of animal and the tail is stiff and straight.

Peteinosaurus

This was a contemporary of *Eudimorphodon*, a little smaller and probably an insect-eater rather than a fish-eater. It shows that the newly-evolved pterosaurs had already diversified into a variety of lifestyles and pursued several different types of food in the variety of habitats that existed around the shores of the Tethys Ocean. It may have been the ancestor of the better-known early Jurassic *Dimorphodon*.

Features: *Peteinosaurus* is a little more primitive in construction than *Eudimorphodon* the teeth are mostly the same size, apart from two pairs of small fangs in the front lower jaw, and all have single points. The wings are about twice as long as the hind legs (this is short for a primitive pterosaur). Most rhamphorhynchoids (a main group of the pterosaurs) had wings that were at least three times as long as the hind legs.

Distribution: Northern Italy.
Classification: Pterosauria, Rhamphorhynchoidea.
Meaning of name: Winged lizard.
Named by: Wild, 1978.
Time: Norian stage of the late Triassic.
Size: 60cm (2ft) wingspan.
Lifestyle: Flying insectivore.
Species: *P. zambelli*.

Above: Peteinosaurus *probably weighed as little as 100g (3½oz).*

EVOLUTION OF THE PTEROSAURS

Above: Heleosaurus, *a primitive archosaur and possible pterosaur ancestor.*

The big mystery of the pterosaurs is that they spring fully evolved into the fossil record in the rocks formed in late Triassic times. There is no record of any undisputed ancestral forms. The evolution of their powers of flight is unknown. This is not really surprising. The odds are against any animal becoming fossilized, and we are lucky to have any examples, especially of lightweight flying animals whose delicate bones would rarely fossilize anyway.

It is possible that their ancestors lie among the primitive archosaurs, probably lizard-like animals that may have run on hind legs. Some of these may have taken to trees in pursuit of tree-living insects, and developed long limbs and lightweight bodies as an adaptation to tree-living life. The next stage may have been the development of gliding patagia between the limbs, to allow travel from tree to tree. How such a form developed into the warm-blooded active flying animal, with muscular manoeuvrable wings, is still to be discovered.

Preondactylus

This pterosaur is known from two partial skeletons, and some bits of forelimbs. One interesting fossil is jumbled up and packed together. The inference is that this was a fish pellet. A dead *Preondactylus* was swallowed by a massive fish and after it was digested the hard parts were regurgitated in a lump that was subsequently fossilized.

Features: This is the earliest pterosaur known. Only part of the skull has been found, but what there is shows the animal to have been very similar to the Jurassic *Dorygnathus*. The tail is well preserved and is similar to that of all other rhamphorhynchoids, in that it consists of long vertebrae lashed together into a stiff column by tendons that have become solidified into bone. This forms a balancing and steering organ.

Distribution: Northern Italy.
Classification: Pterosauria, Rhamphorhynchoidea.
Meaning of name: Finger from the Preone Valley.
Named by: Wild, 1983.
Time: Norian stage of the late Triassic.
Size: 1.5m (6ft) wingspan.
Lifestyle: Fish-eater.
Species: *P. buffarinii*.

Left: Dorygnathus *for comparison.*

Above left: Computer imagery shows the partial skull of Preondactylus *to have straight jaws with different-sized teeth, big fangs at the front and also about halfway along, and a sclerotic ring (a strengthening ring of fine bones), in the eye socket.*

EARLY MEAT-EATING DINOSAURS

What were the first dinosaurs? The truth is that we are not quite sure. We can only look at the earliest remains found, and speculate that these must be the earliest and most primitive of the dinosaur group. As far as we know the dinosaur dynasty began with the small meat-eaters in South America, but they may have existed in other places as well.

Eoraptor

The Valley of the Moon in north-western Argentina consists of dusty outcrops of sandstone and mudstone laid down in lushly forested river valleys in late Triassic times. These river banks were prowled by the first dinosaurs, including fox-sized *Eoraptor*, and various other reptiles that it hunted for food.

Right: Even the most perfect dinosaur skeleton can say little about the skin or the coloration. As in most restorations, skin colour and pattern are conjectural.

Features: *Eoraptor* is known from a complete skeleton that is lacking only the tail, and in shape and size *Eoraptor* conforms to every idea of the primitive dinosaur. Its lower jaw lacks the bone joint behind the tooth row that is seen in every other meat-eater, and there is more than one kind of tooth, something unusual in a meat-eating dinosaur. However, all other skeletal features such as the shape of the hips, the upright stance and a reduction of the number of fingers on the hand show that this dinosaur is definitely an early theropod.

Distribution: North-western Argentina.
Classification: Saurischia, Theropoda.
Meaning of name: Dawn plunderer.
Named by: Sereno, 1993.
Time: Carnian stage of the late Triassic.
Size: 1m (3ft).
Lifestyle: Hunter.
Species: *E. lunensis*.

Herrerasaurus

Distribution: North-western Argentina.
Classification: Saurischia, Theropoda, Herrerasauridae.
Meaning of name: From Victorio Herrera, its discoverer.
Named by: Sereno, 1988.
Time: Carnian stage of the late Triassic.
Size: 5m (16½ft).
Lifestyle: Hunter.
Species: *H. ischigualestensis*.

A contemporary of *Eoraptor* on the late Triassic riverbanks of Argentina, *Herrerasaurus* was a much bigger and more advanced theropod. Because of the difference in size, it must have hunted different prey from its smaller relative. Its skeleton was found in 1959, although it was several decades before it was scientifically studied. The complete skull was not found until 1988.

Features: A big animal with heavy jaws and 5cm- (2in-) long serrated teeth, giving it the appearance and probable lifestyle of the big theropods to come. It has the hinged lower jaw of other theropods. The foot bones are quite primitive, retaining the first and fifth toes that later theropods were to lose. *Herrerasaurus* has complex ear bones suggesting that it had a keen sense of hearing which would help in hunting.

EARLY HISTORY OF MEAT-EATING DINOSAURS

It is possible that the dinosaurs first evolved in South America. Most of the early dinosaur fossils are found in the Santa Maria Formation in Brazil and the Ischigualasto Formation in Argentina both dating from the Carnian stage of the late Triassic. Certainly the most complete early dinosaurs have been found here, although scattered remains have cropped up in other parts of the world. South America was also home to the lagosuchids, a group of archosaurs very closely related to the dinosaurs. It is likely that their common ancestor lived in this area.

Most of the animals on these two pages were once regarded as too primitive to be allocated to either of the major dinosaur groups – saurischia or ornithischia. However, research in the 1990s, mostly by Paul Sereno of Chicago, shows that they were primitive saurischians, as shown by the structure of the skull and the bones of the hind foot.

These early meat-eaters had certain interesting features. There was a joint in the lower jaw, just behind the tooth row, that absorbed the stresses involved in biting struggling prey, curved claws on the first three fingers for grasping, and hollow limb bones for speed – all features possessed by the later meat-eating dinosaurs.

Below: Marasuchus was an anchosaur close to the ancestry of the dinosaurs.

Staurikosaurus

In 1936, an expedition from the Museum of Comparative Zoology at Harvard, USA, looked at the Santa Maria Formation in southern Brazil. They found the skeleton of a big meat-eating dinosaur, but it was not until 1970 that it was studied scientifically and named by the great American palaeontologist Edwin H. Colbert. Since then opinions have differed widely about the classification of *Staurikosaurus*.

Features: *Staurikosaurus* is very similar to its Argentinian contemporary, *Herrerasaurus*, but more lightly built and with a longer, more slender neck. The head is quite large and constructed for tackling large prey. Like *Herrerasaurus* it has a primitive five-toed foot, but the arm and hand bones are missing from the only known skeleton. Some scientists believe that it is the same animal as *Herrerasaurus*, while others think it is so primitive that it cannot be classed as either a theropod or a sauropod. The classification given here is therefore tentative.

Distribution: South-eastern Brazil.
Classification: Saurischia, Theropoda, Herrerasauridae.
Meaning of name: From the Southern Cross constellation.
Named by: Colbert, 1970.
Time: Carnian stage of the late Triassic.
Size: 2m (6½ft).
Lifestyle: Hunter.
Species: *S. pricei*.

Left: Although it must have had the appearance of a theropod, Staurikosaurus may have belonged to a completely different group. It may not even have been a saurischian but something ancestral to both the saurischian and ornithischian lines.

Chindesaurus

The story of *Chindesaurus* shows the confusion caused by the primitiveness of the early dinosaurs. It was unearthed in the 1980s by Bryan Small of the University of California, Berkeley, USA, and was regarded as a prosauropod, an early plant-eater. The media dubbed it "Gertie" after a cartoon dinosaur from the early days of the cinema. In 1985 it was found to be a herrerasaurid theropod.

Distribution: Arizona and New Mexico, USA.
Classification: Saurischia, Theropoda, Herrerasauridae.
Meaning of name: From Chinde Point.
Named by: Murray and Long, 1985.
Time: Carnian stage of the late Triassic.
Size: 3m (10ft).
Lifestyle: Hunter.
Species: *C. bryansmalli*.

Right: This animal lived in North America in the Carnian stage of the late Triassic. Unfortunately it is difficult to equate the age with that of similar rocks in South America, and so it is unclear whether Chindesaurus predated the South American forms or was later.

Features: *Chindesaurus* is similar to the other early dinosaurs, but its legs are particularly long and its tail very whip-like. It is important because it is the first herrerasaurid found in the Northern Hemisphere. Its presence in the Chinle Formation of Petrified Forest National Park, with the remains of other meat-eaters, shows that a variety of prey animals existed in the river-bank forests of the time.

LITHE MEAT-EATERS

The classic shape of a meat-eating dinosaur lent itself rather well to a fast-moving lifestyle. The powerful hind legs propelled the animal forward into an attack, and the teeth and claws would be used as weapons. Some of the smaller dinosaurs became quite lightweight and were obviously built for speed, developing slim bodies and long, agile legs.

Aliwalia

Distribution: South Africa.
Classification: Saurischia, Theropoda.
Meaning of name: From Aliwalia in South Africa.
Named by: Galton, 1985.
Time: Carnian to Nor an stages of the late Triassic.
Size: 8m (26ft).
Lifestyle: Hunter.
Species: *A. rex*.

Features: Hardly anything is known about this animal except that it is a very large meat-eater, probably the first of the really big flesh-eating dinosaurs. The only remains that we know are parts of the hind leg bones and a jaw fragment. We often think of the very big meat-eating dinosaurs as coming later in the Mesozoic,

The few remains that we have of *Aliwalia* were found in a shipment of prosauropod fossils sent from South Africa to a museum in Austria, in 1873. For a century they were thought to have come from a prosauropod, and led to the idea that prosauropods may have been meat-eating animals as well as the plant-eaters we now know them to have been.

when they would have had time to evolve from the earlier small dinosaurs. However, the presence of animals such as *Aliwalia* shows that they existed very early. They probably preyed on prosauropods or any of the other big plant-eaters, such as the pick-toothed rhynchosaurs or the last of the herbivorous mammal-like reptiles.

Left: Life restorations of animals based on only a handful of bones are difficult. However, we know enough about the general appearance of the early meat-eating dinosaurs to make an educated guess as to the probable appearance of Aliwalia.

Shuvosaurus

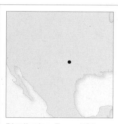

Distribution: Texas, USA.
Classification: Saurischia, Theropoda.
Meaning of name: From Shuvo, the son of the finder.
Named by: Chatterjee, 1993.
Time: Norian stage of the late Triassic.
Size: About 3m (10ft).
Lifestyle: Possibly omnivorous or an egg-stealer.
Species: *S. inexpectatus*.

When the skull of *Shuvosaurus* was discovered by the son of palaeontologist Sankhar Chatterjee in the early 1990s, it caused a surprise. It was suggested that it was the skull of an ostrich-mimic because of its lack of teeth. However, ostrich mimics are only known from Cretaceous rocks, and the Triassic would have been far too early for this group to have lived. The unexpected discovery is the reason for the specific name of the only species found to date.

Features: Nothing is known about this dinosaur except a toothless skull. Its original identification as an ostrich-mimic did not hold water, and soon it was reclassified as a rauisuchian, one of the land-living, crocodile-

like predators that shared the world with the early dinosaurs. On this basis it was assigned to the already known genus *Chatterjeea*, which lacked the skull. Vertebrae and other isolated bones found in the formation known as the Upper Dockum Group of the Norian of Texas may belong to this animal. Current thinking suggests that it is a very specialized coelophysoid, but the matter is not settled.

Above: The toothless jaws for Shuvosaurus *suggest it had a very specialized diet, however there is no evidence as to what that diet was.*

Gojirasaurus

The dinosaur *Gojirasaurus* was found by Kenneth Carpenter of Denver Museum of Nature and Science. As a child he was fascinated by the 1954 Japanese science-fiction film *Godzilla*, which later sparked his interest in palaeontology and dinosaurs. He named his newly found fossil *Gojirasaurus* in honour of his childhood passion.

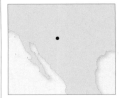

Distribution: New Mexico, USA.
Classification: Saurischia, Theropoda, Ceratosauria.
Meaning of name: From Gojira.
Named by: Carpenter, 1991.
Time: Carnian to Norian stages of the late Triassic.
Size: 5.5m (18ft), from an immature specimen.
Lifestyle: Hunter.
Species: *G. quayi*.

Features: This dinosaur is only known from an associated skeleton, consisting of ribs, bits of backbone, the shoulder girdle, part of the hips and hind leg, and a tooth. What is more, the skeleton is of an individual that was not fully grown, and so we have no good idea of the size of the adult animal. Consequently the restoration here is highly speculative. What we can say is that *Gojirasaurus* was an early meat-eater that attained a respectable size.

Left: The original Japanese name for Godzilla is Gojira, which is itself an amalgam of the Japanese words kujira, meaning whale, and gorira, meaning gorilla.

THE SHAPE OF MEAT-EATERS

The basic shape of the meat-eating dinosaur became established early in the history of the group. The upright posture, with the long hind legs and tail, is familiar to anyone with any knowledge of dinosaurs. Originally scientists believed that a meat-eating dinosaur stood with its back at an angle of 45 degrees with the tail dragging on the ground – the stance of the kangaroo, which is the only large, familiar, tailed bipedal animal of today. No doubt it was the kangaroo that inspired this vision.

About the 1970s the more modern version of the meat-eating dinosaur came to be accepted. The back is held horizontally, the jaws and claws are held out at the front where they can do the most damage, and the whole body is balanced by the heavy tail.

One of the theories for the evolution of this stance was that the dinosaurs' immediate ancestors were crocodile-like semi-aquatic forms, with strong swimming back legs and powerful tails. When they took to the land they walked naturally on their strong hind legs, with their shorter front legs clear of the ground. The heavy muscular tail then balanced the rest of the body quite naturally.

Below: The dinosaur skeleton on the left shows the contemporary understanding of how theropods stood. The skeleton on the right shows the old-fashioned view.

Liliensternus

In Triassic Europe *Liliensternus* was one of the largest meat-eating dinosaurs. In size and appearance it was somewhere between *Coelophysis* and *Dilophosaurus*. It lived in the forests that grew on the banks of the rivers that crossed an otherwise barren continent. It may have hunted the big plant-eating prosauropods that lived in the region.

Distribution: Germany.
Classification: Saurischia, Theropoda, Ceratosauria.
Meaning of name: From Ruhle von Lilienstern.
Named by: von Huene, 1934.
Time: Carnian to Norian stages of the late Triassic.
Size: 5m (16½ft).
Lifestyle: Hunter.
Species: *L. orbitoangulatus, L. airelensis, L. liliensterni.*

Features: *Liliensternus*, in common with the other early meat-eaters, has five fingers on the hand. However, the fourth and fifth are much reduced – seemingly a step on the way to the established three-finger pattern of the vast majority of theropods. It is a very slim animal with a particularly long neck and tail. The crest on the head would probably have been brightly coloured, and would have been used for signalling and communicating with other dinosaurs.

COELOPHYSIDS

The coelophysids seem to have been the most widespread of the meat-eaters in late Triassic times. So successful were they that members of the group survived into the beginning of the Jurassic. They were small, fast-running dinosaurs and probably preyed on the smaller animals of the time, leaving larger prey to the big crocodile-like animals that still existed.

Procompsognathus

One of the earliest, European, meat-eating dinosaurs was *Procompsognathus*. What we know of its skeleton shows that it was an active hunter, but it was not the major hunter of the time because it is clear that much bigger herrerasaurids lived alongside it. Despite the fact that little is known about this animal, and we are uncertain of its appearance or classification, it became widely known as a principal dinosaur in Michael Crichton's 1995 novel *The Lost World: Jurassic Park*, and the subsequent film.

Features: We only know of a single skeleton, and that is badly crushed, with most of the neck, tail, arms and hips missing. The classification, putting it among the Coelophysoidea, is largely based on the long, narrow skull and the teeth. These, however, now seem to have belonged to a different animal altogether, possibly a crocodile. As a result, the restoration shown here and the classification given are rather spurious.

Distribution: Germany.
Classification: Theropoda, Coelophysoidea (unproven).
Meaning of name: Before *Compsognathus*.
Named by: Fraas, 1913.
Time: Norian stage of the late Triassic.
Size: 1.2m (4ft).
Lifestyle: Hunter.
Species: *P. traissicus*.

Camposaurus

Charles Lewis Camp (1893–1975) was an American vertebrate palaeontologist who did a great deal of work on the sites that produced most of our knowledge of the animal life of the Triassic Petrified Forest fossil site in Arizona. One of his finds consisted of the partial remains of this small meat-eater that was named after him long after his death.

Features: We only know of the limb bones and a few vertebrae of this dinosaur. They show it to be very similar to *Coelophysis* in having long running legs and clawed, three-toed feet. However, there are enough differences in the shape and arrangement of bones to show

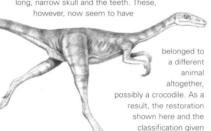

that, although closely related, it is a totally different animal. It would have hunted the smaller animals that lived in the fern thickets and coniferous forests of the Petrified Forest National Park, Arizona.

Distribution: Arizona, USA.
Classification: Theropoda, Coelophysoidea.
Meaning of name: From Charles Lewis Camp.
Named by: Hunt, Lucas, Heckert and Lockley, 1998.
Time: Carnian stage of the late Triassic.
Size: 1m (3ft).
Lifestyle: Hunter.
Species: *C. arizonensis*.

Right: The few bones known of Camposaurus *are just different enough to distinguish it from the slightly later* Coelophysis, *but the two must have been very similar.*

COELOPHYSIS DISCOVERY

The discovery of a pack of *Coelophysis* at the evocatively named Ghost Ranch in New Mexico, in 1947, gives an astonishing insight into the lives and deaths of the early meat-eating dinosaurs. Some were complete and fully articulated. They all had a typical death pose, with the head and neck pulled back, indicating that the body twisted as the tendons dried out in the open air. Inside the rib cages of two were the bones of *Coelophysis* youngsters. The immediate interpretation was that the pack had been driven to cannibalism. However, more recent work shows that in one, the bones of the youngster were below the adult skeleton not within it, and in the other the stomach contents consisted of a small crocodile, not a young *Coelophysis*.

Coelophysis had been named in 1889, and the Ghost Ranch skeletons were identified as the same animal. Later work, however, showed that they were different enough to be a new genus, and they were renamed *Rioarribasaurus*. By this time the site was famous, and the dinosaurs found there had been accepted as *Coelophysis*, resulting in the International Commission of Zoological Nomenclature (ICZN) – the body that regulates the naming of new animals – ruling that the name *Coelophysis* be applied to the Ghost Ranch animals rather then Cope's original find. It is unusual for the ICZN to do this.

Coelophysis

This is without doubt the best known of the coelophysid group and hundreds of skeletons, some complete and articulated, provide our knowledge of the group in general. A dozen skeletons were found at Ghost Ranch in New Mexico, in 1947, indicating that they died of starvation around a drying water hole. In 1998 the space shuttle 'Endeavor' took a skull of *Coelophysis* to the Mir space station, making it the first dinosaur in space.

Features: *Coelophysis* is a slimly built running hunter with a long skull and neck, a lightweight body and a long tail. The premaxilla – the front bone of the skull – is loosely articulated to the rest of the jaw mechanism, suggesting that it could be moved to manipulate small prey. Its fossil remains show that it hunted in packs. Two different weights of skeleton suggest that both males and females travelled together.

Distribution: Arizona, New Mexico and Utah, USA.
Classification: Theropoda, Coelophysoidea.
Meaning of name: Hollow form.
Named by: Cope, 1889.
Time: Carnian and Norian stages of the late Triassic.
Size: 2.7m (9ft).
Lifestyle: Hunter.
Species: *C. bauri*. Classification is in dispute, but some regard *Podokesaurus* and *Syntarsus* as species of *Coelophysis*, and give the name *Rioarribasaurus* to the Ghost Ranch remains.

Left: Coelophysis *is named after its hollow bones, a feature shared with modern birds. They made for a very light, fast-moving animal.*

Eucoelophysis

It used to be thought that there were very few species of hunting dinosaurs living during Triassic times. The discovery of *Eucoelophysis* and *Camposaurus* shows a range of animals all closely related in the coelophysoid group, living in the same area at more or less the same time. *Eucoelophysis* was found in the Petrified Forest beds, slightly older than those containing the remains of *Coelophysis*.

Features: What we know of this animal comes from the discovery of leg bones and other parts of the skeleton in New Mexico in the early 1980s. Their shape shows that it was closely related to *Coelophysis* but different enough to be regarded as a separate genus. Based on this it seems that some of the original *Coelophysis* material found by amateur fossil collector David Baldwin in the 1880s actually belongs to this genus.

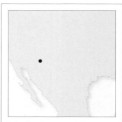

Distribution: New Mexico, USA.
Classification: Theropoda, Coelophysoidea.
Meaning of name: True *Coelophysis*.
Named by: Sullivan and Lucas, 1999.
Time: Carnian to Norian stages of the late Triassic.
Size: 3m (10ft).
Lifestyle: Hunter.
Species: *E. baldwini*.

Right: *The original remains of* Coelophysis *found by Cope, along with some other putative* Coelophysis *species, may actually belong to the genus* Eucoelophysis.

PRIMITIVE SAUROPODOMORPHS

The sauropodomorphs were the first plant-eating dinosaurs. The earliest, probably evolving from the primitive meat-eaters, were small, rabbit-sized beasts but, as the Mesozoic advanced, they evolved into the biggest land animals that have ever existed. They can be divided into two main groups, the earlier prosauropods and the sauropods. There were also animals that were too primitive to be classed in either.

Saturnalia

As an early dinosaur species *Saturnalia* gives us a taste of things to come in the dinosaur world. It is the earliest known plant-eating dinosaur, and was originally thought to have been an early prosauropod. It is now thought to have been more primitive than the prosauropods, and we can only classify it as an early member of the Sauropodomorpha.

Features: From three partial skeletons we can construct the appearance of this elegant, rabbit-sized animal, with a long neck and tail. The head is small and the teeth coarsely serrated, in keeping with a vegetable diet. The body and legs are quite slender. The hip bones are quite primitive, making it a rather borderline dinosaur, but its ankle bones are similar to those of the contemporary meat-eating dinosaurs. It was very close to the ancestry of the prosauropods and the sauropods.

Distribution: Brazil.
Classification: Sauropodomorpha.
Meaning of name: After the Roman solstice festival.
Named by: Langer, Abdala, Richter and Benton, 1999.
Time: Carnian stage of the late Triassic.
Size: 1.5m (5ft).
Lifestyle: Low browser.
Species: *S. tupiniquim*.

Right: The Saturnalia was the Roman winter solstice festival, and the name was given to this dinosaur as that was the time of year when it was found. The species name means "native" in the local Portuguese.

Thecodontosaurus

An early discovery, *Thecodontosaurus* was only the fourth dinosaur to have been named. It was named from a jaw bone with teeth, which was found in limestone quarries near Bristol, England. It was not originally included in Owen's classification of the Dinosauria, possibly because it was so small compared with *Iguanodon* and *Megalosaurus*. It was only drawn into the dinosaur fold by Thomas Huxley in 1870.

Features: *Thecodontosaurus* is known from hundreds of fossils, both juvenile and adult, including partial skeletons. Unfortunately, the earliest specimens were lost in the bombing of Bristol City Museum, England, during World War II. As with all other prosauropods, it has a small head with plant-shearing teeth, and front legs that are shorter than the hind. It probably moved about mostly on all fours, but could spend much of its time on its hind legs.

Distribution: England; Wales.
Classification: Sauropodomorpha, Prosauropoda.
Meaning of name: Socket-toothed lizard.
Named by: Riley and Stutchbury, 1836.
Time: Norian to Rhaetian stage of the late Triassic.
Size: 2m (6½ft).
Lifestyle: Low browser.
Species: *T. antiquus, T. minor, T. caducus*.

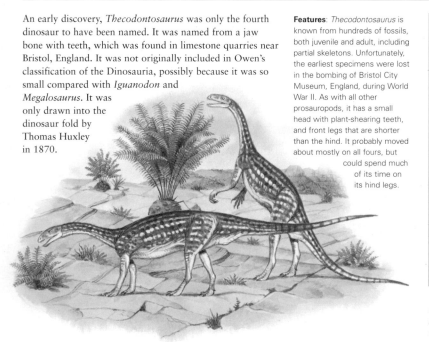

Left: Thecodontosaurus probably ran on two legs and browsed on four.

Efraasia

The only remains of this animal were discovered in 1909 by German dinosaur pioneer Eberhard Fraas. The animal was named after him in 1973. The skeleton shows the distinctive nature of the prosauropod's hand. This dinosaur is now thought by some to be the juvenile form of the larger *Sellosaurus*, although there is a suggestion that it may be a very primitive sauropodomorph, like *Saturnalia*.

Features: As with other prosauropods, *Efraasia* has multipurpose hands with long, grasping fingers and mobile thumbs. The wrist joint is well-developed and flexible, and the palm could be turned towards the ground allowing the animal to walk on all fours. There are only two sacral vertebrae joining the backbone to the hips, however, which is quite a primitive feature because most other saurischian dinosaurs have three. This may be the sign of a young animal.

Distribution: Germany.
Classification: Sauropodomorpha, Prosauropoda.
Meaning of name: After Eberhard Fraas, the German dinosaur pioneer.
Named by: Galton, 1973.
Time: Norian stage of the late Triassic.
Size: 2.4m (8ft).
Lifestyle: Low browser.
Species: *E. diagnostica*, *E. minor*.

Right: The original skeleton of Efraasia was associated with a set of crocodile jaws, and the whole thing was restored with a meat-eater's head and a plant-eater's body. Peter Galton's research in 1973 established the current rendering of the dinosaur.

PLACING DINOSAURS IN THEIR CORRECT PERIOD

Left: The skull of Thecodontosaurus.

The prosauropods, a group of sauropodomorphs, comprise one of the first dinosaur groups to be studied, but we still do not know a great deal about them compared with other dinosaurs. This is probably because they lived at the beginning of the Age of Dinosaurs, and their fossil record is so old that it is fragmentary and incomplete.

The first prosauropod remains to be found were those of *Thecodontosaurus*. Its fragmentary fossils were found in Triassic infillings of Carboniferous limestones near Bristol in southern England. The seeming paradox of Triassic fossils found in Carboniferous rocks can be explained by the geography of the time. In the Triassic period this area was an arid limestone upland, formed from rock laid down during the Carboniferous period. It was riddled with caves and chasms which may have provided shelter for animals. Swallow holes would have supported vegetation around the rim, encouraged by moist air rising from the caves below. Plant-eating animals such as *Thecodontosaurus* would have been attracted to this richer vegetation, and inevitably some fell to their deaths in the caves. The rubble and bones on the cave floor eventually solidified, resulting in fossils of Triassic animals surrounded by Carboniferous rock.

Sellosaurus

One of the prosauropods of the European Triassic deserts, *Sellosaurus* would have fed on the primitive monkey-puzzle-like conifers that grew in the moister areas. The presence of stomach stones inside some of the skeletons suggests that they had quite a sophisticated digestive system, possibly with a big gut full of fermenting bacteria to help break down the tough plant material.

Features: *Sellosaurus* is a typical, medium-sized prosauropod. It has a heavy body, a long neck and a small head. Its hind legs are bigger and stronger than its arms, which it would use to pull down twigs and leaves from trees. It differs from other prosauropods by the shape of the tail vertebrae – the neural spine on each has a saddle-shaped structure, hence the name. It may have been ancestral to the later *Anchisaurus* but this would be difficult to prove. It is known from over 20 partial skeletons, three of them with skulls. Gastroliths, or stomach stones, have been found associated with some of the skeletons.

Distribution: Germany.
Classification: Sauropodomorpha, Prosauropoda, Plateosauridae.
Meaning of name: Graceful saddle lizard.
Named by: von Huene, 1908.
Time: Norian stage of the late Triassic.
Size: 6.5m (21ft).
Lifestyle: Low browser.
Species: *S. gracilis*.

Left: Originally the name Sellosaurus *encompassed two other animals, Efraasia diagnostica and Plateosaurus gracilis. The research of Adam Yates, an Australian palaeontologist, in 2002 indicated that they were three separate genera.*

TRUE PROSAUROPODS

The prosauropods, a group of sauropodomorphs, rapidly developed into the main plant-eaters of late Triassic times, when the landmass of the world was still joined together. Consequently, the remains of similar animals have been found all over the world. Although Europe seems to have been the centre for prosauropod discoveries, they have also been found in North and South America, Africa and China.

Euskelosaurus

The first dinosaur to be discovered in Africa was *Euskelosaurus*. It was known from a set of leg bones found in South Africa in 1866, and was sent to London for study. Since then, bones of this animal have been found all over southern Africa, suggesting that it had been a very common plant-eater of the time. It appears to have lived at a time when dry climates were spreading across the southern continents, and many remains have been found in the Lower Elliot formation in South Africa, entombed in sandstone formed from soft river sands in which they had become mired – rather like the Swiss occurrences of *Plateosaurus*.

Features: This is one of the biggest of the prosauropods, and as such it looks very much like one of the true sauropods that succeeded the group. It supports its bulky body by moving on all fours. While we have many bones of this animal, we do not have the skull, and its relationship with the rest of the prosauropod group is unclear.

Distribution: Lesotho, South Africa; Zimbabwe.
Classification: Sauropodomorpha, Prosauropoda.
Meaning of name: Well-limbed lizard.
Named by: Huxley, 1866.
Time: Carnian to Norian stages of the late Triassic.
Size: 9–12m (30–39ft).
Lifestyle: High browser.
Species: *E. browni, E. africanus, E. capensis, E. fortis, E. molengraafi.*

Right: Euskelosaurus *resembles both the advanced prosauropods such as* Melanorosaurus, *and the primitive sauropods like* Antetonitrus. *However, it is the shape of the thigh bone that curves out that shows it is a traditional prosauropod.*

Blikanasaurus

Evolution is always developing new animal shapes as a response to new conditions. The variation seen in the prosauropods includes a movement towards bigger, heavier animals. One of these lines developed into the sauropods, but some became merely evolutionary sidelines that happened to look like sauropods but developed no further. *Blikanasaurus* was one of these.

Features: All that we know of *Blikanasaurus* is the stoutly built hind limb. It shows the very short foot of an animal that would have spent all its time on all fours, and as such it would have resembled the later sauropods. However, its fifth toe (that is present in all sauropods) is tiny, showing that *Blikanasaurus* represented a side-branch of the prosauropods rather than being part of the evolutionary line to the sauropods themselves.

Distribution: Lesotho.
Classification: Sauropodomorpha, Prosauropoda.
Meaning of name: Lizard from Blikana.
Named by: Galton and van Heerden, 1985.
Time: Carnian to Norian stages of the late Triassic.
Size: 5m (16½ft).
Lifestyle: High browser.
Species: *B. comptoni.*

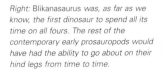

Right: Blikanasaurus *was, as far as we know, the first dinosaur to spend all its time on all fours. The rest of the contemporary early prosauropods would have had the ability to go about on their hind legs from time to time.*

Plateosaurus

Distribution: Germany; Switzerland; and France.
Classification: Sauropodomorpha, Prosauropoda, Plateosauridae.
Meaning of name: Broad lizard.
Named by: von Meyer, 1837.
Time: Norian stage of the late Triassic.
Size: 8m (26ft).
Lifestyle: High browser.
Species: *P. engelhardti*, *P. gracilis*.

Below: The teeth of Plateosaurus *were leaf-shaped and coarsely serrated. They were well suited to shredding the tough leaves of the tree ferns and primitive conifers.*

One of the best known of the prosauropods is *Plateosaurus*. It was one of the most common of all European dinosaurs with fossils found in over 50 localities. The fact that so many skeletons have been found together suggests that they may have been herding animals that migrated across the arid plateaux and basins that formed the landscape of Europe in late Triassic times. The concentration of *Plateosaurus* remains in a few areas has led palaeontologists to suggest that frequent disasters overtook the herds, but it is just as likely that these accumulations of bones took place over long periods of time, highlighting the fact that the herds gathered at specific spots at particular times of the year.

Interesting remains have been found in Switzerland, and consist of the leg bones of *Plateosaurus* standing vertically in river sandstones, while the rest of the skeleton is scattered over a wide area, mixed with the teeth of meat-eating dinosaurs and crocodile-like animals. The interpretation is that the *Plateosaurus* became bogged down in quicksands and was torn apart by hunting animals. *Euskelosaurus* remains showing the same mode of preservation have been found in South Africa.

PROSAUROPOD LINEAGE

The prosauropods are sometimes given the name palaeopods. They evolved from two-footed meat-eating ancestors, and their early forms evolved into the later sauropods. The smaller prosauropods were active little two-footed animals which perhaps were not quite able to run. Even so, they would have been faster than any four-footed animal at the time. The biggest were totally quadrupedal animals. Those in between would have spent most of their time on all fours, but would have raised themselves up to reach the conifer foliage and fruits of the cycad-like plants on which they fed. The hind legs had heavy, five-toed feet, while the front legs were relatively short. They had five fingers and, a characteristic of the group, a big claw on the thumb. This may have been used as a weapon or to help them dig up roots or tear down branches. When the prosauropod walked on all fours, a specialized joint held the thumb claw up and out of the way.

So distinctive are the feet that the footprints in Triassic and early Jurassic sandstone can be readily identified as those of prosauropods. A set of footprints of the ichnogenus *Navajopus* has been linked to the prosauropod *Ammosaurus* – one of the few instances when this can be done.

Right: A prosauropod hand and clawed thumb.

Features: The general understanding of prosauropods is derived from the many remains of *Plateosaurus*. However, new analyses are constantly being made of the material. The traditional view is of an animal walking in a digitigrade stance – on the tips of its toes like a bird. More recent studies, however, suggest that it would have spent most time in a flat-footed plantigrade stance with the weight on the whole of the foot, like a bear.

MORE PROSAUROPODS

As the Triassic period progressed, the prosauropods evolved from small and medium-sized animals into big beasts that took on the appearance of the later sauropods with their four-footed stance, massive bodies, long necks and tiny heads. Most of the big ones fall into the melanorosaurid group of prosauropods.

Lessemsaurus

Similar prosauropods lived in South Africa and South America. In late Triassic times the Atlantic Ocean did not exist, and South America and Africa were part of one continent. Since environmental and climatic conditions were continuous across the two, it is not surprising that the remains of similar animals are found on both continents. Prosauropods were the major plant-eating animals there.

Features: The only thing known of *Lessemsaurus* is the partially articulated spinal column. This shows tall spines sticking upwards that would have produced a prominent ridge along the back. Perhaps this was some kind of signalling device and

Right: This prosauropod is named after Don Lessem, a prolific American writer who popularized dinosaurs.

was brightly coloured, or maybe it was something to do with temperature regulation. From the structure of the backbone it is assumed that the rest of the animal conformed to the typical prosauropod shape.

Distribution: Argentina.
Classification: Sauropodomorpha, Prosauropoda, Melanorosauridae.
Meaning of name: Lessem's lizard.
Named by: Bonaparte, 1999.
Time: Norian stage of the late Triassic.
Size: 9m (30ft).
Lifestyle: High browser.
Species: *L. sauropoides*.

Riojasaurus

The best known South American prosauropod is *Riojasaurus*. However, although the skeleton has been known from the 1960s, the skull has only recently been discovered. It was known that the neck was long and slender, and this suggested that the head was small like that of other prosauropods. The discovery of the skull confirmed this.

Features: *Riojasaurus* is a member of the melanorosaurid family – the big prosauropods that tended to walk on all fours. More than 20 skeletons at different stages of growth give us a good idea of what this animal looked like. The back legs are only slightly larger than the front legs, suggesting a quadrupedal stance. The limb bones are thick and solid, but the backbone is hollow. *Riojasaurus* probably weighed something in the region of one tonne.

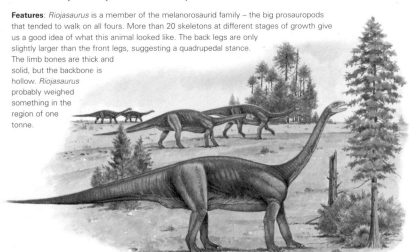

Distribution: Argentina.
Classification: Sauropodomorpha, Prosauropoda.
Meaning of name: Lizard from Rioja province.
Named by: Bonaparte, 1969.
Time: Norian stage of the late Triassic.
Size: 11m (36ft).
Lifestyle: High browser.
Species: *R. incertus*.

Left: Riojasaurus *was probably the heaviest land animal that existed before the evolution of the sauropods. All the skeletons have been found in the foothills of the Andes.*

Camelotia

There are not many differences between the big, advanced prosauropods and the more primitive of the sauropods, and those differences lie mostly in the arrangement of foot bones and the curvature of the leg bones. This can lead to confusion when only part of a skeleton is found. *Camelotia* is a dinosaur that seems to combine features of both groups. It was first found by Seeley in the 1890s and given the name *Avalonia*. There were some teeth associated with the skeleton that did not seem to fit

Features: The vertebrae, hip bones and hind leg bones – the only elements that are known for *Camelotia* – pose a bit of a mystery. The curve of the thigh bone and the areas of leg muscle attachment are similar to those of a sauropod, which has led to the suggestion that *Camelotia* is actually a primitive sauropod. However, the rest of the bones indicate that it is a member of the melanorosaurid prosauropods.

Distribution: England.
Classification: Sauropodomorpha, Prosauropoda, Melanosauridae.
Meaning of name: From Camelot, the legendary castle of King Arthur.
Named by: Galton, 1985.
Time: Rhaetian stage of the Triassic and Jurassic boundary.
Size: 9m (30ft).
Lifestyle: High browser.
Species: *C. borealis*.

with the accepted idea of prosauropod teeth. These have since been identified as the teeth of an ornithopod, called *Avalonianus*.

Left: The name *Avalonia was pre-occupied and Peter Galton renamed this dinosaur when he redescribed it a century after it was found and first named, separating the spurious ornithopod teeth from the rest of the skeleton.*

PROSAUROPOD EVOLUTION
The classification of the prosauropods, as with all other dinosaurs, is continually undergoing revision. The main primitive groups are the Plateosauridae and the Melanosauridae. The line leading to the sauropods branched off early, and the prosauropods themselves continued to evolve into the more advanced forms. The Massospondylidae then continued into the Jurassic.

Below: An advanced prosauropod above and a primitive prosauropod below.

Melanorosaurus

The big prosauropods from the southern continents, such as *Melanorosaurus* and *Euskelosaurus* from South Africa, and *Riojasaurus* from South America, were very similar in appearance, leading to the suggestion that they are all members of the same genus. Certainly they all belonged to the melanorosaurid family, and it would probably be difficult to tell apart in life.

Features: This was a large prosauropod that established the melanorosaurid family. It is the heaviest known, with massive front and hind legs, and was probably the biggest land animal at the time. Unlike some of its relatives, this prosauropod was unlikely to have been able to move on its hind legs. Its shape and lifestyle, using its neck to reach the branches of trees in which to browse, anticipated the big sauropods to come.

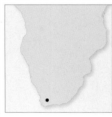

Distribution: South Africa.
Classification: Sauropodomorpha, Prosauropoda, Melanosauridae.
Meaning of name: Black Mountain lizard (from the site of discovery).
Named by: Haughton, 1924.
Time: Carnian to Norian stages of the late Triassic.
Size: 15m (50ft).
Lifestyle: High browser.
Species: *M. readi*, *M. thabanensis*.

Left: The huge size of Melanorosaurus *and its relatives may have been an early defensive adaptation. The sheer size of such an animal would have discouraged attacks by any of the smaller meat-eaters that existed at the time.*

HERBIVOROUS MISCELLANY

At the end of the Triassic period a totally new group of plant-eating dinosaurs, the ornithischians, which were distinguished from the saurischians by their bird-like hips, began to appear among the prosauropods and early sauropods. The most widespread ornithischians were the ornithopods, those with bird-like feet, which first appeared in Triassic times as small, two-footed running animals.

Pisanosaurus

Pisanosaurus is a dinosaur oddity. The ornithopods were thought to have evolved from animals like *Lesothosaurus*, a primitive group from the early Jurassic period in South Africa. They had not evolved the chewing mechanisms and the cheeks that are so distinctive of the group as a whole. Yet *Pisanosaurus* had these features, 25 million years before *Lesothosaurus* existed.

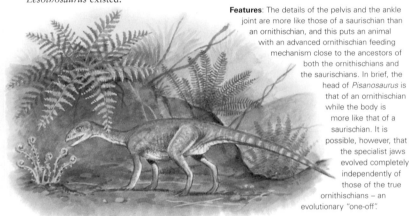

Features: The details of the pelvis and the ankle joint are more like those of a saurischian than an ornithischian, and this puts an animal with an advanced ornithischian feeding mechanism close to the ancestors of both the ornithischians and the saurischians. In brief, the head of *Pisanosaurus* is that of an ornithischian while the body is more like that of a saurischian. It is possible, however, that the specialist jaws evolved completely independently of those of the true ornithischians – an evolutionary "one-off".

Distribution: Argentina.
Classification: Ornithopoda (unproven).
Meaning of name: Pisano's lizard.
Named by: Casamiquela, 1976.
Time: Carnian stage of the late Triassic.
Size: 1m (3ft).
Lifestyle: Low browser.
Species: *P. merti*.

Technosaurus

The origin and evolution of the plant-eating ornithopod dinosaurs is unclear. The discovery of a single jaw bone and teeth in Triassic rocks of Texas, USA, suggests that they evolved quite early, but more evidence is needed before a worthwhile timetable can be established. *Technosaurus* would have foraged in the undergrowth of the coniferous forest that became Petrified Forest National Park, and was probably preyed upon by *Coelophysis* and its relatives.

Distribution: Texas, USA.
Classification: Ornithischia.
Meaning of name: Texas University lizard.
Named by: Chatterjee, 1984.
Time: Carnian stage of the late Triassic.
Size: 1m (3ft).
Lifestyle: Low browser.
Species: *T. smalli*.

Above: The species name T. smalli *is in honour of Bryan Small, a palaeontologist who has done much work in the southern states of the USA.*

Features: Palaeontologists only know of a single jawbone of this dinosaur, but it seems to pose the same kind of problems as *Pisanosaurus* – the presence of a sophisticated, plant-eating dinosaur long before the ornithischian dinosaurs are supposed to have evolved their advanced chewing mechanisms. The jawbone suggests that *Technosaurus* would have been a small, swift two-footed animal, walking on hind legs balanced by a heavy tail.

Antetonitrus

A specimen of the prosauropod *Euskelosaurus* lay on a shelf in the University of Witwatersrand after being unearthed near Bloemfontein, South Africa, in 1981. Then, in 2001, Australian palaeontologist Adam Yates looked closely at the bones and realized that they actually belonged to a sauropod – the earliest ever discovered. He and the original discoverer, James Kitching, re-examined the fossil and established the new genus.

Features: Though considered an early sauropod, *Antetonitrus* still has a prosauropod hand, with the ability to grasp with its fingers. However the hind legs, with the straight thigh bone and the short foot bones – adaptations to carrying heavy weights permanently on all fours – show that this is a primitive sauropod. It is the earliest true sauropod so far known. It was found in rocks only slightly older than those in which *Isanosaurus* (below) was found.

Distribution: South Africa.
Classification: Sauropodomorpha, Sauropoda.
Meaning of name: Before the thunder, referring to a later group of sauropods called the brontosaurs or 'thunder lizards'.
Named by: Yates and Kitching, 2003.
Time: Norian stage of the late Triassic.
Size: 10m (33ft).
Lifestyle: High browser.
Species: *A. ingenipes*.

Right: Although Antetonitrus *is small in relation to the later sauropods (weighing about two tonnes) it is still bigger than any land-living animal that exists today.*

DINOSAUR TEETH

Technosaurus is not the only dinosaur that is based on the fossils of puzzling teeth in the Triassic beds of North America. Carnian rocks reveal the teeth of other plant-eating dinosaurs.

There are primitive teeth that are curved and armed with coarse serrations, looking as if they have come from a meat-eating ornithischian – a dinosaurian absurdity (since ornithischians don't eat meat) but quite plausible considering the evolutionary types that were being 'tried out' in the early days. Some teeth have coarse serrations, which themselves have finer serrations, pointing to a complex feeding strategy.

The teeth of a dinosaur called *Pekinosaurus* have been found in New Jersey and New Mexico, USA, showing that these early plant-eaters were quite widespread in their range. *Tecovasaurus* is another dinosaur known from two kinds of teeth: one from the cheek and the other, longer and curved, from the front of the mouth. The teeth of *Tecovasaurus* are quite common, and useful as an index fossil providing proof that a particular geological bed is the same age as another with similar fossils.

All in all, the abundance and diversity of these teeth seem to suggest that ornithischians evolved in the early Carnian period, and had evolved rapidly into a number of different forms during late Carnian times. The ornithischians appear to have been more diverse in south-western North America than the prosauropods of the time. One day we may have the fossils of the rest of the skeletons. These may give us a better idea of the appearance of the animals that we now know from teeth.

Isanosaurus

For years it was suspected that the sauropods had a history that stretched back into the Triassic period. There has been footprint evidence to suggest this, and the complexity of sauropod types that suddenly appeared in the early Jurassic period indicated a long history. The discovery of *Isanosaurus* in Thailand, and then the discovery of *Antetonitrus* (above) in South Africa, prove that sauropods actually existed so early in history.

Features: *Isanosaurus* is known only from parts of a backbone, a shoulder blade, ribs and a femur. The femur is intermediate between the curved shape of a prosauropod femur and the straight sauropod femur, and the sites of muscle attachment are intermediate between the two. The tall backbones, however, mark this as a sauropod rather than a prosauropod. The lack of fusion between the bones of the backbone show that it was probably a youngster, only half-grown.

Distribution: Thailand.
Classification: Sauropodomorpha, Sauropoda.
Meaning of name: Lizard of Isan (in north-eastern Thailand).
Named by: Buffetaut, Suteethorn, Cuny, Tong, Le Loeuff, Khansubha and Jongauthariyakui, 2000.
Time: Norian and Rhaetian stages of the late Triassic to Jurassic boundary.
Size: 6m (18ft), but not fully grown.
Lifestyle: High browser.
Species: *I. attavipachi*, *I. russelli*.

Left: The species I. attavipachi *honours P. Attavipach, a former director of the Thai Department of Mineral Resources, a supporter of palaeontological research.*

MAMMAL-LIKE REPTILES

It was the preceding Permian period that had been the heyday of the synapsids, the mammal-like reptiles, and although they almost became extinct at the late Permian mass-extinction, enough of them survived and re-established the group in the Triassic. However, they were not quite as important as they had once been.

Lystrosaurus

The dicynodont lineage of mammal-like reptiles that developed towards the end of the Permian, continued into the Triassic. *Lystrosaurus* was a member of this group. Its dicynodont tusks and its general hippopotamus-like proportions have led to suggestions that this was a shallow-water amphibious animal, foraging in the mud at the bottom of lakes for roots.

Features: The front of the skull has a short and deep snout, sloping downwards and ending in a pair of tusks. The nostrils are immediately in front of the eyes, suggesting that it lived in water. An alternative interpretation is that the strangely-shaped skull held powerful biting muscles to tackle tough fibrous desert plants. The animals seem to have lived in areas of dry flood plain rather than swamp land.

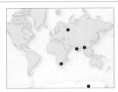

Distribution: South Africa; Antarctica; India; China; and Russia.
Classification: Dicynodontia, Lystrosauriidae.
Meaning of name: Spoon lizard.
Named by: Cope, 1870.
Time: Late Permian and early Triassic.
Size: 1.2m (4ft).
Lifestyle: Low browser.
Species: *L.murrayi, L. hedini, L. curvatus, L. weidenrichi, L. shichanggouensis, L. georgi, L. platycopes, L. declivis, L. mccaigi, L. rajurkari, L. robustus, L. oviceps.*

Left: Lystrosaurus fossils have been found on all the southern continents suggesting that the land was one single landmass.

Kannemeyeria

This huge-headed, ox-sized animal is typical of the kannemeyeriids, a group of cow-sized animals whose fossils are found throughout the world. This was the last of the dicynodonts – those with a pair of dog-like teeth – to exist in South Africa but a similar North American genus *Placerias* is known from the late Triassic beds of Petrified Forest National Park in Arizona, USA.

Features: The massive body is supported by stout legs and limb girdles formed by huge plates of bone. The skull is big, but made quite light by the large openings in it for the eyes, the nostrils and to accommodate the strong jaw muscles. A strong beak at the front is flanked by the two canine teeth, but the backs of the jaws are toothless and must have worked with a scissor action.

Distribution: South Africa; South America; India; Australia (possibly).
Classification: Dicynodontia, Kannemayeriidae.
Meaning of name: From Kanne.
Named by: Pearson, 1924.
Time: Olenekian to Anisian stages of the early and middle Triassic
Size: 3m (10ft).
Lifestyle: Rooter or low browser.
Species: *K. jenseni, K. erithrea, K. simocephalus.*

Left: The kannemeyeriids were the cows of the early Triassic.

Ericiolacerta

Distribution: South Africa; Antarctica.
Classification: Theriodontia, Ericiolacertidae.
Meaning of name: Eric's little lizard.
Named by: Watson, 1931.
Time: Induan stage of the early Triassic.
Size: 20cm (8in).
Lifestyle: Insect-hunter.
Species: E. parva.

The therocephalians – reptiles with mammal-like heads – were quite abundant in Permian times, but only a few made it through to the Triassic. *Ericiolacerta* was one of those. It is possible that they gave rise to the cynodonts, the only mammal-like reptile group to exist into late Triassic times. It was these that gave rise to the mammals.

Features: Little holes in the snout area of the skull suggest that perhaps the snout was extended as a small trunk or had highly developed sense organs, like whiskers. There is a palate in the roof of the mouth, separating the breathing passage from the eating area. This suggests an efficient eating mechanism that may indicate a warm-blooded lifestyle, at odds with its otherwise lizard-like appearance.

Left: An active little hunter, Ericiolacerta would have lived like a modern shrew.

CYNODONTS

With the arrival of the cynodonts (the dog-toothed reptiles), the mammal-like reptiles were becoming even more mammal-like. We can make certain assumptions about their appearance and lifestyle, going beyond the structures that are physically fossilized.

How do we know that they were covered in fur? Some of the more advanced forms have tiny pits in the bones of the snout and the lower jaw. These are identified as the attachment of vibrissae, the sensitive whiskers that we see on a typical mammal. As vibrissae are specialized hairs, there must have been generalized hairs present from which to evolve, and so we can deduce that these animals had hair, and hence a furry coat.

What makes us think that they may have suckled their young like mammals? When we find the young of these creatures they usually lack the teeth at the front. This could well be an adaptation to suckling, as any front teeth would just get in the way when sucking on a nipple.

Below: The most noticeable thing about the skull of a cynodont are the teeth. At the front are nipping incisor teeth, at the side are killing canine teeth, and at the back are meat-shearing molars – just as we find in a dog.

Thrinaxodon

The cynodonts (dog-toothed) as distinct from the dicynodonts (two-dog-toothed) were the mammal-like reptiles that were important in the late Triassic. They evolved in Permian times, but flourished in the late Triassic and gave rise to the mammals themselves. *Thrinaxodon* is a well-known member of the group. It probably lived in families in burrows, and scuttled after its prey on its short, but agile, legs.

Features: The mammal-like skeleton shows *Thrinaxodon* to have been a fast runner. Its mammal-like features include the division of the body into a chest encased in ribs, and a lower back where the backbone supported no ribs. The teeth of the lower jaw are set into a single bone which has grown at the expense of the other jaw bones. This makes the jaw strong. In the later mammals the other bones were to become part of the ear structure.

Distribution: South Africa; Antarctica.
Classification: Theriodontia, Cynodontia.
Meaning of name: Palm tree tooth.
Named by: Seeley, 1894
Time: Induan stage of the early Triassic.
Size: 50cm (20in).
Lifestyle: Hunter.
Species: T. liorhunus.

Below: Thrinaxodon was about the size of a badger, and probably looked rather like one too.

TOWARDS MAMMALS

The Triassic plains of the supercontinent of Pangaea supported a number of animals that would have looked like mammals. The members of the cynodont group, the dog-toothed reptiles, were the most mammal-like of the mammal-like reptiles. However, towards the end of the period the earliest representatives of the mammals lived alongside them.

Cynognathus

Cynognathus had running legs held beneath the body; teeth that divided into incisors at the front, canines at the side and molars at the back; a whiskered snout, covered in fur; and it possibly gave birth to live young... so why is it a reptile rather than a mammal? The distinctive features lie in the rear of the jaws and skull.

Features: *Cynognathus* has a jaw that consists of several bones (although only one is really important), whereas a mammal's jaw is a single bone with the rest accommodated into the structure of the ear. The eye socket and the synapsid gap behind are still two separate openings in *Cynognathus*, whereas in a mammal they have merged into a single space.

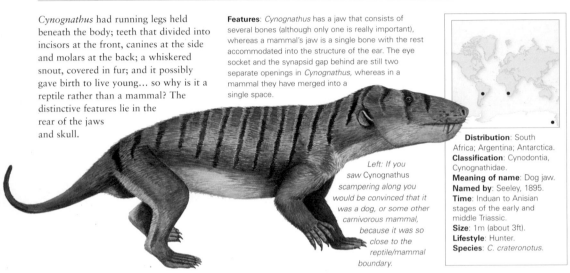

Left: If you saw Cynognathus *scampering along you would be convinced that it was a dog, or some other carnivorous mammal, because it was so close to the reptile/mammal boundary.*

Distribution: South Africa; Argentina; Antarctica.
Classification: Cynodontia, Cynognathidae.
Meaning of name: Dog jaw.
Named by: Seeley, 1895.
Time: Induan to Anisian stages of the early and middle Triassic.
Size: 1m (about 3ft).
Lifestyle: Hunter.
Species: *C. crateronotus*.

Massetognathus

The cynodonts are regarded as being mostly meat-eaters. However, the traversodontid subgroup – those that are defined by having much broader hind teeth than the other cynodonts – consists of plant-eating types. They had broad jaws, suggesting that they may have had cheeks to help with chewing tough plants. As the Triassic continued they became smaller, ending up no bigger than the earliest mammals that replaced them. Several species of *Massetognathus* were first identified but these were found to be different growth stages of the same species.

Features: The younger individuals have larger brain cases and smaller snouts, as is usual in animals. The important observation is that the number of teeth increases as the animal becomes older, and that the teeth in the older individuals are proportionally smaller. This is a mammal-like feature. The rear teeth are broad for grinding plant material. There is a gap between them and the nipping front teeth, to hold the food while it is being processed.

Right:
Massetognathus
was one of the few
herbivorous cynodonts.

Distribution: Argentina.
Classification: Cynodontia, Traversodontidae.
Meaning of name: Chewing muscle jaw.
Named by: Romer, 1967.
Time: Ladinan stage of the middle Triassic.
Size: 50cm (20in).
Lifestyle: Low browser and forager.
Species: *M. ochagaviae*, *M. pascuali*.

Chiniquodon

Distribution: Brazil;
Argentina.
Classification: Cynodontia,
Chiniquodontoidea.
Meaning of name:
Chiniquá tooth.
Named by von Huene, 1936
(but originally Belesdodon).
Time: Ladinian to the Carnian
stages of the middle and
late Triassic.
Size: 60cm (2ft).
Lifestyle: Hunter.
Species: C. magnificus,
C. theotonicus.
C. sanjuanensis.

Belesodon and *Probelesodon* are names that have been attached to this animal, but there is a distinct possibility that they are all the same beast. This indicates the ongoing flux in the study of mammal-like reptiles. The chiniquodonts are a group of medium-sized carnivorous cynodonts from the middle Triassic of South America.

Features: In appearance and lifestyle *Chiniquodon* has been compared to a modern jaguar. The head, however, is very much longer than that of any cat, and longer even than that of the other cynodonts. From the point of view of classification its important features lie in the number of teeth that it has – ten rear teeth in each jaw – which is more than its relatives have, and the greater breadth of the skull bones.

Above: Coprolites or fossil dung, have been found in late Triassic rocks of Argentina. These contain bits of undigested bone. They would have been deposited either by Chiniquodon *or one of the early dinosaurs.*

OTHER EARLY MAMMALS

The classification of the earliest mammals is unclear. Most of the science is based on finds of teeth, which are harder and more easily fossilized than bone. Also the earliest mammals were tiny and their delicate bones were not given to easy fossilization. Nevertheless several Triassic mammals have been identified.

Adelobasileus – From the Carnian stage of the late Triassic of Texas, USA. It is only known from the rear portion of a skull.

Gondwanadon – From the Carnian stage of the late Triassic of India. It is the oldest-known mammal in India.

Tikitherium – From the Carnian stage of the late Triassic of India. It is known from a single tooth that has quite advanced features.

Brachyzostrodon – From the Norian to Rhaetian stages of the late Triassic of France. The identification is based on lower molar teeth.

Erythrotherium – Possibly late Triassic or early Jurassic of Lesotho. It has teeth like those of *Eozostrodon*.

Halluatherium and **Helvetiodon** – Both from the Rhaetian stage of the late Triassic of Switzerland.

Haramaiya – From the late Triassic and early Jurassic of southern England. The teeth show it was a mouse-like plant-eater.

Holwelloconodon – From the Rhaetian stage of the late Triassic of southern England.

Rostrodon – From the Rhaetian stage of the late Triassic to the early Jurassic of Lufeng in China. Possibly a misprint for *Eozostrodon* while translating from a Chinese paper.

Eozostrodon

Morganucodon and *Eozostrodon* are little mammals from the Triassic of the UK. They may be the same animal; the *Morganucodon* specimen is much more complete. However, *Eozostrodon*, having been scientifically described first, has the accepted name for both. It is one of the Triassic animals found in the cavities of Carboniferous limestones in southern England and Wales that formed upland caves in Triassic times, and trapped little animals in them.

Features: The body of *Eozostrodon* or *Morganucodon* is tiny and shrew-like. It has short legs, a long narrow snout, sharp nipping teeth at the front and grinding molars at the back. The molars have three cusps to their grinding surface – hence the family name of Triconodonta. Its large eyes suggest that it was active at night. It would have been warm-blooded, but it is not known whether it bore its young alive or laid eggs.

Distribution: Southern
England and Wales.
Classification: Triconodonta,
Morganucodontiade.
Meaning of name: Early
girdle tooth.
Named by Parrington, 1941
Time: Rhaetian stage of the
late Triassic.
Size: 12cm (5in).
Lifestyle: Insectivore.
Species: E. parvus.

Left: Eozostrodon, or Morganucodon, is the earliest mammal for which we can make a reliable restoration.

THE
JURASSIC
PERIOD

During the Jurassic period the supercontinent of Pangaea began to break up. Rifts, similar to the Great Rift Valley of modern Africa, split across the landmass, and filled with seawater to form embryo oceans. Shallow seas spread across the continental margins and shelves. The result was moist equable climates, and the inland areas became habitable.

The plant-eating dinosaurs fed on conifer trees and in turn were hunted by fierce meat-eating dinosaurs. In the sky the pterosaurs flew, accompanied by the birds, which were just beginning to evolve. In the shallow shelf seas the unrelated sea reptiles fed on fish and invertebrates that inhabited the warm waters.

The Jurassic period is split by scientists into three distinct time bands. These are referred to as early Jurassic, middle Jurassic and late Jurassic.

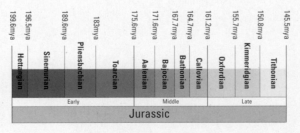

Above: The timeline shows the different chronological stages of the Jurassic period of Earth's history. (mya = million years ago)

1 *Angustineriptus.*
2 *Mamenchisaurus.*
3 *Omeisaurus.*
4 *Tuojiangosaurus.*
5 *Xiasaurus.*

ICHTHYOSAURS

The wide variety of forms of whale-like or eel-like ichthyosaur that existed in Triassic times were whittled down at the beginning of the Jurassic to a more dolphin-like shape. Most Jurassic ichthyosaur families thrived in the early part of the period, with a few important genera remaining at the end. Only one, Platypterigius, *is known to have survived into Cretaceous times.*

Temnodontosaurus

The evolutionary trend among the Jurassic ichthyosaurs was towards a more compact, fish-shaped or dolphin-shaped type. However, the tradition of the whale-sized elongated beasts of the Triassic continued in the various species of *Temnodontosaurus*. The stomach contents show that they dined mainly on squid and the shelled cephalopods of the shallow continental seas.

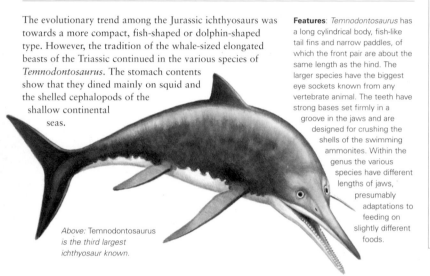

Features: *Temnodontosaurus* has a long cylindrical body, fish-like tail fins and narrow paddles, of which the front pair are about the same length as the hind. The larger species have the biggest eye sockets known from any vertebrate animal. The teeth have strong bases set firmly in a groove in the jaws and are designed for crushing the shells of the swimming ammonites. Within the genus the various species have different lengths of jaws, presumably adaptations to feeding on slightly different foods.

Above: Temnodontosaurus is the third largest ichthyosaur known.

Distribution: Southern England to Germany.
Classification: Ichthyosauria, Leptopterygiidae.
Meaning of name: Cutting toothed lizard.
Named by: Conybeare, 1882.
Time: Early Jurassic.
Size: 10m (35ft).
Lifestyle: Marine hunter.
Species: *T. platydon, T. eurycephalus, T. longirostris, T. trigonodon.*

Ichthyosaurus

This is the creature that we think of when we hear the term ichthyosaur. Its dolphin-like appearance is familiar from all kinds of popular images. Its shape was a perfect adaptation to swift hunting in open waters. Propulsion was by powerful thrusts of the tail and the animal was stabilized by the paddle limbs and the dorsal fin.

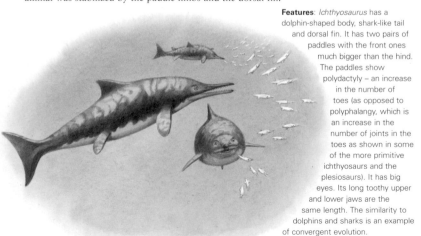

Features: *Ichthyosaurus* has a dolphin-shaped body, shark-like tail and dorsal fin. It has two pairs of paddles with the front ones much bigger than the hind. The paddles show polydactyly – an increase in the number of toes (as opposed to polyphalangy, which is an increase in the number of joints in the toes as shown in some of the more primitive ichthyosaurs and the plesiosaurs). It has big eyes. Its long toothy upper and lower jaws are the same length. The similarity to dolphins and sharks is an example of convergent evolution.

Distribution: Southern England.
Classification: Ichthyosauria, Ichthyosauridae.
Meaning of name: Fish lizard.
Named by: Conybeare, 1821.
Time: Sinemurian stage of the early Jurassic.
Size: 2m (6ft).
Lifestyle: Ocean hunter.
Species: *I. communis, I. intermedius, I. conybeari, I. breviceps.*

Excalibosaurus

A group of ichthyosaurs called the eurhinosaurids were the veritable swordfish of their day. The very long upper jaw, about four times the length of the lower, may have been used as a weapon, or it may have meant that, with the resulting downward-directed mouth, the animal foraged on the seabed probing in the sand for likely food.

Right: A specimen of Excalibosaurus *was found and described in 2003. Rather than being a different species, the specimen found is probably an adult version of the already known species, and it is this find on which the 7m (23ft) length is based.*

Features: The feature that distinguishes the members of the Eurhinosauridae is, of course, the remarkable upper jaw. Where it projected beyond the mouth the teeth that were present extended sideways, rather like the teeth of a sawfish. *Excalibosaurus*, at 7m (23ft) long, is the giant of the group. Most others, including *Eurhinosaurus* itself – a similar animal after which the group is named – usually reached 3–4m (10–13ft). The paddles are long and narrow, suggesting manoeuvrability rather than fast swimming.

Distribution: Southern England.
Classification: Ichthyosauria, Eurhinosauridae.
Meaning of name: Excalibur lizard.
Named by: McGowan, 1986.
Time: Sinemurian stage of the early Jurassic.
Size: 7m (23ft).
Lifestyle: Ocean hunter or scavenger.
Species: *E. costini*.

ICHTHYOSAUR FOSSILS

The ichthyosaurs were familiar to nineteenth-century naturalists long before dinosaurs were. This is understandable since sea-going animals were more likely to have their remains preserved as fossils. Their dead bodies would have sunk to the sea bed where sediments were gathering and would eventually, through a process of lithification, become sedimentary rock.

At first ichthyosaurs were thought of as some kind of ancient crocodile, and their remains were avidly collected in the quarries and coastal cliffs of southern England. The first professional woman fossil collector, Mary Anning, made important finds.

Complete skeletons all seemed to have had a kink in the tail, with the end of the tail bent downwards. This was thought to have been some kind of damage, and the first restorations of ichthyosaurs showed straightened crocodile-like tails. Full-sized statues of these animals built in the 1850s and still to be seen in the grounds of the Crystal Palace in south London, England, show this straightened tail.

In the 1880s ichthyosaur fossils were found in the slate quarries at Holzmaden in Germany. These were so well preserved that the soft tissues were still present as a thin film of carbon surrounding the skeletons. It was now obvious that the bent tail supported a fluke, like the tail of a shark but the other way up – a shark's tail has the backbone stiffening the upper fluke. A shark-like dorsal fin was also evident for the first time. Some of these skeletons also showed females in the act of giving birth, the first indication that ichthyosaurs were viviparous.

Ophthalmosaurus

The big eyes, up to 10cm (4in) in diameter, suggest that *Ophthalmosaurus* was a deep-water or night-time feeder. Its toothless jaws may indicate that it fed on soft-bodied prey such as squid. It has been estimated that *Ophthalmosaurus* could dive to depths of about 500m (1,640ft). The disc-like vertebrae made a fairly rigid body shape.

Features: This is the most streamlined of the ichthyosaurs, with a body that was almost teardrop-shaped. The tail fin is big and half-moon-shaped, with muscles for fast propulsion. The eyes fill almost the whole of the side of the head and each contains a sclerotic ring (a ring of bone) to stop it from collapsing under pressure. The front paddles are bigger than the hind, suggesting that these did most of the steering.

Below: Ophthalmosaurus *seems to have been prone to the "bends", the non-technical name for the injury that divers sustain when they surface too quickly. The condition is caused by the formation of nitrogen bubbles in the bloodstream and can lead to bone damage. Such damage has been found in fossils of* Ophthalmosaurus.

Distribution: Southern England; northern France; western North America; Canada; Argentina.
Classification: Ichthyosauria, Ichthyosauridae.
Meaning of name: Eye lizard.
Named by: Seely, 1874.
Time: Late Jurassic (but included here because of its importance).
Size: 3.5m (11ft).
Lifestyle: Fish hunter.
Species: *O. discus, O. icenius*.

EARLY JURASSIC PLESIOSAURS

The dawn of the Jurassic saw the dominance of two main lines of swimming reptiles, the ichthyosaurs and the plesiosaurs. Other swimming reptiles existed as well, such as marine crocodiles and turtles, but it was these two groups that were most abundant. Among the plesiosaurs there were two groups – the long-necked, small-headed plesiosauroids, and the short-necked, big-headed pliosauroids.

Macroplata

The plesiosaur *Macroplata* appears to have been quite a long-lived genus. The two species known span 15 million years at the beginning of the Jurassic, an unusual time span that has led some to believe that they are two different genera. The structure of the shoulder girdle indicates that the front paddles could produce a powerful forward stroke for fast swimming.

Features: As in all pliosaurs, *Macroplata* has a long head, in this case half the length of the neck. The neck vertebrae are numerous, about 29 of them, but short and compact, producing a rigid structure. The skull tapers to a needle-like snout with the front of the two lower jaws fused for most of their length, not just at the tip. The length of this symphysis is used by palaeontologists to determine the identity of pliosaurs.

Distribution: England.
Classification: Plesiosauria, Pliosauroidea.
Meaning of name: Big plate (referring to the broad shoulder bones).
Named by: Swinton, 1930.
Time: Hettangian to Toarcian stages of the early Jurassic.
Size: 5m (16ft).
Lifestyle: Swimming hunter.
Species: *M. tenuiceps*, *M. longirostirs*.

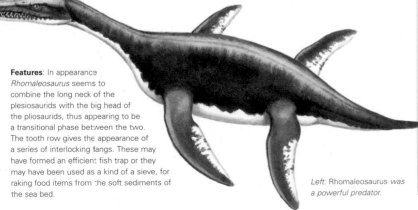

Above: Usually it is easy to tell the difference between a plesiosauroid and a pliosauroid – the former has a long neck while the latter has a big head. Macroplata appears to have possessed both.

Rhomaleosaurus

The nostrils of *Rhomaleosaurus* were not used for breathing (they were too small to allow for the passage of enough air), but for hunting prey. Water that passed into the open mouth while the animal was swimming was channelled out through the nostrils, which were situated on the top of the head. This water would have been tasted or smelled by *Rhomaleosaurus*, and used to analyse the surrounding environments.

Features: In appearance *Rhomaleosaurus* seems to combine the long neck of the plesiosaurids with the big head of the pliosaurids, thus appearing to be a transitional phase between the two. The tooth row gives the appearance of a series of interlocking fangs. These may have formed an efficient fish trap or they may have been used as a kind of a sieve, for raking food items from the soft sediments of the sea bed.

Distribution: Central England; and Germany.
Classification: Plesiosauria, Pliosauroidea.
Meaning of name: Strong lizard.
Named by: Stutchbury, 1846.
Time: Hettangian stage of the early Jurassic.
Size: 6m (20ft).
Lifestyle: Fish-hunter.
Species: *R. cramptoni*, *R. megacephalus*, *R. victor*, *R. zetlandicus*, *R. propinguus*.

Left: Rhomaleosaurus was a powerful predator.

Attenborosaurus

The original specimen of *Attenborosaurus*, in Bristol City Museum, England, was regarded as a species of *Plesiosaurus* (*P. conybeari*), but it was destroyed during World War II. A good plastercast survived for later study to be carried out, and to allow for its description as a new genus. Unfortunately the important evidence of the skin was lost.

Features: The original fossil of *Attenborosaurus* carried traces of the skin as a thin brownish film. This showed the skin to have been a continuous membrane with no sign of scales. However there were small oblong bones in the hip region that may have been dermal plates. The skull is quite large and pliosaur-like, but the neck is very long, and the pelvic girdle is quite primitive for a plesiosaur of this type. The snout is tapered. *Attenborosaurus* has fewer teeth than normally found in plesiosaurs, and these have massive crowns.

Distribution: England.
Classification: Plesiosauria, Plesiosauroidea.
Meaning of name: Attenborough's lizard (after Sir David Attenborough, the natural historian).
Named by: Bakker, 1993.
Time: Sinemurian stage of the early Jurassic.
Size: 5m (16ft).
Lifestyle: Fish hunter.
Species: *A. conybeari*.

Above: The genus was named after the naturalist and journalist Sir David Attenborough, whose childhood interests in plesiosaurs eventually led him to a distinguished career in natural history journalism.

SWIMMING ACTION OF THE PLESIOSAUR

The paddles of a plesiosaur are stiff and relatively inflexible. The bones consist of five toes, which are very elongated. Each toe has more than the usual number of bones or phalanges, up to about 20, which compares with the four or five normally found in a tetrapod. This condition is known as polyphalangy. All these bones are tightly lashed together with gristle to produce a powerful swimming organ.

A plesiosaur swam by means of a flying action, with an undulating body movement very much like that of a modern penguin or sea lion. The hydrodynamics suggest that in most plesiosaurs the rear paddles were used for power with the downward stroke producing the propulsive force. The front paddles were used mainly for stabilization and to alter the angle and the direction of swimming. There may have been a vertical diamond-shaped fin on the tail for steering, but palaeontologists are not all in agreement about this. The body was solid and compact, with the shoulder and pelvic girdles occupying about half the area of the undersurface. This formed a firm base for the swimming muscles.

Gastroliths, or stomach stones, have been found in association with plesiosaur skeletons. Swallowing stones helped the animal to adjust its buoyancy; this is found in contemporary sea lions and penguins, modern marine animals that swim by using a similar flying action.

"Plesiosaurus"

This was the first of these plesiosaurs to have been described, and hence has given its name to the whole group. It is also something of a wastebasket taxon in that all kinds of specimens have been referred to it. "A snake threaded through the body of a turtle", was the description given by Dean Buckland, one of the Victorian naturalists who first studied it.

Features: The skull is small and the neck very long, about half as long again as the body, and composed of about 40 cervical vertebrae. The paddles are long, each consisting of five toes but with many more joints than usual. The forelimbs are slightly longer than the hind, although in the youngsters this is reversed. The teeth are long and sharp, and interlock when they close to form a fish trap.

Distribution: England.
Classification: Plesiosauria, Plesiosauroidea.
Meaning of name: Almost a lizard.
Named by: de la Beche and Conybeare, 1821.
Time: Sinemurian and Toarcian stages of the early Jurassic.
Size: 3.5m (11ft).
Lifestyle: Swimming hunter.
Species: *P. dolichodeirus*, *P. guilelmiimperatoris*, *P. macrocephalus*, *P. megdeirus*, *P. winspitensis*, *P. eurymerus*.

JURASSIC CROCODILES

The crocodiles that represented one of the major groups of land-living tetrapods in the Triassic period, were now becoming less important. They were adopting the semi-aquatic freshwater lifestyles that we know of in today's representatives. However, the early and middle Jurassic threw up a few specialist forms, mostly adapted for a marine fish-hunting environment.

Metriorhynchus

The sea crocodile group represented by *Metriorhynchus* were probably ambush hunters. At rest they would have drifted just below the surface, at an angle of 45 degrees, with just the nostrils at the tip of the snout showing. They would have darted after prey with a sudden lunge of the flexible body and finned tail, as do most sharks today.

Features: The head is broad and flat, suggesting that it could strike to the side without much water resistance. The skull is quite lightweight and porous, moving the centre of gravity back and giving rise to the 45-degree resting position. There are no armour scutes that we would expect on the skin of a crocodile, and this appears to have increased the flexibility of the body. However, the paddle fins look as if they evolved for sustained swimming.

Distribution: England; France; Chile.
Classification: Crocodylomorpha, Thalattosuchia.
Meaning of name: Moderate snout.
Named by: Meyer, 1830.
Time: Middle to late Jurassic.
Size: 3m (10ft).
Lifestyle: Fish hunter.
Species: *M. moreli*, *M. superciliosus*.

Above: Metriorhynchus may have swum for long distances between feeding grounds but hunted as an ambush predator once it was there.

Geosaurus

Along with *Metriorhynchus*, *Geosaurus* had a tail fin. We know this because the backbone of the tip of the tail is bent downwards like that of the ichthyosaurs, as though it had been snapped. In *Geosaurus* this bending is accompanied by a lengthening of the spines above each vertebra, indicating the attachment of a broad fin that must have produced a powerful swimming organ. At the time that *Geosaurus* lived the seas were full of other swimming reptiles, particularly the long-necked and short-necked plesiosaurs and the fish-shaped *Ichthyosaurus*. Food in the sea was evidently varied enough to support all these groups.

Features: The skeleton of *Geosaurus* is very similar to that of *Metriorhynchus*, but is slimmer, and more eel-like in profile, with tiny front paddles. The jaws have a significantly smaller number of teeth. Otherwise, it is like its relative in having a tail fin above the tip of the backbone (a "hypocercal caudal fluke" in technical parlance) and a lack of armour to increase the flexibility of the body. Rings of bone – sclerotic rings – are present in the eyes of *Geosaurus*. These are an adaptation for withstanding pressure. The jaws contain relatively fewer teeth than the other marine crocodiles.

Below: The skull of one specimen of Geosaurus *shows bite marks, evidence of an attack that must have killed it.*

Distribution: England; Germany; Argentina; Mexico.
Classification: Crocodylomorpha, Thalattosuchia.
Meaning of name: Earth lizard.
Named by: Cuvier, 1842.
Time: Jurassic.
Size: 3m (10ft).
Lifestyle: Fish hunter.
Species: *G. vignaudi*, *G. gracilis*, *G. giganteus*, *G. araucanensis*, *G. suevicus*.

Teleosaurus

Distribution: Southern England.
Classification: Crocodylomorpha.
Meaning of name: Completed lizard.
Named by: Deslongchamps, 1866.
Time: Early Jurassic.
Size: 3m (10ft).
Lifestyle: Fish and cephalopod hunter.
Species: *T. hastifer*, *T. mandelslohi*, *T. chapmanni*.

The teleosaurids were a family of marine crocodiles that existed from early Jurassic to the middle Cretaceous times. They were less well adapted to permanent sea life than their cousins, the thalattosuchians, in lacking the tail fin, the flipper limbs, and in retaining the body armour of regular crocodiles. They were widespread, ranging across the shallow seas in the region of modern Europe and North America. A close relative, but much larger, was *Steneosaurus*. Remains have been found in Europe. Crocodile teeth are quite common in marine sediments of the time. This is because the teeth fell out and were replaced throughout the life of the crocodile, and the hard enamel coating means that they were easily preserved.

Features: The jaws of *Teleosaurus* and the other teleosaurids are long and very narrow, lined with interlocking teeth – an ideal arrangement for seizing slippery fish and cephalopods like belemnites. The shape of the vertebrae suggest that in life it had a very flexible body, capable of much more bending than in modern crocodiles. The body and tail are long and slender, and the front legs are only half the length of the hind. The presence of armour, both on top of and beneath the body, indicates that *Teleosaurus* was not a fast hunter. It must have hunted for food in areas avoided by more agile aquatic reptiles like ichthyosaurs or plesiosaurs with which they would have competed.

Below: The tiny front legs of Teleosaurus *would have been held against the body as the crocodile swam with a sinuous eel-like action.*

SWIMMING CROCODILES

Above: Dakosaurus.

Perhaps the most bizarre of the swimming crocodiles was *Dakosaurus*, which lived in the late Jurassic and early Cretaceous seas of South America. Its enormous teeth and massive jaws led to its nickname "Godzilla". Even in the most aquatic of the sea crocodiles, the hind limbs were not as fully converted into flippers as their fore feet were. In fact, there has been an observation that the hind feet of *Metriorhynchus* did not form paddles at all. The possibility is that these animals retained some kind of ability for moving on land in order to return to the beach now and again to lay their eggs.

Another general observation is that all the marine genera illustrated here have been found with stomach stones in their body cavities. They presumably swallowed stones to help to adjust their buoyancy and help them to swim and hunt at different layers of the ocean.

Calsoyasuchus

Crocodiles with a modern aspect established themselves in the middle of the Mesozoic. One such group consisted of the goniopholids, which resembled modern crocodiles except that their backbones were more primitive and they did not have the advanced structure to the roof of the mouth. They were very close to modern crocodiles but became extinct by the end of the Cretaceous. The discovery of *Calsoyasuchus*, in 2002, in early Jurassic deposits showed that this was a much older line than had previously been thought.

Features: In appearance *Calsoyasuchus* would have looked much like a modern crocodile, and the complex system of nasal cavities that it has is very similar to that in modern crocodile forms. Only the skull is known and the interior of the snout is better preserved than in any other crocodile specimen from the time, and so it is difficult to make comparisons with its contemporaries. Like a modern crocodile it would have lived as an ambush hunter in rivers and freshwater lakes.

Distribution: Arizona, USA.
Classification: Crocodylomorpha.
Meaning of name: Calsoyas' crocodile (after the chief of the local Navajo Nation who encouraged the research on their land).
Named by: Tykoski, Rowe, Ketcham and Colbert, 2002.
Time: Sinemurian or Pliensbachian stage of the early Jurassic.
Size: 2m (6½ft).
Lifestyle: Fish hunter or scavenger.
Species: *C. valliceps*.

Left: Calsoyasuchus *would have lurked in shallow waters waiting to ambush small dinosaurs when they came to drink.*

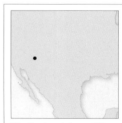

RHAMPHORHYNCHOID PTEROSAURS

Much of what we know about early Jurassic flying reptiles, or pterosaurs, comes from the remains found in southern England and Germany. In early Jurassic times much of Europe was covered in a shallow shelf sea, and the fine mud of the sea bed resulted in deposits of heavy shale. All these early pterosaurs with tails belong to the suborder Rhamphorhynchoidea, itself divided into a number of families.

Dimorphodon

Distribution: Dorset, England.
Classification: Pterosauria, Rhamphorhynchoidea.
Meaning of name: Two types of teeth.
Named by: Owen, 1859.
Time: Hettangian stage of the early Jurassic.
Size: 1.4m (4½ft) wingspan.
Lifestyle: Fish hunter.
Species: *D. macronyx.*

The first specimen of *Dimorphodon* was found by the famous fossil collector Mary Anning in the cliffs of Lyme Regis, in Dorset, in 1828. Only a few specimens have been found and these are all from the Dorset coast, although another possible specimen has been collected from the banks of the River Severn in Gloucestershire, England.

Features: The two types of teeth implied in the name consist in each jaw of a set of 30–40 tiny pointed teeth, with four large teeth at the front. They are set in a very deep, narrow skull, made up of flimsy struts of bone surrounding large cavities, keeping the body weight to a minimum. The legs are very long and powerful and the claws on both the feet and the hands are strong. This is seen as an adaptation to clinging to cliffs and rocks.

Below: As in all fossil animals, the coloration is speculative. It does seem possible that the deep face of such pterosaurs as Dimorphodon could have been brightly coloured for signalling, like modern deep-beaked birds.

Campylognathoides

This pterosaur is known from several skeletons, some complete, from the shale deposits of Holzmaden in Germany. One species, *C. indicus*, is known from India. The site at Holzmaden is famous for the complete ichthyosaur skeletons. This genus was originally named *Campylognathus* but the scientific description was not valid for technical reasons, and so it was redescribed and renamed later. The name refers to the slight upward hook to the front of the jaw.

Features: The head of *Campylognathoides* is large with a pointed snout. It has very large eye sockets, which suggests it had acute eyesight, and possibly a nocturnal habit. The hip bones of a pterosaur that may be *Campylognathoides* show that the legs were held sideways and upwards, suggesting an inability to walk on its hind legs while on the ground. It is not known if this applies to all others of the rhamphorhynchoid group.

Distribution: Germany and India.
Classification: Pterosauria, Rhamphorhynchoidea.
Meaning of name: Like the curved jaw (it was originally named *Campylognathus* or curved jaw).
Named by: Wild, 1971.
Time: Pliensbachian stage of the early Jurassic.
Size: 1.75m (5¾ft) wingspan.
Lifestyle: Fish-hunter.
Species: *C. liasicus, C. indicus, C. zitteli.*

Above: The quarries of Holzmaden in east Germany, where Campylognathoides was found, produced many fossils of early Jurassic animals, including the pterosaurs that hunted for fish in the shallow sea that then covered this area.

Dorygnathus

Some very fine skeletons of *Dorygnathus* have been uncovered from the slate quarries of Holzmaden. In the early 19th century it was originally classed as a species of *Pterodactylus*, but that was before a distinction was made between the long-tailed rhamphorhynchoids and the short-tailed pterodactyloids.

Features: The wings are relatively short for a rhamphorhynchoid. The foot has a long, narrow fifth toe, which may have supported a web of skin for use in manoeuvring, possibly compensating for the short wings. The very long front teeth intermesh when the jaw closes, making an ideal device for catching fish. As in all rhamphorhynchoids, the tail is long and stiff, the vertebrae lashed together by tendons.

Below: Dorygnathus *is similar to a new pterosaur,* Cacibupteryx, *which was found in Cuba in 2004, extending the range of rhamphorhynchoids.*

Distribution: Germany.
Classification: Pterosauria, Rhamphorhynchoidea.
Meaning of name: Spear jaw.
Named by: Wagner, 1860.
Time: Pliensbachian stage of the early Jurassic.
Size: 1m (3ft) wingspan.
Lifestyle: Fish-hunter.
Species: *D. banthensis. D. mistelgauensis.*

WING STRUCTURE

The rhamphorhynchoid wing (below top) differed from that of the pterodactyloid, principally in the structure of the supporting limb. The wrist bone was short, and so the weight of the wing was carried mostly on the elongated fourth finger. The bones of the finger were as thick as those of the arm. A rhamphorhynchoid wing was generally longer and narrower than that of a pterodactyloid.

In a pterodactyloid's wing (below bottom) the wrist bones were often as long as the humerus. This put the fingers of the hand further along the leading edge of the wing. In both there was an extra bone at the base of the wrist, which appears to have supported a membrane that stretched in front of the arm bones between the wrist and the base of the neck. This would have been used to adjust the flow of air over the wing and give some manoeuvrability.

The other main difference between the two pterosaur groups was the presence of a long tail in the rhamphorhynchoids.
This consisted of long vertebrae lashed together into a stiff rod by tendons and bearing a diamond-shaped fin at the end. It would have been used in flight for steering.

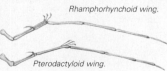

Rhamphorhynchoid wing.

Pterodactyloid wing.

Parapsicephalus

This pterosaur is known only from fragmentary skulls from Whitby, England, although some bones from Germany, found in 1970, may represent a more complete skeleton. It was a significant find. This is the first pterosaur skull that was discovered preserved in three dimensions allowing a study of the brain cavity. Since the original find, however, there have been better brain cavities found in specimens of *Rhamphorhynchus* and *Anhanguera*.

Features: *Parapsicephalus* appears to be a typical rhamphorhynchoid pterosaur, so typical that it was originally classed as a species of *Scaphognathus*, and even now is thought by some to be a specimen of *Dorygnathus*. The skull is longer than those usually found in the group. The brain is much smaller than that of a bird, but larger than that of any terrestrial reptile. The angle of the ear structure suggests that the head was held horizontally in flight.

Distribution: North-east England; possibly Germany.
Classification: Pterosauria, Rhamphorhynchoidea.
Meaning of name: Almost an arched head.
Named by: Newton, 1888.
Time: Toarcian stage of the early Jurassic.
Size: 1m (3ft) wingspan.
Lifestyle: Fish hunter.
Species: *P. purdoni, P. mistelgauensis.*

Left: The cast of the brain shows that the semicircular canals in the ear region are well developed. This implies that Parapsicephalus *had a good sense of balance while in flight.*

COELOPHYSID MEAT-EATERS

The coelophysids, the most extensive hunters of Triassic times, continued into the Jurassic period and were the main predators of the early part of the period before being replaced by all sorts of new theropods that were beginning to evolve. The coelophysids existed all over the world from North America to Africa and China.

Dilophosaurus

Familiar from its appearance in the *Jurassic Park* film, unfortunately *Dilophosaurus* was inaccurately portrayed, being much too small and having poison glands and an erectile neck frill, all from the film designer's imagination. Three skeletons were found together, suggesting that it may have hunted in packs. Another species was found in China, showing that the genus was quite widespread.

Distribution: Arizona, USA; China.
Classification: Theropoda, Coelophysoidea.
Meaning of name: Lizard with two crests.
Named by: Welles, 1954.
Time: Sinemurian to Pliensbachian stages of the early Jurassic.
Size: 6m (19½ ft).
Lifestyle: Hunter.
Species: *D. wetherilli, D. breedorum, D. sinensis.*

Features: The significant distinguishing feature of *Dilophosaurus* is the pair of semi-circular crests that runs the length of the skull. They are quite thin and were probably used for signalling or display. The articulated tip to the snout, as seen in other coelophysids, is present and very marked. It was probably used for poking small prey out of tight corners, but the dinosaur's size suggests that it would have hunted bigger animals, even though its jaws and teeth were quite weak.

Right: The crests of Dilophosaurus have never been found attached to the skull. However, the standard restoration seems the most convincing.

Segisaurus

Lacking hollow, bird-like bones *Segisaurus* was originally thought to have been different from other meat-eating dinosaurs of the time. This later proved not to have been the case, and it is currently classified among the coelophysids. Like the other coelophysids it was built for running and probably jumping, making it an active hunter. It would have been a desert-living animal with the ability to survive sand storms and droughts.

Features: The single, incomplete skeleton of *Segisaurus*, consisting of backbones, ribs, shoulder, pelvic girdles and limb bones, found in desert sandstone, unfortunately does not reveal much about the dinosaur. It seems to have a collar bone, which is unusual in such an early dinosaur, and its hands are long and slender. This suggests that the arms were strong and used for catching prey, probably the small prosauropods that lived in the deserts.

Distribution: Arizona, USA.
Classification: Theropoda, Coelophysoidea.
Meaning of name: Lizard from the Segi Canyon.
Named by: Camp, 1936.
Time: Toarcian stage of the early Jurassic.
Size: 1m (3ft).
Lifestyle: Hunter.
Species: *S. halli.*

Right: The lack of a skull and teeth means that we cannot be sure about the diet of Segisaurus. However, its fast-runner build and clawed hands suggest that it was a meat-eater

Podokesaurus

The only remains of this dinosaur were destroyed in a museum fire in 1917, so all recent work has been carried out on casts. The Connecticut valley, USA, where it was found, is famed as the site of extensive dinosaur tracks found in the early nineteenth century. The fossil footprints named *Grallator* may have been formed by *Podokesaurus*.

Features: In its build *Podokesaurus* is very similar to the smaller specimens of the late Triassic *Coelophysis*, from the other side of the North American continent, with its slim body, long hind legs, long tail and long, flexible neck. In fact the two are so similar that many palaeontologists think that they are the same genus. Unfortunately, since the original specimen was destroyed, this can never be proved because the surviving casts are too poor to give much information. The specimen appears to have been a juvenile, incompletely grown, and so the small size quoted could be an underestimation.

Distribution: Massachusetts, USA.
Classification: Theropoda, Coelophysoidea.
Meaning of name: Swift foot lizard.
Named by: Talbot, 1911.
Time: Pliensbachian to Toarcian stages of the early Jurassic.
Size: 1m (3ft).
Lifestyle: Hunter.
Species: *P. holyokensis*.

Right: Podokesaurus is the earliest dinosaur to have been found in the eastern United States. It was discovered in 1910 by local fossil hunter Dr Mignon Talbot, nicknamed the "dinosaur lady."

Syntarsus

More than 30 *Syntarsus* skeletons were found in a bonebed in Zimbabwe, suggesting that a pack was overwhelmed by a disaster such as a flash flood. The cololites, or stomach contents, suggest that they preyed on smaller vertebrates. Another species of *Syntarsus*, found in Arizona, has a pair of crests on the head, rather like those of *Dilophosaurus* but smaller.

Features: The close similarity between *Syntarsus* and *Coelophysis* has led many to suggest that the two are actually the same genus, with three nimble fingers, a long neck and tail, strong hind legs and a slim body. There seem to have been two sizes of adult – the larger was probably the female and the smaller the male, judging by the size ranges in modern bird flocks. Computer reconstructions of the braincase of *Syntarsus* show it to be quite large compared with that of the earlier herrerasaurids, indicating an increase in intelligence. A bird-like cunning seems to have been evolving at this time.

Right: There is a possibility that the name Syntarsus is pre-occupied by an insect. If this proves to be the case the generic name will be changed to Megapnosaurus, and the two known species will be M. rhodesiensis and M. kayentakatae.

Distribution: Arizona, USA; Zimbabwe.
Classification: Theropoda, Coelophysoidea.
Meaning of name: Fused ankle.
Named by: Raath, 1969.
Time: Hettangian to Pliensbachian stages of the early Jurassic.
Size: 2m (6½ft).
Lifestyle: Hunter.
Species: *S. rhodesiensis*, *S. kayentakatae*.

DIVERSIFYING MEAT-EATERS

In early Jurassic times life began to diversify. Different plant-eating animals began to appear in different areas and, as a result, different meat-eating dinosaurs evolved to take advantage of this new variety of food. The coelophysids, who had been the main meat-eaters up to this point, were beginning to be replaced by a variety of other carnivorous hunter families.

Sarcosaurus

Early Jurassic Britain consisted of a scattering of low islands across a shallow sea on the northern edge of the supercontinent. The islands were wooded and would have supported plenty of wildlife. *Sarcosaurus* was the big hunter of the time, and its remains are found in the marine lias beds – beds of alternating shale and limestone – having been washed out to sea.

Right: Knowledge of another species, S. andrewsi, *also from the early Jurassic period in England, is based on an isolated femur. That now seems to have belonged to a coelophysid.*

Features: All we know of this dinosaur is a partial pelvis, a femur and some vertebrae. It is a lightly built, two-footed predator. The state of the bones shows it to have been an adult. Its pelvis is remarkably similar to the later *Ceratosaurus*, and some palaeontologists think it is actually an early species of that genus. Officially, though, it is classed in the neoceratosaurids, a group placed between the coelurosaurids and the ceratosaurids on the evolutionary chart. The bones show similarities with *Ceratosaurus* and also with *Dilophosaurus* and *Liliensternus*.

Distribution: Leicestershire, England.
Classification: Theropoda, Neoceratosauria.
Meaning of name: Flesh-eating lizard.
Named by: Andrews, 1921.
Time: Sinemurian stage of the early Jurassic.
Size: 3.5m (12ft).
Lifestyle: Hunter.
Species: *S. woodi.*

"Saltriosaurus"

Saltriosaurus was found in a limestone quarry in northern Italy, near the Swiss border, in 1996. At the time of writing it has not been scientifically described, and so its name is unofficial. It is important, though, because preliminary studies show it to have been the earliest known member of the Tetanurae, a name that means "stiffened tail", the group to which all the advanced theropods belong.

Right: As well as the way the tail vertebrae are bound together, the Tetanurae are characterized by the presence of no more than three fingers, and an extra opening in the skull.

Features: Just one example of *Saltriosaurus* has been found. It consists of about 100 bone fragments, which is about 10 per cent of the skeleton. It seems to have had a long neck and would have weighed more than one tonne. The fact that it is a hunter was proved by the presence of a sharp cutting tooth which is 7cm (2¾in) long. The bones include a wishbone and three-fingered hands, both features of hunting dinosaurs that came in later times. Although it was found in marine limestone, its sheer size suggests that there must have been big landmasses close by.

Distribution: Northern Italy.
Classification: Theropoda, Tetanurae.
Meaning of name: Lizard from Saltrio.
Named by: Dal Sasso, 2000 (but this is unofficial).
Time: Sinemurian stage of the early Jurassic.
Size: 8m (26ft).
Lifestyle: Hunter.
Species: None allocated.

Cryolophosaurus

Distribution: Antarctica.
Classification: Tetanurae,
Carnisauria.
Meaning of name: Frozen
crested lizard.
Named by: Hammer and
Hickerson 1994.
Time: Pliensbachian stage of
the early Jurassic.
Size: 6m (19½ft).
Lifestyle: Hunter.
Species: *C. elliotti*.

The name *Cryolophosaurus*, meaning "frozen crested lizard", is a bit of a misnomer. Although its skeleton was found in the Antarctic, this was not a frozen continent in early Jurassic times. The area was much further north, and on the edge of the southern supercontinent of Gondwana – one of the two landmasses formed as the ancient Pangaea split in two. The environment may not have been tropical, but it was certainly much more temperate than today. Other remains in the area consist of prosauropods, pterosaurs and smaller meat-eaters, and there seems to have been quite a breadth of animal life.

The remoteness of the site means that very little field work has been done there, but an expedition in 2004, led by William Hammer of Augusta College, Georgia, USA, who found the original specimen, has yielded even more bones.

Features: *Cryolophosaurus* is a big meat-eating dinosaur with a peculiar crest on its head. In most crested theropods the crest runs fore-and-aft. In *Cryolophosaurus* it runs crossways, turning up in a distinctive quiff that earned the animal the unofficial name "Elvisaurus" before it was formally described. This crest must have been used for display. The skull is tall and narrow, and the lower jaw deep and strong.

DINOSAUR DIVERSIFICATION
The sudden expansion of different types of meat-eating dinosaur in the early Jurassic period must have mirrored the expansive evolution of different types of plant-eating animals. The more types of prey there are, the more kinds of predator evolve to hunt them.

A reason for this expansion may be found in the geography of the time. The single great landmass of Pangaea, which had existed through most of Permian and Triassic times, was beginning to split up. Two main supercontinents were beginning to appear – Laurasia in the north and Gondwana in the south. It would be a long time before they were totally separate, but the cracks were beginning to open.

The rift valleys that formed along the cracks produced arms of the ocean that penetrated deep into the landmasses. The Red Sea, in Egypt, today is an example. At the same time a rising sea level spread shallow seas over much of the continents, spreading moister climates into the interiors, where there had formerly been desert. Lush vegetation began to grow, and this encouraged the development of widespread herbivorous animals, and subsequently carnivorous forms.

The splitting of the landmass was ultimately leading to the formation of new continents, and the spreading of seas on to the continental shelves resulted in the formation of many strings of islands across shallow areas of sea. Isolation is always a spur to evolution, with new species appearing on remote islands, separate from, and different from, living things elsewhere.

THE LATE PROSAUROPODS

Although the sauropods made their presence felt in the early Jurassic period, the more primitive prosauropods continued to survive in many parts of the world. Their remains have been found in early Jurassic rocks of North America and southern Africa, but China seems to be the last bastion of this group, with several prosauropod genera known from the depths of the Asian continent.

Anchisaurus

Though discovered in 1818 *Anchisaurus* was not recognized as a dinosaur until 1885. The two known species may actually represent the male and female versions of the same, with the larger being the female. It was originally thought to have been a Triassic contemporary of *Plateosaurus*, but in the 1970s the New England sandstones in which the fossil was found were proved to be Jurassic in age.

Features: If *Plateosaurus* is the classic example of the biggest medium-size prosauropod, then *Anchisaurus* is the best-known of the smallest medium-size prosauropods. It has the long neck and tail, and a slim body, with long hind legs that allowed it to walk on all fours or with its forelimbs off the ground. The teeth are bigger and the jaw mechanism stronger than those of *Plateosaurus*, suggesting that it ate tougher food.

Left: Between its discovery in 1818 and its proper identification in 1885, this animal lived under a number of names, including Megadactylus and Amphisaurus, both of which were pre-occupied.

Distribution: Connecticut and Massachusetts, USA.
Classification: Sauropodomorpha, Prosauropoda.
Meaning of name: Near lizard.
Named by: Marsh, 1885.
Time: Pliensbachian to Toarcian stages of the Jurassic.
Size: 2.5m (8ft).
Lifestyle: Low browser.
Species: *A. major,*
A. polyzelus.

Jingshanosaurus

Distribution: China.
Classification: Reptile, Sauropodomorpha.
Meaning of name: Lizard from Golden Hill.
Named by: Zhang and Young, 1995.

Time: Early Jurassic.
Size: 9.8m (32ft).
Lifestyle: High browser.
Species: *J. dinwaensis.*

This is one of the best-known of the Chinese prosauropods, and is known from a complete skeleton. It is named after the town of Jingshan, or Golden Hill, in Lufeng province, China, close to where the skeleton was found. This town is the site of the Museum of Lufeng Dinosaurs, one of several big dinosaur museums in China.

Features: The complete skeleton shows *Jingshanosaurus* to be a typical large prosauropod, with a heavy body, long neck and tail, and legs that would have allowed a bipedal or quadrupedal mode of travel. It has the characteristic big sauropod claw on the thumb. The skeleton looks like a particularly large version of *Yunnanosaurus*: it may well be the same animal.

Left: Although Jingshanosaurus was a typical prosauropod with the typical prosauropod dentition and presumably vegetarian diet, its describers have suggested that it may have eaten molluscs as well.

Yunnanosaurus

This dinosaur is a well-known Chinese prosauropod. There have been about 20 skeletons found, two of which have skulls. One specimen has more than 60 teeth preserved, giving an insight into the feeding mechanisms of the big prosauropods. However, there is a suggestion that the teeth actually come from a different animal, a sauropod.

Features: The distinctive feature of *Yunnanosaurus* is the teeth. They are very similar to those of one of the later sauropods and, had they been found on their own, would have been identified as part of a sauropod dentition. The wear on them is the same as that of sauropods, and shows that the advanced prosauropods had the same eating strategies as the sauropods. The rest of the skeleton is a typical, medium-size prosauropod.

Distribution: China.
Classification: Sauropodomorpha, Prosauropoda.
Meaning of name: Lizard from Yunnan.

Named by: Young, 1942.
Time: Hettangian to Pliensbachian stages of the early Jurassic.
Size: 7m (23ft).
Lifestyle: High browser.
Species: *Y. huangi*.

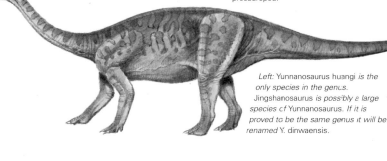

Left: Yunnanosaurus huangi *is the only species in the genus.* Jingshanosaurus *is possibly a large species of* Yunnanosaurus. *If it is proved to be the same genus it will be renamed Y.* dinwaensis.

LUFENG BASIN

The Lufeng Basin in south-western China consists of continental sediments up to 1,000m (3,280ft) thick. The lower part of it was originally thought to have been Triassic in age, but now the whole sequence is known to be early Jurassic. The sediments were deposited in rivers and lakes, and show a picture of a varied flora and fauna of the interior of the continent in early Jurassic times.

Animals found there consist of small mammals, mammal-like reptiles, crocodiles and dinosaurs, particularly prosauropods. The mammal-like reptiles are unusual since they mostly existed in the Permian period, dying out elsewhere in Triassic times. Everywhere else they were replaced by their descendants, the mammals proper. Their continued existence into the early Jurassic parts of the Lufeng Basin led to the idea that these beds were Triassic in age.

With late surviving prosauropods and mammal-like reptiles, the Lufeng Basin could almost be considered as a kind of 'lost world' with animals surviving into early Jurassic times that had died out elsewhere in the previous Triassic period. Environmental conditions may have remained stable there while they changed in other parts of the world. Such an occurrence is biologically known as a "refugium". A modern example would be mountain-tops that still sustain arctic fauna that has been extinct in the lowlands since the end of the Ice Age.

Massospondylus

The first *Massospondylus* specimen to be found consisted of a few broken vertebrae shipped to Sir Richard Owen, in London, from South Africa in 1854. Since then the skeletons of more than 80 individuals have been found across southern Africa. There has even been a nest of six eggs found that have been attributed to *Massospondylus*. Another possible specimen has been found in Arizona, which may indicate that this was a very widespread animal.

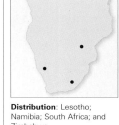

Distribution: Lesotho; Namibia; South Africa; and Zimbabwe.
Classification: Sauropodomorpha, Prosauropoda.
Meaning of name: Massive vertebra.
Named by: Owen, 1854.
Time: Hettangian to Pleinsbachian stages of the early Jurassic.
Size: 4m (13ft).
Lifestyle: Low browser.
Species: *M. carinatus*, *M. hislopi*.

Features: The teeth are large, some serrated and some flat. It is usually shown as a much more slender animal than other prosauropods of the same size. It has five fingers, but the fourth and fifth are very small. The huge, clawed first finger could be curved over, thumb-wise, across the second and third, making this a versatile hand.

LAST PROSAUROPODS AND PRIMITIVE SAUROPODS

In early Jurassic times the prosauropods and the newly evolving sauropods all shared the same features –
a big body, long neck and tiny head, with some minor differences. In view of their similarities, there is
still some disagreement over the classification of some of these genera.

Gongxianosaurus

The dinosaur *Gongxianosaurus* presented its Chinese finders with a puzzle when it was excavated close to the village of Shibei in Gongxian County, China. Was it the largest of the prosauropods, or a primitive sauropod? Most scientists agree on the latter, and there is enough of the skeleton to show an interesting mixture of the two. The find consisted of three disarticulated skeletons, one of which was a juvenile. There were problems working out which bones went with which individual.

Features: The heavily built forelimbs of *Gongxianosaurus* were about three-quarters the length of the hind, but they had a very short forearm. The hind leg was massive. The hind feet were like those of the heaviest prosauropods but had short foot bones, like a sauropod. There is a collection of belly ribs, which is usually a sign of a prosauropod, but what is known of the skull

Distribution: China.
Classification: Sauropodomorpha.
Meaning of name:

Right: Whether Gongxianosaurus was a prosauropod or a sauropod it must have had a typical sauropod lifestyle – a four-footed browser that fed from trees.

looks distinctly like that of a sauropod, particularly in the large spoon-shaped teeth.

Lizard from Gonxian County.
Named by: He, Wang, Liu S., Zhou, Liu T., Cai and Dai, 1998.
Time: Early Jurassic.
Size: 14m (46ft).
Lifestyle: High browser.
Species: *G. shibeiensis*.

Yimenosaurus

Ten partial skeletons of this animal are known, one of which includes part of the skull. Its presence in the early Jurassic rocks of China shows that the prosauropod group continued to develop and survive, even though the sauropods were beginning to take over as the main plant-eaters in the forests of conifers that thrived at the time.

Features: *Yimenosaurus* is a large prosauropod which the finders classed as a plateosaurid. Others disagree, as the skull seems to be too short and too deep to have been a member of the more primitive prosauropod group. In fact it resembles the skull of an early sauropod. This would make *Yimenosaurus* one

Distribution: China.
Classification: Sauropodomorpha, Prosauropoda.
Meaning of name: Lizard from Yimen.
Named by: Bai, Yang and Wang, 1990.
Time: Early Jurassic.
Size: 9m (30ft).
Lifestyle: High browser.
Species: *Y. youngi*.

Right: Ten tonnes of rock yielded the remains of ten individuals when the site in Yimen County, China, was excavated in 1987.

of the more advanced prosauropods in its resemblance to the sauropod group. The later prosauropods tended to look more and more like their successors.

Lufengosaurus

All sorts of names have been given to *Lufengosaurus* by people who thought they were finding a completely new animal, because it is known from so many remains. They include:

• *Gyposaurus*, found by Robert Broom in 1911 in South Africa. This was described as a small prosauropod 1.5m (5ft) long. Later, more *Gyposaurus* material was found in China. The South African specimen is now regarded as a subadult *Anchisaurus*, while the Chinese fossils are thought to be a subadult *Lufengosaurus*.

• *Tawasaurus* – this was found, in 1982, by the famous Chinese palaeontologist Young Chung Chien (who has contributed greatly to the knowledge of Chinese prosauropods). It is now thought to be a juvenile *Lufengosaurus*.

• *Fulengia* – Carroll and Galton found this tiny animal in 1977, and thought it was the fossil of a lizard. It turned out to be a hatchling prosauropod and is very likely *Lufengosaurus*. (Incidentally, the original genus name was to have been *Lufengia*, commemorating the area in which it was found, but that name already existed, so they just swopped round the 'l' and the 'f'.)

Features: *Lufengosaurus* is the most familiar of the Chinese prosauropods. There have been more than 30 skeletons found, and mounted skeletons grace a number of Chinese museums. Its large size meant that it would have been a quadrupedal animal.

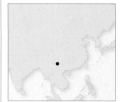

Distribution: China.
Classification: Sauropodomorpha, Prosauropoda.
Meaning of name: Lizard from the Lufeng Series.
Named by: Young, 1941.
Time: Hettangian to Pliensbachian stages of the early Jurassic.
Size: 6m (19½ft).
Lifestyle: High browser.
Species: *L. changhduensis*, *L heunei*.

Tazoudasaurus

This early sauropod was discovered by accident in 1998 when the authorities were investigating the illegal traffic in fossil bones. Among the fossil bones of this animal was a jawbone with 17 teeth, which prove it to have been a sauropod. The only earlier sauropod – *Antetronitus* – was found in Triassic rocks at the other end of Africa.

Right: The Tazoudasaurus remains include the oldest sauropod skull yet found. The oldest sauropod skeleton, that of Antetonitrus, lacked the head and neck.

Features: *Tazoudasaurus* is known from the head, jaw and some vertebrae. The jaw and teeth show it to have been a primitive sauropod rather than an advanced prosauropod, an identification that would have been assumed had the teeth not have been found. The typical spoon-shaped sauropod teeth would have been used for raking the leaves from branches that it reached with its long neck.

Distribution: Morocco.
Classification: Sauropodomorpha, Sauropoda.
Meaning of name: Lizard from Tazouda village.
Named by: Allain, 2003.
Time: Toarcian stage.
Size: 10m (33ft).
Lifestyle: Browser.
Species: *T. naimi*.

PRIMITIVE SAUROPODS

The primitive sauropods, the big plant-eating dinosaurs with the long necks, originated in Triassic times, but came to prominence during the early Jurassic period. The Victorian palaeontologists named the group "sauropod" because they saw a resemblance in the arrangement of the foot bones to that of a lizard, as the sauropod had five toes compared with three of the contemporary theropods.

"Kunmingosaurus"

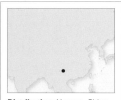

Distribution: Yunnan, China.
Classification:
Sauropodomorpha,
Sauropoda, maybe
Volcanodontidae.
Meaning of name: Lizard
from Kunming.
Named by:
Chao, 1985.
Time: Early Jurassic.
Size: 11m (36ft).
Lifestyle: Herbivore.
Species: "*K. wudingensisu*".

A name that has been given to a fossil animal that has not been formally described in a scientific paper is a *nomen dubium*, a doubtful name. This does not mean that the animal did not exist, but shows that we cannot say too much about it with confidence since other scientists have not had the opportunity to check the original discoverer's research. This is the case with "*Kunmingosaurus*", and the custom is to put the name in inverted commas to show that it is a *nomen dubium*.

Features: "*Kunmingosaurus*" has not been scientifically described. However, there has been an impressive full-body skeleton constructed. All that we can say about the animal is that it is a primitive sauropod and walked on all fours.

Left: The mounted skeleton is assembled from disarticulated bones found in the same quarry in 1954. At first the teeth were thought to have been from Yunnanosaurus *and the jaw from* Lufengosaurus – *both prosauropods.*

Kotasaurus

Known from a single bone bed in Andhra Pradesh in India, the remains of 12 individuals of various sizes of *Kotasaurus* lay in river sandstones. They are possibly the remains of a herd that had been drowned during a river crossing. A mounted skeleton now stands in the Science Centre in Hyderabad.

Features: *Kotasaurus* seems to be intermediate between the prosauropods and the sauropods. Some of the hip bones are prosauropod-like, but others are definitely those of a sauropod, as are the vertebrae and teeth. Studies carried out in 2001 show that its simple vertebrae, its narrow shoulder bone and slim leg bones are all signs of a primitive sauropod. Like other early sauropods, it is a heavy, four-footed animal with a long neck and small head.

Distribution: India.
Classification: Sauropoda,
Sauropodomorpha,
Vulcanodontidae.
Meaning of name: Lizard
from Kota Formation.
Named by: Yadagiri,
1988.
Time: Toarcian stage of
the early Jurassic.
Size: 9m (30ft).
Lifestyle: Herbivore.
Species: *K. yamanpalliensis*.

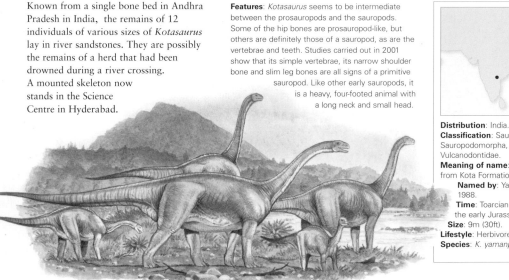

Vulcanodon

When it was discovered in 1972, *Vulcanodon* was regarded as a prosauropod. The teeth were definitely those of a meat-eater, which fitted in with the current thinking that the prosauropods were omnivorous. It has now been established that these teeth, for which the animal was named, belonged to an unidentified theropod that had scavenged the carcass. The true teeth of this animal, as well as the skull and neck, are unknown.

Features: The status of *Vulcanodon* as a sauropod was established in 1975 on the basis of the shape of the toe bones. These are broader than they are deep – a sauropod feature – although they seem to be even broader than those of most other sauropods. Another distinguishing feature of *Vulcanodon* is the length of the front legs, which are comparatively long compared to other sauropods. The ichnogenus *Deuterosauropodopus* from Lesotho probably represents the footprints of *Vulcanodon*.

Distribution: Zimbabwe.
Classification: Sauropoda, Sauropodomorpha, Vulcanodontidae.
Meaning of name: Volcano tooth, from the volcanic deposits in which it was found.
Named by: Raath, 1972.
Time: Hettangian stage of the early Jurassic.
Size: 6.5m (21ft).
Lifestyle: Herbivore.
Species: *V. karibaensis*.

Left: The restoration of Vulcanodon is made with difficulty – the only skeleton found had been eaten by a theropod of some kind.

Barapasaurus

Barapasaurus seems to have been quite a common early Jurassic sauropod in India. More than 300 specimens representing parts of about six individuals have been found. However, as is common in sauropod dinosaurs, no skulls are known, and neither have the foot bones been found, which poses further difficulties for precise classification.

Features: Although *Barapasaurus* is defined, and has been scientifically described, based on the hip bone, many of the other bones are known as well. The vertebrae in the back have a characteristically deep cleft for the spinal cord, a feature that distinguishes this animal from other sauropods. It appears to have had rather long and slim legs. The teeth are spoon-shaped, like many later sauropods, and used for raking leaves from branches. The narrowness of the hip bones suggest that this dinosaur belongs to the Cetiosauridae, rather than the more primitive Vulcanodontidae.

Distribution: India.
Classification: Sauropodomorpha, Sauropoda, Cetiosauridae.
Meaning of name: Big legged lizard.
Named by: Jain, Kutty, Roy-Chowdry and Chatterjee, 1975.
Time: Toarcian stage of the early Jurassic.
Size: 18m (59ft).
Lifestyle: Herbivore.
Species: *B. tagorei*.

Right: Barapasaurus was the first of the really big sauropods, comparable in length to the giants that appeared later in the Jurassic. A mounted skeleton stands in the Geological Studies Unit of the Indian Statistical Institute, Calcutta, India.

LITTLE ORNITHOPODS

The early plant-eating ornithopods were small, beaver-sized animals. By the early Jurassic period some had not yet evolved the cheek pouches and the chewing teeth of the later forms, but one group, the heterodontosaurids, had developed a very strange, almost mammal-like arrangement of differently sized teeth. The remains of these animals are known from desert deposits of southern Africa.

Lesothosaurus

The most primitive ornithischians, such as *Lesothosaurus*, had not evolved the complex chewing mechanism that was to characterize the later forms. Instead, they would have crushed their food by a simple up and down chopping action of the jaws. This is quite an unspecialized feeding method, and these animals may well have eaten carrion or insects as well as plants in order to survive.

Features: *Lesothosaurus* is one of the most primitive of the ornithischians, and as such it is difficult to put into a strict classification. It is a small, two-footed plant-eater, built for speed. The head, on the end of a flexible neck, is short, triangular in profile, with big eyes. The teeth are arranged in a simple row and, unlike all other ornithopods, the mouth does not seem to have cheeks. The jaw action is one of simple chopping. The snout ends with a horn-covered, vegetation-cropping beak.

Right: Lesothosaurus is very similar to the earlier-discovered Fabrosaurus. *However the Fabrosaurus material is so poor it is impossible to make direct comparisons. If they are the same genus, then the name Fabrosaurus would have to take precedence, being applied first.*

Distribution: Lesotho; South Africa.
Classification: Ornithopoda, Fabrosauridae.
Meaning of name: Lizard from Lesotho.
Named by: Galton, 1978.
Time: Hettangian to Sinemurian stages of the early Jurassic.
Size: 1m (3ft).
Lifestyle: Low browser.
Species: *L. diagnosticus.*

Abrictosaurus

The name of the "wide-awake lizard" derives from a dispute between palaeontologists. Tony Thulborn proposed that the heterodontosaurids slept away the hot desert summer as some modern animals do. J. A. Hopson disagreed with this theory, based on the study of the growth of the teeth, and celebrated his notion of a year-round active animal by giving this new dinosaur an appropriate name.

Features: Our knowledge of *Abrictosaurus* is based on two skulls and some fragmentary pieces of the skeleton. The skulls are almost identical to those of *Heterodontosaurus*, with a similar varied arrangement of teeth, except that they lack the prominent tusks. The remains may, in fact, represent female specimens of *Heterodontosaurus*. It is possible that only the males had the tusks and used them for display, as with many modern animals. On the other hand, many modern animal groups, such as the pigs, have prominent tusks in some genera but not in others. A wild boar's differ from those of a warthog.

Right: Only the skull and a few bones of Abrictosaurus *are known, and so the remainder of the restoration is based on the skeleton of* Heterodontosaurus.

Distribution: Lesotho; South Africa.
Classification: Reptile, Ornithopoda, Heterodotosauridae.
Meaning of name: Wide-awake lizard.
Named by: Hopson, 1975.
Time: Hettangian to Sinemurian stages of the early Jurassic.
Size: 1.2m (4ft).
Lifestyle: Herbivore.
Species: *A. consors.*

Lanasaurus

This animal – the "woolly lizard" – was named in honour of the noted palaeontologist A. W. Crompton, whose nickname was "Fuzz". As with most of the heterodontosaurids it is known only from the jaw and teeth. It is possible that it is another specimen of *Lycorhinus*, or one of the other genera of heterodontosaurids that seemed to have abounded in southern Africa in early Jurassic times.

Below: Lanasaurus is closely related – or indeed identical – to Lycorhinus, which is known only from the jaw bone. It was so mammal-like that it was first identified as an early mammal. The name means "wolf snout."

Distribution: South Africa.
Classification: Ornithopoda, Heterodontosauridae.
Meaning of name: Woolly lizard.
Named by: Gow, 1975.
Time: Hettangian to Sinemurian stages of the early Jurassic.
Size: 1.2m (4ft).
Lifestyle: Herbivore.
Species: *L. scalpridens.*

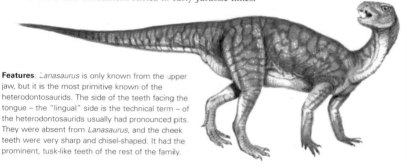

Features: *Lanasaurus* is only known from the upper jaw, but it is the most primitive known of the heterodontosaurids. The side of the teeth facing the tongue – the "lingual" side is the technical term – of the heterodontosaurids usually had pronounced pits. They were absent from *Lanasaurus*, and the cheek teeth were very sharp and chisel-shaped. It had the prominent, tusk-like teeth of the rest of the family.

HETERODONTOSAURID TEETH

So, what was the purpose of the strange dentition of the heterodontosaurids? The chisel-like teeth at the back were obviously for chewing. The muscular cheeks would have held the plant material as it was worked over again and again by the chewing action. The cheeks themselves have not been fossilized, of course, but we surmise that they were there because the tooth rows were set in from the side of the skull, leaving a gap at each side of the mouth that must have been covered over with something. Also, the chopping action of the teeth would have meant that half the food would have escaped the mouth if there had been nothing there to catch it.

The big tusks may have been used for digging up tubers that were part of the desert vegetation. They may also have been used as defensive weapons, protecting the family or herd against the attacks of meat-eating dinosaurs. A third purpose could have been display, being used in ritual combat and competition in the herd hierarchy, as we see in modern warthogs. If this were the case it could explain why only some specimens have been found with the very big tusks. Perhaps only the males sported them.

Below: Heterodontosaurus teeth and skull.

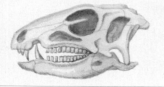

Heterodontosaurus

The original description of *Heterodontosaurus* in 1962 was based on a single skull, but in 1976 it was backed up by the discovery of one of the most complete and well-preserved dinosaur skeletons known – almost caught in an action pose as if it had been running away and frozen in time. It shows a swift-footed animal with long front legs and strong hands bearing five fingers.

Features: The typical heterodontosaurid tooth pattern consists of three pairs of sharp stabbing teeth at the front of the mouth, the rearmost of which are big like canine tusks. It also has a set of closely packed shearing and grinding teeth further back. The rear teeth would have been enclosed in cheeks to hold the food while it was being chewed. The front teeth worked against a horny beak on the lower jaw.

Distribution: South Africa.
Classification: Ornithopoda, Heterodontsauridae.
Meaning of name: Differently-toothed lizard.
Named by: Crompton and Charig, 1962.
Time: Hettangian to Sinemurian stages of the early Jurassic.
Size: 1m (3ft).
Lifestyle: Herbivore.
Species: *H. tucki.*

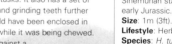

PRIMITIVE ARMOURED DINOSAURS

In early Jurassic times, a new group of dinosaurs began to appear. They were distinguished by the presence of armour on their backs, as bony plates covered in horn, embedded in the skin. Armoured dinosaurs were all plant-eaters and evolved from the ornithischian, or the bird-hipped, line. The evolutionary spur would have been the presence of so many big, contemporary, meat-eating dinosaurs.

Scutellosaurus

This plant-eating dinosaur that would have sought refuge from its enemies in two different ways, either by running away, or by squatting down and letting the armour deter enemies. It had long hind legs and a very long tail for balance that would have enabled it to run. Its strong front legs would have carried its body as it hunkered down in passive defence.

Features: If it were not for the armour plates, this animal would have been regarded as a fabrosaurid, like *Lesothosaurus*. It has the simple dentition, and the long hind legs and tail. The presence of this armour suggests that *Scutellosaurus* was close to the ancestry of the big plated and armoured dinosaurs to come. It would have evolved from purely bipedal ancestors.

Distribution: Arizona, USA.
Classification: Thyreophora.
Meaning of name: Little shield lizard.
Named by: Colbert, 1981.
Time: Hetangian stage of the early Jurassic.
Size: 1.2m (4ft).
Lifestyle: Low browser.
Species: *S. lawleri*.

Left: The armour consists of more than 300 little shields in six different types, ranging from tiny lumps to big stegosaur-like plates.

Scelidosaurus

The first almost complete dinosaur skeleton to be discovered and named was *Scelidosaurus*. As an armoured dinosaur it would have been a slow-moving creature, but the length of its hind legs and weight of its tail seem to suggest that it may have been able to run on two legs for short distances.

Features: This very primitive animal belonged to the group known as the thyreophorans. The group consists of the plated dinosaurs (the stegosaurs) and the armoured dinosaurs (the ankylosaurs). In the past it has been regarded as ancestral to the stegosaurs, but now it is thought to be closer to the ancestors of the ankylosaurs. Its back, sides and tail are covered by an arrangement of armoured scutes with a mosaic of small armoured scales between them.

Distribution: Dorset, England.
Classification: Thyreophora, Ankylosauria.
Meaning of name: Leg lizard.
Named by: Owen, 1868.
Time: Sinemurian to Pliensbachian stages of the early Jurassic.
Size: 4m (13ft).
Lifestyle: Low browser.
Species: *S. harrisonii*.

Right: Two almost complete skeletons of Scelidosaurus have been found in southern England, and it is from these skeletons that we can reconstruct the likely appearance of related, less complete, animals.

OWEN'S MISSED CHANCE

The early evolution of the armoured dinosaurs has always been the subject of much debate, conflict and confusion. This is despite the fact that the best-preserved early armoured dinosaur, *Scelidosaurus*, was one of the first dinosaurs to have been discovered and has been known since 1868.

It was studied and named by none other than Sir Richard Owen, the most prominent British scientist of the day, and the man who actually invented the term 'dinosaur'. He had been involved in the scientific study of the first dinosaurs discovered – meat-eating *Megalosaurus* and plant-eating *Iguanodon* – in the 1830s and 1840s. His scientific studies of them were insightful and thorough, and laid the foundation of modern dinosaur research. However, by the time that *Scelidosaurus* was discovered, Owen's career had moved on. By then he was deeply involved in the politics of the science, working on projects such as the establishment of the British Museum (Natural History) – now the Natural History Museum in London – and on the social impact of Charles Darwin's newly published theories. The scientific description of *Scelidosaurus*, although workmanlike, did not display the originality and vigour of his earlier work, and did not have the same impact in the scientific world.

The chance of establishing the significance of the armoured dinosaurs at this early stage was missed, and it was not until the *Scelidosaurus* skeleton was re-interpreted a century later that the modern ideas about the evolution of this important dinosaur group were aired.

Emausaurus

The scientific description of *Emausaurus* is based on an almost complete skull and some skeletal parts found in northern Germany. The skull is about half the size of that of the better-known *Scelidosaurus* but does not seem to have been as heavily armoured, and probably represents an early stage in the evolution of the armoured dinosaurs.

Features: The skull and teeth of *Emausaurus* are very similar to those of *Scelidosaurus*, but the head seems to be wider towards the rear and narrows towards the snout. The jaw joint is very simple, which suggests that the mouth worked in a scissor-like action, but there does not seem to have been any wear on the teeth. Perhaps *Emausaurus* just tore at the plants with its mouth, and swallowed without chewing.

Distribution: Germany.
Classification: Thyreophora, Ankylosauria.
Meaning of name: Acronym of Ernst-Moritz-Arndt Universität lizard.
Named by: Haubold, 1990.
Time: Toarcian stage of the early Jurassic.
Size: 2m (6ft).
Lifestyle: Low browser.
Species: *E. ernsti*.

Left: Despite its placing with Scelidosaurus *and its relatives here, certain aspects of the skull are stegosaur-like – very close to* Huayangosaurus.

"Lusitanosaurus"

French Jesuit priest and palaeontologist Albert-Felix de Lapparent (1905–75), who hunted dinosaurs all over Europe and North Africa was the finder of "*Lusitanosaurus*". The specimen consisted of a fragment of jawbone a few centimetres (1–2in) long. It appears to be rather like that of *Scelidosaurus* and is about the same age, hence it belongs to the same family.

Features: "*Lusitanosaurus*" is something of a *nomen dubium*. Only parts of the upper jaw and an array of eight teeth are known. These differ from those of *Scelidosaurus* in being taller, and in lacking the scalloped edges of the English specimen. There is no guarantee that the rest of the animal looked anything like *Scelidosaurus*, but it is the best guess until more remains come to light that may confirm or disprove this suggestion.

Distribution: Portugal.
Classification: Thyreophora, Ankylosauria.
Meaning of name: Lizard from Lusitania (an old name for Portugal).
Named by: de Lapparent and Zbyszewski, 1957.
Time: Sinemurian stage of the early Jurassic.
Size: 4m (13ft).
Lifestyle: Low browser.
Species: "*L. liasicus*".

Right: Albert de Lapparent, who discovered and named "Lusitanosaurus", made many dinosaur discoveries between the late 1940s and the early 1960s. Most of his work was done in the Sahara Desert, Africa.

MAMMALS AND THE LAST MAMMAL-LIKE REPTILES

The mammal-like reptiles were almost gone by the beginning of the Jurassic. Only the tritylodonts survived into this period, and these would have been almost indistinguishable from the mammals proper of the time. The mammals were all small shrew-like creatures that lived as a minor part of the ecosystem.

Oligokyphus

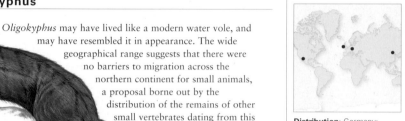

Oligokyphus may have lived like a modern water vole, and may have resembled it in appearance. The wide geographical range suggests that there were no barriers to migration across the northern continent for small animals, a proposal borne out by the distribution of the remains of other small vertebrates dating from this time. Two sizes of *Oligokyphus* have been found in the same fissure in southern England. They were named as different species – *O. major* and *O. minor*. These may represent male and female members of the same species.

Features: The teeth of *Oligokyphus* are very mammalian, with large incisors separated by a gap from a row of multi-cusped grinding teeth. There are no canine teeth. The backbone is also very mammal-like, and would have enabled the animal to curl up into a ball to keep warm. It was probably able to suckle its young as well. Despite these mammal-like features, *Oligokyphus* is regarded as a reptile because the jaw- and ear-bones are so similar to those of the advanced cynodonts. It retains the quadrate bone in the jaw. In true mammals this bone has evolved into the incus bone of the middle ear.

Above left: The limb and body movements of Oligokyphus *would have made it look just like a mammal scurrying away.*

Distribution: Germany; England; China; Arizona, USA.
Classification: Therapsida, Tritylodontidae.
Meaning of name: Small and curved.
Named by: Hennig, 1922.
Time: Rhaetian to Pleinsbachian stages of the early Jurassic.
Size: 30cm (12in).
Lifestyle: Small omnivore.
Species: *O. triserialis, O. major, O. lufengensis.*

Kuehneotherium

The remains of *Kuehneotherium* were found in fissure deposits in south Wales, alongside the remains of other early mammals such as *Morganucodon*. It seems that the environment was such as to support more than one early mammal. The name honours Walter Kühne, a German refugee who worked extensively in the early Mesozoic remains of that area during the War years and immediately after. Unfortunately the remains consist only of the teeth, and so the appearance and the size of the body are quite conjectural. However, the tooth shapes are distinctive enough to make this a significant animal.

Features: The molar teeth – the defining elements of the early mammals – have three tall cusps. and three roots. The middle cusp is taller than the others, and they are arranged in a triangle. This tooth structure is similar to that of modern opossums. Their arrangement suggests that there was some degree of lateral movement of the jaw. The lower jaw itself is quite narrow and lightly built – more so than in *Morganucodon*.

Distribution: Southern Wales.
Classification: Mammal, order uncertain.
Meaning of name: Kühne's beast.
Named by: Kermack, Kermack and Musset, 1968.
Time: Early Jurassic.
Size: 10cm (4in).
Lifestyle: Insectivore.
Species: *K. praecursoris.*

Right: Kuehneotherium and the other early mammals co-existed, along with lizards and all kinds of other small vertebrates on the limestone uplands of the time.

Hadrocodium

Distribution: China.
Classification: Mammal, Symmetrodonta.
Meaning of name: Big head.
Named by: Luo, Crompton and Sun, 2001.
Time: Early Jurassic.
Size: 4cm (1¾in).
Lifestyle: Insectivore.
Species: H. wui.

A complete skull of this little mammal was found in China, and raised some excitement. Its state of preservation allowed palaeontologists to see that the ear bones were separated from those of the lower jaw – a true distinguishing feature that separates the true mammals from the most mammal-like reptiles. The advanced mammalian features – the combination of the big brain and the separate ear bones – were about 40 million years in advance of anything found up to the time of its discovery. It is, however, unclear what the rest of the body looked like.

Right: Hadrocodium *is the smallest mammal so far known. Its diet must have been restricted to the smallest invertebrates around, and it must have kept feeding all the time to fuel its high metabolism.*

Features: The skull, the only thing known about this animal, has space for a bigger brain than any of the early Mesozoic mammals so far found. In the larger brain the areas devoted to hearing and smell seem to have been well developed, indicating fine hunting skills. The space in the skull for housing the hearing mechanism is quite expanded too, something that seems to have evolved along with the larger brain. *Hadrocodium* may have been on the line that led to most modern mammals.

THE DIFFERENCES BETWEEN MAMMALS AND MAMMAL-LIKE REPTILES

1. Mammals have a much larger brain – 3–4 times as big.
2. They also have a simplified jaw articulation, involving only one jaw bone.
3. Their teeth are of varied sizes and shapes.
4. Mammals have a specialized chewing mechanism.
5. They also have shearing back teeth.
6. Big ears are a feature of mammals.
7. They have a small body size (weighing around 20–30g (1oz)).
8. The legs of mammals are held beneath the body.
9. The skull and neck joint allow the head to move up and down and rotate.
10. They have a more specialized vertebrae in the back.
11. They have a palate present in the roof of the mouth.
Other early mammals from the time that are known from only fragmentary remains include *Sinoconodon*, *Kayentatherium*, *Indotherium*, *Erythrotherium*, and *Dianzhongia*.

Below: Lower jaw of an early mammal, consisting of a single bone and teeth of varying shapes and sizes.

Megazostrodon

Megazostrodon is accepted as one of the first of the true mammals, although it retained many reptile characteristics. It would have suckled its young, like all mammals do, but it probably laid leathery eggs like its reptile ancestors. The placenta, that allows for live birth, would not have evolved until later. It would have hunted insects and maybe small lizards.

Features: A complete skeleton from Lesotho shows *Megazostrodon* to have a long snout, a long body and a long tail. It has a bigger brain than its reptile ancestors and in life it found its prey at night with the senses of smell and hearing. There is no direct evidence for the external ear flaps, shown on all restorations, but the rest of its ear structure is built the same way as in modern mammals.

Distribution: Lesotho, South Africa.
Classification: Mammal.
Meaning of name: Large girdle tooth.
Named by: Crompton and Jennings, 1968.
Time: Late Triassic to early Jurassic.
Size: 10cm (4in).
Lifestyle: Insectivore.
Species: M. rudnerae.

Left: Megazostrodon *was a shrew-like mammal that probably hunted by night.*

MEGALOSAURUS AND ITS CLONES

The big, meat-eating dinosaurs whose fossils have been found in the middle Jurassic rocks, especially those from Europe, have traditionally been regarded as megalosaurs. This classification is undergoing revision. Although these big beasts seem similar to one another, being quite conservative in their build and appearance, they seem to have represented quite a range of animals.

Megalosaurus

The first dinosaur to be studied scientifically and named goes to *Megalosaurus*. It has always been the source of some confusion, however, since for a long time any early or middle Jurassic theropod found in Europe was assigned to the genus *Megalosaurus* without too much study. The scientific term "wastebasket genus" is applied to such a genus.

Features: Although the original *Megalosaurus* remains consist of nothing but a jawbone and teeth, and a few scraps of bone, subsequent discoveries have shown that it gives quite a good impression of a generalized, large meat-eating dinosaur. It has the big head with the long jaws and sharp teeth which are to be expected. It ran on powerful hind legs with the smaller front legs held clear of the ground. The tiny hands have three fingers and the feet have four toes, three of which reach the ground.

Distribution: England; France; and perhaps Portugal.
Classification: Theropoda, Tetanurae.
Meaning of name: Big lizard.
Named by: Ritgen, 1826.
Time: Aalenian to Bajocian stages of the middle Jurassic.
Size: 9m (30ft).
Lifestyle: Hunter or shoreline scavenger.
Species: *M. bucklandii*, "*M. dabukaensis*," "*M. phillipsi*", "*M. tibetensis*", although these are *nomen dubia*.

Above: About 50 genera have been misassigned to Megalosaurus, *including* Dryptosaurus, Proceratosaurus, Eustreptospondylus, Magnosaurus, Iliosuchus, Metriacanthosaurus, Carcharodontosaurus, Dilophosaurus *and even the prosauropod* Plateosaurus.

Eustreptospondylus

Distribution: England.
Classification: Theropoda, Tetanurae.
Meaning of name: Well-curved vertebrae.
Named by: Walker, 1964.
Time: Callovian stage of the middle Jurassic.
Size: 9m (30ft), but only known from the skeleton of a 6m (19½ft) juvenile.
Lifestyle: Hunter.
Species: *E. oxoniensis*.

This dinosaur was one of those theropods that was casually assigned to *Megalosaurus* until it was found to have been something completely different. The vertebrae are more curved than those of *Megalosaurus*, hence its name. It is known from an almost complete skeleton found in 1871. This skeleton is now mounted and displayed in the University Museum in Oxford, England.

Features: *Eustreptospondylus* is a medium to large hunter which seems to have been more closely related to the ancestors of the later *Allosaurus* or of the spinosaurids than to its contemporary *Megalosaurus*. In the one skeleton found, the skull is fragmentary and difficult to reconstruct. The vertebrae are mostly missing their upper parts, suggesting that the skeleton is of a juvenile, and that its bones had not yet had time to knit together properly. Casts made from the disarticulated skull are often used as the basis for the skull of *Megalosaurus* in museum displays.

Above: The single skeleton of Eustreptospondylus *was found in marine clays. It had died on land and had been washed out to sea. Once the decaying body had released all its gases it sank to the bottom and was buried in the fine sediment.*

Poekilopleuron

We cannot say too much about *Poekilopleuron*, since most of its remains were destroyed during World War II when the Musée de la Faculté des Sciences de Caen, France, was bombed. What we do know is based on casts that were subsequently found in the Museum National d'Histoire Naturelle in Paris, France. An almost complete skull and other fragments assigned to *P. valesdunensis* were excavated in Normandy in 1994.

Features: The front limbs – almost all that is known of the original specimen – are very short but remarkably strong, with attachment areas for powerful muscles. From what is known of the rest of the skeleton, this animal, along with *Eustreptospondylus*, seems to belong among the ancestors

of the spinosaurid group of theropods, judging from the length of spines on the bones of the middle neck. The skull, from the more recently found material, is quite low and has no ornamentation of any sort.

Distribution: France.
Classification: Theropoda, Tetaneurae.
Meaning of name: Varying side.
Named by: Eudes-Deslongchamps, 1838.
Time: Bathonian stage of the middle Jurassic.
Size: 9m (30ft).
Lifestyle: Hunter.
Species: *P. bucklandii*, *P. valesdunensis*, "*P. schmidtii*" – the *nomen dubium* of some unidentifiable bones from Russia.

A NEW CLASS OF REPTILES

For a long time the quarries in the so-called Stonesfield slate, actually a thinly bedded limestone, yielded fossil remains. The quarries were close to Oxford in England, and were worked for building materials. In about 1815 some bones, including a large jawbone with teeth in place, were discovered and passed to William Buckland, the first Professor of Geology at the University of Oxford. One of his colleagues, James Parkinson, published a drawing of the tooth, which was named *Megalosaurus* in 1822. As there was no concept of a dinosaur at that time, it was assumed that the jawbone and teeth came from some kind of gigantic lizard.

However, other big animals were coming to light, and in 1841 Sir Richard Owen established the classification "dinosauria" to include these discoveries. Even then they were regarded as giant lizards, and were represented as big, dragon-like, four-footed animals.

The most famous of these early restorations are the full-sized statues that were constructed for the Crystal Palace park in south London, in 1854. They are there to this day, erroneous but impressive for the insight they give to the

scientific understanding of the time.

Left: The first impressions of Megalosaurus.

Magnosaurus

This is another of those theropods that had originally been assigned to *Megalosaurus*. When it was found to be a different genus it was renamed *Magnosaurus* in view of its similarity to it. The bones seem to have belonged to a juvenile animal. They are now in Oxford University, England, the last resting place of many British megalosaurs.

Features: *Magnosaurus* is not a very well-known animal. The only remains are jaws, teeth and bits of bone. Some of the limb bones are known only from the casts of the internal cavities. The teeth are thicker than those of *Megalosaurus*, and there are fewer in the jaws. It lived some time earlier, and there is still some argument over whether *Magnosaurus* is indeed just an early, small species of *Megalosaurus*.

Distribution: Dorset, England.
Classification: Theropoda.
Meaning of name: Huge lizard.
Named by: von Huene, 1932.
Time: Bajocian stage of the middle Jurassic.
Size: 4m (13ft).
Lifestyle: Hunter.
Species: *M. nethercombensis*, *M. lydekkeri*.

Left: Magnosaurus *was a swift, medium-size hunter. However we have no direct evidence as to its prey – there are hardly any other land-living vertebrate remains found in that particular horizon of the middle Jurassic.*

CARNOSAURS

The term carnosaur was once used to refer to any big, meat-eating dinosaur, as distinguished from the small meat-eaters, which were called coelurosaurs. Nowadays it is restricted to the family of meat-eaters related to Allosaurus. This family is quite primitive compared with the spectacular and highly adapted carnivorous dinosaurs to come, but they were the main predators of middle Jurassic times.

Xuanhanosaurus

Distribution: China.
Classification: Theropoda, Tetanurae, Carnosauria.
Meaning of name: Lizard from Xuanhan County.

Named by: Dong, 1984.
Time: Bathonian to Callovian stages of the middle Jurassic.
Size: 6m (19½ft).
Lifestyle: Hunter.
Species: *X. quilixiaensis.*

Many of the middle Jurassic carnosaurs have been found in China, and their names are based on the place names of the areas where they were found. As a result many of their names seem difficult for western tongues to pronounce. *Xuanhanosaurus*, from Xuanhan county, is known only from forelimbs, shoulders and some pieces of backbone.

Features: The distinctive feature of this middle Jurassic carnosaur is the extraordinary length of the forelimbs, which are much longer than the limbs of other theropods. When it was discovered there was speculation as to whether it may have walked on all fours, but this is not now thought to be the case. The muscle attachments at the shoulders show that the big arms were very strong and bore four powerful fingers. The humerus is bigger and heavier than the lower arms, and there is a clavicle – a collar bone that is not usually preserved in dinosaurs.

Above: The Xiashaximiao Formation, in which Xuanhanosaurus *was found, also yielded specimens of sauropods such as* Shunosaurus, Datousaurus *and* Omeisaurus, *which may have represented its prey.*

Gasosaurus

The Dashanpu dinosaur quarries in Sichuan, in central China, are one of the world's most extensive dinosaur graveyards. One of the many dinosaurs found there is *Gasosaurus*. Unlike the other dinosaurs found there, it was incomplete when extracted. Its name derives from the fact that it was discovered by the workers of the local oil and gas extraction company.

Features: *Gasosaurus* is one of the smallest of the carnosaurs found so far. About the size of a lion, it would have been a fierce hunter and probably the major predator of the Dashanpu fauna. We only know it from the pelvis and parts of the front and back legs, yet, despite this, the reconstruction of a whole skeleton has been on display in the Zigong Dinosaur Museum, and was part of an exhibition that toured the West in the late 1980s.

Left: Gasosaurus *lived in the same time and the same locality as* Xuanhanosaurus *(above). Its smaller size suggests that it hunted a different, probably smaller, prey.*

Distribution: Sichuan, China.
Classification: Theropoda, Tetanurae, Carnosauria.
Meaning of name: Gas company lizard.
Named by: Dong and Tang, 1985.
Time: Bathonian to Callovian stage of the middle Jurassic.
Size: 3.5m (11½ft).
Lifestyle: Hunter.
Species: *G. constructus.*

Monolophosaurus

The almost complete skeleton of this medium-sized meat-eating dinosaur was found in 1984, and was originally referred to by the genus name *Jiangjunmiaosaurus*, after the region in which it was discovered. Its more descriptive name, referring to the single spectacular crest on the head, was awarded after it was scientifically described by a joint Chinese-Canadian study ten years later.

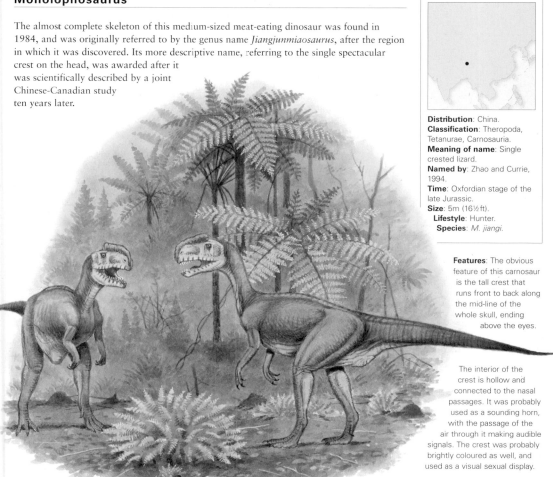

Distribution: China.
Classification: Theropoda, Tetanurae, Carnosauria.
Meaning of name: Single crested lizard.
Named by: Zhao and Currie, 1994.
Time: Oxfordian stage of the late Jurassic.
Size: 5m (16½ ft).
Lifestyle: Hunter.
Species: *M. jiangi*.

Features: The obvious feature of this carnosaur is the tall crest that runs front to back along the mid-line of the whole skull, ending above the eyes.

The interior of the crest is hollow and connected to the nasal passages. It was probably used as a sounding horn, with the passage of the air through it making audible signals. The crest was probably brightly coloured as well, and used as a visual sexual display.

TETANURAE CLASSIFICATION

The word Tetanurae means 'stiffened tail'. This refers to the fact that the tail was not the flexible swinging structure that we have seen in its sister group, the Ceratosauria – the group that contained the earlier coelophysids – but was held out stiff and straight behind, balancing the jaws, teeth and claws as they were held out in front during an attack.

The Tetanurae comprise three main groups:

The **Carnosauria** tend to be the big ones, some of them becoming the biggest meat-eating land animals known.

The **Coelurosauria** are generally smaller (although they do contain the tyrannosaurs) and even encompass the modern birds.

The third group consists of the **Spinosauria**, from Cretaceous times.

Another feature that palaeontologists use to define the Tetanurae is the presence of an extra hole in the side of the skull between the eyes and the nostrils. This classification has only been in place since the early 1990s, and constant updates and revisions mean that it is always changing. As far as current dinosaur classification goes, we can regard the theropods as consisting of the two groups, the Tetanurae and the Ceratosauria.

Below: A coelurosaur.

Left: A spinosaur.

Left: A carnosaur.

VARIOUS THEROPODS

The meat-eating dinosaurs ranged from little chicken-sized animals to great dragon-like monsters. There were also many that fell between the two extremes. From China to Australia, South America and Great Britain, the medium-size, meat-eating dinosaurs were widespread in middle Jurassic times. At that time the continents were beginning to split up, but the same kinds of animal still existed everywhere.

Ozraptor

This is the only Jurassic theropod to be found in Australia so far. In middle Jurassic times Australia was at the far reaches of the southern arm of the supercontinent Pangaea. It was still part of the major landmass and so, despite its remoteness, there is no reason why it should not have supported a similar fauna to the rest of the world.

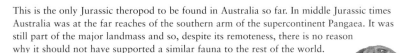

Features: *Ozraptor* is only known from a part of a leg bone, but from it palaeontologists can infer that the animal was a swift-footed carnosaur. The joint at the ankle end is unique among theropods, and seems to have been adapted for fast running. In this respect it resembles that of a basal dromaeosaur, one of the swift hunters that existed in North America in Cretaceous times. There are probably many more Jurassic theropods to be discovered in Australia.

Right: The leg bone of Ozraptor *was found in 1966 by college students and was originally thought to have belonged to a turtle. It was not until 30 years later that it was removed from its matrix and seen to be the bone of a dinosaur.*

Distribution: Western Australia.
Classification: Theropoda, Tetanurae.
Meaning of name: Oz (a colloquial name for Australia) lizard.
Named by: Long and Molnar, 1998.
Time: Bajocian stage of the middle Jurassic.
Size: Maybe 3m (10ft).
Lifestyle: Fast hunter.
Species: *O. subotaii.*

Kaijiangosaurus

This is an obscure theropod from the famous Dashanpu fossil beds of China. As with many potentially interesting dinosaurs it is only known from a partial skeleton from which we can deduce very little. Like its contemporary, *Gasosaurus*, it would have preyed on the vast range of plant-eating sauropod dinosaurs that inhabited this area during middle Jurassic times.

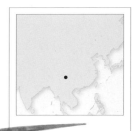

Features: All that is known of *Kaijiangosaurus* is a group of seven neck vertebrae. They are definitely the vertebrae of a carnosaur but are very primitive, lacking the ball-and-socket articulations that we usually find. Their unspecialized nature suggests that *Kaijiangosaurus* may be very close to the ancestors of the carnosaur group. There is even a suggestion that it is the same animal as *Gasosaurus*, which comes from the same time and place.

Above: Kaijiangosaurus *and* Gasosaurus *(inset) would have been the main predators of the sauropods found in the Dashanpu quarries.*

Distribution: China.
Classification: Theropoda, Tetanurae.
Meaning of name: Lizard from the Kaijiang River.
Named by: He, 1984.
Time: Bathonian to Callovian stages of the middle Jurassic.
Size: 6m (19½ft).
Lifestyle: Hunter.
Species: *K. lini.*

THE BODY SHAPE OF THE MEAT-EATERS

Plant-eating dinosaurs developed all kinds of different shapes during the 160-million year tenure of the Age of Dinosaurs. There were the big, elephantine, long-necked sauropods, the two-footed ornithopods that were either small enough to sprint away from danger or found safety in herds, the heavy thyreophorans with their decoration of plates or their defence of armour, and the later ceratopsians with their shields and horns.

For all that, the meat-eating dinosaurs seemed to have adopted a single shape and stuck with it for the entire Mesozoic. It was a shape that worked. The simple carnivorous digestive system did not need a big body. The resulting animal was light enough to be carried on two legs. The killing mechanisms – the jaws, teeth and claws – could be carried well forward, balanced by the tail behind. The typical theropod was a splendidly designed killing machine.

Enough complete skeletons of theropods are known to tell us that few of them departed from this basic shape. Accordingly when new animals turn up, and they consist only of a few bones and teeth, we can be confident in the prediction that the whole animal would not have differed much from this basic shape.

Right: The shape of the meat-eating dinosaur.

Piatnitzkysaurus

The discovery of this carnosaur in the 1970s was the first indication that relatives of the North American dinosaurs actually existed in South America in middle Jurassic times. The implication is that North and South America were connected at that time, and the seaway between them that was known to exist later had not yet formed, isolating South America as an island continent. It is known from two partial skulls and some skeletal material. It is regarded as having the dinosaur name that is most difficult to spell!

Distribution: Argentina.
Classification: Theropoda, Tetanurae, Carnosauria.
Meaning of name: Piatnitzky's lizard, after a friend of the discoverer.
Named by: Bonaparte, 1986.
Time: Callovian to Oxfordian stages of the middle to late Jurassic.
Size: 4.3m (14ft).
Lifestyle: Hunter.
Species: *P. floresi*.

Features: This early ancestor of the *Allosaurus* group differed from its later relative in the more primitive set of hip bones, a longer arm bone and more powerful shoulders. It has a very similar build to *Allosaurus*, and presumably pursued the same hunting lifestyle. As it is much smaller than the local sauropods, it may have concentrated on hunting juveniles or the weak and ageing members of the plant-eater herds.

Proceratosaurus

When first studied, by noted British palaeontologist Sir Arthur Smith Woodward in 1910, this dinosaur was thought to have been yet another *Megalosaurus*. German dinosaur expert Friedrich von Huene just as erroneously attributed it to the *Ceratosaurus* line when he studied the skull 15 years later. Another specimen of what could be the same genus was found in France in 1923.

Distribution: Gloucestershire, England.
Classification: Theropoda, Tetanurae, Carnosauria.
Meaning of name: Before Ceratosaurus.
Named by: von Huene, 1926.
Time: Bathonian stage of the middle Jurassic.
Size: 2m (6½ft).
Lifestyle: Hunter.
Species: *P. bradleyi*, *P. divesensis*.

Features: *Proceratosaurus* is known only from a fragmentary skull, which shows the base of a horn on the nose. This led von Huene to surmise that it was an ancestor of the later horned *Ceratosaurus*. Although there is not much in the way of skull material to study, it certainly seems to be a carnosaur rather than a ceratosaur, which it would be if it really were related to *Ceratosaurus*.

Right: Despite the paucity of remains, it seems as if Proceratosaurus can be placed in the carnosaur family. That makes it the earliest well-known member of the group.

THE SAUROPODS OF THE DASHANPU QUARRY, CHINA

The Dashanpu quarry, near Zigong in Sichuan Province, China, is the world's most famous middle Jurassic dinosaur site. In recent decades more than 40 tonnes of fossils, or more than 8,000 dinosaur bones, have been excavated here. In 1987 the Zigong Dinosaur Museum opened, celebrating this richness.

Omeisaurus

The first skeleton of this dinosaur was unearthed in 1939 by Chinese palaeontologist Young Chung Chien (or more correctly, Yang Zhongjian) and American Charles L. Camp. As with *Mamenchisaurus*, the length of the neck was not obvious at first. This fact only came to light with the discovery of a more complete skeleton in the 1980s. The long necks may have been used to reach food high in the trees, or over large areas on the ground. Remains of tail clubs found at the same site have been attributed to *Omeisaurus*. However, it is more likely that they belong to large specimens of *Shunosaurus*.

Features:
Omeisaurus, with its extremely long neck, short body and stocky limbs, must have looked rather like the slightly earlier *Mamenchisaurus*. Indeed, some specimens of *Omeisaurus* have been mis-identified as *Mamenchisaurus*. The difference lies in the shape of the vertebrae in the back – the spines are divided in *Mamenchisaurus*, but in *Omeisaurus* they are not. Small features like these help palaeontologists to distinguish one genus from another.

Left: Like most sauropods the hips of Omeisaurus *were higher than the shoulders. The nostrils were well forward on the skull – an unusual feature.*

Distribution: China.
Classification: Sauropoda.
Meaning of name: Lizard from Mount O-mei.
Named by: Young, 1939.
Time: Kimmeridgian to Tithonian stages of the late Jurassic.
Size: 15m (50ft).
Lifestyle: Browser.
Species: *O. junghsiensis*, *O. tianfuensis*, *O. luoquanensis*.

Shunosaurus

Sometimes known as *Shuosaurus*, this is the best-known of the Chinese dinosaurs. There have been about 20 individual skeletons found, some of them complete. Unusually for a dinosaur, every single bone of *Shunosaurus* is known. Palaeontologists regard this as a very common dinosaur of middle Jurassic times, and probably the most abundant in the eastern part of the Laurasian continent.

Features: The remarkable feature of *Shunosaurus* is the spiked club on the tail, evidently used as a defensive weapon. When first found, in 1979, it was thought to have been an abnormal growth at the site of an injury. However, several fully articulated skeletons have been excavated with the club in place. The skull is relatively long and low with the nostrils pointing sideways. In sauropods the teeth are usually spoon-shaped or pencil-shaped; in *Shunosaurus* they seem to be between the two.

Distribution: China.
Classification: Sauropoda.
Meaning of name: Shu (the old local name for Sichuan Province) lizard.
Named by: Dong, Zhou and Chang, 1983.
Time: Bathonian to Callovian stages of the middle Jurassic.
Size: 9m (30ft).
Lifestyle: Browser.
Species: *S. lii*, *S. ziliujingensis*.

Mamenchisaurus

When it was discovered in 1952, the nature of the *Mamenchisaurus constructus* vertebrae was unclear. The vertebrae were so delicate and poorly preserved that they could not be excavated undamaged. Early restorations of this animal showed only a moderate length of neck. It was with the discovery of *M. hochuanensis*, in 1957 that the fantastic length of the neck was appreciated.

Features: *Mamenchisaurus* has the longest neck of any known dinosaur. It consists of 19 vertebrae – the greatest number so far found – and is about 14m (46ft) long, taking up about two-thirds of the length of the entire animal. The vertebrae are very thin and lightweight, made up of fine struts and sheets, rather like the later diplodocids. However, the short, deep skull shows that it belongs to the more primitive euhelopid group.

Distribution: China; Mongolia.
Classification: Sauropoda, Euhelopodidae.
Meaning of name: Lizard from Mamen Brook.
Named by: Young, 1954.
Time: Tithonian stage.
Size: 21m (69ft).
Lifestyle: Browser.
Species: *M. sinocanadorum, M. hochuanensis, M. youngi, M. jingyaninsis, M. anyuensis, M. constructus.*

Left: The extreme length of its neck may have enabled Mamenchisaurus to reach in between closely-spaced trees to eat undergrowth in dense woodland.

THE DASHANPU SITE

Middle Jurassic dinosaur-bearing rocks are not common. Practically the only site that has shown a good variety of middle Jurassic dinosaurs is the Dashanpu mudstone quarry, in the Zigong region of Sichuan Province, in central China. The site was discovered in the 1960s by a construction crew installing a gas pipeline. The work was halted while the site was excavated, with the eminent Chinese palaeontologist Dong Zhiming of Beijing leading the work.

In Jurassic times this area was a well-watered lowland, with rivers and lakes across it supporting dense forests of conifers, cycads and ferns. There was a vast river delta in the area, which existed for many millions of years, from early, through middle, to late Jurassic times, depositing its muds and silts very gently all the time. It is not usual for the geographic conditions to remain constant in a particular place for such a long time.

Dead dinosaurs were embedded in the mud and their skeletons preserved without being broken up. Fossils of pterosaurs, mammals and amphibians have also been found here. The mudstones contain the fossils of the trunks of huge conifer trees, and the skeletons of a wide selection of dinosaurs, including theropods, stegosaurs and a particularly wide array of long-necked sauropods.

Left: An open plain enviroment.

Datousaurus

This dinosaur is known from two incomplete skeletons. The skull that has been attributed to *Datousaurus* was actually found some distance away from the skeletons, and so there is some uncertainty as to whether it actually belongs to this animal: the Dashanpu quarry is full of all sorts of sauropods. Hence there is some confusion over its classification.

Features: *Datousaurus* has a neck that is longer than that of most early sauropods, but not as long as those to come. There are 13 vertebrae in the neck. The skull is large and heavy for a sauropod, with the nostrils at the front. The jaws have spoon-shaped teeth. The skeleton is *Diplodocus*-like, but the skull, if it is indeed the skull of the same animal, suggests that *Datousaurus* belongs to a different sauropod line altogether.

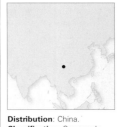

Distribution: China.
Classification: Sauropoda, Cetiosauridae or possibly Euhelopidae.
Meaning of name: Chieftain lizard.
Named by: Dong and Tang, 1984.
Time: Bathonian to Callovian stages of the middle Jurassic.
Size: 15m (50ft).
Lifestyle: Browser.
Species: *D. bashanensis.*

Right: Datousaurus is regarded as one of the least abundant of the sauropods that existed in central China in the middle Jurassic.

CETIOSAURS

The cetiosaurs were the most primitive family of the sauropod dinosaurs. They were discovered in Europe and were among the first dinosaurs to be recognized – but now they are also known from fossils in Australia, Africa, North America and, significantly, South America. As a rule, their front and hind legs were more or less the same length, unlike most sauropods in which the front legs tend to be shorter.

DEVELOPMENT OF THE EVOLUTIONARY LINE

The sauropods were the natural successors to the prosauropods of the Triassic and early Jurassic periods. They had the same big bodies with the huge plant-food processing gut, and the long necks that enabled the little heads to reach high into trees or in a wide arc on the ground. In fact, many of the more advanced prosauropods looked very much like primitive sauropods, and there is often confusion in the classification about where the line should be drawn between the two.

South America had a wide variety of prosauropods in early times. In the later Cretaceous period there was a large number of sauropods, especially in the advanced titanosaurid family. The late sauropods flourished in South America towards the end of the age of dinosaurs, even though they were on the wane in the rest of the world at the time. However, in between, there is little evidence of what was evolving. The few middle Jurassic sauropods that we know seem to be closely related to the cetiosaurids that were more typical of Europe at that time.

Cetiosaurus

This was the first of the sauropods to be discovered – in 1825 – and described. As such it suffered the same fate as its contemporary theropod, *Megalosaurus*: for a long time any sauropod remains found in Europe were attributed to it.

Sir Richard Owen, who named it in 1842, the same year that he invented the name "dinosaur", did not recognize it as one of this group of animals. He thought that he was studying the remains of a gigantic crocodile. He saw similarities in the backbones to those of a whale, hence the name he gave it, and assumed that it was an aquatic animal. It was dinosaur pioneer Gideon Mantell who recognized these bones as dinosaur bones in 1854.

Cetiosaurus remains have mostly been found in marine deposits, suggesting that this dinosaur lived close to the sea, or at least close to rivers where the remains could be washed out to sea. The best *Cetiosaurus* skeleton was found in 1968 by workmen digging in a clay pit in the English Midlands. This skeleton is now on display in the Leicester Museum and Art Gallery.

Distribution: England; Portugal; and perhaps Morocco.
Classification: Sauropoda, Cetiosauridae.
Meaning of name: Whale lizard.
Named by: Owen, 1842.
Time: Bajocian to Bathonian stages of the middle Jurassic.
Size: 14m (46ft).
Lifestyle: Browser.
Species: *C. mogrebiensis, C. medius, C. conybearei, C. oxoniensis.*

Features: The *Cetiosaurus* skeleton is that of a typical sauropod, with a small head, long neck and tail, and a heavy body supported by elephantine legs. It is quite primitive because the vertebrae are not hollowed out as a weight-saving measure as they are in more advanced members of the group. Instead they are spongy and coarse – the whale-like feature that was noted by Owen.

Amygdalodon

This is the earliest-known South American sauropod. It predates, by tens of millions of years, the varied titanosaur sauropods that were to dominate that continent in Cretaceous times. It is possible that *Amygdalodon*'s ancestors are to be found among the many prosauropods that existed in South America, or it may show that the primitive cetiosaurid group had spread worldwide from its European origin. The name derives from the shape of the teeth, which are almond-like. This is the most primitive sauropod known from South America.

Features: It is known from the teeth, parts of the vertebrae, some ribs and hip bones and a fragment of limb from two individuals. As usual, when only part of a skeleton has been found and identified, we have to assume that the whole animal followed the long-necked sauropod form. Its teeth suggest that it is a member of the cetiosaurid group, mostly known from Europe. If this is the case it indicates how widespread this early line of the sauropods was in middle Jurassic times, when all the continents were still joined as a single landmass.

Distribution: Argentina.
Classification: Sauropoda, Cetiosauridae.
Meaning of name: Almond-shaped tooth.
Named by: Cabrera, 1947.
Time: Bajocian stage of the middle Jurassic.
Size: 13m (43ft).
Lifestyle: Browser.
 Species: *A. patagonicus*.

Right: Although we presume Amygdalodon *to have been long-necked and long-tailed, it seems likely that the neck and tail were not quite as long as those in later sauropods.*

Patagosaurus

In the late 1970s the skeletons of about a dozen *Patagosaurus* were found together, five of them with skulls, suggesting that they moved about in herds. This discovery makes *Patagosaurus* the best-known of the early South American sauropods. Their presence, along with that of theropods such as *Piatnitzkysaurus*, shows that South America was connected to the other landmasses at that time.

Features: The teeth are similar to those of the earlier *Amygdalodon*. The rest of the skeleton, however, is more advanced. The skeleton resembles that of *Cetiosaurus* because the neck vertebrae have the same kind of undivided spines, and the back vertebrae have shallow cavities rather than the deep hollows of later sauropods. These simple backbones are largely what define the cetiosauridae family. However, the hip bones and the tail are different from its European relative. A fine mounted skeleton of *Patagosaurus* stands in the Argentine Museum of Natural Sciences in Buenos Aires, and is dedicated to the pioneer Argentinian scientist Bernardino Rivadavia.

Distribution: Argentina.
Classification: Sauropoda, Cetiosauridae.
Meaning of name: Lizard from Patagonia.
Named by: Bonaparte, 1979.
Time: Callovian stage of the middle Jurassic.
Size: 18m (60ft).
Lifestyle: Browser.
Species: *P. fariasi*.

Right: Patagosaurus *moved about in groups. The finding of two adults and three juveniles together supports this.*

SAUROPOD MISCELLANY

Although widespread, the cetiosaurids were not the only sauropods of the middle Jurassic period. By this time the main families of the sauropod clan (the diplodocids, cararsaurids and the brachiosaurids) seem to have evolved. The fact that all the landmasses were still together as a single continent meant that these dinosaurs became quite wide-ranging.

Rhoetosaurus

On the sheep farm, or station, at Durham Downs in Queensland, the manager, Arthur Browne, found some fossil bone fragments in 1924. He sent them to the Museum of Queensland where Dr. Heber Longman identified them as part of a sauropod dinosaur. The rest of the skeleton was not uncovered until 1975 when Dr. Mary Wade continued excavating there.

Features: *Rhoetosaurus* is one of only two well-known sauropods from Australia, the other being the early Cretaceous *Austrosaurus*. It is certainly the biggest, with a thigh bone that is 1.5m (5ft) long. It is probably a cetiosaurid, but it is difficult to be sure. The tail seems to have been

Distribution: Queensland.
Classification: Sauropoda, Cetiosauridae.
Meaning of name: Browne's Rhoetos lizard (a giant from Greek mythology).
Named by: Longman, 1925.
Time: Aalenian to Bajocian stages of the middle Jurassic.
Size: 12m (39ft).
Lifestyle: Browser.
Species: *R. brownei*.

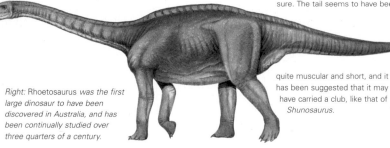

Right: Rhoetosaurus *was the first large dinosaur to have been discovered in Australia, and has been continually studied over three quarters of a century.*

quite muscular and short, and it has been suggested that it may have carried a club, like that of *Shunosaurus*.

Cetiosauriscus

Cetiosauriscus was one of those many sauropods that was assumed to be a species of *Cetiosaurus* when it was discovered. It was described by Owen in 1842. Von Huene realized it was not quite right for *Cetiosaurus* and renamed it *Cetiosauriscus* in 1927. Other species have since been found, the most significant being C. stewarti.

Features: Far from being a primitive cetiosaurid, this was probably an early genus of the more advanced diplodocids. The only part of the skeleton that is well known is the rear half and the legs. The main diplodocid-like feature is the long, whip-like tail. No other diplodocid is known from Europe – they are mostly found in North America and Africa. As usual it is difficult to fix the classification without the skull.

Distribution: England; Switzerland.
Classification: Sauropoda, Diplodocidae.
Meaning of name: Like Cetiosaurus.
Named by: Charig, 1980.
Time: Bajocian to Tithonian stages of the middle to late Jurassic.
Size: 15m (49ft).
Lifestyle: Browser.
Species: *C. stewarti*, *C. glymptonensis*, *C. greppini*, *C. longus*.

SAUROPOD CLASSIFICATION

The sauropods – the big long-necked plant-eaters – were very varied. As the Mesozoic period progressed they split into a number of different lines, most of which can be traced back into middle Jurassic times.

The actual classification is, as usual, constantly undergoing revision, as is the evolutionary sequence that brought these groups about. However, several main groups can be easily recognized. They are:
- Cetiosaurids – with the middle Jurassic Chinese sauropods, they are regarded as fairly primitive. They tended to have front and back legs about the same length.
- Diplodocids – the long, low-slung sauropods were the diplodocids. They had narrow heads and were lightly built for sauropods. Nicely balanced about the hips, they may have been able to rise to their hind legs from time to time for high feeding.
- Camarasaurids – the big, boxy skulls with the huge nasal openings and the spoon-shaped teeth distinguished the camarasaurids.
- Brachiosaurids – they were the big, heavy sauropods, with front legs longer than the hind, and the long necks that took the heads high into the trees.
- Titanosaurids – the last of the sauropods were the titanosaurids. When all other groups had died out by the early Cretaceous period, they held on until the end, many of them sporting a distinctive back armour.

Volkheimeria

Not much of this animal is known, just part of the backbone and hips, and a hind limb. What can be deduced, though, is quite interesting. It appears to be a brachiosaurid, although the brachiosaurid features are not very pronounced and the backbone is quite primitive, giving further evidence for South America, and indeed the rest of the southern continents, being attached to the northern landmasses at that time.

Features: The spines protruding from the vertebrae in the hip region are low and quite flat. If it is a brachiosaur it is a very primitive one. The nearest relative seems to be the Madagascan *Lapparentosaurus* from the same time. The southern portion of the supercontinent Pangaea consisted of Madagascar, Africa and South America, as well as India, Australia and Antarctica, and so it is possible that the same kinds of animals were widespread over this region.

Distribution: Argentina.
Classification: Sauropoda, Macronaria.
Meaning of name: Wolfgang Volkheimer's (an Argentinean palaeontologist) thing from Chubut Province.
Named by: Bonaparte, 1979.
Time: Callovian stage of the middle Jurassic.
Size: 9m (30ft).
Lifestyle: High browser.
Species: V. chubutensis.

Left: Volkheimeria was found and named by José Bonaparte, who found more than 20 new types of dinosaur in South America in the 1970s and 1980s and encouraged a whole new generation of South American palaeontologists.

Lapparentosaurus

Lapparentosaurus was first attributed to the European wastebasket genus *Bothriospondylus* but now, after further study, it appears to be very close to the ancestor, if not the actual ancestor, of *Brachiosaurus* that did not appear for another 20 million years. It is difficult to know, though, since the skeletons found are of youngsters only about 1.8m (6ft) long.

Features: This animal is known from a few, reasonably complete skeletons of juveniles, unfortunately lacking the heads. The vertebrae, although distinctly brachiosaurid in shape, are more primitive than those of other brachiosaurids. Studies by Rimblod-Baly, de Ricqles and Zylberberg of the bones of these youngsters show that they grew quickly while they were very young, and reached their full size within a few years. This is now accepted as the normal growth pattern among sauropods.

Distribution: Madagascar.
Classification: Sauropoda, Macronaria.
Meaning of name: Albert F. de Lapparent's lizard from Madagascar.
Named by: Bonaparte, 1986.
Time: Bathonian stage of the middle Jurassic.
Size: 18m (59ft) maybe.
Lifestyle: High browser.
Species: L. madagascariensis.

Left: Lapparentosaurus is known only from juvenile skeletons, so its final size at maturity is open to dispute.

MORE SAUROPODS

As the Jurassic period developed, the sauropods expanded and spread over the world. Different families evolved, having their heydays at different times. They all had the same general shape – big bodies supported on elephantine legs, long necks and tails, and tiny heads. The end of the Jurassic saw the different families reach their greatest abundance and diversity.

Ferganasaurus

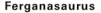

In 1966 an expedition from the Leningrad State Museum, led by N. N. Verzin, found the first sauropod skeleton to be unearthed in the then USSR, in the Chatkal Range of Kyrgyzstan. It was not formally described but it was given the name *Ferganasaurus*. The site was re-excavated by Alexander O. Averianov in 2000, and his team found more bones.

Features: *Ferganasaurus* is a medium-sized sauropod that has many rather primitive features. Unfortunately, the original skeleton found in 1966 is lost, and today's palaeontologists only have drawings of the bones to work with. We do not know what the skull or the feet are like – a severe handicap when determining its relationships with other sauropods.

Left: Without the skull, which has never been found, and with only a few bones of the feet preserved, the restoration shown here is speculative.

Distribution: Kyrgyzstan.
Classification: Sauropoda.
Meaning of name: Lizard from the Fergana Valley.
Named by: Alifanov and Averianov, 2003.
Time: Callovian stage of the middle Jurassic.
Size: 9m (30ft).
Lifestyle: Browser.
Species: *F. verzilini*.

Bellusaurus/Klamelisaurus

Distribution: China.
Classification: Sauropoda, Klamelisaurinae.
Meaning of name: (*Klamelisaurus*) Klameli lizard. (*Bellusaurus*) Beautiful lizard.
Named by: (*Klamelisaurus*) Zhao Xijin, 1993; (*Bellusaurus*) Dong, 1990.
Time: Oxfordian stage of the late Jurassic.
Size: 17m (56ft).
Lifestyle: Browser.
Species: *B. sui, K. gobiensis*.

Right: Klamelisaurus had spoon-shaped teeth and tall spines on the backbone.

Bellusaurus is known from a skull and parts of the skeletons of 17 or so individuals – all small – found in 1954, and jumbled up together in a bone bed in the Junggar Basin in north-western China. The deposit seems to have been formed when a herd was caught in a flood and the smaller ones died.

Features: The complete skeleton of *Klamelisaurus*, found later than the *Bellusaurus* bone bed, is regarded as the adult form of this animal. Palaeontologists regard *Klamelisaurus* as an intermediate form in sauropod evolution – it was a fairly advanced sauropod that still retained primitive features. Palaeontologists have proposed a new sauropod family, Klamelisaurinae, which is related to the Brachiosauridae, to accommodate it.

Europasaurus

Europasaurus is a sauropod example of the results of island dwarfism – the evolution of a small body size as a response to the limited resources found on islands. Other dinosaur examples include the ankylosaur *Struthiosaurus* and the hadrosaur *Telmatosaurus*. Once the ancestors of *Europasaurus* had established themselves on the island chain, with an estimated area of 2000 sq km/1,242sq miles, that stretched across northern Europe at that time they would quickly have evolved the small size that enabled them to survive.

Below: Apart from the small size, Europasaurus resembled the other macronarians.

Features:
The most obvious feature of *Europasaurus* is its small size, with a shoulder height no greater than the shoulder height of an average human. Otherwise it is very similar to other macronarian sauropods like *Camarasaurus*, but distinguishable from then by subtle differences in the bones of the skull and the shape of the bones of the forelimb.

Distribution: Northern Germany.
Classification: Sauropoda, Macronaria.
Meaning of name: Lizard from Europe.
Named by: Mateus, Laven and Knötschke, 2006
Time: Kimmeridgian stage of the late Jurassic.
Size: 6m (20ft)
Lifestyle: Island-based browser.
Species: *E. holgeri*.

Bothriospondylus

Bothriospondylus has been something of a wastebasket genus, with several incomplete specimens attributed to it over the last century. The fact that its remains have been found as far apart as England and Madagascar is an indication of this. There has been no modern work done on *Bothriospondylus*, and some specimens have been reassigned to the genus *Pelorosaurus*.

Features: *Bothriospondylus* was based on only four vertebrae and part of the hip bone of a juvenile sauropod, studied by Sir Richard Owen in the early days of dinosaur science. Owen found the hollows in these vertebrae to be unusual and named it accordingly. We now know that most of the sauropods had deep hollows in their backbones – a weight-saving device. It seems likely that the original

Bothriospondylus was an early member of the brachiosaurids despite the fact that the group is usually characterized by vertebrae made of lightweight struts and plates.

Right: The specimens of Bothriospondylus from France and from Madagascar consist of nothing but a few vertebrae. These have the same distinctive massive form as the original English specimens.

Distribution: England; France; Madagascar.
Classification: Sauropoda, Macronaria.
Meaning of name: Furrowed or hollowed vertebra.
Named by: Owen, 1875.
Time: Bathonian to Kimmeridgian stages of the middle and late Jurassic.
Size: 15–20m (49–66ft).
Lifestyle: Browser.
Species: *B. suffosus*, *B. madagascariensis*, *B. robustus* (*nomena dubia*).

ORNITHOPODS

*The ornithopods were a group of plant-eating dinosaurs, part of the much larger ornithischian order.
Their remains have been found in locations from the prolific quarries of Dashanpu, in China, to the
coast of Portugal. However, in middle Jurassic times they were not as abundant or as varied as their
sauropod cousins within the dinosaur group. Their time was yet to come.*

Yandusaurus

Known from two almost complete skeletons including the skulls, we have a good idea of
what *Yandusaurus hongeensis* looked like. Another species of *Yandusaurus*,
Y. multidens, is probably really *Agilisaurus*. It is one of the dinosaurs from the
Dashanpu quarries, which are famed for their
sauropod remains.

Features:
Yandusaurus is a
typical member of the
hypsilophodont family. It is a small, two-footed
plant-eater with very long legs, and was built for
speed. It has a small head and chopping teeth. The
food was held in cheek pouches while it was being
chewed. Hypsilophodonts were more common in
Cretaceous times but their ancestry can be traced
back to the middle Jurassic. *Yandusaurus* is
distinguished by the ridges on the teeth.

*Above: The short forearms of
Yandusaurus had the full
complement of five fingers.*

Distribution: China.
Classification: Ornithopoda,
Hypsilophodontidae.
Meaning of name: Lizard
from Yandu.
Named by: He, 1979.
Time: Bathonian to Callovian
stages of the middle Jurassic.
Size: 1.5m (about 6ft).
Lifestyle: Low browser.
Species: *Y. hongheensis*,
Y. multidens (possibly).

Agilisaurus

Despite the fact that *Agilisaurus* is known from an almost
complete skeleton, its actual position in the dinosaur
evolutionary hierarchy is still unclear. Paul Barrett, of the
Natural History Museum, London, England, suggests that it
is too primitive to be a conventional ornithopod. Perhaps it
is a fabrosaurid, one of the very primitive ornithopods that
had not yet evolved the cheek pouches, or perhaps it is a
new family
altogether.

Features: The skull of *Agilisaurus*
is small, with big eyes, and its
teeth are leaf-shaped. The teeth
are larger and pointed at the
front. The hind limbs are long,
much longer than the front limbs.
The thigh bone is particularly
short compared with the rest of
the leg bones – a sign of a fast-
running
animal with
a lightweight
foot and all
the muscle
concentrated
near the hip.
The tail is
long and
used for
balance
while
running.

Distribution: China.
Classification: Ornithopoda.
Meaning of name: Fast
lizard.
Named by: Peng, 1992.
Time: Bathonian to Callovian
stages of the middle Jurassic.
Size: 1.2m (4ft).
Lifestyle: Low browser.
Species: *A. louderbecki*,
Yandusaurus multidens may
well be a species of
Agilisaurus.

*Left: Despite the similarities between the skeletons
of Agilisaurus and Yandusaurus (above), these are
thought to have been quite different animals.*

"Xiaosaurus"

This animal is known only from the teeth and jawbone, and a few bones from the rest of the skeleton including the hind leg. So little is known about this small dinosaur that "*Xiaosaurus*" is really a *nomen dubium*. It may even be a species of *Agilisaurus*. It does show, however, that there was quite a range of basal ornithopods in middle Jurassic China.

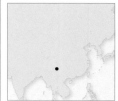

Distribution: China.
Classification: Reptile, Ornithopoda.
Meaning of name: Small lizard.
Named by: Dong, Tang, 1983.
Time: Bathonian stage of the middle Jurassic.
Size: 1m (3ft).
Lifestyle: Low browser.
Species: "*X. dasanpensis*".

Right: Even if "Xiaosaurus" is found to be ancestral to the horned dinosaurs, it would still have the general appearance shown here

Features: The remains, found in the Dashanpu quarries, are so fragmentary that very little can be deduced about this animal. However, an odd thing about the femur is the arrangement of leg muscle attachments. They seem to be very similar to those of the primitive horned dinosaurs, and this has led to speculation that *Xiaosaurus* may be close to the ancestry of such dinosaurs as *Triceratops* that did not flourish until late Cretaceous times.

ESSENTIAL DIFFERENCES

From a distance a small ornithopod would have looked very similar to a small theropod – both would have been standing on hind legs and balanced by a heavy tail. If we visited Mesozoic times we would have to know the difference, as one would be dangerous.

The most obvious difference between the two types is the size of the body. The more complex, plant-eating digestive system of the ornithopod would have meant that its body would be bigger. Then there is the shape of the head. The theropod had long jaws and sharp teeth, and probably eyes pointing forward. The ornithopod would have a smaller head, big eyes on the side, usually cheeks at the side of the mouth and a beak at the front.

The hands would be different, too. The theropod would have three, or even two fingers, and be sporting big, curved, claws. The ornithopod would have four or five fingers, with blunt claws. Finally, the colour. This is something we know nothing about, but it would seem likely that theropods would be brightly coloured, like birds or tigers, while ornithopods would be camouflaged, with greens and browns.

Below: Theropod (left) and ornithopod (right) for comparison.

"Alocodon"

All we know about this dinosaur, unfortunately, are a few distinctive teeth. This is a usual state of affairs in vertebrate palaeontology. Sometimes the teeth can be used to reconstruct the whole animal, but more often they pose more questions than they answer. These teeth were found near Pedrógão, Portugal, one of the few dinosaur sites in that country.

Features: The teeth of *Alocodon* have vertical grooves in them, hence the name. The only other dinosaur teeth that they resemble are those of the hypsilophodont *Othnielia*, which came later in the evolutionary line. Some palaeontologists regard this animal as being intermediate between the lesothosaurids and the hypsilophodonts. A set of sharper teeth furnished the front of the upper jaw. There has even been a suggestion that these teeth come from a primitive thyreophoran, one of the plated or armoured dinosaurs. Whatever classification is finally decided upon, "Alocodon" is featured as a primitive ornithopod here.

Distribution: Portugal.
Classification: Ornithopoda.
Meaning of name: Furrowed tooth.
Named by: Thulborn, 1973.
Time: Middle or late Jurassic.
Size: 1m (3ft).
Lifestyle: Low browser.
Species: "*A. kuehni*".

Left: An early mammal discovered in 1977 and named Alocodon had to have its name changed to Alocodontulum in 1978 as it was found that the name Alocodon had already been given to this dinosaur.

EARLY THYREOPHORANS

Heavy, four-footed, bird-hipped dinosaurs, with ornamentation and armour on their backs and tails, appeared in middle Jurassic times. These were the thyreophorans, part of the ornithischian order. The earliest to come to prominence, in the late Jurassic period, were the stegosaurs, but they were replaced in Cretaceous times by the ankylosaurs. The early forms of both were present in the middle Jurassic period.

Huayangosaurus

This primitive stegosaurid is known from complete adult skeletons found in the Dashanpu quarries, China, in the early 1980s. It was re-described in 1992. The arrangement of teeth, and the fact that its front legs are long for a stegosaurid put it so far from later stegosaurids that it is placed in a family of its own.

Features:
Huayangosaurus has a double row of heart-shaped plates on the neck that are replaced by long, narrow, spine-like plates on the back. They become smaller on the tail, ending about half-way down. The tip of the tail is furnished with two pairs of spines. The skull has teeth at the front of the mouth – something that was lost in later stegosaurids – and there is a pair of horns near the eyes.

Distribution: China.
Classification: Thyreophora, Stegosauria.
Meaning of name: Lizard from Huayang.
Named by: Dong, Tang and Zhou, 1982.
Time: Bathonian to Callovian stages of the middle Jurassic.
Size: 4m (13ft).
Lifestyle: Low browser.
Species: *H. taibaii*.

Lexovisaurus

An early stegosaur that roamed the islands that dotted the shallow sea covering Europe in middle Jurassic times was *Lexovisaurus*. Since it was discovered in the 1880s, several different names have been given to it, including *Omosaurus* and *Dacenturus*. At times it was regarded as a species of *Stegosaurus* and *Kentrosaurus*, which it resembled. The definitive study was done by R. Hoffstetter in 1957.

Features: The plates of *Lexovisaurus* are narrow and short on the neck and back, and there are several pairs of long spines on the tail. There is also a pair of long spines jutting sideways from the shoulders. Old restorations show these spines jutting from the hips.

Distribution: England; France.
Classification: Thyreophora, Stegosauria.
Meaning of name: Lizard of the Lexovi tribe (a tribe of ancient Gaul).
Named by: Hulke, 1887.
Time: Callovian to Kimmeridgian stages of the middle and late Jurassic.
Size: 5m (16½ft).
Lifestyle: Low browser.
Species: *L. vetustus*, *L. duobrivensis*.

Right: According to modern theory, it makes sense for the shoulder spines to be in a position where they could damage an enemy, using a thrusting motion, rather than at the hip.

Tianchisaurus

When discovered *Tianchisaurus* was named *Jurassosaurus* because it was so unusual for an ankylosaur to be found in rocks as early as the Jurassic period. Its cumbersome species name is derived from the initials of the cast of *Jurassic Park* – Neill, Dern, Goldblum, Attenborough, Peck, Ferrero, Richards and Mazello. Steven Spielberg, the film's director, named it because he financed the research.

Features: This small primitive ankylosaur, found in 1974, has well-developed shoulder armour and its back is covered in scutes. The head is quite heavy and there may or may not be a club on the end of its tail. The jawbone is more stegosaur-like than ankylosaur-like. In the future, it may prove to be a nodosaurid rather than an ankylosaurid. It is certainly the earliest known of the ankylosaur group, most of which lived in late Cretaceous times.

Distribution: China.
Classification: Thyreophora, Ankylosauria, Ankylosauridae.
Meaning of name: Tian Chi (Heavenly Pool) lizard.
Named by: Dong, 1993.
Time: Bathonian stage of the middle Jurassic.
Size: 3m (10ft).
Lifestyle: Low browser.
Species: *T. nedegoapeferima*.

ARMOURED DINOSAUR CLASSIFICATION

The first breakdown of Sir Richard Owen's 1842 classification of the Dinosauria came in 1870 when Thomas Henry Huxley separated off all forms of dinosaur with armour into the Scelidosauridae family grouping. This was based on the well-preserved skeleton of *Scelidosaurus*. When Harry Govier Seely divided the Dinosauria into the lizard-hipped Saurischia and the bird-hipped Ornithischia, a classification still used today, the armoured dinosaurs fell into the latter group. O. C. Marsh proposed the Stegosauria in 1896 as the group to which the armoured dinosaurs belonged, based on the shared arrangement and shape of the teeth.

The stegosaur group was split into two in 1927 by Alfred Sherwood Romer who was the first to separate the plated dinosaurs (the Stegosauria family) from the armoured dinosaurs (the Ankylosauria family). *Scelidosaurus* was grouped with the Stegosauria in Romer's classification.

Franz Baron Nopcsa in 1915 had proposed grouping them all together with the horned dinosaurs in a group called the Thyreophora. This never really caught on, but the concept was resurrected in the 1980s by Paul Sereno who used this name (but without the horned dinosaurs) to combine both the plated and the armoured dinosaurs once more. Under the current scheme *Scelidosaurus* is more closely related to the ankylosaurs than to the stegosaurs. This is how the classification stands today.

"Sarcolestes"

Sarcolestes is known only from the left half of the lower jaw. The animal is so obscure that it was first thought to have been a flesh-eating dinosaur, hence the meaning of the name. For almost a century it was regarded as the oldest known ankylosaur, but that was until the discovery of *Tianchisaurus* in China. It is often regarded as a *nomen dubium*.

Features: The jaw is very similar to that of the nodosaurid *Sauropelta*. Its ankylosaur features are the small teeth, extending all the way along the lower jaw bone with the little denticles along the edge, and an armour plate welded to the jawbone. The area that joins the two lower jaws together at the front is also typically ankylosaurid. It was probably a heavily built animal, armoured at the sides and shoulders and lacking a tail club.

Distribution: Cambridgeshire, England.
Classification: Thyreophora, Ankylosauria, Nodosauridae.
Meaning of name: Flesh thief.
Named by: Lydekker, 1893.
Time: Callovian stage of the middle Jurassic.
Size: 3m (10ft).
Lifestyle: Low browser.
Species: "*S. leedsi*".

Left: The name of this animal, meaning "flesh thief," could not be more inappropriate. Far from being carnivorous, it was a slow-moving plant-eating animal.

LATE ICHTHYOSAURS

The fish-lizards were abundant in late Jurassic oceans. Perfect specimens preserving the outline of the soft anatomy as a carbon film are known from the shale of Holzmaden, Germany. But as the period drew to a close and the Cretaceous began, their importance waned and the group died away. Only one group survived into Cretaceous times, to be placed in the wastebasket taxon "Platypterygius".

Nannopterygius

Nannopterygius is not the most common ichthyosaur in the fossil record but its remains have been found in the lithographic limestone quarries in Solnhofen in Bavaria, Germany, and also at Kimmeridge and Lyme Regis on the south coast of England. The individual vertebrae, with their distinctive pentagonal shape and their deep depressions front and back, are easily recognized when found loose. A specimen from Germany shows gastroliths – stomach stones – in its stomach area. While gastroliths are common in plesiosaurs, they are very rare in ichthyosaurs.

Features: *Nannopterygius* has a very streamlined body, with the head continuous with the shoulders and no discernable neck. It also has a longer tail than would be expected from the more teardrop-shaped later ichthyosaurs. The paddles are very small – proportionally the smallest of any ichthyosaur. The teeth are small, circular in cross section and with bulbous roots. The skull has a long snout and very large eye sockets, with rings of bone to guard against outside pressure.

Distribution: Southern England; southern Germany.
Classification: Ichthyosauria, Opthalmosauridae.
Meaning of name: Little wing.
Named by: von Huene, 1922.
Time: Kimmeridgian stage of the late Jurassic.
Size: 2m (6½ft).
Lifestyle: Fish-hunter.
Species: *N. entheciodon.*

Above: The powerful tail would have made Nannopterygius *a fast swimmer.*

"Platypterygius"

An interesting specimen of this ichthyosaur was found in Queensland, Australia, with the contents of the stomach preserved. The animal's last meal had consisted of turtle hatchlings and birds. Until then it had been assumed that ichthyosaurs fed almost exclusively on cephalopods (the tentacled group of invertebrates to which the modern octopus and squid belong), due to the presence of hooks like those of squid suckers in the stomachs of other specimens. Perhaps *Platypterygius*, the last of the ichthyosaurs, had evolved into an opportunistic feeder.

Below: "Platypterygius" is a wastebasket taxon in which all Cretaceous ichthyosaurs are placed.

Distribution: Queensland, Western Australia, Northern Territory; Wyoming, USA; possibly the Isle of Wight.
Classification: Ichthyosauria, Opthalmosauridae.
Meaning of name: Flat wing.
Named by: von Huene, 1922.
Time: Early Cretaceous.
Size: 7m (23ft).
Lifestyle: Opportunistic feeder.
Species: *P. longmani, P. americanus* and about ten others.

Features: Although the ichthyosaurs were mostly Jurassic animals, this was one of the few that survived into the Cretaceous period, after which they were replaced by the mosasaurs. Study of the head of a well-preserved specimen shows that it must have been deaf. It probably relied on other senses, such as detection of vibrations, to find its prey.

Brachypterygius

Distribution: Southern and eastern England.
Classification: Reptile, Ichthyosauria.
Meaning of name: Broad wing.
Named by: von Huene, 1922.
Time: Kimmeridgian stage of late Jurassic.
Size: 5m (16ft).
Lifestyle: Fish-hunter.
Species: *B. extremus*, *B. mordax*.

Brachypterygius was one of the ophthalmosaurs – a group of ichthyosaurs that had particularly big eyes. The eyes were braced by a circle of bone, known as the sclerotic ring, within the eyeball. This was probably to resist the pressure of deep diving. There is a species of *Plesiosaurus*, *P. brachypterygius*. The similarity of names and of lifestyle often causes confusion.

Below: Like all ophthalmosaurs, Brachypterygius had a distinctive teardrop shape.

Features: As the genus name suggests, *Brachpterygius* has a short paddle that is almost disc shaped. This is made up of a veritable pavement of little bones in which both the number of digits and the number of bones in those digits are increased. These bones would have been lashed together with gristle to form a rigid structure. The way in which this paddle articulates to the rest of the body is the distinguishing feature between this ichthyosaur and the others. There is currently a great deal of work being done on the classification of ichthyosaurs. It may be that several genera that were thought to have been distinct will soon be regarded as species of *Brachypterygius*.

THE SWIMMING ICHTHYOSAUR

We see restorations of ichthyosaurs leaping above the waves, curving gracefully like dolphins and plunging back into the water. It is a tempting image, given the resemblance of the typical ichthyosaur to a dolphin. However, the graceful arching leap of a dolphin is achieved by the fact that its backbone articulates in an up and down plane. The flukes of the tail are horizontal, producing a motion in which the animal is driven along in a vertical undulating manner.

The principal articulation of the backbone of an ichthyosaur, however, is from side to side, like other reptiles. This is reflected in the fact that the tail fin is set vertically. An ichthyosaur would have moved by a horizontal undulating action. Studies by Emily Bucholtz of Wellesley College, Massachusetts, indicate that there was little flexibility in the body, and all the motion was provided by the tail, with the flippers used only for stabilizing and steering.

If an ichthyosaur leapt from the surface of the sea it would have done so, not like a dolphin, but more like a swordfish or a tarpon – by a powerful sideways thrust of the tail fin.

Top: A dolphin with horizontal tail flukes.
Below: Ichthyosaur, with vertical tail flukes.

Aegirosaurus

This is the first ichthyosaur to have been found in the lithographic limestones of Bavaria, Germany, famous for their fossils of pterosaurs and the first bird *Archaeopteryx*. Many marine animal fossils have been found here, especially of animals that have swum into the poisonous waters of a lagoon from the open ocean and perished. This is probably what happened to *Aegirosaurus*.

Features: *Aegirosaurus* has a long and slender snout with many small, tightly packed, delicate teeth. The eye is medium size. The hind paddle is particularly small. The details of the hip bones – broad towards the rear – and the arrangement of the bones in the front paddle – with three bones articulating with the humerus rather than the two we see in most tetrapods – are enough for palaeontologists to distinguish it as a different genus from other ichthyosaurs.

Below: Aegirosaurus shared the shallow waters of the northern Tethys with the bodies of dead pterosaurs and early birds.

Distribution: Southern Germany.
Classification: Ichthyosauria, Opthalmosauridae.
Meaning of name: Aegir's lizard (after the Nordic god of the oceans).
Named by: Bardet and Fernandez, 2000.
Time: Kimmeridgian stage of the late Jurassic.
Size: 2m (6½ft).
Lifestyle: Fish-hunter.
Species: *A. leptospondylus*.

LATE JURASSIC PLESIOSAURS

By the middle and end of the Jurassic the plesiosaurs had really become established. The difference between the long-necked plesiosauroids and the big headed pliosauroids had become quite marked. The shallow shelf seas of the time provided plenty of different ecological niches, and different food sources to exploit, and could support a wide variety of marine animals.

Muraenosaurus

Although *Muraenosaurus* is known principally from fossils found in England, very similar remains have been found in Russia, and in North and South America. It seems likely that an ocean-going animal like this would exist throughout the seas of the world. Although it was quite a large plesiosaur, one species, *M. beloclis*, was a veritable dwarf at only 2.5m (8ft) long.

Features: The neck of *Muraenosaurus* is as long as the body and the tail together and is supported by 44 vertebrae. The tiny head is only about a sixteenth of the length of the whole animal and is quite broad with a short snout. The teeth, 19–22 pairs of them in each jaw, become gradually larger towards the front of the mouth.

Above: Muraenosaurus *was quite widespread with possible specimens known as far apart as Wyoming, USA, South America and Russia as well as Europe. Andrews had reconstructed the skull as rather flat and snake-like. However, a new reconstruction by Mark Evans of Leicester Museum, England shows it to have been high and domed.*

Distribution: England; France.
Classification: Plesiosauria, Plesiosauroidea.
Meaning of name: Eel lizard.
Named by: Andrews, 1910.
Time: Late Jurassic.
Size: 6m (20ft).
Lifestyle: Fish-hunter.
Species: *M. leedsii, M. beloclis, M. purbecki, M. elasmosauroides.*

Cryptoclidus

This was the typical long-necked plesiosauroid of the middle and late Jurassic seas. Numerous fossils have been found that bear the tooth marks of something big like *Liopleurodon*, indicating that it was not only the hunter of the seas but also the prey. Sometimes the fossils of isolated paddles are discovered, as if the owner had been crippled by an attack before being eaten.

Features: The typical solid body and long pointed paddles are here, as are the small head and the long neck. However, the neck is not quite as long as in some of the other plesiosauroids and it does not seem to be particularly flexible. The teeth are all about the same length and resemble thin curved needles. The eyes are on the top of the skull and point upwards.

Distribution: England.
Classification: Plesiosauria, Plesiosauroidea.
Meaning of name: Hidden collar bone.
Named by: Phillips, 1871.
Time: Late Jurassic.
Size: 4m (13ft).
Lifestyle: Fish-hunter.
Species: *C. eurymerus, C. richardsoni.*

Liopleurodon

This was the top predator of the Jurassic seas. It was the middle and late Jurassic equivalent of the modern great white shark, a veritable aquatic *Tyrannosaurus*. The fossils of half-chewed *Ichthyosaurus* skeletons have been found, as have plesiosaur limb bones with tooth marks corresponding to the bite of *Liopleurodon*, all attesting to the diet of this gigantic animal. The huge specimens that are being found suggest a revised length of about 25m (80ft) and make it the biggest flesh-eating vertebrate known.

Features: *Liopleurodon* has the typical pliosaurid shape, with the huge head, the short neck and the compact streamlined body. The teeth are 20cm (8in) long, a good three quarters of which consists of root embedded deeply into the jaws giving a very strong bite. The strongest are arranged in a rosette at the front of the mouth. The paddles are large and worked on the underwater flight principle found in all plesiosaurs, using the front flippers as wings in the manner of modern sea lions.

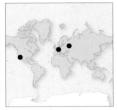

Distribution: England; France; Germany; Russia; and possibly Mexico.
Classification: Plesiosauria, Pliosauroidea.
Meaning of name: Smooth-sided tooth.

Above: A skull recently found in Mexico suggests that they may have been even bigger than their officially recorded length. The arrangement of nostrils in the skull suggests that they were used for sensing the water rather than for breathing. Prey could have been detected many kilometres away.

Named by: Sauvage, 1873.
Time: Late Jurassic.
Size: 25m (80ft).
Lifestyle: Ocean hunter.
Species: *L. macromerius, L. ferox, L. pachydeirus, L. grossouveri, L. rossicus.*

PLESIOSAUR MANOEUVRABILITY

There has always been speculation about the flexibility of the neck of plesiosaurs. The classic restoration shows the animal swimming close to the surface of the ocean, with the head held high, on top of a neck held in a graceful swan-like S-shape.

Study of the bones of the neck suggests that this may not have been possible. Both the long-necked genera featured here, *Muraenosaurus* and *Cryptoclidus*, may have been unable to lift the main part of their necks above the horizontal, with the head end flexible enough just to peek upwards. They may, however, have been able to lower their necks to about 45 degrees below the horizontal. This suggests that they swam close to the surface and hunted for fish below them. Their necks seem to have been more flexible sideways than in the vertical plane.

Below: Cryptoclidus, showing the lowered neck.

Peloneustes

With its short, streamlined body and long skull, *Peloneustes* must have looked like a giant diving bird, like a penguin or a gannet. The long jaws of *Peloneustes* made up for its short neck in catching fast-moving prey. Stomach contents analysis has revealed a preponderance of horny hooks from the tentacles of squid and other cephalopods.

Features: The long snout and the relatively few teeth of this pliosaur indicate a specialized diet. The big hind paddles suggest a fast and manoeuvrable swimming action, allowing it to pursue the swift soft-bodied squid and belemnites that flourished in the shallow seas of late Jurassic Europe. Head and neck, with 20 vertebrae, are approximately the same length. The underside is solid, formed by a fusion of the shoulder and hip girdles and the belly ribs, giving a completely inflexible body.

Distribution: Europe.
Classification: Plesiosauria, Pliosauroidea.
Meaning of name: Mud swimmer.
Named by: Seeley, 1869.
Time: Late Jurassic.
Size: 3m (10ft).
Lifestyle: Swimming hunter.
Species: *P. philarchus.*

Right: Early restorations showed Peloneustes with quite a thin and flexible neck. It now seems more likely that the neck was stiff and made a streamlined shape with the head and body.

RHAMPHORHYNCHOID PTEROSAURS

The late Jurassic was a rich time for pterosaurs. The two main types existed together: the short-tailed
pterodactyloids, abundant at the end of the era, were establishing themselves, while the more primitive
rhamphorhynchoids (illustrated here) were still abundant and diverse. None of the rhamphorhynchoid
line survived into the Cretaceous. They were warm-blooded and covered in fur.

Rhamphorhynchus

Distribution: Germany;
England; and Tanzania.
Classification: Pterosauria,
Rhamphorhynchoidea.
Meaning of name: Beak
snout.
Named by: von Meyer, 1847.
Time: Oxfordian to
Kimmeridgian stages of the
late Jurassic.
Size: 1.75m (5¾ft) wingspan.
Lifestyle: Fish-hunter.
Species: *R. intermedius, R.
gemmingi, R. jessoni, R.
longicaudus, R. longiceps, R.
muensteri. R. tendagurensis.*

This is the animal that gives its name to the
whole group. It is the most common genus
found in the Solnhofen deposits in Germany.
Rhamphorhynchus is what most people
think of as the typical pterosaur, with its
leathery wings and its long tail ending in a
vertical diamond paddle.

Features: The jaws of *Rhamphorhynchus* combine a
set of very long fangs and a
pointed beak, hence the
name. The upper jaw has
ten pairs of these teeth while
the lower jaw has seven, and they project
forwards and outwards. The breast bone is broad
and strong and carries a forward-pointing crest,
giving a wide attachment for strong wing-muscles.
The neck is short and compact, holding the head
straight out and not at an angle as with birds. The
wings are stiffened with fine struts of gristle that
radiated from the arm bones in the same pattern as
the flight feathers of a bird.

Below: Rhamphorhynchus
probably hunted fish by
skimming close to the
surface of the water. It is
possible that it had a
throat pouch, like a
pelican, for holding
its prey.

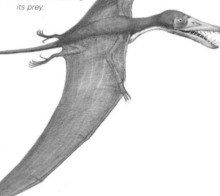

Batrachognathus

Most rhamphorhynchoids were adapted to hunting fish. *Batrachognathus*, however, had a
completely different diet. Its blunt teeth were ideal for crushing insect carapaces, and the
broad jaws formed a kind of scoop. Its tiny body would have made it very manoeuvrable
in flight. It must have flown over lake surfaces catching insects in flight, as modern
swallows do.

*Left: The range of pterosaur
types must reflect a range of
feeding strategies, as with the
range of types of modern birds.*

Features: *Batrachognathus* has a
broad frog-like mouth with little
peg-like teeth. It has a high
short skull, unlike that of other
rhamphorhynchoids. The tail is
short. The tail vertebrae are fused
together, quite unlike what would
be expected from a
rhamphorhynchoid and more
like that of a pterodactyloid. The
wing structure with its short
wrists is, however, very much
that of its long-tailed relatives.

Distribution: Kazakhstan.
Classification: Pterosauria,
Rhamphorhynchoidea.
Meaning of name: Frog face.
Named by: Riabinin, 1948.
Time: Late Jurassic.
Size: 0.5m (1½ft) wingspan.
Lifestyle: Insectivore.
Species: *B. volans.*

Jeholopterus

The extreme specialization of this rhamphorhynchoid has led to active speculation as to its lifestyle. One suggestion by independent palaeontologist David Peters is that the large claws allowed it to cling to the sides of big dinosaurs, while the wide-opening mouth and the pair of strong buttressed fangs at the front could be driven into thick skin to reach blood vessels. *Jeholopterus* was a vampire!

Above: Jeholopterus is known from a complete articulated skeleton, found in fine lake deposits. The deposits are so delicately preserved that they show the hairy covering and also the wing membrane of this pterosaur.

Features: The skull is broad and flattened at the front, giving it a rather cat-like appearance. The jaws are articulated to open wider than those of any other pterosaur. All the teeth are reduced except for a pair in the upper jaws at the front, and these are particularly strong and protruding, deeply rooted in a very strong palate. The claws are large and sharper than those of other pterosaurs, while the tail is short for a rhamphorhynchoid. It is closely related to other short-tailed rhamphorhynchoids like *Batrachognathus*.

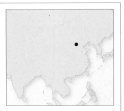

Distribution: Northeastern China.
Classification: Pterosauria, Rhamphorhynchoidea.
Meaning of name: Wing from the Jehol geological formation.
Named by: Wang, Zhou, Zhang and Xu, 2002.
Time: Late Jurassic or early Cretaceous.
Size: 1m (3ft) wingspan.
Lifestyle: Insectivore or perhaps blood-sucker.
Species: *J. ninchengensis.*

OTHER RHAMPHORHYNCHOIDS

Scaphognathus has a shorter head than *Rhamphorhynchus* and has long teeth that are set upright rather than pointing forward. It also lacks the beak at the front of the jaws.
Anurognathus is similar to *Batrachognathus*, but with a wingspan of 50cm (1½ft) it is the smallest known pterosaur. Like *Batrachognathus* it has a short head and no tail to speak of, and may have caught insects on the wing.

OTHER PTERODACTYLOIDS

The pterodactloids below were located in the Solnhofen lagoons of Germany.
Germanodactylus has a long, straight, narrow beak with widely-spaced teeth, and a long crest running down the head's midline.
Diopecephalus is a large pterodactyloid, with a wingspan of about 1.45m (5ft) and a particularly long neck and head.

Across the world, in the famous Morrison Formation of late Jurassic North America, rhamphorhynchoids and pterodactyloids also lived together. The dinosaur beds of Colorado and Utah yielded **Comodactylus**, a

rhamphorhynchoid, and **Mesadactylus**, a pterosauroid. These are only known from individual bones, but they show that the two groups were cosmopolitan.

Sordes

For a long time palaeontologists debated about whether or not the pterosaurs were warm-blooded, like birds or bats. The argument for being warm-blooded was very strong, as a warm-blooded metabolism would have been needed for active aerial hunting. It was the discovery of *Sordes*, with its coat of hair, a sure sign of warm-bloodedness, which finally settled the argument.

Features: *Sordes* is a small rhamphorhynchoid, similar to the general form presented by *Rhamphorhynchus* itself. The first fossil of *Sordes* found is so well-preserved that the fur-like covering is visible. Short hairs about 6mm (¼in) long cover the whole body but leave the tail naked. There is even hair on the wing membrane and between the toes although it is much sparser here. There seems to be no diamond fin on the tail, but the tail itself is flattened and paddle-like at the end.

Distribution: Kazakhstan.
Classification: Pterosauria, Rhamphorhynchoidea.
Meaning of name: Filth.
Named by: Sharov, 1971.
Time: Late Jurassic.
Size: 60cm (2ft) wingspan.
Lifestyle: Insectivore or fish-hunter.
Species: *S. pilosus.*

Above: The hairy body covering would have helped to regulate the body temperature during active flying, but it would also have helped to reduce the noise of flying, useful for an aerial hunter.

PTERODACTYLOID PTEROSAURS

The pterodactyloids (illustrated here) became the dominant pterosaur group from the late Jurassic onwards. They had short stubby tails and longer necks than the more primitive rhamphorhynchoids, and generally had broader, more manoeuvrable wings. They carried their heads at more of a bird-like angle. Like the rhamphorhynchoids they were warm-blooded and covered in fur.

Ctenochasma

With its array of long fine teeth, *Ctenochamsa* could have been nothing but a filter feeder. It would have rested on all fours in shallow ponds and skimmed tiny animals, such as crustaceans or invertebrate larvae, from the surface, rather like the modern flamingo does. Recent studies, by Stephane Jouve, from the Museum National d'Histoire Naturelle, Paris, have suggested that some juvenile pterosaurs, once regarded as *Pterodactylus*, were actually *Ctenochasma*.

Left: The many teeth of Ctenochasma were not present throughout its life. The hatchling possessed about 60 of them and they developed to the full complement as the pterosaur grew towards maturity.

Distribution: Germany; France.
Classification: Pterosauria, Pterodactyloidea.
Meaning of name: Comb mouth.
Named by: von Meyer, 1852.
Time: Late Jurassic.
Size: 1.2m (3.9ft) wingspan.
Lifestyle: Plankton feeder.
Species: *C. porocristata*, *C. gracilis*.

Features: The long jaws contain more than 250 fine needle-like teeth that fan outwards at the tip. The two species of the *Ctenochasma* genus are distinguished from each other by the presence of a crest on the skull of *C. porocristata*, which s absent in *C. gracilis*. This crest probably forms the base of a horny display structure. The crest is very lightweight, consisting of very porous bone, to present as little impediment to flight as possible.

Gnathosaurus

The masses of needle-like teeth in the long jaws of this pterosaur were so reminiscent of the jaws of some aquatic-feeding crocodiles that it was classed as a crocodile when first discovered, and remained so until a second specimen was found in 1951. Like *Ctenochasma* it would have been a filter-feeder living off small invertebrates in shallow waters. *Gnathosaurus* was found in the lithographic limestone deposits of southern Germany, where *Ctenochasma* also sieved for food. There must have been a big enough range of food to support at least two different filter feeders.

Features: The skull and lower jaw of *Gnathosaurus* are very similar to those of *Ctenochasma*, but there are fewer teeth – about 130 in all – and those that it does have are thicker. Unlike *Ctenochasma* the snout is expanded and spoon-shaped at the end. A crest runs down the mid-line of the skull for about three quarters of its length. Only the skull and jaw have been found in Germany; we know nothing of the rest of the skeleton.

Distribution: Germany.
Classification: Pterosauria, Pterodactyloidea.
Meaning of name: Jaw lizard.
Named by: von Meyer, 1833.
Time: Late Jurassic.
Size: 1.7m (5½ft) wingspan.
Lifestyle: Filter feeder.
Species: *G. sublatus*.

Right: Only the jaws and a scrap of bone are known from Gnathosaurus. However, the body shape of the pterodactyloids is so conservative that we can restore the appearance of the whole animal with confidence.

SOLNHOFEN FOSSIL BEDS

In late Jurassic times part of the northern shoreline of the Tethys Ocean became cut off as a series of quiet lagoons. The ocean was blocked by reefs of coral and sponges. These lagoons became toxic and anything that swam in them or died and fell into in them was preserved in the very fine limestone that was deposited on the lagoon bed. Nowadays these limestones outcrop at Solnhofen in southern Germany, where they have been quarried since Roman times, and latterly for the printing industry (their fine grain making them ideal).

The fossils preserved are exquisitely detailed. They consist of fish complete with scales, horseshoe crabs that have dropped dead at the end of their tracks (also preserved), lizards with completely articulated skeletons, fine sea lilies that are still intact, and even the first bird, *Archaeopteryx*, complete with its feathers. The many pterosaurs, both rhamphorhynchoid and pterodactyloid types, that lived on the arid shorelines and islands of the lagoons are preserved along with imprints of their wing membranes. It is because of these

deposits that we know so much about late Jurassic pterosaurs.

Left: A detailed Solnhofen fossil of a pterosaur.

Pterodactylus

The word pterodactyl is widely used as a popular term for any of the pterosaur group. The word, originally written as "pterodactyle", was made up by Cuvier when the fossils were first studied. At first this creature was thought to have been a mammal like a bat, and there were even suggestions that it was a swimming creature like a penguin, with the elongated finger supporting a paddle.

Features: *Pterodactylus* is the typical pterodactyloid pterosaur, with a short tail, long wrist bones, a longer neck than that of a typical rhamphorhynchoid, formed by elongation of the individual neck vertebrae, and a head held at an angle, like that of a bird, rather than in line with the neck as in a rhamphorhynchoid. The wing membrane is strengthened by fibres that radiate away from the arm and wrist, rather like the orientation of the flight feathers of a bird.

Right: Many Pterodactylus species have been found, and reassigned to other genere.

Distribution: Germany; France; England; Tanzania.
Classification: Pterosauria, Pterodactyloidea.
Meaning of name: Wing finger.
Named by: Soemmering, 1812 (and Cuvier in 1809).
Time: Kimmeridgian to Tithonian stages of the late Jurassic.
Size: Up to 2.5m (8ft) wingspan.
Lifestyle: Fish hunter.
Species:
P. antiquus,
P. arningi,
P. grandis,
P. manseli,
P. maximus,
P. pleydelli,
P. micronyx,
P. kochi,
P. cerinensis,
P. grandipelvis,
P. suprajurensis.

Gallodactylus

This pterosaur was described from a specimen found in Var in southern France in 1974, but since then a species of *Pterodactylus* found in Germany 120 years earlier has been assigned to the genus. It is also known from the site of Solnhofen, Germany where, along with *Pterodactylus*, it is one of the largest of the Jurassic pterosaurs.

Features: *Gallodactylus* shows another in the range of head shapes that appeared in the pterodactyloids as soon as they evolved. The long jaws are almost toothless except the tips where the long and fairly slender teeth are bunched together. This is another of the fish-traps evolved in the pterosaur line. The skull is extended back into a crest, presumably used as a display structure. The rest of the body is similar to that of all other pterodactyloids.

Distribution: France and Germany.
Classification: Pterosauria, Pterodactyloidea.
Meaning of name: Finger from Gaul (the Roman name for France).
Named by: Fabre, 1974.
Time: Late Jurassic.
Size: 1.35m (4½ft) wingspan.
Lifestyle: Fish eater.
Species: G. canjuersensis.

Above: Gallodactylus differed from Pterodactylus in the arrangement of teeth in the jaws and the presence of the crest at the rear of the skull. Apart from these features it was very similar to members of that genus.

MORRISON FORMATION MEAT-EATERS

In the second half of the nineteenth century the pioneering palaeontologists made most of their dinosaur discoveries in the vast Morrison Formation of the American Mid-west. The Jurassic rocks here are widespread. Even today new finds are being made, both in the well-worked quarries and in new areas where no-one has looked before. Meat-eating theropods form a spectacular part of the Morrison fauna.

THE MORRISON FORMATION

These animals, and many of those in the pages to follow, were found in the famous Morrison Formation of North America, named after a small town to the west of Denver where the formation was first identified.

At the end of the Jurassic period, a shallow seaway spread southwards across the middle of the North American continent. To the west the ancestral Rocky Mountains were rising. The rock fragments that washed off this crumbling mountain range were brought down to the sea and were deposited to form broad deltas and well-watered plains. The sand, silt, mud and pebbles that formed the soil of this plain are now found today as a vast swathe of sandstone and mudstone that stretches from New Mexico far north into Canada. These rocks form what is known as the Morrison Formation.

The dinosaurs that lived on the plains stayed mostly by the water. That is where the thickest vegetation grew. Occasionally they would fall into the rivers or lakes and become buried by the sediment. Sometimes many dead dinosaurs would be washed down together, and the result would be one or another of the famous dinosaur quarries of the Mid-west. The modern climate in this area is quite arid, and with only sparse vegetation masking the rocks, eroding dinosaur bones can be quite easily seen by prospectors. Most of our early knowledge of late Jurassic dinosaurs was obtained in the 1880s by the pioneering, and often rival, palaeontologists who excavated the Morrison Formation.

Ceratosaurus

One of the last of the wiggly-tailed ceratosaurids was *Ceratosaurus* itself, and what a spectacular animal it was! It was not the biggest meat-eating dinosaur of the time, but it was certainly a beast with character. If any dinosaur looked like a medieval dragon, this was it. With horns on the head, a jagged crest right down its back and teeth that seemed too big for its mouth, it must have presented a fearsome aspect to the plant-eaters of the late Jurassic period. It was one of the most common hunters of the time, and may have hunted down its prey in packs or in family groups.

Features: The most obvious feature of this dinosaur is the big horn on the nose and the two smaller ones above the eyes, probably used for display. It has four fingers on the hands – a rather primitive condition – and the arms, although short, are quite strong. One species, *C. ingens*, known only from the teeth, is much larger than the others and may have been one of the biggest theropods.

Distribution: Colorado, Utah, USA; and perhaps Tanzania.
Classification: Theropoda, Neoceratosauria.
Meaning of name: Horned lizard.
Named by: Marsh, 1884.
Time: Tithonian stage of the late Jurassic.
Size: 6m (17½ft).
Lifestyle: Hunter.
Species: *C. nasicornis, C. dentisulcatus, C. ingens, C. magnicornis, C. meriani, C. roechlingi.*

Torvosaurus

Although all that has been found officially of this dinosaur is the forelimb, other bones, such as the jawbone, part of the skull, the neck bones and parts of a hip have been attributed to it. Put together these give the picture of a very big, meat-eating animal. It was found in the same quarry, the Dry Mesa Quarry in the mountains of Colorado, as the biggest of the late Jurassic North American sauropods.

Features: This dinosaur is actually regarded by some palaeontologists as a species of *Megalosaurus*. It is as big as *Allosaurus* – the best-known of the late Jurassic meat-eating dinosaurs – but it seems to be more massive. The short arms are very powerful and carry big claws. Some palaeontologists surmise that it may have been too heavy to hunt and was merely a scavenger, eating the bodies of dinosaurs that had already been killed

Distribution: Colorado, USA; and possibly Portugal.
Classification: Theropoda, Tetanurae.
Meaning of name: Savage lizard.
Named by: Galton and Jensen, 1979.

Time: Tithonian stage of the late Jurassic.
Size: 10m (33ft).
Lifestyle: Hunter or scavenger.
Species: *T. tanneri*.

Left: Dry Mesa quarry, as it appears today – a ledge high on an arid hillside overlooking a deep valley. The fossils formed from a build-up of corpses in a Jurassic river.

Left: A leg bone found in 1998 in Portugal may belong to Torvosaurus. *However, it has also been suggested that it belongs to* Allosaurus, *or even the distinctly Portuguese* Lourinhanosaurus.

Edmarka

Known only from parts of a skull, some ribs and a shoulder bone from three individuals, one of which is a juvenile, *Edmarka* is not the best-known of the Jurassic theropods. In fact, it probably represents a species of *Torvosaurus* (or *Megalosaurus*, if its taxonomy can be correctly established). It was found at the famous dinosaur site of Como Bluff in Wyoming, USA, where it would have preyed upon the big sauropods of the time, or scavenged from their dead bodies.

Distribution: Wyoming, USA.
Classification: Theropoda, Tetanurae.
Meaning of name: Named after Dr William Edmark, the

Features: *Edmarka* may have been the heaviest meat-eating dinosaur of the Jurassic period, weighing maybe two tonnes. The presence of *Edmarka* in the late Jurassic rocks of Wyoming seems to uphold Robert T. Bakker's idea that each area and time period produced one particularly large meat-eating dinosaur that was the dominant predator, and the presence of this one prevented the evolution of any other.

Left: As with all other big heavy meat-eating dinosaurs there is much debate as to whether Edmarka was an active hunter or a slow-moving scavenger, or even both.

Colorado State University scientist.
Named by: Bakker, Krails, Siegwath and Filla, 1992.
Time: Kimmeridgian stage of the late Jurassic.
Size: 11m (36ft).
Lifestyle: Hunter.
Species: *E. rex*.

THE SINRAPTORIDS

While Allosaurus *and its kin represented the big meat-eaters of the Western Hemisphere, China had its own group of fearsome hunters. Closely related to the allosaurids these were the sinraptorids, and most have been found in China. They carried their heads higher than their western relatives, and may have been more adapted to active hunting.*

Szechuanosaurus

The original specimens of this obscure beast were some teeth discovered in 1915 at Wuijaba in Zigong. Later, a headless skeleton was discovered, but it is uncertain that the bones and the teeth come from the same animal.

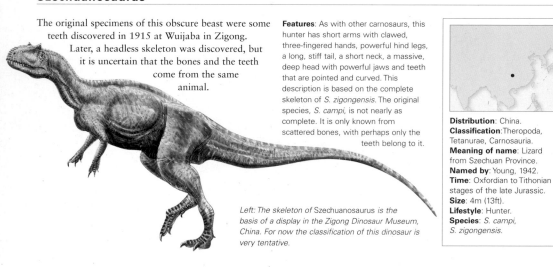

Features: As with other carnosaurs, this hunter has short arms with clawed, three-fingered hands, powerful hind legs, a long, stiff tail, a short neck, a massive, deep head with powerful jaws and teeth that are pointed and curved. This description is based on the complete skeleton of *S. zigongensis*. The original species, *S. campi*, is not nearly as complete. It is only known from scattered bones, with perhaps only the teeth belong to it.

Left: The skeleton of Szechuanosaurus *is the basis of a display in the Zigong Dinosaur Museum, China. For now the classification of this dinosaur is very tentative.*

Distribution: China.
Classification: Theropoda, Tetanurae, Carnosauria.
Meaning of name: Lizard from Szechuan Province.
Named by: Young, 1942.
Time: Oxfordian to Tithonian stages of the late Jurassic.
Size: 4m (13ft).
Lifestyle: Hunter.
Species: *S. campi*, *S. zigongensis*.

Sinraptor

The nearly complete skeleton of *Sinraptor* was found in 1987 during a joint Canadian and Chinese dinosaur project. It was evidently a carnosaur, but appeared to be more primitive those of North America. Other Chinese carnosaurs seemed to show some similarity to it, and this was the basis for erecting a new family, the Sinraptoridae.

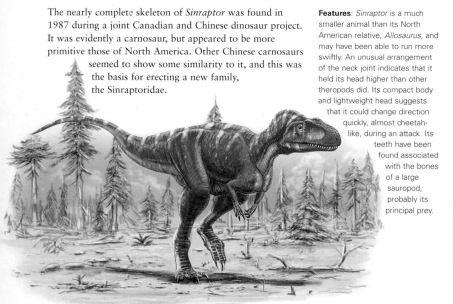

Features: *Sinraptor* is a much smaller animal than its North American relative, *Allosaurus*, and may have been able to run more swiftly. An unusual arrangement of the neck joint indicates that it held its head higher than other theropods did. Its compact body and lightweight head suggests that it could change direction quickly, almost cheetah-like, during an attack. Its teeth have been found associated with the bones of a large sauropod, probably its principal prey.

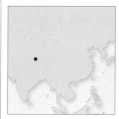

Distribution: China.
Classification: Theropoda, Tetanurae, Carnosauria.
Meaning of name: Chinese hunter.
Named by: Currie and Zhao, 1994.
Time: Kimmeridgian to Tithonian stages of the late Jurassic.
Size: 7m (23ft).
Lifestyle: Hunter.
Species: *S. dongi*, *S. hepingensis*.

Yangchuanosaurus

The almost complete skeleton of *Yangchuanosaurus*, lacking only the forelimbs, the hind feet and the tip of the tail, was found in Szechuan, eastern China, in the 1970s. It was one of the biggest and most complete theropods found in China. A mounted skeleton is now on display at the Beijing Natural History Museum. The genus was named after Yangchuan county and the main species was named after the Shangyou reservoir, where it was found during construction work.

Features: *Yangchuanosaurus* is a big, powerful meat-eater, not quite as big as its relative *Allosaurus* but very similar in build. However, it has more teeth than *Allosaurus*, and has a ridge or crest along the top of the skull and a bony knob on the snout. There is a suggestion, from the shape of the vertebrae, that there may have been a narrow ridge or even a low fin down its back. Such a fin would be brightly coloured and used for display or signalling.

Distribution: China.
Classification: Theropoda, Tetanurae, Carnosauria.
Meaning of name: Lizard from Yangchuan District.
Named by: Dong, Chang, Li and Zhou, 1978.
Time: Kimmeridgian to Tithonian stages of the late Jurassic.
Size: 10m (33ft).
Lifestyle: Hunter or scavenger.
Species: *Y. shangyouensis*, *Y. longqiaoensis*, *Y. magnus*, *Y. yandonensis*.

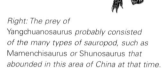

Right: The prey of Yangchuanosaurus *probably consisted of the many types of sauropod, such as* Mamenchisaurus *or* Shunosaurus *that abounded in this area of China at that time.*

CARNOSAURS

The Carnosauria are divided into three main families – the Sinraptoridae, the Allosauridae and the Carcharodontidae. The first of these represents the most primitive, and the ancestors of the other two groups may well be found among the sinraptorids.

Taking its name from *Sinraptor* – the "Chinese hunter" – the group is almost totally restricted to China. This may well be due to geographical factors. Eastern China in Jurassic times was separated from the other landmasses of the world, and so evolution would be expected to produce a fauna that would be different from the rest of the world. As a group, their bodies were quite primitive, but were distinguished by ornamental crests on their heads. So far *Lourinhanosaurus* is the only sinraptorid to have been found outside China. The allosaurids were the largest hunters of the Jurassic, while the carcharodontids filled this role in the Cretaceous. Two other families, the megalosaurids and spinosaurids are now considered of dubious merit.

Lourinhanosaurus

The single skeleton found of this theropod consists of much of the backbone, some ribs and the hind legs. An interesting find was the presence of gastroliths, or stomach stones, in the body cavity. Normally only plant-eating dinosaurs have them, using them to help grind up tough plant material. Palaeontologists are still puzzling over their function here.

Features: What distinguished *Lourinhanosaurus* from the other carnosaurs was the shape of the vertebrae which were longer than they were tall. Unfortunately we do not know anything about the skull. However, the arrangements of the neck vertebrae and the hip bones, particularly the expanded end of the pubis bone, were distinctly carnosaurian. The primitiveness of the skeleton suggests that it was one of the Sinraptoridae, rather than one of the more advanced Allosauridae, and as such, the first to be found outside China.

Distribution: Portugal.
Classification: Theropoda, Tetanurae, Carnosauria.
Meaning of name: Lizard from the Lourinhã Formation.
Named by: Mateus, 1998.
Time: Kimmeridgian to Tithonian stages of the late Jurassic.
Size: 4.5m (15ft).
Lifestyle: Hunter.
Species: *L. antunesi*.

Left: A series of eggs, each about 13cm (5in) long, found near the skeleton of Lourinhanosaurus *may have been laid by this animal. The embryos suggest that the hatchlings were adapted for catching insects.*

THEROPODS OF CLEVELAND-LLOYD

One of the most prolific of the quarries in the Morrison Formation is Cleveland-Lloyd in Utah. More than 12,000 bones have been recovered here, representing 11 species of animal. It seems likely that complete skeletons are present here, but they are all disarticulated and jumbled. Most of the specimens that have been identified are of meat-eating dinosaurs.

Marshosaurus

This theropod was discovered in Dry Mesa quarry in Colorado, and the Cleveland-Lloyd quarry in Utah. The latter is famous for the large number of skeletons of the bigger theropod *Allosaurus*. Whatever lifestyle *Marshosaurus* pursued, it must have been in the presence of these enormous *Allosaurus* meat-eaters. Perhaps *Allosaurus* scavenged from the kills of a more active *Marshosaurus*.

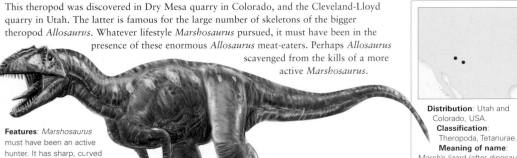

Features: *Marshosaurus* must have been an active hunter. It has sharp, curved teeth with serrated edges. What we know of the skull suggests that the head was quite long. The short and massive upper arm bone shows that it had small but powerful arms. At this time it is difficult to say what kind of theropod it was – some palaeontologists suggest it was a carnosaur, like *Sinraptor*, while others say that it may be an early dromeosaur.

Left: The species name, M. bicentisimus, commemorates the fact that the United States of America was celebrating 200 years of its foundation in the year that this dinosaur was discovered.

Distribution: Utah and Colorado, USA.
Classification: Theropoda, Tetanurae.
Meaning of name: Marsh's lizard (after dinosaur pioneer Othniel Charles Marsh).
Named by: Madsen, 1976.
Time: Tithonian stage of the late Jurassic.
Size: 5m (16½ft).
Lifestyle: Hunter.
Species: *M. bicentisimus*.

Stokesosaurus

Distribution: Utah, USA.
Classification: Theropoda, Tetanurae.

Meaning of name: Lee Stokes's (an American palaeontologist) reptile.
Named by: Madsen, 1974.
Time: Tithonian stage of the late Jurassic.
Size: 4m (13ft).
Lifestyle: Hunter.
Species: *S. clevelandi*. *Iliosuchus incognitus* may be the same genus.

Another of the theropods found in the Cleveland-Lloyd quarry, *Stokesosaurus* is based on very few scattered remains – mostly a hip bone, jaw bone and braincase – that are very different from the other theropod remains found there. Some of the bones seem to resemble those of late Cretaceous theropods, and this has led some scientists to surmise that it may be a primitive tyrannosaurid.

Features: *Stokesosaurus* seems to have been very similar to another dinosaur called *Iliosuchus*, found in England in the 1930s, and these are now regarded as the same animal. The braincase that has been found has many tyrannosaurid features,

Right: A jawbone and a braincase found in South Dakota may also have belonged to Stokesosaurus.

but there is still some uncertainty as to whether or not this actually belonged to *Stokesosaurus* – the bones are so scattered. Verification would have an effect on the classification of this animal. If it is a tyrannosaurid, it is the earliest that has been found.

Allosaurus

The most well-known of the late Jurassic theropods must be *Allosaurus*. It was unearthed by Othniel Charles Marsh during the "bone wars", and since then many specimens have been found. The Cleveland-Lloyd quarry alone has yielded more than 44 individuals. Species attributed to *Allosaurus* have been found as far away as Tanzania and Portugal. Sauropod bones have been found bearing marks gouged by *Allosaurus* teeth.

Features: *Allosaurus* is a familiar animal, with its massive hind legs, its strong S-shaped neck, its huge head with jaws that could bulge sideways to bolt down great chunks of meat, sharp, serrated, steak-knife teeth with 5cm- (2in-) long crowns, and its short, heavy arms with three-

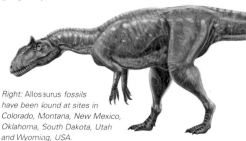

fingered hands bearing ripping claws that were up to 15cm (6in) long. This enormous carnivore would have hunted the biggest plant-eaters of the time including the massive sauropods, the remains of which were found in the Cleveland-Lloyd quarry.

Right: Allosaurus fossils have been found at sites in Colorado, Montana, New Mexico, Oklahoma, South Dakota, Utah and Wyoming, USA.

Distribution: USA; possibly Portugal and Tanzania.
Classification: Theropoda, Tetanurae, Carnosauria.
  Meaning of name: Different lizard.
Named by: Marsh, 1877.
Time: Tithonian to Kimmeridgian stages of the late Jurassic.
Size: 12m (39ft).
Lifestyle: Hunter or scavenger.
Species: *A. tendagurensis, A. fragilis, A. amplexus, A. trihedrodon, A. whitei* and *A. (Saurophaganax) maximus?*

CLEVELAND-LLOYD QUARRY

The Cleveland-Lloyd quarry lies 38km (30 miles) south of Price, Utah, close to the community of Cleveland. Dinosaur excavations started on the site in 1929 and over the next 10 or 12 years they continued thanks to financing by Malcolm Lloyd, a Philadelphia lawyer.

The University of Utah began a thorough excavation in 1960, headed by William Lee Stokes, and another in 2001 with Utah state geologist James H. Madsen in charge. The site was designated a National Natural Landmark in 1966. The excavations are open to view by the public, and skeletons from the site are on view in more than 60 museums worldwide.

The skeletons found at the site include sauropods and stegosaurs, but most appear to be the bones of meat-eaters. It is possible that the site represents a watering hole in an abandoned river meander on an arid Jurassic plain. Such a site would have attracted plant-eating animals from all over, and the meat-eaters would have converged, finding easy pickings among the weak and dehydrated sauropods and stegosaurs. The mud of the water hole would have hampered any escape. The skeletons appear to have been torn apart on the spot, and the bones show signs of having been trampled underfoot. Subsequent floods would have buried all this and set the fossilization process in motion.

Saurophaganax

The few bones known of *Saurophaganax* were excavated in the 1930s, but were not studied seriously until the 1990s. It turned out to be very similar to *Allosaurus* but a great deal bigger. After Don Chure's naming of it in 1995, David K. Smith re-analysed it in 1998 and came to the conclusion that it represented a particularly big species of *Allosaurus*.

Features: The description of *Allosaurus* can just as well apply to *Saurophaganax*, so close are the two animals. The differences lie in the sheer size of *Saurophaganax*, and in the shape of the neck and tail vertebrae. A complete mounted skeleton of *Saurophaganax* is on display in the Sam Noble Museum in Oklahoma City, but most of it is made up from sculpted bones scaled up from those of *Allosaurus* from the Cleveland-Lloyd quarry.

Distribution: Oklahoma, USA.
Classification: Theropoda, Tetanurae, Carnosauria.
Meaning of name: The greatest reptile-eater.
Named by: Chure, 1995.
Time: Kimmeridgian stage of the late Jurassic.
Size: 12m (39ft).
Lifestyle: Hunter or scavenger.
Species: *S. maximus.*

SMALLER THEROPODS

*We often think of theropods as being big, fierce animals. However, some of them were quite small.
In modern faunas we have the big carnivores like lions and wolves, but we also find smaller types
such as weasels and raccoons. It was the same in the dinosaur age – various sizes of carnivore
evolved to hunt different-sized prey.*

Compsognathus

One of the first complete dinosaur skeletons ever found was
also one of the smallest. *Compsognathus* was found
perfectly preserved in the fine lithographic limestones of
Bavaria, Germany. In its build and appearance it is very
similar to the first bird, *Archaeopteryx*, which was also
found here, giving an early indication of
the close relationships between the
birds and the dinosaurs.

Right: Compsognathus
*is usually shown with two
fingers on each hand.
However, some of
the finger
bones may
have been
missing from the original
specimen, and it may
have had three.*

Features: Although the length
of *Compsognathus* is given as
1m (3ft), this gives the wrong
impression. Most of that is tail
and neck. The body is about as
big as a chicken. It was evidently
a meat-eater, as the German
specimen has the remains of a
lizard in its stomach cavity, its last
meal before death. A recent
discovery, by Peter
Griffiths, was
the presence of
unhatched eggs
surrounding the fossil. This
specimen had been a female and
the eggs had been ejected from
the body cavity by the trauma of
her death. The French specimen
is of a slightly larger species.

Distribution: Germany
and France.
Classification: Theropoda,
Tetanurae, Coelurosauria.
Meaning of name:
Pretty jaw.
Named by: Wagner, 1859.
Time: Tithonian stage of the
late Jurassic.
Size: about 1m (3ft).
Lifestyle: Hunter.
Species: *C. longipes*,
C. corallestris.

Ornitholestes

The name of this dinosaur suggests that it hunted birds,
and many early restorations show it doing so – invariably
chasing *Archaeopteryx*, a bird that actually lived on the
other side of the world. It is more likely to have hunted
small reptiles and ground-living mammals. Its skeleton
was found in 1900 in the famous Bone Cabin Quarry
that yielded many Morrison
Formation dinosaurs in the
early days.

Features: This fox-sized animal,
with a long neck and tail, hunted
on two legs. It is lightly built with
hollow bones, and its long tail –
making up more than half the
length of the animal – gave it
manoeuvrability while hunting. Its
head is deeper than that of other
small theropods, and has a bony
crest on the snout. Its hands
have four
fingers,
one of
which is tiny
and almost invisible,
while another is
big and could be used
as a thumb.

*Left: The teeth were not
built for snatching but
for cutting.* Ornitholestes
*probably chased its prey
and grasped it with its
versatile hands, using
both teeth and claws
to kill it.*

Distribution: Utah and
Wyoming, USA.
Classification: Theropoda,
Tetanurae, Coelurosauria.
Meaning of name: Bird thief.
Named by: Lambe 1904.
Time: Tithonian stage of the
late Jurassic.
Size: 2m (6½ft).
Lifestyle: Hunter.
Species: *O. hermanni*.

Coelurus

For a long time *Coelurus* was thought to have been another specimen of *Ornitholestes*. However, studies by John Ostrom in 1976 and Jacques Gauthier in 1986 show that the hands are like those of the maniraptorans. In contrast, the neck is nothing like that of a maniraptoran and it is unclear where this animal fits into the dinosaur family tree.

Features: This is another of the small, hunting dinosaurs. It has a strangely down-curved jaw with sharp, curved teeth. The hands are long but not particularly strong, with a wrist joint similar to that of a bird, and very flexible fingers. The "hollow tail" part of the name refers to the deep excavations in the vertebrae of the back and tail, something like those found as a weight-saving measure in sauropods.

Right: Specimens of Coelurus *found in four locations in the same quarry may have come from the one individual. That individual may not even have been fully grown, and so the size estimate here may be on the small side.*

Distribution: Wyoming, USA.
Classification: Theropoda, Tetanurae.
Meaning of name: Hollow tail.
Named by: Marsh, 1879.
Time: Tithonian stage of the late Jurassic.
Size: 2m (6½ft).
Lifestyle: Hunter.
Species: *C. fragilis*.

THEROPOD DIET

What was the prey of the small theropods? Sometimes we have direct evidence, as in the lizard bones found in *Compsognathus's* stomach. For the rest we have to make inferences from the other animals that lived in the area.

We know that small reptiles and amphibians abounded in late Jurassic times. The Morrison Formation has fossils of snakes and lizards, and frogs and salamanders. There were also small mammals, none of which would have been bigger than a rat, although the only physical fossils we have of them are the teeth. All of these animals would have been prey for the small, active theropods. Although doubt has been thrown on the many early illustrations of *Ornitholestes* chasing *Archaeopteryx*, it is likely that there were birds around at that time. Active small dinosaurs may have ambushed them on the ground, as do cats today. The abundant pterosaurs may also have been part of the small theropod diet.

Then there are the insects and other invertebrates. We know the remains of many types of insects from Jurassic rocks, including dragonflies and beetles. The small lightweight dinosaurs would have fed on some of them.

Below: Coelophysis *skeleton with bones of a small crocodile in its stomach region.*

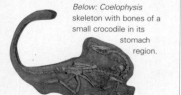

Elaphrosaurus

This theropod is one of the fossils found in the late Jurassic site at Tendaguru, in what is now Tanzania, by the famous German expeditions in the 1920s. Palaeontologists are unsure about the relationships of this animal, but it seems likely to have been a ceratosaurid. It would have been too small to tackle the big sauropods of the area, and probably hunted the small ornithopods.

Features: It looks rather like one of the later ostrich-mimics – and was once classified as an early member of this group – but the legs are shorter in proportion and the body very long and shallow-chested. We have a good idea of what the body is like, but the skull is missing. Small teeth like those of *Coelophysis* were found near the skeleton and seem to have belonged to *Elaphrosaurus*, suggesting that the head also was *Coelophysis*-like. It is the shortest theropod dinosaur in stature, going by the height of the hips compared with the overall length of the animal.

Distribution: Tanzania.
Classification: Theropoda, possibly Neoceratosauria.
Meaning of name: Light lizard.
Named by: Janensch, 1920.
Time: Tithonian to Cenomanian stages of the late Jurassic and early Cretaceous.
Size: 6m (19½ft).
Lifestyle: Hunter.
Species: *E. bambergi*, *E. gautieri*, *E. iguidensis*.

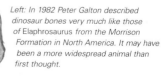

Left: In 1982 Peter Galton described dinosaur bones very much like those of Elaphrosaurus *from the Morrison Formation in North America. It may have been a more widespread animal than first thought.*

BIRDS AND BIRD-DINOSAURS

It has been accepted for some time that the dinosaurs evolved from birds. Since the 1860s Archaeopteryx has been hailed as the earliest bird. However, Archaeopteryx was not the only part-bird, part-dinosaur to exist in late Jurassic times. There were several, which suggests that bird-like features evolved more than once, or else evolved once and then developed along several lines, only one of which became the birds.

Archaeopteryx

Archaeopteryx was the first bird – that is what all the text books say. Indeed, with its mixture of dinosaur and bird features, it is regarded as the perfect transitional form between one group of animals and another. So far there have been ten specimens of *Archaeopteryx* found, one of which is only a feather.

Below: The clarity of the specimens of Archaeopteryx means there is not much doubt about the restoration. However, it may be that the leg feathers were longer than first thought.

Features: The bird-like features of *Archaeopteryx* are the covering of feathers and the wings that are constructed the same way as those of modern birds. The dinosaur-like features are the long stiff tail, the three-clawed hands and the skull with the tooth-filled jaws. Had the skeleton been found without the feathers then undoubtedly it would have been classed as a dinosaur. In fact one of the specimens lay in a collection mis-identified as a small dinosaur for several years before its true identity was revealed.

Distribution: Southern Germany.
Classification: Bird, order uncertain.
Meaning of name: Ancient wing.
Named by: Meyer, 1861.
Time: Kimmeridgian stage of the late Jurassic.
Size: 45cm (1½ft)
Lifestyle: Insectivore.
Species: *A. lithographica, A. recurva, A. bavarica.*

Epidendrosaurus

This is known from the almost complete skeleton of a juvenile found in lake deposits along with pterosaurs and amphibians. It was a small dinosaur that must have lived in trees. Some of its features make it more like a bird than like *Archaeopteryx*, leading to the idea that the adaptations for flight evolved from those for climbing.

Features: We can deduce by the long forelimbs, the long third finger and the curved toe bones that *Epidendrosaurus* lived in trees. The very long third finger is rather like that of the modern aye-aye lemur of Madagascar and was probably used the same way – for winkling insects out of holes in tree trunks. It must have been very close to the transition between dinosaurs and birds.

Distribution: Nei Mongol, China.
Classification: Theropoda, Tetanurae, Coelurosauria.
Meaning of name: Lizard in the tree.
Named by: Zang, Zhou, Xu and Wang, 2002.
Time: Late Jurassic.
Size: 15cm (6in).
Lifestyle: Climbing insectivore.
Species: *E. ningchengensis.*

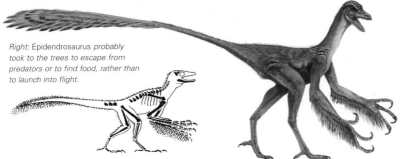

Right: Epidendrosaurus probably took to the trees to escape from predators or to find food, rather than to launch into flight.

Pedopenna

Distribution: Nei Mongol, China.
Classification: Theropoda, Tetanurae, Coelurosauria.
Meaning of name: Feathery foot.
Named by: Xu and Zhang, 2005.
Time: Late Jurassic.
Size: 1m (about 3ft).
Lifestyle: Climbing hunter.
Species: *P. daohugouensis*.

The discovery of the fossil leg of *Pedopenna* – the only part of it to have been found – suggests that bird flight first evolved with a gliding mechanism with wings on all four limbs. The presence of the feathers suggests that it was very close to the ancestry of the birds.

Features: The leg of *Pedopenna* is that of one of the sickle-clawed maniraptorans. However, it does not have the killing claw on the second toe as does its more famous relatives like *Velociraptor*. It has a row of about 50 long feathers sprouting from the long foot bones. Each feather has a distinct vane with branches. The feathers seem quite flexible and are more primitive than those of modern birds and of *Archaeopteryx* as the branches are fewer.

Left: Pedopenna *probably chased small prey through the treetops.*

AN ICONIC FOSSIL

In 1861 workers in a limestone quarry at Solnhofen in Bavaria, Germany, found the fossil of a feather – the first ever. A few months later they found the almost complete skeleton of a feathered animal. The foremost German palaeontologist Hermann von Meyer studied it and named it *Archaeopteryx*, meaning "the ancient wing". This is the fossil that became known as the "London specimen" and found a home in the Natural History Museum, London.

In 1877 a more complete specimen was found, which had the skull and jaws – items that were lacking in the London specimen. This one became known as the "Berlin specimen" and is currently housed in the Humboldt University Museum in Berlin, Germany.

No others were found until 1956, and to date only ten specimens have been found, all in the same lithographic limestone quarries at Solnhofen, Germany. These are the same quarries that have yielded pterosaurs complete with the impressions of wing membranes. Fish and crabs have also been perfectly preserved where they fell into the limy mud of the Jurassic lagoon, and even delicate creatures like jellyfish and sea lilies fossilized.

The idea that birds evolved from dinosaurs dates from the time of the early discoveries, but fell out of favour in the early part of the twentieth century. It was thought that birds evolved from reptiles, but not specifically dinosaurs. The dinosaur ancestry theory resurfaced in the 1970s with the discovery of bird-like dinosaurs such as *Deinonychus* and the other maniraptorans.

Koparion

The troodontids are known from the late Cretaceous, but the discovery of a single troodontid tooth in the late Jurassic Morrison Formation of Dinosaur National Monument put the history of the group back some considerable time. The species name honours Earl Douglass, the notable dinosaur hunter who worked extensively in the Mid-West, particularly in the quarry that became Dinosaur National Monument.

Features: The tooth is small and serrated, adapted for tearing flesh, and came from the upper jaw. It was found by sieving, and represents the only time a tooth has been distinctive enough to identify its owner as a new theropod. Nothing of the rest of the animal is known, but as the later troodontids were feathered animals, and feathers were definitely appearing at this time, it is probable that Koparion had a feathery covering too.

Distribution: Utah, USA.
Classification: Theropoda, Tetanurae, Coelurosauria.
Meaning of name: An ancient surgical tool (after the shape of the tooth).
Named by: Chure, 1994.
Time: Kimmeridgian stage of the late Jurassic.
Size: 1.5m (5ft).
Lifestyle: Hunter.
Species: *K. douglassi*.

Left: Koparion *is the oldest-known member of the troodontid group.*

BRACHIOSAURIDS AND CAMARASAURIDS

The sauropods were the big plant-eaters of late Jurassic times. The plains of North America and the forests of Europe would have been alive with herds of these vast animals roaming to and fro, from one feeding place to another. There were several different groups of sauropods at that time, one of which, the Macronaria, consisted of the brachiosaurid and camarasaurid families.

Giraffatitan

Distribution: Tanzania.
Classification: Sauropoda, Macronaria.
Meaning of name: Giraffe giant.
Named by: Olshevsky, 1991 (after Janensch, 1914).
Time: Kimmeridgian to Tithonian stages of the late Jurassic.
Size: 22m (72ft).
Lifestyle: High browser.
Species: *G. brancai*.

Although the name may be unfamiliar, *Giraffatitan* is one of the best-known sauropods in the world. Under the name *Brachiosaurus brancai* it has stood for 80 years as an exhibit in the Humboldt Museum of Berlin, Germany, as the biggest mounted dinosaur skeleton in existence. It was one of the many dinosaurs collected early in the twentieth century at Tendaguru, in what was then German East Africa.

Features: What was originally regarded as *Brachiosaurus* was given a new name after studies by Greg Paul in 1988 and George Olshevsky in 1991 showed that it was different from the original *Brachiosaurus*. The main difference is in the vertebrae of the back which, in *Giraffatitan*, produces a ridge between the shoulders like a horse's withers. To be fair, not all palaeontologists agree with this, and many books will continue to describe this animal as a species of *Brachiosaurus*.

Left: The famous mounted skeleton in the Humboldt Museum, Germany, is a composite of at least five individual skeletons found at the same site. In total 34 individuals were found at Tendaguru.

Lusotitan

It has long been known that brachiosaurids were not restricted to North America and Africa, where the most famous finds were made. Remains have also been found in late Jurassic rocks of Europe. In 1975 a partial skeleton was found in Portugal and named *Brachiosaurus atalaiensis* by Lapparent and Zybszewski. It is now thought to be a separate genus. Specimens of *Lusotitan* consisting of isolated vertebrae have been found in five different localities in Portugal.

Features: *Lusotitan* is known to be a brachiosaurid because of the low spines on the vertebrae and the muscle attachment of the long arm bone. As with all brachiosaurids, *Lusotitan* has very long forelimbs and high shoulders, allowing it to reach up into trees for food. No skull is known, but it would be the typical brachiosaurid pattern, with the very high and open nostrils, and the spoon-shaped teeth.

Left: The spikes on the backs of some restorations are conjectural. There is some evidence to suggest that spines existed on the backs of the other major sauropod group, the diplodocids, and some palaeontologists believe that the macronaria had them as well.

Distribution: Portugal.
Classification: Sauropoda, Macronaria.
Meaning of name: Giant from Portugal.
Named by: Teles, Antunes and Mateus, 2003.
Time: Tithonian stage of the late Jurassic.
Size: 25m (82ft).
Lifestyle: High browser.
Species: *L. atalaiensis*.

MACRONARIA

The macronaria are the sauropods with the boxy heads. The distinctive head shape was due to the fact that the holes in the skull representing the nostrils were much bigger than the eye sockets. And since the nostrils were on top of the head, it is easy to see why the rather strange-looking skull has been the subject of a great deal of scientific speculation.

It was once thought that the high nostril meant that the animal could remain submerged in a lake and breathe without any of its body showing. That idea has been dismissed. Some scientists drew attention to the fact that the nostrils on an elephant's skull are extremely large as well, and that maybe these sauropods had an elephant-like trunk. But this does not make much sense, since, with a long neck, the added reach of a trunk would seem to be superfluous. Anyway, a sauropod's skull lacks the broad plates of bone that would be needed to anchor the muscles that make up a trunk.

The most likely hypothesis is that the big nasal cavities would have been filled with moist membranes in life, and would have kept the interior of the skull and the little brain cool under the hot sun that would have beaten down on the late Jurassic plains of North America.

Brachiosaurus

The best-known *Brachiosaurus* skeleton in the world is now thought to be a different genus – *Giraffatitan* (far left). However, an even bigger animal, *Ultrasaurus*, found in the Dry Mesa Quarry in Colorado, is now regarded as a particularly big specimen of *Brachiosaurus*. The original *Brachiosaurus* was discovered as two partial skeletons in the Morrison Formation near Fruita in Utah in 1900 by Elmer G. Riggs.

Left: The position of the neck whether it was vertical or horizontal – is an on-going debate among scientists.

Distribution: Colorado and Utah, USA.
Classification: Sauropoda, Macronaria.
Meaning of name: Tall-chested arm lizard.
Named by: Riggs, 1903.
Time: Kimmeridgian to Tithonian stages.
Size: 22m (72ft).
Lifestyle: High browser.
Species: B. altithorax.

Features: About half of the height of *Brachiosaurus* is due to the neck. This, with its long front legs and tall shoulders, meant that it could reach high up into the trees to feed. Even its front feet contributed to its high reach – the fingers are long and pillar-like, and arranged vertically in the hand. Despite its fame, it is one of the rarest of the sauropods from the Morrison Formation.

Camarasaurus

This must have been one of the most abundant of the Morrison Formation sauropods, judging by the number of remains found. It often thought of as a small animal. This is because the best skeleton found is of a juvenile, perfectly articulated, lying in the rock of Dinosaur National Monument in Utah, USA, and mounted in Pittsburgh Museum.

Features: The "chambered lizard" in its name refers to the cavities in the backbone, designed to keep down the weight of the animal. Other cavities are present in the skull, which is merely a framework of bony struts, with enormous nostrils and spoon-shaped teeth. The forelimbs and hind limbs are approximately the same length, making the back of the animal horizontal. It is the Morrison Formation sauropod that is less bulky than *Brachiosaurus*, but not as slim as *Diplodocus*.

Distribution: New Mexico to Montana, USA.
Classification: Sauropoda, Macronaria.
Meaning of name: Chambered lizard.
Named by: Cope, 1877.
Time: Kimmeridgian to Tithonian stages of the late Jurassic.
Size: 20m (66ft).
Lifestyle: Browser.
Species: C. supremus, C. grandis, C. lentus, C. lewisi.

DIPLODOCIDS

If the macronarians (the brachiosaurids and camarasaurids) went for height to be impressive, the diplodocids went for length. Among these sauropods we find the longest land animals that ever existed. The slimness of the neck and tail meant that, despite its length, a diplodocid was not a particularly heavy animal. The tail was twice the length of the body and neck together, and ended in a whiplash.

Diplodocus

The familiar long, low sauropod is known as *Diplodocus*. It is well known from the many casts of the graceful skeleton of *D. carnegii*, the second species to be found. The casts, which appear in museums throughout the world, were excavated, reproduced and donated with finance provided by the Scottish-American steel magnate Andrew Carnegie in the early years of the twentieth century.

Features: The neck and tail are finely balanced around the hips and, as a result, *Diplodocus* could probably have raised itself on to its hind legs to reach high into the trees. The wear on the teeth shows that it could browse high in the treetops or among the undergrowth.

Finds in the 1990s have led American palaeontologist Steven Czerkas to suggest that there may have been a row of horny spines down the neck, back and tail.

Distribution: Colorado, Utah and Wyoming, USA.
Classification: Sauropoda, Diplodocidae.
Meaning of name: Double beam.
Named by: Marsh, 1878.
Time: Kimmeridgian to Tithonian stages of the late Jurassic.
Size: 27m (89ft).
Lifestyle: Low or high browser.
Species: *D. longus*, *D. carnegiei*, *D. hayi*.

Supersaurus

The dinosaur *Supersaurus* was one of the big sauropods found in the Dry Mesa Quarry, Colorado, by dinosaur hunter Jim Jensen. Unfortunately that site is such a jumble – probably representing a log-jam of bones in a Morrison Formation river – that the skeletons are all mixed up. One of Jensen's giants (then called *Ultrasaurus*) actually consisted of a shoulder blade of *Brachiosaurus* and ribs from *Supersaurus*.

Features: The vertebrae and the partial shoulder and hip that have been found of this animal show that it is closely related to *Diplodocus*. It may even be a large species of *Diplodocus* that stood 8m (26ft) high at the shoulder. The tallest vertebrae are as tall as a standing child. There are even bigger bones from Dry Mesa in the basement of Brigham Young University, in Salt Lake City, USA, that are yet to be examined and could alter our interpretation of what this dinosaur looked like.

Distribution: Colorado, USA.
Classification: Sauropoda, Diplodocidae.

Meaning of name: Super lizard.
Named by: Jensen, 1985.
Time: Kimmeridgian to Tithonian stages of the late Jurassic.
Size: 30m (98ft).
Lifestyle: Low or high browser.
Species: *S. vivianae*.

Right: The body proportions are based on those of Diplodocus *to which* Supersaurus *is obviously related. However, the neck may be longer than that of* Diplodocus.

Seismosaurus

Currently holding the record for the longest dinosaur known, *Seismosaurus* has taken 13 years to excavate, largely because of its size. It is only known from a single specimen, consisting of most of the vertebrae, part of the pelvis, the ribs and some stomach stones. Some palaeontologists regard it as a species of *Diplodocus*. The two are certainly closely related.

Features: The amazing length of this animal is entirely due to the length of the neck and tail. The body is not particularly big for a diplodocid, and the legs are quite stubby in comparison. Chevron-shaped protrusions beneath the backbones probably supported the muscles that helped the mobility of the neck and tail. The characteristic whip-like diplodocid tail has a downward curve not far from the hips, suggesting that the distant part of it trailed on the ground.

Distribution: New Mexico, USA.
Classification: Sauropoda, Diplodocidae.
Meaning of name: Earthquake lizard.
Named by: Gillette, 1991.
Time: Kimmeridgian stage of the late Jurassic.
Size: 40m (131ft).
Lifestyle: Low or high browser.
Species: *S. hallorum*.

Right: Seismosaurus was not the only huge sauropod that inhabited the river plains that produced the Morrison Formation. Others included Supersaurus *and* Diplodocus *itself. Obviously the range of food was enough to support several related big animals.*

DIPLODOCID TEETH

Above: The skull of Diplodocus.

The teeth of the diplodocids were distinctive. They had pencil-like teeth arranged like the teeth of a rake in front of the long jaws. There were no teeth at the back. The wear on these teeth suggests that the animals could browse from high up in the trees or from the ground. (The term graze nowadays refers to eating grass, but there was no grass in Mesozoic times so the term is used to mean low browsing.) The articulation of the neck bones and the musculature suggests that diplodocid necks were usually held horizontally, which would suggest that low browsing was the normal mode of feeding. The long neck would provide access to a large area of ferns or horsetails without moving the big body too far.

This tooth arrangement meant that these animals could not chew food. They would have spent their time raking and swallowing to obtain enough to feed. To help digestion they swallowed stones. These gathered in a gizzard and ground up the tough plant material. When the stones became smooth, they regurgitated them and swallowed fresh ones. Little piles of smooth stones are known from the Morrison Formation.

Barosaurus

This species is known from five partial skeletons from the Morrison Formation, three of them in Dinosaur National Monument in Utah, USA. The African species, until recently known as *Gigantosaurus*, was part of the Tendaguru fauna, and known from four skeletons. The rearing skeleton of *Barosaurus* in the American Museum of Natural History, at a height of 15m (49ft), is the tallest mounted skeleton in the world, only made possible by modern techniques of producing casts of fossil bones in lightweight materials.

Distribution: South Dakota and Utah, USA; and Tanzania.
Classification: Sauropoda, Diplodocidae.
Meaning of name: Slow heavy lizard.
Named by: Marsh, 1890.
Time: Kimmeridgian to Tithonian stages of the late Jurassic.
Size: 27m (89ft).
Lifestyle: Low or high browser.
Species: *B. lentus, B. africanus.*

Left: The way that Barosaurus, and the other diplodocids, was balanced at the hips suggests that it could rear up on its hind legs for feeding.

Features: *Barosaurus* is very much like *Diplodocus* – indeed the limb bones are indistinguishable between the two genera – but its tail bones are shorter and its neck bones at least one-third longer, one of which is 1m (3ft) long. The length of the neck has led some palaeontologists to suggest that there were several hearts along its length to enable the blood to reach the brain when it was feeding from high trees.

MORE DIPLODOCIDS

The diplodocid family seems to have evolved first in Europe with animals such as Cetiosauriscus *from middle Jurassic times. They reached their heyday in North America in the late Jurassic period; the Morrison Formation is full of them. They continued into the early Cretaceous period and by that time were spreading into the Southern Hemisphere.*

Apatosaurus

One of the most popular dinosaurs, *Apatosaurus*, keeps changing its identity. For almost a century it went by the evocative name *Brontosaurus*, the "thunder lizard". In 1877 Othniel Charles Marsh discovered *A. ajax* and named it. Two years later he found a more complete animal which he named *Brontosaurus excelsus*. It was not until the twentieth century that it was realized that these were actually two species of the same genus. When there is confusion of this kind, with one animal given two names, it is the first name given that is deemed to be the valid one, in this case *Apatosaurus*. The official change took place in 1903.

Features: *Apatosaurus* is a heavily built diplodocid. The vertebrae have a groove along the top. This held a strong ligament that supported the weight of the neck and the tail, like the cables on a suspension bridge.

The "deceptive lizard" of the name refers to the fact that the chevron bones attached to the vertebrae look confusingly like those of the aquatic reptile *Mosasaurus*. Although the head is about the size of that of a horse, the brain is only as big as that of a cat. The whole skeleton is similar to that of *Diplodocus*, but is much more stocky and massive, going for weight rather than length.

Distribution: Colorado, Oklahoma, Utah and Wyoming, USA.
Classification: Sauropoda, Diplodocidae.
Meaning of name: Deceptive lizard.
Named by: Marsh, 1877.
Time: Kimmeridgian to Tithonian stages of the late Jurassic.
Size: 25m (82ft).
Lifestyle: Low or high browser.
Species: *A. ajax, A. excelsus, A. louisae, A. montanus.*

Brachytrachelopan

Found in Patagonia by a farmer out looking for lost sheep, this dinosaur surprised everybody because it was a sauropod – one that actually had a neck that was shorter than its back. Evidently *Brachytrachelopan* evolved its short neck to browse vegetation that grew close to the ground – a food supply unavailable to its long-necked relatives.

Features: The most amazing feature of *Brachytrachelopan* is the extremely short neck – more like that of a stegosaur than a diplodocid. Like other members of the dicraeosaur group it has high spines along the back and hips, presumably supporting a kind of sail structure in life. The spines towards the front curve forward, indicating that the head would have been held low and

Distribution: Argentina.
Classification: Sauropoda, Dicraeosauridae.
Meaning of name: Short-necked shepherd.
Named by: Rauhut, Remes, Fechner, Cladera and Puerta, 2005.
Time: Tithonian stage of the late Jurassic.
Size: 10m (33ft).
Lifestyle: Low browser.
Species: *B. mesai.*

Left: Brachytrachelopan *would have looked like a long-tailed stegosaur, but without the armour plates.* that the animal fed from the ground. Unfortunately the skull and tail are missing from the single skeleton known.

Eobrontosaurus

Many palaeontologists, including Robert T. Bakker, were unhappy with the renaming of the original *Brontosaurus* genus to *Apatosaurus*, and so when a recently discovered species of *Apatosaurus*, named *A. yahnahpin* by Filla, James and Redman in 1994, was found to be different enough to merit its own genus name, Bakker resurrected a form of the name *Brontosaurus* and named it *Eobrontosaurus*.

Below: Unusually, the single specimen of Eobrontosaurus *that has been found contained the gastralia, the belly ribs. These are not usually preserved in a sauropod.* Eobrontosaurus *seems to be a more primitive animal than any of the accepted species of the* Apatosaurus *genus.*

Distribution: Wyoming, USA.
Classification: Sauropoda, Diplodocidae.
Meaning of name: Early Brontosaurus.
Named by: Bakker, 1998.
Time: Kimmeridgian to Tithonian stages of the late Jurassic.
Size: 20m (66ft).
Lifestyle: Browser.
Species: *E. yahnahpin*.

Features: *Eobrontosaurus* is distinguished from *Apatosaurus* by two specific features. The first is the slightly thicker neck produced by the bigger ribs, which is supported by the neck vertebrae. The second is that *Eobrontosaurus* has a slightly different arrangement of shoulder bones – a little more like those of a macronarian.

APATOSAURUS

Above: The skull of Apatosaurus *is long and narrow.*

The shape of the head also created uncertainty. A restoration of *Apatosaurus* (or *Brontosaurus*) carried out in the early twentieth century shows a dinosaur with a short boxy head like that of *Camarasaurus*. This is because the skull of *Apatosaurus* had not been found at the time, and when the skeleton was mounted in the American Museum of Natural History in New York, USA, the technicians built a *Camarasaurus*-type skull for it.

A long and narrow skull, like that of *Diplodocus*, had actually been found close to one of the skeletons at Dinosaur National Monument, but only the director of the Carnegie Museum in Pittsburgh, W. J. Holland, thought that this was the true *Apatosaurus* skull. He was overruled by the much more influential H. F. Osborne. The correct skull was finally established by John McIntosh and David Berman in the 1970s.

Dicraeosaurus

The only late Jurassic diplodocid found in Africa was *Dicraeosaurus*. It was a member of the Tendaguru fauna and, along with the other animals, showed that the same families of dinosaurs existed in North America and Africa at that time. However, *Dicraeosaurus* was so different from the North American forms that it has been given its own family, Dicraeosauridae.

Features: For a diplodocid, *Dicraeosaurus* has a strangely short neck with only 12 vertebrae, far fewer than any of the other late Jurassic diplodocids except for *Brachytrachelopan*. The vertebrae have extremely long spines that are deeply cleft in the neck and form a kind of low sail over the back. These features would have made it look bigger in profile and would have helped to deter predators. The tail has the typical diplodocid whiplash that would have been used as a weapon.

Distribution: Tanzania.
Classification: Sauropoda, Dicraeosauridae.
Meaning of name: Two forked lizard.
Named by: Janensch, 1914.
Time: Kimmeridgian stage of the late Jurassic.
Size: 20m (66ft).
Lifestyle: Low or high browser.
Species: *D. hansemanni, D. sattleri*.

Left: A full skeleton of Dicraeosaurus *is mounted in the Humboldt Museum in Berlin, Germany, beside that of its Tendaguru neighbour Giraffatitan.*

ORNITHOPODS

The two-footed ornithopod dinosaurs were present at the same time as the sauropods of the late Jurassic and were also plant-eaters. The ornithopods, however, were mostly quite small animals, scampering about at the feet of their big lizard-hipped cousins, and presumably exploiting food sources that were not available to bigger animals. They belonged to the ornithischian order.

Camptosaurus

The original species of *Camptosaurus* is based on ten partial skeletons, ranging from juveniles to adults. The species is well known from the mounted skeletons of a juvenile and an adult collected by Fred Brown and William H. Reed in the 1880s in Wyoming, USA, and put on display in the Smithsonian Institution in Washington D.C. The English species is a *nomen dubium* and may not even be an ornithopod.

Features: *Camptosaurus* is very similar to its cousin *Iguanodon*, a genus which was a characteristic feature of the landscape of the later early Cretaceous period in Europe. However, its head is longer and lower, and it has four toes on the back foot rather than three. Its heavy body can be carried on four legs or on two, both its front five-

Right: It is possible that the original species of Camptosaurus is the only true one. The others may be species of Iguanodon.

fingered feet and its hind feet carrying weight-bearing hooves. Its long mouth contains hundreds of grinding teeth and it has a beak at the front. Its food would be kept in cheek pouches while chewed.

Distribution: Colorado, Oklahoma, Utah and Wyoming, USA; and England.
Classification: Ornithopoda, Iguanodontidae.
Meaning of name: Flexible lizard.
Named by: Marsh, 1885.
Time: Kimmeridgian to Tithonian stages of the late Jurassic.
Size: 3.5–7m (11½–23ft).
Lifestyle: Low browser.
Species: *C. dispar, C. leedsi, C. depressus, C. prestwichii.*

Othnielia

Othniel Charles Marsh found this little dinosaur in 1877 and gave it the name *Nanosaurus*, meaning "little lizard". It was renamed after its original finder by Peter Galton exactly 100 years later. It is known from the almost complete skeletons of two individuals, but the only parts of the skull known consist of a few pieces of bone and teeth. It is said that the skull of one skeleton was stolen by collectors before it could be excavated.

Features: This small herbivore was a very fast runner, with compact thighs to hold the leg muscles, and easily moved lightweight legs with long shins and toes. It has five-fingered hands and four-toed feet. Its tail is stiff and straight, with the bones lashed together with tendons, to give balance while running. It is similar to the better-known *Hypsilophodon* with its big eyes and beak, but differs from it by having enamel on both sides of its teeth.

Distribution: Colorado, Utah and Wyoming, USA.
Classification: Ornithopoda, Hypsilophodontidae.
Meaning of name: Othniel's one.
Renamed by: Galton, 1977 (originally named in 1877).
Time: Kimmeridgian to Tithonian stages of the late Jurassic.
Size: 1.4m (4½ft).
Lifestyle: Low browser.
Species: *O. rex.*

Right: The tiny ornithopods like Othnielia must have scampered about among the feet of the huge Morrison Formation sauropods, eating the ground-hugging plants that were not available to their big neighbours.

Drinker

If *Othnielia* was named after one of the greatest American dinosaur pioneers of the nineteenth century, Othniel Charles Marsh, then the dinosaur *Drinker* was named after the other, namely Edward Drinker Cope. It is known from skeletons of an adult and a juvenile. Although there are quite a few remains known, not much has been published about this dinosaur.

Features: *Drinker* differs from its close relative *Othnielia* by having a more flexible tail, and its teeth seem to have more complex crowns. The chewing action that they produced would have worked on a ball of food held in the cheeks. Its toes are long and spreading, suggesting that it may have lived in swampy conditions. However, the differences between the two dinosaurs are so slight that instead of two different dinosaur species we may just be looking at two species of *Othnielia*.

Distribution: Wyoming, USA.
Classification: Ornithopoda, Hypsilophodontidae.
Meaning of name: Charles Drinker Cope's one.
Named by: Bakker, Galton, Siegwarth and Filla, 1990.
Time: Kimmeridgian to Tithonian stages of the late Jurassic.
Size: 2m (6ft).
Lifestyle: Low browser.
Species: *D. nisti*.

LUSITANIAN BASIN

In late Jurassic times the area of Portugal was a broad, shallow plain called, by geologists, the Lusitanian Basin. It had a warm climate, and rivers flowed into the basin from the nearby mountains. The resulting vegetation supported a large number of late Jurassic dinosaurs.

The fossils found at the site represent brachiosaurs, camarasaurs, diplodocids, stegosaurs and meat-eating theropods. There are also fossils of crocodiles, turtles, pterosaurs and mammals. The fauna is very similar to that of the North American Morrison Formation, indicating that the two areas, which are now far apart, were very close geographically at that time.

Studies of the rich fossil fauna from the site began in 1982 when a farmer found a dinosaur bone in his field. The fossil belonged to the carnosaur *Lourinhanosaurus*. The Museum of Lourinhã, Portugal, is now a world-famous institution, housing the dinosaurs found in the local area. In 1993 local scientist Isabel Mateus found fossilized dinosaur eggs in a sea cliff just to the north of the Lusitanian Basin. These eggs were identified by Philippe Taquet of the National Museum of Natural History in Paris, France, as theropod eggs. There is speculation that these were the eggs of *Lourinhanosaurus*. Since then much of the local dinosaur fauna, including the ornithopod *Draconyx*, has been collected and studied by Octavio Mateus, Isabel's son.

Draconyx

The dinosaur *Draconyx* is known from a scattering of bones from a single individual. The remains include teeth, vertebrae, leg bones and finger and toe bones, including the claws, after which it was named. The best material comes from the hind limbs. It is now housed in the Museum of Lourinhã, the principal centre for Portuguese dinosaur study.

Features: *Draconyx* is an early iguanodont of medium size, seemingly closely related to *Camptosaurus*. It is about the same size, and has the same heavy hind legs and short front legs. It differs from *Camptosaurus* by the shape of the thigh bone and the arrangement of toe bones – it has a vestigial first toe whereas *Camptosaurus* does not, and it lacks the fifth toe that *Camptosaurus* possesses. Otherwise the beasts are very similar.

Distribution: Portugal.
Classification: Ornithopoda, Iguanodontidae.
Meaning of name: Dragon claw.
Named by: Mateus and Antunes, 2001.
Time: Late Kimmeridgian to early Tithonian stages of the late Jurassic.
Size: 7m (23ft).
Lifestyle: Low browser.
Species: *D. loureiroi*.

Left: Draconyx *is another dinosaur excavated from the prolific Lourinhã dinosaur quarries of Portugal.*

STEGOSAURS

The stegosaurs were the plated dinosaurs of Jurassic and early Cretaceous times. They appear to have evolved in Asia – most of the primitive forms are found in China – and then migrated to North America and to Africa. They probably evolved from the same ancestors that gave rise to the armoured dinosaurs. The Stegosauria family is today grouped in the Thyreophora group of the Ornithischia order.

Stegosaurus

Although *S. armatus* was the first stegosaurus species to be found, *S. stenops*, found by Othniel Charles Marsh, is the more familiar species. The back plates were once thought to have been

Features: As well as the plates that *Stegosaurus* has along its back, it also has two pairs of spikes on the end of the tail to use as weapons. Recent studies show that these spikes stick out sideways. A mass of little bony ossicles protect the throat. The brain is the smallest, when compared with the bulk of the animal, for any dinosaur.

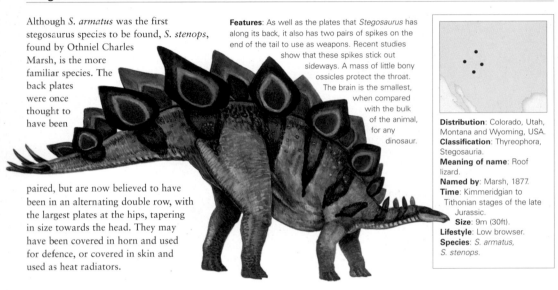

paired, but are now believed to have been in an alternating double row, with the largest plates at the hips, tapering in size towards the head. They may have been covered in horn and used for defence, or covered in skin and used as heat radiators.

Distribution: Colorado, Utah, Montana and Wyoming, USA.
Classification: Thyreophora, Stegosauria.
Meaning of name: Roof lizard.
Named by: Marsh, 1877.
Time: Kimmeridgian to Tithonian stages of the late Jurassic.
Size: 9m (30ft).
Lifestyle: Low browser.
Species: *S. armatus, S. stenops*.

Tuojiangosaurus

The best-known of the many Chinese stegosaurs, and the first to have been discovered is *Tuojiangosaurus*. It is known from two partial skeletons, one of which is 50 per cent complete. Its similarity to *Stegosaurus* shows that the two animals obviously had the same lifestyle and ate the same food – leafy plants growing low on the ground.

Features: *Tuojiangosaurus* has 15 pairs of small, pointed plates running down the neck, back and tail, as well as two pairs of spikes on the tail. As with all stegosaurs, it has a long head, with spoon-shaped teeth and a toothless beak. The teeth are very like those of *Stegosaurus* – small, coarsely serrated and vertically grooved.

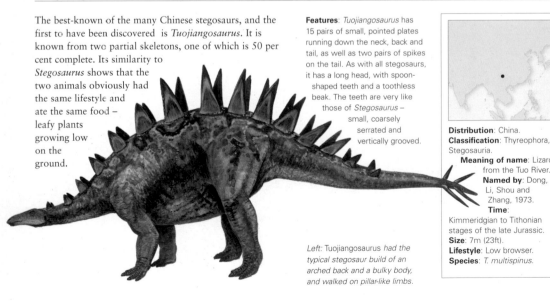

Left: Tuojiangosaurus *had the typical stegosaur build of an arched back and a bulky body, and walked on pillar-like limbs.*

Distribution: China.
Classification: Thyreophora, Stegosauria.
Meaning of name: Lizard from the Tuo River.
Named by: Dong, Li, Shou and Zhang, 1973.
Time: Kimmeridgian to Tithonian stages of the late Jurassic.
Size: 7m (23ft).
Lifestyle: Low browser.
Species: *T. multispinus*.

Chungkingosaurus

This stegosaur is known from several partial skeletons. It was one of the smaller, more primitive members of the stegosaur group. On the evolutionary scale it seems to lie somewhere between *Kentrosaurus* and *Stegosaurus*. The incomplete remains suggest that it has more than the usual two pairs of spikes on the end of the tail. The head is deep and narrow, and it has large nostrils.

Features: The distinguishing feature of *Chungkingosaurus* is the shape of the upper arm bone, which seems to be very primitive but with very broad ends, which is unusual. The hip bones are also primitive. Over the back the plates do not have strong bases, and may not have been firmly embedded in the skin. The tail plates appear to have been stronger.

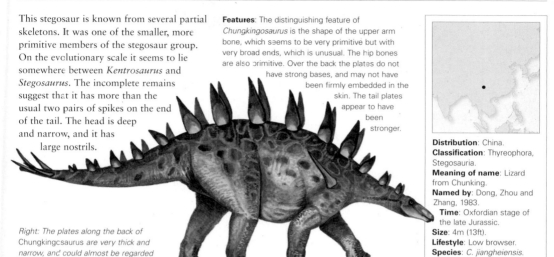

Right: The plates along the back of Chungkingosaurus *are very thick and narrow, and could almost be regarded as spikes.*

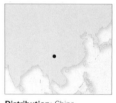

Distribution: China.
Classification: Thyreophora, Stegosauria.
Meaning of name: Lizard from Chunking.
Named by: Dong, Zhou and Zhang, 1983.
Time: Oxfordian stage of the late Jurassic.
Size: 4m (13ft).
Lifestyle: Low browser.
Species: *C. jiangheiensis*.

DINOSAUR PLATES

Ever since the first *Stegosaurus* was found, there has been much scientific discussion as to how the plates were arranged. None has ever been found in fossil form attached to the skeleton of a dinosaur.

An early idea of their arrangement was that they lay horizontally over the body, overlapping like the scales of a pangolin, and forming a protective shield. However, it was soon realized that they must have stuck up vertically.

For a long time the perceived image was of the plates being arranged in pairs, forming a symmetrical double row down the neck, back and tail. Now the accepted arrangement is of a double alternating row. Other interpretations abound, however. There is a suggestion that they formed a single row, but overlapped one another.

Scientific understanding of the function of the plates has always been subject to conjecture too. The obvious interpretation is that they were covered in horn, and formed an armour to protect the neck and back. Robert T. Bakker even suggests that their bases were muscular, and the points could have been turned towards any attacker.

In the 1970s it was suggested that the plates were covered in skin rather than horn, and used as heat exchangers, warming the blood in the morning sun and dissipating the heat at midday. If this worked for broad-plated stegosaurs like *Stegosaurus*, it would not have done so for the narrow-plated stegosaurs like *Chungkingosaurus*, since they would not have provided the surface area needed to make the system work.

Chialingosaurus

The first important stegosaur to have been found in China was *Chialingosaurus*. It is known from only one partial skeleton. Like others of the family it would have fed close to the ground, from the ferns and cycads that abounded at the time. Many stegosaurs were well-balanced at the hips, and seem to have been able to rise to their hind legs to feed from low-growing branches as well.

Features: This is a medium-size stegosaur, with a high, narrow skull. Its plates are small. They appear to be almost disc-shaped over the neck, but tall and spike-like over the hips and tail. They run in two rows from the neck to the tail. Its front legs are long for a stegosaur. The tall extensions on the tail vertebrae suggest muscular hips and perhaps an ability to rear on its hind legs.

Distribution: China.
Classification: Thyreophora, Stegosauria.
Meaning of name: Jia-ling River lizard.
Named by: Young, 1959.
Time: Middle to late Jurassic.
Size: 4m (13ft).
Lifestyle: Low browser.
Species: *C. kuani*, *C. guangyuanensis*.

MORE STEGOSAURS

The most familiar image we have of the stegosaurs is of Stegosaurus *itself, with the big, broad back plates and the spiked tail. However,* Stegosaurus *represents something of an extreme. Most stegosaurs had much smaller back structures, and many of them had spines rather than plates. The more primitive forms even had spines on their shoulders.*

Kentrosaurus

This stegosaur was excavated between 1909 and 1912 from the Tendaguru site by a team from Germany. Several hundred *Kentrosaurus* bones were found, suggesting that something like 70 individuals died there. The group find suggests that it may have been a herding animal. Two mounted skeletons were prepared for the Humboldt Museum in Berlin, Germany, but one was destroyed by bombing during World War II.

Features: The nine pairs of plates on the neck and back are very much narrower than those of *Stegosaurus,* and the five pairs of spines run in a double row right down the tail. Another pair of spines projects sideways from the shoulders. Unlike more advanced stegosaurs, it seems *Kentrosaurus* did not have ossicles across its body embedded in the skin. The little skull contains a tiny brain with well-developed olfactory bulbs. This suggests *Kentrosaurus* had a very good sense of smell, which would have aided food gathering.

Distribution: Tanzania.
Classification: Thyreophora, Stegosauria.
Meaning of name: Pointed lizard.
Named by: Henning, 1915.
Time: Kimmeridgian stage of the late Jurassic.
Size: 5m (16½ft).
Lifestyle: Low browser.
Species: *K. aethiopicus, K. longispinus.*

Dacentrurus

Remains of *Dacentrurus* are found throughout the late Jurassic period, and they seem to represent individuals ranging in size from 4–10m (13–33ft), making it one of the biggest of the stegosaurs. The youngest remains are common in the

Features: This was the first stegosaur to be discovered, by Owen in 1875, and was named *Omosaurus*. The forelimbs are longer than those of other stegosaurs, and the back is lower. The pieces of armour that we know suggest that it may have had spines rather than plates. In build it is a very primitive stegosaur, but the vertebrae and the hips have some quite advanced features. It is possible that the remains found represent more than one species.

Lourinhã site in Portugal. The French specimens were lost when the museum at Le Havre was destroyed during World War II.

Right: The remains of Dacentrurus *cover about 10 million years. This is a long period and may indicate that they consist of more than one genus.*

Distribution: England; France; and Portugal.
Classification: Thyreophora, Stegosauria.
Meaning of name: Very pointy lizard.
Named by: Lucas, 1902.
Time: Oxfordian to Kimmeridgian stages of the late Jurassic.
Size: 10m (33ft).
Lifestyle: Low browser.
Species: *D. armatus, D. phillipsi.*

TENDAGURU FORMATION

The Tendaguru Formation represents the richest deposits of late Jurassic fossils in the whole of Africa. It lies in Tanzania, north-west of the town of Lindi.

In the early years of the twentieth century, the German palaeontologist Eberhard Fraas brought back to Germany an account of fossil bones from what was then German East Africa. In 1909 the Natural History Museum of Berlin, Germany, mounted an expedition led by Werner Janensch and Edwin Hennig, during which they excavated about 225,000kg (4,432 cwt) of fossil bones in four years. The excavation involved the hiring of several hundred local people as labourers and porters to carry the finds down to the coast.

The beds consist of marls laid down by rivers and lakes, interspersed with marine sandstone. The area in late Jurassic times was a semi-arid plain crossed by rivers emptying into a shallow sea in the east, behind an offshore sand barrier. The riverbanks were thickly forested and attracted many types of herbivorous dinosaurs.

The fauna of the Tendaguru Formation is similar to that of the Morrison Formation of North America, which was laid down under similar circumstances and thriving at the same time. There were big sauropods, small ornithopods, stegosaurs, and large and small theropods. *Kentrosaurus* is the African equivalent of the North American *Stegosaurus* in this fauna.

"Yingshanosaurus"

The dinosaur known as *"Yingshanosaurus"* seems to be a *nomen dubium* – a name that has no scientific backing – and there has been little published work on the specimen. However, an almost complete skeleton was excavated from the Sichuan basin in China in the 1980s. It is a stegosaur with a double row of plates, two pairs of tail spines and a huge pair of shoulder spines.

Features: The features that make this animal unique are the wing-like shoulder spines, known scientifically as "parascapular spines". Other stegosaurs possess them, but none are as large. It seems that shoulder spines are only present in the primitive stegosaurs; the more advanced types such as *Stegosaurus* and *Wuerhosaurus* show no evidence of them.

Distribution: China.
Classification: Thyreophora, Stegosauria.
Meaning of name: Lizard from Yingshan.
Named by: Zhou, 1984.
Time: Late Jurassic.
Size: 5m (16½ft).
Lifestyle: Low browser.
Species: 'Y. jichuanensis'

Left: Note the large-scale spines on the shoulders. Stegosaurs used to be restored with these spines on the hips, but it makes more sense from a defensive angle to have them on the shoulders.

Hesperosaurus

The skeleton of this stegosaur was found in 1985, not far above base of the Morrison Formation, during the excavation of a *Stegosaurus* skeleton. It was complete, only missing the limbs, which were probably lost by erosion as it lay near the surface. It is now on display in the Denver Museum of Natural History, USA. *Hesperosaurus* is the oldest known stegosaur from North America.

Features: *Hesperosaurus* is distinguished by its short and wide skull, and deep lower jaw. Its armour consists of at least 10 plates, probably 14, which are oval, and the traditional two pairs of defensive spikes on the end of the tail. Unlike those of *Stegosaurus*, the plates are longer than they are high, and the spikes point in more of a backward direction. Its closest relative seems to have been *Dacentrurus* of contemporary Europe.

Distribution: Wyoming, USA.
Classification: Thyreophora, Stegosauria.
Meaning of name: Western lizard.
Named by: Carpenter, Miles and Cloward, 2001.
Time: Kimmeridgian to Tithonian stages of the late Jurassic.
Size: 6m (17½ft).
Lifestyle: Low browser.
Species: H. mjosi.

Left: The species name, H. mjosi, honours Ronald G. Mjos who collected and prepared the specimen and mounted a cast of it for display in the Denver Museum of Natural History, USA.

ANKYLOSAURS AND CERATOPSIANS

Both ankylosaurs and ceratopsians are typical of the Cretaceous fauna. However, there are early, possibly ancestral, examples that have been found in late Jurassic rocks, suggesting that the ankylosaur and ceratopsian families evolved slowly at first, biding their time, while the stegosaurs were the main, slow-moving ground feeders of the late Jurassic period.

Mymoorapelta

Until the discovery of *Mymoorapelta* in the 1980s, there were no ankylosaurids known from the Morrison Formation. The skeleton is more primitive than any of the other members of the polacanthid group. The skeleton remains were found as the scattered and trampled bones of a small animal beside a late Jurassic water hole. It is the smallest adult dinosaur found in the Morrison Formation. The remains were found in the Mygatt-Moore Quarry.

Features: The armour of *Mymoorapelta* is similar to that of other polacanthids with spines on the shoulders, sharp plates projecting sideways from the tail, and a solid mass of tiny bones across the hips. However, it is much more primitive than others of the group. A fossil egg found with the bones suggests that this skeleton might have been that of an adult female. If this is the case, then it will be the first recorded find of an ankylosaur egg.

Distribution: Colorado, USA.
Classification: Ankylosauria, Polacanthidae.
Meaning of name: Mygatt-Moore lizard – after its finders Peter and Marilyn Mygatt and John and Vanetta Moore.
Named by: Kirkland and Carpenter, 1994.
Time: Kimmeridgian stage of the late Jurassic.
Size: 2.7m (9ft).
Lifestyle: Low browser.
Species: *M. maysi.*

Right: Mymoorapelta was found in the Grand Junction area of Colorado, but other bones that may be from the same genus have been found in the Garden Park Quarry, along with bones of Camarasaurus and Othniela.

Dracopelta

From the late Jurassic period of Portugal comes perhaps the best-known of the early nodosaurids. It is known from the ribcage and a collection of armour plates. A small but heavy animal, *Dracopelta* would not have escaped predators by running. Instead, it would have hunkered down and relied on the solid pavement of back armour to protect it. As with most ankylosaurs, there is no evidence of any armour on the underside.

Features: The rib cage and armour show the typical curved ribs of the ankylosaur, giving a very broad, barrel-shaped body, probably containing a huge fermenting gut. It has the characteristic protection of at least five types of horn-covered bony plates, ranging from very small coin-sized pieces of armour to plates the size of CDs. Several rows of

Distribution: Portugal.
Classification: Ankylosauria, Nodosauridae.
Meaning of name: Dragon's armour.
Named by: Galton, 1980.
Time: Kimmeridgian stage of the late Jurassic.
Size: 2m (6½ft).
Lifestyle: Low browser.
Species: *D. zybszewskii.*

cone-shaped plates along the back are separated by a mass of smaller scutes, and there are spines along the flanks. The neck is protected by broader plates. We do not know what the head is like.

Gargoyleosaurus

This primitive ankylosaur from the Morrison Formation has features shared by both main ankylosaur families – the ankylosaurids and the nodosaurids. They include the fact that the armoured covering of the skull is fused to the skull bones rather than being set into the skin. It is possible that *Gargoyleosaurus* is close to the ancestors of both groups. Its species name honours its discoverers, J. Parker and T. Pinegar.

Features: This primitive ankylosaur has a long, narrow snout. The front of the mouth has a beak, with front teeth in both the upper and lower jaw, and that is unusual for an ankylosaur which usually has teeth only in the lower beak. Unlike other ankylosaurs there is no palate separating the mouth from the nasal passages, and in this case the nasal

passages are straight rather than contorted. Its shoulders are not as heavily armoured as the rest of the group.

Distribution: Wyoming, USA.
Classification: Thyreophora, Ankylosauria.
Meaning of name: Hideous statue lizard.
Named by: Carpenter, Miles and Cloward, 1998.
Time: Kimmeridgian to Tithonian stages of the late Jurassic.
Size: 3m (10ft).
Lifestyle: Low browser.
Species: *G. parkpinorum*.

Chaoyangsaurus

Ceratopsians – the horned dinosaurs – are usually regarded as purely Cretaceous animals, particularly from the late Cretaceous period. However, the earliest of the group hails from the late Jurassic beds of Liaoning province in China. Discovered in 1983, it was not sufficiently studied at the time, but a new description published in 1999 shows it to have been an early ceratopsian.

Features: The specimen consists only of the front part of the skull and lower jaw, and part of the backbone. The lower jaw is very similar to that of *Psittacosaurus*, hitherto the most primitive type of ceratopsian, and the chopping teeth are set in from the edge of the skull showing the presence of cheeks. By comparison with other primitive ceratopsians, we can assume that *Chaoyangsaurus* is a lightweight, two-footed, fast runner.

Below: The appearance of Chaoyangsaurus gives little indication of the spectacular types to follow, such as Protoceratops (below right) and Triceratops (above).

Distribution: China.
Classification: Marginocephalia, Ceratopsia.
Meaning of name: Lizard from Chaoyang County.
Named by: Zhao, Zheng and Xu, 1999.
Time: Middle to late Jurassic.
Size: 1m (3ft).
Lifestyle: Low browser.
Species: *C. youngi*.

Right: The name Chaoyangsaurus first appeared in a guidebook for an exhibition of fossils in 1981. It remained a nomen dubium until a scientific description was published in 1999.

THE CRETACEOUS PERIOD

By the Cretaceous period the supercontinent Pangaea had split apart into individual continents that could be recognizable today. In the Southern Hemisphere, though, the landmasses of Australia and Antarctica were still joined – in a supercontinent called Gondwana. With isolated continents, different animals evolved, so that the dinosaurs of South America differed from those of North America and Europe. Tree ferns and conifers were beginning to give way to flowering plants, with the conifers being pushed to high latitudes and mountainous areas. However, the inhabitants of these forests were quite remarkable – for this was the climax of the Age of Dinosaurs.

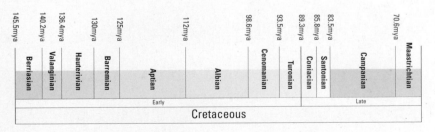

Above: The timeline shows the different chronological stages that make up the Cretaceous period. (mya = million years ago)

1 *Quetzalcoatlus.*
2 *Euoplocephalus.*
3 *Albertosaurus.*
4 *Corythosaurus.*
5 *Chasmosaurus.*
6 Lizard.

LIZARDS AND LIZARD-LIKE ANIMALS

The Cretaceous period was not just the time of the giant dinosaurs. At the feet of the mighty ran all sorts of little animals mostly feeding on the insects and other tiny invertebrates of the undergrowth. Many of these belonged to the small reptile groups that we would recognize today, but some were quite specialized. As with most reptile groups, some took up an aquatic mode of life.

Hyphalosaurus

The skeletons of small land-living animals are so delicate that they are rarely fossilized. Only with exceptional fossilization conditions, such as in the lake deposits of Liaoning, where the waters were poisoned by volcanic eruptions and anything that died in them was covered in ash, do we get good fossils. There are skeletons of little flying lizards such as *Xianglong* complete with rib-supported wings, and many swimming types. *Hyphalosaurus* was one of those.

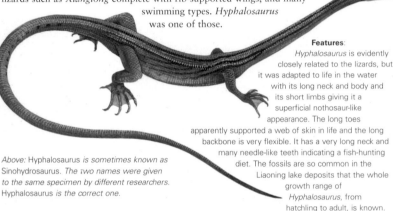

Features:
Hyphalosaurus is evidently closely related to the lizards, but it was adapted to life in the water with its long neck and body and its short limbs giving it a superficial nothosaur-like appearance. The long toes apparently supported a web of skin in life and the long backbone is very flexible. It has a very long neck and many needle-like teeth indicating a fish-hunting diet. The fossils are so common in the Liaoning lake deposits that the whole growth range of *Hyphalosaurus*, from hatchling to adult, is known.

Above: Hyphalosaurus is sometimes known as Sinohydrosaurus. The two names were given to the same specimen by different researchers. Hyphalosaurus is the correct one.

Distribution: Liaoning, China.
Classification: Lepidosauria, Squamata.
Meaning of name: Thread lizard.
Named by: Gao, Tang and Wang, 1999.
Time: Barremian stage of the early Cretaceous.
Size: 1.5m (5ft).
Lifestyle: Fish-hunter.
Species: *H. lingyuanensis*.

Pleurosaurus

The sphenodonts, the almost extinct order to which the modern tuatara belongs, were a diverse bunch in Mesozoic times. One group, the pleurosaurs, became highly adapted to life in the sea, reducing their limbs and developing long eel-like bodies and whip-like tails. There are two known genera – *Palaeopleurosaurus*, and two species of *Pleurosaurus*. The former lived at the end of the Jurassic and differed in appearance by its hooked beak-like jaw.

Below: Pleurosaurus moved through the shallow waters with an eel-like motion.

Features: *Pleurosaurus* has a large number of vertebrae – about 57, twice as many as is found in other sphenodonts. Many of these go into the very long tail. The limbs do not show any particular adaptation to the marine environment – they are just small. The teeth are attached to the edge of the jaws, not in sockets, and the nostrils are set far back on the snout, close to the eyes.

Distribution: France; Germany.
Classification: Lepidosauria, Sphenodontia.
Meaning of name: Sideways lizard.
Named by: Meyer, 1831.
Time: Early Jurassic to late Cretaceous.
Size: 60cm (2ft).
Lifestyle: Fish-hunter.
Species: *P. ginsburgi*, *P. goldfussi*.

Pachyrachis

Distribution: Israel.
Classification: Lepidosauria, Squamata.
Meaning of name: Thick ribs.
Named by: Haas, 1979.
Time: Early Cretaceous.
Size: 1m (about 3ft).
Lifestyle: Fish-hunter.
Species: *P. problematicus*.

It is not quite clear how the first snakes evolved from the lizards. There are several known from Cretaceous times and it is possible that they developed their distinctive shape and lack of limbs as an adaptation to a burrowing or a swimming mode of life. Certainly *Pachyrachis* with its tiny limbs, seems close to the ancestry.

Features: The front limbs and the shoulder girdles have completely disappeared, with the body beginning immediately behind the skull, and no neck vertebrae whatsoever. However, part of the hind limb structure remains. It was at first thought to have had a head like that of a modern monitor lizard. Another snake with hind legs, *Haasiophis*, was found in the same area in 2000. Studies on this and *Pachyrachis* show that the skull is less lizard-like than first thought – more like the skull of a modern python.

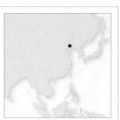

Right: The thick ribs of Pachyrachis show that it must have lived in the water and swam by snake-like undulations.

SNAKEBITE

Until 2006 scientists thought that the venom of the three modern poisonous snake families – the Viperidae (vipers), the Elapidae (cobras) and the Atractaspididae (stiletto snakes) – evolved independently after these families had split apart. The fourth family, the Colubridae (snakes, such as pythons and anacondas that kill by constriction), consisting of about half the living snake species, never had poison glands.

Genetic analysis of snake venom by Bryan Greg Fry of the University of Melbourne, Australia, indicates otherwise. It seems as if poison evolved in the ancestors of both lizards and snakes as long ago as the Triassic period. As the groups evolved and diversified some of them lost the potency of their venom and turned to other ways of killing prey – like constriction. This accounts for the fact that, as well as snakes, a few modern lizards, such as the Gila monster, still have a poisonous bite. The so-called "non-poisonous" modern snakes do produce venom but, as far as humans are concerned, they just lack a way of administering it that would be dangerous to us.

Monjurosuchus

Distribution: Liaoning, China; Japan.
Classification: Lepidosauria, Squamata.
Meaning of name: Manchu crocodile.
Named by: Endo, 1940.
Time: Barremian stage of the early Cretaceous.
Size: 1.2m (4ft).
Lifestyle: Hunter.
Species: *M. splendens*.

Monjurosuchus – another of the well-preserved small animals from Liaoning – is another of the diapsid group that seems to resemble both the lizards and the sphenodonts. It was probably a semi-aquatic animal that fed on small fish, amphibians and invertebrates – a good specimen studied in the year 2000 showed stomach contents consisting of parts of arthropod shells.

Features: The "crocodile" part of the name refers to the bony scutes in the skin, like those of the modern Chinese crocodile lizard. These scales are small and overlapping, bigger on the upper surface than underneath, and with a double row of large oval scales running down the back from the neck to the tail. The feet are webbed, showing it to have been at least partly aquatic. The eyes are quite large.

Below: Monjurosuchus *must have spent some of its time in the water and some of its time on land, hunting the smaller animals around the Liaoning lake.*

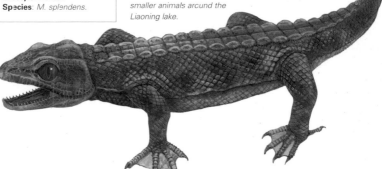

EARLY CRETACEOUS CROCODILES

The crocodiles of the early Cretaceous were, by and large, similar in appearance to those that live today. They played an important part in the ecology of the shallow waters of the rivers and swamps of the low lands of the time, feeding on small and large animals. Some, however, retained the terrestrial lifestyle of their ancestors.

Sarcosuchus

Hailed as "super croc" when the full skeleton was uncovered in the Sahara desert and announced in 2000 by Paul Sereno of Chicago, *Sarcosuchus* is one of the biggest crocodiles ever. It lurked in the meandering waterways that covered a large proportion of North Africa in early Cretaceous times. Since the 1940s it was known only from isolated teeth and pieces of armour, but the 2000 expedition unearthed about half the skeleton. Growth lines on the bony plates suggest that *Sarcosuchus* would have been fully grown at about 50 years of age – about twice the lifespan of a modern crocodile.

Features: Apart from its extreme size – weighing about ten tonnes, as long as a bus and twice the size of any living crocodile – the main feature of *Sarcosuchus* is its head with its long snout, representing about 75 per cent of the skull's length, expanded at the front. The teeth do not interlock but are stout, rounded and smooth, suitable for grabbing and crushing large prey rather than shearing meat or snapping at fish. The eyes and nostrils are on the top of the skull – a feature of a water lurker. Its back is armoured from the neck to half way down its tail.

Below: Growth rings on the armour scutes suggest that Sarcosuchus would have taken 50 or 60 years to grow to full size.

Distribution: Niger.
Classification: Crocodylomorpha, Pholidosauridae.
Meaning of name: Flesh crocodile.
Named by: Broin and Tarquet, 1966
Time: Early Cretaceous.
Size: 12m (40ft).
Lifestyle: Fish-hunter or scavenger.
Species: *S. imperator*.

Araripesuchus

The presence of *Araripesuchus* in both South America and Africa had always been held up as proof of continental drift and was used to determine the timing of the split up of Pangaea. However, recent research suggests that the African form, *A. wegeneri*, may not have been the same genus, although it was closely related. Another close relative has now been found in Madagascar. This indicates that all the southern continents were a single landmass at that time.

Features: This is a rather dog-like crocodile in that it hunted on land rather than in the water, unlike its relatives. Its longer legs, held in an almost mammal-like stance, indicate this. The head is quite short and broad compared with that which we usually find in crocodiles, showing a completely different hunting and feeding strategy. It has been found in the same formation that yields the abundant South American pterosaur fauna, and perhaps these represented its main prey. It may have raided the pterosaur nests along the lake sides.

Below: Araripesuchus was probably a fast hunter of the lakeshores.

Distribution: Argentina; Brazil; possibly Niger; Madagascar.
Classification: Crocodylomorpha, Mesoeucrocodilia.
Meaning of name: Crocodile from Araripe.
Named by: Price, 1959.
Time: Albian to Cenomanian stages of the Cretaceous.
Size: 2m (6½ft).
Lifestyle: Hunter.
Species: *A. gomesii*, *A. patagonicus*, *A. buitreraensis*, "*A*." *wegeneri*.

Bernissartia

Distribution: Southern England; Belgium; Spain.
Classification: Crocodylomorpha, Pholidosauridae.
Meaning of name: From Bernissart in Belgium.
Named by: Buffetaut and Ford, 1979.
Time: Berriasian stage of the early Cretaceous.
Size: 60cm (2ft).
Lifestyle: Fish- or shellfish-eater, or scavenger.
Species: B. fagesii.

This tiny crocodile lived in the swamps and creeks that spread across northern Europe from England to Belgium, sharing the environment with such dinosaurs as *Iguanodon* and *Neovenator*. It was one of the smallest of the Cretaceous crocodiles. The environment was inhabited by other crocodiles that were more similar in size to those of the present day. Such a variety of crocodiles indicates a range of different types of prey.

Features: There are two kinds of teeth in the jaws of *Bernissartia*. Those at the front are sharp and pointed, like normal crocodile teeth, while those at the back are rounded and button-shaped and situated at the point where greatest pressure could be exerted. It is possible that it varied its diet between fish, caught with the front teeth, and shellfish, crushed with the rear teeth. It may even have used its rounded teeth for crushing the bones of scavenged corpses. As in other crocodiles, the back and tail were covered with little bony plates that supported thick horny scales.

Below: Bernissartia *looked just like a miniature version of a modern crocodile.*

OTHER EARLY CRETACEOUS CROCODILES

Goniopholis – a larger crocodile than *Bernissartia*, from the same area and at the same time. It had two prominent rows of broad scutes down its back.
Sichuanosuchus – a very primitive form, with a flat triangular-shaped head from the early Cretaceous of China.
Malawisuchus – a small terrestrial crocodile from Africa that may have been a burrower.
Pachycheilosuchus – from Texas, USA, this swimming crocodile had strong bones in its upper jaw and a neck that was armoured like one of the ankylosaurs.
Susisuchus – found in north-east Brazil, this large swimming crocodile had armour that was almost identical to the armour of modern types.

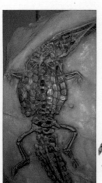

Left: When a crocodile is well-preserved we can even see the scutes in the skin.

Chimaerasuchus

This was the first vegetarian crocodile to be discovered. The only specimen is of a partial skull and some body parts including a limb, discovered in the 1960s and kept in the Institute of Vertebrate Palaeontology and Palaeoanthropology in Beijing, China. The skull was so unusual that it was not at first recognized as a crocodile, but regarded as a mammal or a mammal-like reptile.

Features: The significant feature of *Chimaerasuchus* is the appearance of the teeth. There are two pairs of sharp canine-like teeth at the front, but the back teeth are blunt, solid and mammal-like, rather than the usual array of pointed gripping fangs we would expect to see in the mouth of a crocodile. The jaw has a complex hinge that allows a fore-and-aft grinding action, in keeping with the broad grinding surfaces of the back teeth.

Below: The foot bones and the presence of armour scutes in the skin suggest that the body of Chimaerasuchus *was similar to that of a conventional crocodile.*

Distribution: Hubei, China.
Classification: Crocodylmorpha, Mesoeucrocodylia.
Meaning of name: Chimaera (monster from Greek mythology).
Named by: Wu and Suess, 1995.
Time: Aptian to Albian stages of the early Cretaceous.
Size: 1m (about 3ft).
Lifestyle: Browser.
Species: C. paradoxus.

DIVERSE PTEROSAURS

In the early Cretaceous the pterodactyloid pterosaurs evolved into all kinds of different types. The body and wing arrangement were quite conservative, but there were great differences in the shape of the head, indicating different food types and lifestyles. We have good evidence of pterosaurs that lived by the sea, lakes and rivers. There must have been more from inland and mountain areas that were never fossilized.

Dsungaripterus

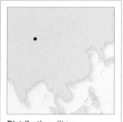

Distribution: China.
Classification: Pterosauria, Pterodactyloidea.
Meaning of name: Wing from the Junggar Basin.
Named by: Young, 1964.
Time: Early Cretaceous.
Size: 3m (10ft) wingspan.
Lifestyle: Shellfish-eater.
Species: *D. brancai, D. weii.*

The strange mouth has been interpreted as an adaptation to feeding on shellfish, with the narrow beak used for prizing the shells from their substrate and the bony knobs used for crushing them. Since the specimens found are all a long way from any sea, the shellfish must have been freshwater types, from rivers or lakes.

Features: The remarkable feature of this large pterodactyloid is the head. The jaws are pointed and turned upwards at the tips, like specialist forceps. The sides of the mouth are armed with bony knobs in the place of teeth. The eye socket is surprisingly small, but much of the side of the deep skull is taken up by a cavity that combines the nostril with the other holes usually found there. A thin crest sticks up along the mid-line of the skull, and another juts out behind.

Left: The body of Dsungaripterus is very strong, with the back vertebrae fused together in the shoulder region, giving a structure similar to that of a bird.

Plataleorhynchus

Nicknamed the Purbeck spoonbill after the area in southern England where it was found, this creature must have worked the shallow waters of the swampy lowlands that existed at the time that the limy sediment of the famous Purbeck building stone was deposited. Like its namesake bird, the spoonbill, it would have filtered the water for organisms such as crustaceans and insect larvae.

Features: The only part of this pterosaur known is the tip of the long narrow jaws, in a specimen in which the top side is exposed. This is flat with a spoon-shaped end, and is lined with small pointed teeth that curve outwards. This is superficially similar to the structure in *Gnathosaurus*, but the arrangement of jaw bones is different; the maxilla at the front is very small, suggesting that it evolved independently. This is an example of convergent evolution. The rest of this pterosaur is unknown, but as with the other little-known examples, we can restore the living appearance of the animal by reference to the other pterosaurs in the group.

Below: At the end of the Jurassic and beginning of the Cretaceous southern England was a wide swampy plain, covered in slow rivers and lakes – an ideal environment for animals that fed on shallow-water creatures.

Below: Gnathosaurus for comparison.

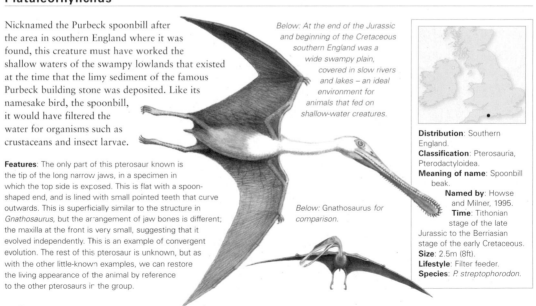

Distribution: Southern England.
Classification: Pterosauria, Pterodactyloidea.
Meaning of name: Spoonbill beak.
Named by: Howse and Milner, 1995.
Time: Tithonian stage of the late Jurassic to the Berriasian stage of the early Cretaceous.
Size: 2.5m (8ft).
Lifestyle: Filter feeder.
Species: *P. streptophorodon.*

Criorhynchus

A number of pterosaurs had crests on the tips of their jaws, both the upper and the lower. *Criorhynchus* was the first to be found. The crests were probably used for cleaving the water as the pterosaur skimmed across at wave height, making it easier to catch fish just under the surface.

Features: A semicircular crest at the tip of the upper jaw is matched by a similar one on the lower, giving the impression of a continuous vertical disc when the jaws are closed. The first specimen was so fragmentary that it was not until the discovery of a more complete specimen of the closely related *Tropeognathus* more than a century later, in 1987, that the arrangement of crests on the skull was determined.

Distribution:
England.

Classification:
Pterosauria, Pterodactyloidea.
Meaning of name: Battering ram snout.
Named by: Owen, 1874.
Time: Cenomanian stage of the middle Cretaceous.
Size: 5m (16ft) wingspan.
Lifestyle: Fish-hunter.
Species: *C. simus*.

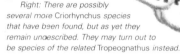

Right: There are possibly several more Criorhynchus *species that have been found, but as yet they remain undescribed. They may turn out to be species of the related* Tropeognathus *instead.*

OTHER PTEROSAUR GENERA

It is impossible to cover all the different genera of pterosaur here, even the ones that we know from good fossils. Other early cretaceous types include:
Phobetor A close relative of *Dsungaripterus*, but only about half the size and with straighter forceps jaws.

Domeykodactylus Another of the *Dsungaripterus* group but from South America – the first to be found outside China.

Doratorhynchus Known from a single neck vertebra which is particularly long, suggesting that it is an early member of the group that covered the giant pterosaurs of the late Cretaceous, such as *Quetzalcoatlus*.

Istiodactylus A large British pterosaur with a long rounded duck-like beak lined with short teeth. It is quite unlike any other known pterosaur and is classed in a family of its own.

Santanadactylus A Brazilian form whose discovery opened the study of pterosaurs in South America.

Pterodaustro

Filter feeding was taken to an extreme in this pterosaur, the first to be found in South America. It must have rested close to the water and swept the surface with its bristle-combed lower jaw, scooping out small invertebrates, rather like a modern flamingo does. It has been suggested that its fur was pink – a pigment imparted by the crustaceans on which it fed, as in the plumage of modern flamingos.

Features: The jaws are very long and slim, and sweep upwards in a bow-shape. There are no teeth in the lower jaw, instead there is a brush-like arrangement of long elastic bristles, up to 500 of them, into which the upper jaw fits when closed. The upper jaw has many short blunt teeth that comb food from the bristles and into the mouth. The hands are small and the feet big for a pterosaur.

Distribution: Argentina, Chile.
Classification: Pterosauria, Pterodactyloidea.
Meaning of name: Southern wing.
Named by: Bonaparte, 1969.
Time: Early Cretaceous.
Size: 1.3m (4¼ft).
Lifestyle: Filter feeder.
Species: *P. guinzaui* and possibly two others, unnamed.

Left: Pterodaustro *was the first pterosaur to have been found in South America. Several others have been discovered since.*

BIG PTEROSAURS

The pterosaurs continued to evolve, reaching their greatest diversity in early Cretaceous times. After this they began to decline, as the birds evolved and took over their niches. This competitive evolutionary pressure encouraged the pterosaurs to adopt more and more bizarre forms as they adapted to very restricted lifestyles.

Tapejara

Distribution: Araripe Plateau, north-eastern Brazil.
Classification: Pterosauria, Pterodactyloidea.
Meaning of name: Old one.
Named by: Kellner, 1989.
Time: Aptian stage of the Cretaceous.
Size: Probably 5m (16ft) wingspan (but only skulls are known).
Lifestyle: Fruit-eater, fish-eater or scavenger.
Species: *T. wellnhoferi, T. imperator, T. navigans.*

The big lightweight crest of *Tapejara* is a mystery. The obvious explanation is that it is a display structure, but it is possible that it may have had something to do with aerodynamics, and was used as a windsurfer sail. The short mouth, likewise, is a puzzle. It may have been a fruit-eater, like modern heavy-billed birds such as hornbills or toucans, or it may have lived on fish or carrion.

Right: If Tapejara had been a fruit-eater, then it probably served an important ecological purpose, helping to spread seeds over a wide area, just like modern fruit-eating birds.

Features: The distinctive feature of *Tapejara* is the extraordinary head crest. Two bony struts protrude from the skull, one broad and pointing directly upwards and one narrow and sweeping back. In the species *T. navigans* the back section is almost non-existent, but in *T. imperator* it is long and slim. In life these would have been joined by a flap or a sail, giving a broad display structure. The front of the short snout is turned down as a strong beak.

Tupuxuara

This pterosaur was found in the same area and from the same period as the other spectacularly crested pterosaur, *Tapejara*. The skin-covered bony crest of *Tupuxuara* was well-endowed with blood vessels, which suggests that the crest could change colour, or at least the intensity of the colour, according to the animal's mood and activity. This would have made for an elaborate communication system for threat or mating displays. Perhaps only the males possessed the crest.

Right: The name Tupuxuara means "familiar spirit" in the language of the Tupi, the original inhabitants of the area of Brazil in which it was found. Likewise the name Tapejara means "old one" in the same language, and refers to a being in local mythology.

Features: The crest sweeps upwards and backwards in a semicircular plate, making the skull longer than the body itself. The jaws are narrow and toothless. Because of the spectacular crest, *Tupuxuara* was thought to have belonged to the same family as *Tapejara*, but detailed study of the skeleton suggests that it is closer to the giants like *Quetzalcoatlus*. Blood vessels in the crest indicate that it was covered by skin. It may even have had a keratinous extension, that would have made it even bigger.

Distribution: Ceará, Brazil.
Classification: Pterosauria, Pterodactyloidea.
Meaning of name: Familiar spirit of local Tupi Indian mythology.
Named by: Kellner and Campos, 1988.
Time: Aptian stage of the Cretaceous.
Size: 5.5m (17ft) wingspan.
Lifestyle: Fish-eater or meat-eater.
Species: *T. leonardii, T. longicristatus.*

Anhanguera

The shape of the brain in one particularly well-preserved specimen of *Anhanguera* shows that it could co-ordinate information from all parts of the body. The wing was used as a sensory organ and the brain was able to control the position and attitude of the body while keeping its eyes focused on its prey. This ability probably applies to the rest of the pterodactyloids.

Features: This large pterosaur has very long wings and crests on both the upper and lower jaws, about halfway along. The legs are quite small, and the hip bones relatively weak. Their articulation indicates that it is impossible to bring the legs vertically under the body like a bird, and so when at rest the legs must have been splayed out at the side.

Distribution: North-eastern Brazil. Possibly also Argentina; Isle of Wight; and Queensland, Australia.
Classification: Pterosauria, Pterodactyloidea.
Meaning of name: Old devil.
Named by: Campos and Kellner, 1985.
Time: Aptian stage of the Cretaceous.
Size: 4m (13ft) wingspan.
Lifestyle: Fish-eater.
Species: *A. santanae*, *A. blittersdorffi*.

Left: The skull of Anhanguera *is as long as the body.*

PTEROSAUR EGGS

It had always been assumed that pterosaurs laid eggs. However, when a fossil egg is found it is difficult to tell what kind of animal laid it.

In early 2004 a tiny egg, no more than 5cm (2in) in diameter, was found in the famous late Jurassic and early Cretaceous lake deposits in Liaoning province in China. It was the fossil of a soft egg, like that of a snake or a turtle, rather than one with a hard shell as we find in birds and dinosaurs. So well preserved was this egg that when it was dissected it was found to contain a young pterosaur with long wings tightly folded. The baby would have had a 27cm (11in) wingspan when it hatched.

The long wing bones had solidified in the egg before hatching, which suggested that as soon as the youngster hatched it would have been ready for flight. This means that the youngsters would have had to fend for themselves as soon as they were free from the shell, like crocodiles and turtles, and would have had little in the way of parental attention. This is in keeping with the observation that pterosaurs grew slowly, as all their youthful energy would have gone into finding food rather than growing.

Cearadactylus

This genus is known only from a single damaged skull with the jaws still articulated. The resemblance of the head to that of some fish-eating dinosaurs and crocodiles suggests that *Cearadactylus* hunted fish. The bunch of fine teeth at the end of the jaw would have been ideal for holding on to slippery prey which it would then swallow whole.

Features: *Cearadactylus* has long jaws, with the long needle-like teeth restricted to the tip and forming a kind of a rosette. The teeth of the upper and lower jaw tip interlock when the jaws close. Apart from these, the teeth in the rest of the jaw are quite small and conical. The upper jaw has a notch close to the end. In fact the whole skull is reminiscent of that of the fish-catching dinosaur *Baryonyx*.

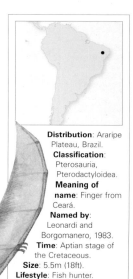

Distribution: Araripe Plateau, Brazil.
Classification: Pterosauria, Pterodactyloidea.
Meaning of name: Finger from Ceará.
Named by: Leonardi and Borgomanero, 1983.
Time: Aptian stage of the Cretaceous.
Size: 5.5m (18ft).
Lifestyle: Fish hunter.
Species: *C. atrox*.

EARLY CRETACEOUS CARNOSAURS

In early Cretaceous times the theropods began to diversify, and all kinds of new meat-eating dinosaurs started to evolve. However, the traditional line of theropods, the carnosaurs, which were the main hunters of the big plant-eaters of the Jurassic period, continued to flourish, and to prey on the plant-eating dinosaurs of the time.

Acrocanthosaurus

The presence of the fin on this animal originally led scientists to think that it must have been related to the spinosaurs. However, the fin seems different in construction from that of these later dinosaurs, and the rest of the animal is definitely carnosaurian. Footprints found in Texas, USA, may be those of *Acrocanthosaurus* and, if so, show that it hunted in packs.

Above: Acrocanthosaurus hunted in packs.

Left: Allosaurus.

Features: Imagine an *Allosaurus* with a low fin down the middle of its back. That is what the *Acrocanthosaurus* looks like. The fin is supported by spines from the vertebrae, 50cm (20in) over the back and progressively lower down the neck and tail. This fin was probably thicker and fleshier than that of the spinosaurs. It would have been bright, and used for identification and signalling. Its 68 thin, sharp, serrated teeth show that it was a hunter. The feet are quite small and adapted to walking on firm dry ground. The hand has three grasping fingers.

Distribution: Oklahoma, Texas, Utah and perhaps Maryland, USA.
Classification: Theropoda, Carnosauria.
Meaning of name: High-spined lizard.
Named by: Stovall and Langston, 1950.
Time: Aptian to Albian stages of the Cretaceous.
Size: 8m (26ft).
Lifestyle: Hunter.
Species: *A. atokensis*.

Neovenator

Judging by the number of remains found close to one another on the Isle of Wight, *Neovenator* may have been a pack hunter. As an agile and fast hunter, it would have been the main predator of the herds of *Iguanodon* and *Hypsilophodon* that grazed the horsetail-choked marshes that existed in the area.

Features: About 70 per cent of the skeleton of *Neovenator* has been found. It is a very sleek and streamlined animal. The skull is narrow and has a particularly large nostril cavity. The arrangement of five teeth on the premaxilla, the front bone of the lower jaw, shows that it is related to *Allosaurus*. Its skeletal features suggest that it may have been on the ancestral line to the big theropods, such as *Carcharodontosaurus* and *Giganotosaurus*, that were to come. *Neovenator* is the first allosaurid to have been found in Europe, and apart from *Baryonyx* is the biggest meat-eater from the area and time.

Below: The one species of Neovenator found is named after the family Salero, on whose land the fossil was found.

Distribution: Isle of Wight.
Classification: Theropoda, Carnosauria.
Meaning of name: New hunter.
Named by: Hutt, Martill and Barker, 1996.
Time: Barremian to Aptian stages of the Cretaceous.
Size: 6–10m (19½–33ft).
Lifestyle: Hunter.
Species: *N. salerii*.

Afrovenator

By the 1990s this was the only almost complete skeleton found of an African theropod. It was uncovered by a team from the University of Chicago, USA, led by Paul Sereno. The only parts missing were the lower jaw, some ribs and vertebrae, and the toe bones. Its similarity to the North American *Allosaurus* suggests that the two continents were still united at that time.

Features: The strong hind legs show that *Afrovenator* was built for active hunting, and the strong arms, which are longer than those of its relative, *Allosaurus*, with the big curved claws, were perfectly designed for catching and holding prey. The skeleton is

Distribution: Niger.

Classification: Theropoda, Carnosauria.
Meaning of name: African hunter.
Named by: Sereno, J. A. Wilson, Larsson, Dutheil and Suess, 1994.
Time: Hauterivian to Barremian stages of the late Cretaceous.
Size: 8–9m (26–30ft).
Lifestyle: Hunter.
Species: *A. abakensis*.

quite lightweight for the size of animal, and the tail is stiffened by overlapping bony struts, all features of a fast-moving beast. The skull is low, and did not have much in the way of crests.

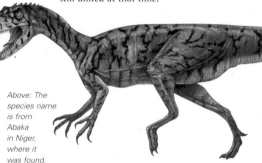

Above: The species name is from Abaka in Niger, where it was found.

CHARACTERISTICS OF THE TETANURAE

The Tetanurae, the group to which the carnosaurs and all the other advanced theropods belong, is characterized by the presence of three fingers or fewer, an extra opening in the side of the skull behind the nostril, and the stiffened tail.

The stiffening of the tail was brought about by a series of overlapping, bony struts, formed by bony material built up along the tendons that lashed the vertebrae of the tail together. The purpose was to hold the tail stiffly behind, while balancing the teeth and the jaws at the front. In later tetanurans, such as the maniraptorans, this was taken to an extreme. The bony struts could envelop up to ten vertebrae behind the one from which they protruded. This made the tail stiff and inflexible like a tightrope walker's balancing pole.

An even greater extreme is found in the tetanuran's present-day representatives. Our modern birds do not appear to have a bony tail at all. The original long dinosaurian tail has now evolved into a bony lump called the pygostyle, in which all the vertebrae are fused together. This is used as a base from which the lightweight feather tail sprouts, and anchors the muscles that control the feathered fan in flight.

Fukuiraptor

The "raptor" part of the name demonstrates that when this dinosaur was originally named it was thought to have been one of the maniraptorans, with the big killing claw on the second toe of the hind foot. Instead this claw was found to be a claw from the hand, and the animal was reclassified as a primitive carnosaur. Other Japanese theropods, *Kitadanisaurus* and *Katsuyamasaurus*, are now thought to have been *Fukuiraptor*.

Features: The big claws on the hand are distinctive, and led to the original mis-identification of the family. The jawbones into which the teeth are set are also similar to those of maniraptorans, and added to the initial confusion. *Fukuiraptor* seems to have been related both to *Sinraptor* from China, and the allosaurs found in Australia.

Distribution: Fukui Prefecture, Japan.
Classification: Theropoda, Carnosauria.
Meaning of name: Plunderer from Fukui.
Named by: Azuma and Currie, 2000.
Time: Albian stage of the Cretaceous.
Size: 4.2m (14ft).
Lifestyle: Hunter.
Species: *F. kitadaniensis*.

Above: The Fukuiraptor skeleton that was uncovered is of an immature individual, and the adult would have been bigger than the 4.2m (14ft) length given here.

MEDIUM-SIZE THEROPODS

At the beginning of the Cretaceous period, new types of plant-eating dinosaur began to evolve. The late Jurassic period, which preceded the Cretaceous, was the heyday of the plant-eating sauropods, but now the plant-eating ornithopods were coming into their own. Many were much smaller animals, and as a result smaller theropods, described here, evolved to prey on them.

Huaxiagnathus

For more than 150 years, the smallest theropods were assumed to be the compsognathids. *Compsognathus* and *Sinosauropteryx*, the only two compsognathids hitherto known, were no bigger than chickens. So it was a surprise when the complete articulated skeleton of a compsognathid as big as a cassowary came to light in the bone-littered lake beds of Liaoning, China, in 2004.

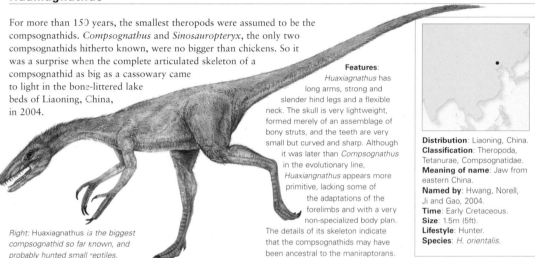

Features: *Huaxiagnathus* has long arms, strong and slender hind legs and a flexible neck. The skull is very lightweight, formed merely of an assemblage of bony struts, and the teeth are very small but curved and sharp. Although it was later than *Compsognathus* in the evolutionary line, *Huaxiangnathus* appears more primitive, lacking some of the adaptations of the forelimbs and with a very non-specialized body plan. The details of its skeleton indicate that the compsognathids may have been ancestral to the maniraptorans.

Distribution: Liaoning, China.
Classification: Theropoda, Tetanurae, Compsognatidae.
Meaning of name: Jaw from eastern China.
Named by: Hwang, Norell, Ji and Gao, 2004.
Time: Early Cretaceous.
Size: 1.5m (5ft).
Lifestyle: Hunter.
Species: *H. orientalis*.

Right: Huaxiagnathus is the biggest compsognathid so far known, and probably hunted small reptiles.

Dilong

The earliest indisputable tyrannosaurid was found in 2004 as a semi-articulated skeleton, along with scattered bones of three other specimens in the lake deposits of the Yixian Formation, in Liaoning, China. The imprints of the feathers were clearly seen, and this has led to renewed speculation that even the biggest of the Cretaceous meat-eating dinosaurs were feathered, at least in their small juvenile stages.

Features: Although the skeleton is similar to that of primitive coelurosaurs, the skull is definitely that of a tyrannosaurid, with its strongly bonded front bones and the front teeth with the characteristic D-shaped cross-section. The arms are longer than those of later tyrannosaurids, and it has three fingers. The entombing rocks where *Dilong* was found are so fine-grained that the covering of primitive feathers was preserved, the first indisputable proof we have that the early tyrannosaurids were covered with feathers.

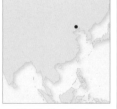

Distribution: Liaoning, China.
Classification: Theropoda, Tetanurae, Compsognatidae.
Meaning of name: Emperor dragon from Chinese mythology.
Named by: Xu, Norell, Kuang, Wang, Shao and Jia, 2004.
Time: Early Cretaceous.
Size: 1.5m (5ft).
Lifestyle: Hunter.
Species: *D. paradoxus*.

Above: The feathers, or more accurately protofeathers, of Dilong consisted of branched structures about 3cm (1¼in) long. They appear to have covered the entire animal and would have been used for insulation.

Nqwebasaurus

The earliest coelurosaurids to be found on any of the southern continents was *Nqwebasaurus*. The find suggests that the family spread throughout the world before the southern supercontinent of Gondwana split from northern Laurasia. It is known from a skeleton that is 70 per cent complete, and was given the nickname "Kirky", from the Kirkwood Formation in which it was found.

Features: The body of *Nqwebasaurus* is similar to that of a conventional coelurosaurid. It has a small body, a long flexible neck and long running legs. It has a very long first finger, partially opposable like a thumb, that bore a particularly large claw. Stomach stones were found with the skeleton. They usually indicate a vegetarian diet, but *Nqwebasaurus* was definitely a meat-eater. The stones may have come from the stomach of a plant-eating animal which it had killed and eaten.

Distribution: Eastern Cape Province, South Africa.
Classification: Theropoda, Tetanurae, Coelurosauria.
Meaning of name: Lizard from Nqweba, the local name for the Kirkwood District of South Africa.
Named by: de Klerk, Forster, Sampson, Chinsamy and Ross, 2000.
Time: Early Cretaceous.
Size: 0.8m (2½ft).
Lifestyle: Hunter.
Species: *N. thwazi*.

THE EVOLUTIONARY LINE OF BIRDS

It is now accepted that birds are descended from the tetanuran theropod dinosaurs. In fact, some scientists insist that birds must be referred to as dinosaurs, and the animals that are more traditionally referred to as dinosaurs should be called "non-avian dinosaurs".

The anatomy of birds and dinosaurs is so similar that there is hardly a doubt about the connection. However, one anomaly persists. The Tetanurans had a maximum of three fingers on the hand, with some, such as the later tyrannosaurids, having only two, and the alvarezsaurids only one. Birds' wings are likewise made up of three fingers, but they are a different three fingers. In Tetanurans the fingers are the first, second and third, and they have lost the equivalent of the human little finger and ring finger. Studies of bird embryos show that a bird's three fingers are the second, third and fourth, missing the thumb and little finger. If these animals had common ancestors, then we would expect the same fingers to be involved. This matter has yet to be resolved.

Eotyrannus

Amateur fossil collector Gavin Leng discovered *Eotyrannus* preserved in hard mudstone high on the cliffs of the coast of the Isle of Wight, in 1996. The skeleton was about 40 per cent complete, and there was enough of it for it to be identified as one of the earliest tyrannosaurids known. It would have preyed on some of the smaller herbivores of the time.

Features: *Eotyrannus* is evidently an early tyrannosaurid judging by the heavy skull, the D-shaped front teeth and the arrangement of the shoulder and limb bones. As in other early tyrannosaurids the hands are long compared with those of later members of the group, and in fact the second finger is almost as long as the forearm. The only skeleton found is a juvenile, and so the adult may have been longer than the 4.5m (15ft) stated.

Distribution: Isle of Wight, England.
Classification: Tetanurae, Tyrannosauroidea.
Meaning of name: Dawn tyrant.
Named by: Hutt, Naish, Martill, Barker and Newbery, 2001.
Time: Barremian stage of the early Cretaceous.
Size: 4.5m (15ft).
Lifestyle: Hunter.
Species: *E. lengi*.

Left: Eotyrannus *is very similar to* D long *from the other side of the world, and it has been suggested that they are actually the same genus.*

LITTLE FEATHERED DINOSAURS

The fine-grained lake deposits from the early Cretaceous period of Liaoning, in China, have long been famous for their exquisite fossils. However, it was not until the 1990s that they became well-known to Western scientists. Since this date a steady stream of beautifully preserved dinosaur material has been removed from the site, much of it seeming to show transitional stages between dinosaurs and birds.

Microraptor

The feathers of this little dinosaur show that it was a glider, and it probably represented an intermediate stage between ground- or tree-dwelling dinosaurs and birds with a flapping flight. It could spread out its arms and legs, and form a gliding surface enabling it to fly from tree to tree.

Features: The remarkable feature of this dinosaur is the distribution of feathers. Like a bird, it has flight feathers along the arms, but unlike a bird it also has them along the legs. Other adaptations to flight include a very short body with few vertebrae in the back, making the body stiff and strong. It is possibly a dromaeosaurid, but has similarities to the troodontids and also the birds.

Left: Microraptor had long feathers on both the arms and the legs. When spread, these would have provided an effective gliding surface. The tail also was feathered, presumably forming a steering organ.

Distribution: Liaoning, China.
Classification: Theropoda, Tetanurae, Coelurosauria.
Meaning of name: Small plunderer.
Named by: Xu, Zhou, Wang, Kuang, Zhang and Du, 2003.
Time: Barremian stage of the early Cretaceous.
Size: 40–60cm (16–24in).
Lifestyle: Insectivore.
Species: *M. zhaoianus, M. gui.*

Sinosauropteryx

This was the first of the Liaoning dinosaurs discovered with a covering of feathers or feather-like structures. The presence of feathers gives strength to the theory that all small meat-eating dinosaurs had feathers, or were at least insulated in some manner. This would support an active lifestyle for a warm-blooded predator. The bones of a mammal were found in the stomach of one specimen.

Features: Apart from the feathers, this is a typical small meat-eating dinosaur, similar to *Compsognathus*, with the short arms and long tail. The feathers are not like the branched feathers of modern birds, but would have formed more of a fuzzy or downy covering. Each filament is up to 3cm (1⅛in) long. *Sinosauropteryx* has a very long tail, with 64 vertebrae, unlike the stumpy tail of a modern bird. In the first skeleton found, the feathers were only obvious along the back. Some scientists, unable to accept the idea that dinosaurs had feathers, interpreted this as a continuous crest of skin. Skeletons with clearer feather impressions have since been found.

Above: There is no evidence as to the colour or pattern of the plumage of feathered dinosaurs, but it seems likely that they were as varied as modern birds.

Distribution: Liaoning, China.

Classification: Theropoda, Tetanurae, Coelurosauria.
Meaning of name: Chinese bird with feathers.
Named by: Ji Q. and Ji S., 1996.
Time: Barremian stage of the early Cretaceous.
Size: 1.3m (4ft).
Lifestyle: Hunter.
Species: *S. prima.*

ARCHAEORAPTOR

In 1999 an interesting fossil came to the West from the prolific fossil site at Liaoning, in China. It appeared to be a feathered bird but with the tail of a dinosaur.

It came to light through a devious route, the finder having circumvented the legal process of exporting fossils from China. It was sold to an American museum for $80,000. The purchaser, dinosaur researcher and museum-owner Stephen Czerkas, began to prepare a scientific paper on it, but potential co-authors were unwilling to help because of its murky history. To legitimize the procedure, Chinese palaeontologist Xu Xing was sent to America to work on the specimen, in anticipation of its being returned to China.

The resulting paper did not meet the stringent requirements for scientific publication, but nevertheless the popular magazine *National Geographic* publicly announced the name of this new specimen as *Archaeoraptor*. In fact the specimen was discovered to be a fake. A slab of stone containing the body fossil of the bird *Yanornis* had been glued to a fossil containing the tail of a *Microraptor*, before it had reached the dealer in China. The finder thought that he could make the fossil look more interesting, and hoodwinked the palaeontologists of America and China for months.

Caudipteryx

Distribution: Liaoning, China.
Classification: Theropoda, Tetanurae, Coelurosauria.
Meaning of name:
Feathered tail.
Named by: Ji Q., Currie, Norell and Ji S., 1998.
Time: Barremian stage of the early Cretaceous.
Size: 70–90cm (27½–35in).
Lifestyle: Hunter.
Species: *C. zoui*, *C. dongi*.

Caudipteryx is known from several almost complete skeletons that include the feathers. The feathers of *Caudipteryx* show that it was a close relative of birds, but not the ancestor – many types of fully formed birds existed at that time in the Liaoning area. It may have waded at the edge of a lake on its long legs, or even perched on branches.

Features: The distribution of feathers on this animal is quite distinctive. As well as a covering of fine insulating feathers, it also has long branching feathers on the arms and the end of the tail. Since these are symmetrical they are not flight feathers, and were probably used for display. Dark bands in the fossil feathers are probably remains of the colour pattern. It has a short face with big eyes and long, sharp front teeth.

Left: The species C. zoui is named after Sou Jiahua, vice-president of China at the time of its discovery.

Protarchaeopteryx

This dinosaur is known from two specimens found that show it to have been a small, feathered theropod. It is less well known than the other feathered dinosaurs of the area, but like its contemporary, *Caudipteryx*, it has long, symmetrical feathers on the arms and the tip of the tail which were used for display rather than for flight. It would have chased small prey along the ground by running after it on its long legs.

Below: The lakeside environments of early Cretaceous China must have been very colourful, with so many plumed and feathered animals, presumably all displaying to one another to attract mates or warn off rivals, as modern birds do.

Features: Long legs and a long neck, as well as long, clawed fingers on its hands, show that this is a running animal that grabbed at swift animals on the ground. The wrists are jointed so that the long hands could shoot out forwards to grab prey. The feathers on the arms, and on the short tail, would have been used for display, or even as aerodynamic structures that help steer at speed. It probably fed on insects and small vertebrates that it would be able to run down and snatch.

Distribution: Liaoning, China.
Classification: Theropoda, Tetanurae, Coelurosauria.
Meaning of name: Before Archaeopteryx.
Named by: Ji Q. and Ji S., 1997.
Time: Barremian stage of the early Cretaceous.
Size: 70cm (27½in).
Lifestyle: Hunter.
Species: *P. robusta*.

THEROPOD DINO-BIRDS

During early Cretaceous times birds continued to evolve from the small theropod dinosaurs. The traditional bird lines had not become established at this time, and there were a number of evolutionary branches that may even be regarded as experiments. Many animals existed that combined the features of both groups, usually with feathers and sometimes with wings as well.

Falcarius

At the time the birds were evolving from the dinosaurs, all sorts of other transitions were taking place. *Falcarius* was an early representative of the therizinosaurs – a group of dinosaurs from the meat-eating theropod line that developed into plant-eaters. Its close resemblance to the feathered Chinese form *Beipiaosaurus* indicates that it would have been feathered too.

Features: The teeth are distinctly those of a plant-eater – widely spaced and with shredding edges. The hips are broader and the pubis bone is swept back to give enough room for a big plant-eating gut. The legs are also shorter than those of other theropods and built for moving slowly, not for chasing after prey. The shoulder joints would have allowed the arms to reach up into the branches to pull down the twigs and leaves on which it fed. It retains big claws on its particularly prehensile forelimbs, hence the name, but these would have been used for tearing down leaves from trees.

Above: Falcarius shows the perfect transition between a meat-eating and a plant-eating dinosaur.

Distribution: Utah, USA.
Classification: Theropoda, Tetanurae, Therizinosauria.
Meaning of name: Sickle bearer.
Named by: Kirkland, 2005.
Time: Aptian to Albian stages of the early Cretaceous.
Size: 4m (13ft).
Lifestyle: Browser.
Species: *F. utahensis*.

Nuthetes

Below: The presence of a deinonychosaur in European deposits shows that this was a widespread group of animals.

The dromaeosaurids, a subgroup of the Deinonychosauria to which *Velociraptor* and *Deinonychus* belong, were thought to have existed only in Asia and North America. Then, in 2002, Angela Milner of the Natural History Museum in London, England, studied some remains that had been classified as *Megalosaurus* by Sir Richard Owen 150 years earlier. She found that the remains were actually those of one of the dromaeosaurids.

Features: The remains consist only of pieces of jaw bone and isolated teeth. The teeth were found to have the same characteristics as those of the deinonychosaurs, having the correct number and pattern of denticles along the cutting edges. This specimen was known as "fuzzy raptor" until it was properly identified. A number of other bones found beside the original remains were thought by Owen to have been armour plates that had been embedded in the skin of this dinosaur. Analysis showed they were pieces of armour from turtles of the time.

Distribution: Southern England.
Classification: Theropoda, Coelurosauria, Deinonychosauria.
Named by: Owen, 1854.
Time: Berriasian stage of the early Cretaceous.
Size: 1.8m (6ft).
Lifestyle: Hunter.
Species: *N. destructor*.

Mei

This is the shortest genus name of all dinosaurs, shorter even than that of the ankylosaur *Minmi* that held the record before. When the fossil was found in 2004 it caused a sensation. It had evidently died in its sleep and been preserved just as it lay. A sudden fall of volcanic ash buried and preserved it. Another troodontid, *Sinornithoides*, has been found in China in a similar curled up sleeping pose.

Features: *Mei* is a typical small troodontid, albeit rather an early one. Its main fame is the condition in which it was found – curled up with its tail wrapped around and its head tucked underneath its arm. This is precisely the position that a bird adopts when asleep, emphasizing the similarity of the small meat-eating dinosaurs to birds. The fact that the bones are not totally fused shows that it was not quite an adult when it died.

Distribution: Liaoning, China.
Classification: Theropoda, Tetanurae, Troodontidae.
Meaning of name: Sleeping. The whole species name, *Mei long* means "sleeping dragon" in Chinese.
Named by: Xu and Norell, 2004.
Time: Valanginian or early Barremian stage of the early Cretaceous.
Size: 53cm (21in).
Lifestyle: Hunter.
Species: *M. long*.

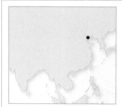

Right: The small skull, long hind limbs, closely packed teeth in the middle of the jawbone, and the U-shaped wishbone show that Mei was a different genus from other troodontids.

FEATHER EVOLUTION

It seems likely that feathers evolved before flight. So many dinosaur remains have been found that have traces of a feathery covering that Mark Norrell from the American Museum of Natural History in New York has speculated that nearly all the theropods would have carried feathers. It is unlikely that the largest would have been covered in feathers throughout life, and they would have lost their youthful plumage as they grew, leaving a soft scaly skin behind.

The most primitive feathers used for insulation, known as protofeathers, would have been merely filaments. These would have been cone-shaped and hollow, growing from a follicle in the skin.

In the next stage the shell of this filament would have split to produce a tuft of unbranched filaments, called barbs. Branches, called barbules, would then have formed along these filaments.

A more advanced structure would see the branching of the barbules becoming more regular, forming symmetrically along a central stem or rachis. These would then become branched to form modern-looking feathers.

Flight feathers are more complex. They tend to be assymetrical, with the arrangement of barbs along one side of the rachis being of a different size from that along the other. The arrangement is structured in such a way that once a whole row of feathers fits together they form an aerodynamic plane, with the feathers overlapping just enough to give strength and to keep the weight to a minimum.

Scansoriopteryx

This sparrow-size, bird-like dinosaur is very similar to another climbing theropod called *Epidendrosaurus*, from the same time and found in the same year. Both were climbing dinosaurs and had a distinctive long third finger. If the two names describe the same genera, then *Epidendrosaurus* is the name that has priority. Its existence could indicate that birds evolved from tree-living animals like this.

Features: What makes this animal stand out among the other small part-bird, part-dinosaur animals of the time is the long hand with the particularly long third finger, almost like that of the modern aye-aye lemur of Madagascar. The long finger was probably used in the same way, for probing holes for insects in tree trunks. The hips and shoulder girdle are more dinosaur-like and less bird-like than those of *Archaeopteryx*, and are less well adapted to flight.

Distribution: Liaoning, China.
Classification: Theropoda, Tetanurae.
Meaning of name: Climbing wing.
Named by: Czerkas and Yuan, 2002
Time: Barremian stage of the early Cretaceous.
Size: 30cm (1ft).
Lifestyle: Climbing insectivore.
Species: *S. heilmanni*.

Left: The trees around the Liaoning lake must have been full of little creatures – an ideal habitat for small insect-eating climbing animals like Scansoriopteryx.

LIAONING DINO-BIRDS

The early Cretaceous lake deposits in Liaoning are famous for their delicate preservation of birds and feathered dinosaurs. The fact that feathers are preserved under the exceptional conditions of these beds suggests that birds were common among dinosaurs elsewhere but were just not fossilized. The dates of these deposits are unclear, but they are definitely early Cretaceous.

Sinovenator

Discovered in the famous Yixian formation of Laioning, China, the presence of *Sinovenator* among the dinosaurs sheds light on the evolution of the troodontids from the dromaeosaurid stock. It was a fast runner, probably hunting the many birds and bird-like dinosaurs of the area at that time. In build it was very bird-like. Unlike other remains from the area, the skeleton of *Sinovenator* was not found articulated.

Features: *Sinovenator* is classed as a troodontid because of the shape of the bones of the lower jaw and the arrangement of skull bones around the eye sockets. It differs from the more advanced troodonts by the fact that the foot bones are not so highly adapted for fast running, and the hip bones are more like those of a dromaeosaurid than a troodontid. All in all, it indicates that bird-like characteristics evolved in the group before the dromaeosaurids and troodontids became specialized.

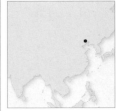

Distribution: Liaoning, China.
Classification: Tetanurae, Troodontidae.
Meaning of name: Chinese hunter.
Named by: Xu, Norell, Wang, Makivicky and Wu, 2002.
Time: Barremian stage of the early Cretaceous.
Size: 1m (about 3ft).
Lifestyle: Hunter.
Species: *S. changii*.

Right: Sinovenator was a turkey-size swift-footed, long-legged, running hunter.

Incisivosaurus

There always has been a mystery surrounding the diet of the oviraptorosaurs. Once thought to have been egg-stealers, they possess short jaws and powerful beaks, the purpose of which are unclear. *Incisivosaurus* is known from only the skull and neck vertebrae and is the earliest-known oviraptorosaur. It was a plant-eater.

Features: The skull and lower jaw – all that has been found – are longer and lower than that of later oviraptorosaurs, and the amazing dentition of the upper jaw consists of a pair of big incisor-like teeth at the front, followed by three pairs of smaller, almost conical teeth. The back teeth are pointed and very worn. The front teeth are almost rodent-like in their relative size and in the way they must have protruded from the mouth when the animal was alive. The back teeth made contact with one another when the mouth was closed, just as they do in herbivorous animals today.

Distribution: Liaoning, China.
Classification: Theropoda, Tetanurae, Oviraptorosauria.
Meaning of name: Incisor lizard.
Named by: Xu, Cheng, Wang and Chang, 2002.
Time: Barremian stage of the early Cretaceous.
Size: 1m (about 3ft).
Lifestyle: Browser.
Species: *I. gautheri*.

Right: Incisivosaurus would have eaten the bark from the trees, and the cones of the conifers that surrounded the early Cretaceous Liaoning lakes.

Cryptovolans

This "hidden flyer" is known from two skeletons, with feather impressions, and some scattered bones. In some respects it could have been a better flier than *Archaeopteryx*, with more bird-like ribs that articulated with the breastbone. It is possibly the same animal as *Microraptor*. The species name honours Greg Paul, the dinosaur writer who championed the bird-dinosaur connection.

Distribution: China.
Classification: Theropoda, Tetanurae, Deinonychosauria.
Meaning of name: Hidden flyer.
Named by: Cxerkas, Zhang, J. Li, and Y. Li, 2002.
Time: Barremian or Aptian stage of the early Cretaceous.
Size: 95cm (3ft).
Lifestyle: Flying hunter.
Species: C. pauli.

Right: In structure Cryptovolans is very similar to all other dromaeosaurids, but has a particularly long tail.

Features: The fossils of *Cryptovolans*, like those of *Microraptor*, show long simple flight feathers on the back legs, the arms and the backs of the hands, and also a fan of long feathers at the tip of the tail. These flight feathers are not as complex as those of modern birds, but they are formed from a central quill which is simply branched. The rest of the body is covered by simple hair-like "protofeathers". Spread out, the flight feathers of the arms and legs may have linked together to form a continuous aerodynamic surface.

PERFECT PRESENTATION

Left: A Liaoning fossil.

Usually palaeontologists are very lucky if they find a good fossil of an animal. Occasionally a fossil site preserves every part of the contemporary ecosystem and gives a unique view of life at a particular time. The name given to such occurrences is lagerstätten – German for "deposition places". Only about a dozen occurrences of such sites are known that merit the term. Yixian limestones of Liaoning province, China, is one of them. At the beginning of the Cretaceous period, this area consisted of a series of lakes in volcanic terrain. Every so often the volcanoes would erupt and ash would rain down, poisoning the air and the waters, and burying the victims before their bodies could decay. The result is the cross-section of the living environment of the time preserved as perfect fossils. The assemblage of animals and plants preserved there is known as the Jehol biota.

Sinornithosaurus

The fabulous feathered fossil of *Sinornithosaurus*, nicknamed "fuzzy raptor". It was one of the perfectly preserved vertebrates found in the early Cretaceous lake deposits in Laioning, China, and one that shows the complete pattern of feathering over the body. The fossil skeleton was lying flat, with its arms held out to the side. This shows the almost wing-like flexibility of the front limbs.

Features: It is the distribution of feathers that is important here. The head is covered with a fine mat of filaments. The filaments form thick sprays over the shoulders and down the backbone. Those on the lower arms are quite long and branched, like the feathers of a bird, and project to the rear as if forming wings. The filaments run down both sides of the tail, sticking out and back at an angle of about 45 degrees. At the tail tip they are longer and form a paddle shape.

Distribution: Liaoning, China.
Classification: Theropoda, Tetanurae, Deinonychosauria.
Meaning of name: Chinese bird-lizard.
Named by: Xu, Wang and Wu, 1999
Time: Barremian stage of the early Cretaceous.
Size: 1m (about 3ft).
Lifestyle: Hunter.
Species: S. milleni, S. haoiana.

Left: Beneath the feathers, "fuzzy raptor" seems to be a normal dromaeosaurid.

MORE DINO-BIRDS

We used to think that Compsognathus *from the late Jurassic of Germany was the smallest of the dinosaurs. Nowadays we find even smaller animals from the early Cretaceous lake beds in Liaoning, China. Do we regard them as dinosaurs or as birds? The answer is unclear as the boundary between the two is becoming less clear with every new discovery.*

Graciliraptor

This is the earliest definitive dromaeosaurid that existed several million years before others of the group. It is known from some jaw bones and teeth, the tail bones and the bones of the front and hind limbs. No feathers are known from this animal, but it is assumed that it had a plumage similar to its relatives.

Below: The earliest dromaeosaurid known, Graciliraptor, *has the long tail and long limbs of its descendants. It was probably closely related to the other small raptors, such as* Microraptor *and* Cryptovolans.

Features: As the name suggests this is a very lightweight theropod, possibly showing an evolutionary trend that ended with flying animals, although there is no suggestion that *Graciliraptor* flew. The teeth that are known are similar to those of other dromaeosaurids in having larger serrations on the rear edge than on the front edge. The foot bones are partially fused – a condition between that of other theropods and that of birds. There are, as usual, large claws on the hands, suitable for gripping prey or tree branches.

Distribution: Liaoning, China.
Classification: Theropoda, Tetanurae, Deinonychosauria.
Meaning of name: Graceful hunter.
Named by: Xu and Wang, 2004.
Time: Valanginian to early Barremian stage of the early Cretaceous.
Size: 1m (about 3ft).
Lifestyle: Hunter.
Species: *G. lujiatunensis*.

Shenzhouraptor

This creature could be a flying dromaeosaurid or a long-tailed bird – the evidence is ambiguous. More than half of its length is made up of the dinosaur-like tail. The rest of the skeleton is that of a bird, with adaptations to perching on branches. The name *Jeholornis* was applied to a specimen of this animal and actually announced a few days before *Shenzhouraptor*, but amongst scientists this does not amount to a scientific description and so the name *Jeholornis* cannot take priority.

Features: The presence of fossilized seeds in the stomach area of *Shenzhouraptor* is the earliest direct evidence of seed-eating in Mesozoic birds. They lie in a big lump, indicating that the bird swallowed them without chewing, and relied on the grinding action of the crop to break them up. The curved toe bones and claws show that it was a tree-living animal, and may have gathered growing seeds from the trees or picked them from the ground.

Distribution: Liaoning, China.
Classification: Bird, order uncertain.
Meaning of name: Hunter from Shenzhou.
Named by: Zhou and Zhang, 2002.
Time: Barremian stage of the early Cretaceous.
Size: 75cm (2½ft).
Lifestyle: Seed eater.
Species: *S. sinensis*.

Left: Despite the teeth in the jaws, Shenzhouraptor *seems to have swallowed its food whole.*

Yixianosaurus

Distribution: Liaoning, China.
Classification: Theropoda, Tetanurae, Deinonychosauria.
Meaning of name: Lizard from Yixian.
Named by: Xu and Wang, 2003.
Time: Barremian stage of the early Cretaceous.
Size: 60cm (2ft).
Lifestyle: Climber.
Species: *Y. longimanus*.

All that exists of *Yixianosaurus* is a pair of forelimbs with the shoulder structure attached, along with a few scraps of ribs and belly ribs. Enough of the animal is preserved to show that it is a dromaeosaurid rather than a more advanced bird, but the length of the hands makes it quite unlike any other found.

Features: As the species name suggests, the hand is very long – one and a half times the length of the forearm. The fingers, too, are very long, much longer than the bones of the hand itself, and have the capability of grasping narrow objects, indicating a tree-climbing way of life. There are traces of feathers with this fossil as well.

Left: The long fingers of Yixianosaurus *show it to have been a nimble tree-climber.*

THE EVOLUTION OF FLIGHT

Flight in birds may have come about from either of two different directions, summarized by ground up or trees down. In the former, the idea is that bird ancestors firstly became adept at running, and their long arms with their plumes were used as aerofoils to help in balance and in turning. Eventually the selective pressure would have been for the arms to become longer and more efficient, and they would have formed wings that allowed the animal to lift off the ground for a time. Certainly the long legs of many bird-like dinosaurs suggest this route of evolution.

Alternatively the bird ancestors could have first been tree dwellers. On escaping from enemies or to reach new trees to exploit food sources they would have become leapers. Insulating fur could then have developed into gliding structures, and these could have evolved to become flapping wings. The long curved claws on many of the earliest birds and the bird-like dinosaurs suggests that they were climbing animals to begin with.

Recent experiments with young birds whose wings have not yet fully grown give evidence for both ideas. Kenneth Dial of the University of Montana, USA, watched partridge chicks as they ran. He found that as they ran up sloping tree trunks they beat their wings giving them an added thrust that increased their grip on the bark. He regarded this behaviour as a throwback to the bird's early pre-flight ancestors, and saw this as a step towards proper flight.

Shenzhousaurus

The ornithomimosaurs were the "ostrich mimics" that strode majestically across the late Cretaceous plains of Asia and North America. However, they had their origins in the little bird-like dinosaurs of Liaoning. *Shenzhousaurus* is known from a single articulated skeleton missing only the forelimbs, shoulder girdle and the end of the tail. The neck is twisted back and the head lies over the torso, as is often seen in dinosaur death-poses. It was found in river sediments rather than in the lake sediments that are more typical of the area.

Left: Shenzhousaurus probably lived in undergrowth, unlike the plains-dwelling lifestyle of the later ornithomimosaurs.

Distribution: Liaoning, China.
Classification: Tetanurae, Ornithomimosauria.
Meaning of name: Lizard from Shenzhou (the old name for China).
Named by: Ji, Q., Norell, Makovicky, Gao, Ji. S. and Yuan, 2003
Time: Barremian stage of the early Cretaceous.
Size: 90cm (about 3ft).
Lifestyle: Insectivore or omnivore.
Species: *S. orientalis*.

Features: The graceful proportions of the later ornithomimosaurs are not yet evident in this early example. The head is still fairly large in relation to the body and the legs are not particularly long compared with other theropods. As in all members of the group the teeth are reduced, but in this instance there are still teeth at the front of the mouth – a primitive feature. The shape of the hip bones, too, distinguishes *Shenzhousaurus* from other primitive ornithomimosaurs.

BIRDS OF LIAONING

Many of the excellent specimens from the Liaoning lake beds have definitely crossed the border from dinosaurs to birds. However, most are primitive forms, representing evolutionary sidelines rather than the direct ancestors of modern birds, and often possessing dinosaur-like teeth and bony tails. Although some have tentatively been grouped into families their relationship to the later orders of bird is uncertain.

Sinornis

Below: Flying through the branches of the lakeside trees, Sinornis would have looked very much like a typical small bird of today.

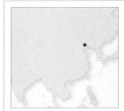

Sinornis was the first of the spectacularly preserved remains from the lake deposits of Liaoning to reach the west, and alert western scientists about the wealth of new material to be found there. It was a sparrow-sized bird that would have been as capable as a modern bird of sustained and manoeuvrable flight. The wrist joint was very flexible allowing the wing to twist in flight and allowing the wings to be held flush against the body when at rest.

Distribution: Liaoning, China.
Classification: Bird, order uncertain.
Meaning of name: Chinese bird.
Named by: Sereno and Rao, 1992.
Time: Barremian stage of the early Cretaceous.
Size: 10cm (4in) – sparrow sized
Lifestyle: Insectivore.
Species: *S. santensis*.

Features: Although it has the appearance and all the flight structure of a modern bird, *Sinornis* still has toothed jaws instead of a beak, a primitive arrangement of hip bones, and fingers on the wings – just like the Jurassic *Archaeopteryx*. However, its perching feet and its tail of long feathers fanning from a stump called a pygostyle are very much like those of modern birds. A similar bird, *Cathayornis*, is now regarded as the same as *Sinornis*.

Confuciusornis

Confuciusornis is one of the earliest birds to have possessed a beak rather than toothed jaws, and a modern type of wing structure. It appears to have been the most common bird in the lake region, judging by the number of fossils found. It lived in large flocks around the water's edge.

Features: The tail of *Confuciusornis* is a short pygostyle, but in the male of the species it is furnished with a pair of extremely long tail feathers. The females are smaller and lack these feathers on the tail. There are still fingers on the wings and there is only a claw on the thumb and middle finger. The index finger has no claw and is adapted to carrying the wing feathers, being made up of broad flat bones.

Left: The fine preservation of feathers in the Liaoning deposits is evident from this fossil of Confuciusornis.

Above: Spectacular plumage evolved early in the history of birds, as is seen in the tail feathers of Confuciusornis.

Distribution: Liaoning, China.
Classification: Bird, Confuciusornithidae.
Meaning of name: Bird of Confucius.
Named by: Hou, 1997.
Time: Barremian stage of the early Cretaceous.
Size: 75cm (2½ft) including the tail feathers – with a pigeon-sized body.
Lifestyle: Insectivore.
Species: *C. chounzohous*, *C. suniae*.

Changchengornis

Distribution: Liaoning, China.
Classification: Bird, Confuciusornithidae.
Meaning of name: Great Wall of China bird.
Named by: Ji and Chiappe, 1999.
Time: Barremian stage of the early Cretaceous.
Size: 75cm (2½ ft) including the tail feathers with a starling-sized body.
Lifestyle: Aerial insect-hunter.
Species: C. hengdaoziensis.

Unlike *Confuciusornis*, *Changchengornis* is quite a rare bird, known so far only from a single starling-sized specimen. The structure of the tail suggests that it may have been a manoeuvrable bird, able to catch insects on the wing as swallows do.

Features: There are two long display feathers on the tail of *Changchengornis*, just as on *Confuciusornis*, but it is not known if only the males possessed them. The toe bones are more curved than that of its relative and the claws are more hooked. This suggests that it was more adept at perching in the branches. Despite its long tail feathers, it does not have a feathered tail that could be used in flight – a feature that it shares with *Confuciusornis*. The "bastard wing", or alula, had not evolved either.

Below: Clouds of insects would have hovered over the lake surface and would have provided food for a number of aerial predators like Changchengornis.

Liaoxiornis

PART AND COUNTERPART

Split a block of finely-bedded sedimentary rock along its bedding plane. It falls open in two parts. Any fossil that lies along this bedding plane may remain fixed to one part or the other, or else parts of it remain embedded in the one and parts in the other. The result is essentially two fossils, one the mirror image of the other. Palaeontologists call this occurrence the "part and counterpart" of the specimen, and both are important in the study and analysis of the fossil.

Occasionally unscrupulous fossil collectors separate the two and sell them as individual fossils. This can lead to confusion, especially if the part is described scientifically and is given its scientific name, and the counterpart is given another name by another researcher unaware of the circumstances.

Below: The part and counterpart of a fossil, when put together, look like the mirror images of one another. Both are important in the study of the fossil.

Liaoxiornis, the smallest known of all Mesozoic birds, was the size of a wren. It is known from a complete specimen which is regarded as the skeleton of an adult, although some scientists disagree, seeing juvenile characteristics in the skull. It appeared about two million years later than the confuciusornithids. Another fossil was named *Lingyuanornis*, but it turned out to have been the counterpart – the other half of the slab of the rock that contained the fossil – of this specimen, so the two birds are the same.

Features: The skull is tall and short, and has large eye sockets. There are small teeth in the jaw. The neck is long, and so is the tail, although it has a pygostyle that takes up about three quarters of its length. The finger bones are fused and lack claws – an advanced feature – but the leg bones are quite primitive and crow-like. There are long claws on the feet but these are not strongly curved.

Right. Liaoxiornis was a tiny bird, living on insects in the undergrowth around the Liaoxing lakes.

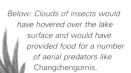

Distribution: Liaoning, China.
Classification: Bird, Enantiornithes.
Meaning of name: Bird of the distant west.
Named by: Hou and Chen, 1999.
Time: Barremian stage of the early Cretaceous.
Size: 8cm (3¼ in) – wren sized.
Lifestyle: Insectivore.
Species: L. delicates.

MORE BIRDS FROM LIAONING

Some Cretaceous birds had long dinosaur-like tails, while others had the more modern pygidium – the fused series of tail bones that supports the fan of tail feathers in modern birds, and sometimes referred to as the "parson's nose". Some had teeth in their jaws while some had modern beaks. Nearly all had claws on their wings. All the birds of Liaoning, China, had mixtures of these features.

Liaoningornis

Liaoningornis is known from a single partial skeleton. It was very much like modern small birds in its anatomy and presumably its physiology and lifestyle. It is regarded as the most primitive of the modern bird group. It existed at the same time as the even more primitive confuciusornithids, and suggests that there was a great radiation of bird types at the time.

Right: The implied complex breathing system indicates that Liaoningornis *had a lifestyle as active as that of modern birds.*

Features: The big keeled breastbone indicates that *Liaoningornis* had strong flying muscles. The articulation of the ribs and breastbone suggests that it had the complex breathing mechanism using air sacs for efficient ventilation and oxygen consumption as modern birds do – something its contemporaries seemed to lack. The legs also are quite modern. However, it retains the primitive feature of having teeth in the jaws even though in life it would also have had a beak. The length of its toes gives the meaning to its species name, which means "long toed".

Distribution: Liaoning, China.
Classification: Bird, order uncertain.
Meaning of name: Bird from Liaoning.
Named by: Hou, 1997.
Time: Barremian stage of the early Cretaceous.
Size: 10cm (4in) – sparrow sized.
Lifestyle: Seed-eater.
Species: *L. longidigitrus.*

Yanornis

Below: Yanornis *waded around the shores of the lake, dipping for fish and invertebrates*

This fossil is notorious in the field of palaeontology as being part of an elaborate hoax played on the Chinese authorities and American academics by fossil hunters. In 1999 a fossil of a bird with a dinosaur-like tail went on display in the United States. It proved to be a forgery, uniting the fossils of a dinosaur tail and of a bird. The bird part was subsequently studied and named *Yanornis*. Several other specimens have since been found.

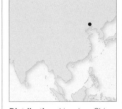

Distribution: Liaoning, China.
Classification: Bird, order uncertain.
Meaning of name: Bird of the Yan Dynasty.
Named by: Zhou and Zhang, 2001.
Time: Barremian stage of the early Cretaceous.
Size: 40cm (16in) – crow sized.
Lifestyle: Fish-hunter.
Species: *Y. martini.*

Features: The head and jaws are quite long and furnished with strong conical teeth. This, along with a long neck and its long legs and toes suggest that it was a wading bird, and fish bones in the stomach show its diet. However stomach stones have been found in the crop of another specimen – an indication of a plant diet – and so *Yanornis* may have changed its diet from season to season.

Cuspirostrisornis

Distribution: Liaoning, China.
Classification: Bird, order uncertain.
Meaning of name: Pointed beak bird.
Named by: Hou, 1997.
Time: Barremian stage of the early Cretaceous.
Size: 12cm (5in) – large sparrow sized.
Lifestyle: Insect-eater.
Species: C. houi.

Another almost complete and articulated bird skeleton from the Liaoning deposits is *Cuspirostrisornis*. Unlike in many other bird fossils the skull is complete, and shows the unusual length of the jaw bones. The variety of fossils that have been found here indicate that either the bird fauna of the area was particularly rich in the early Cretaceous, or that there was abundant bird life everywhere at the time and the unique fossilization processes of the area managed to catch it better than anywhere else.

Right: As is usual we have no evidence for the pattern or colour of the plumage, and so the pattern shown here must be regarded as speculative.

Features: The jaws are long, straight and slender, with five teeth on each side of both upper and lower jaws. The keeled breast bone, the wing structure and the hips are all well developed. The legs are long and the feet evidently evolved for perching with the toe bones carrying very large and curved claws, suitable for gripping small branches.

MORE BIRDS FROM LIAONING

There are a vast number of birds from the Liaoning lake deposits in China. Some have not yet been fully studied, and some are very fragmentary.

Chaoyangia – known from one specimen. It has a remarkable rib structure which is more advanced than that of the other primitive birds.
Eocathayornis – found as a partial skeleton. It had a powerful flapping capability.
Eoenantiornis – a halfway stage in evolutionary terms between *Archaeopteryx* and *Sinornis*. Possessing toothed jaws, a pygostyle and possibly an alula, or bastard wing.
Longchengornis – sparrow sized, toothed and with big claws.
Longipteryx – a kingfisher-like bird with forelimbs about one and a half times as long as the hind, giving long wings. A specialized fish-eater.
Sapeornis – a gull-sized soaring shorebird. The biggest of the early Cretaceous birds.
Songlingornis – about the size of a sparrow, with a compact arrangement of small teeth and a skeletal structure that suggests a water bird.
Vescornis – possessed fingers on the wings but these were much reduced, showing that it did not rely on them for climbing. Instead it had strong perching feet.
Yangadgornis – a ground-living bird, possibly flightless, about the size of a crow, with a long bony tail, but a lightweight skull and toothless, beaked jaws.
Yixianornis – a powerful flying bird of modern appearance but still with teeth and clawed wings. About the size of a jay.

Boluochia

The birds of the Liaoning lakeside were evidently not all delicate little seed-eaters or insectivores. The discovery of *Boluochia* in 1995 with a hawk-like beak shows that they had diversified into most of the bird niches known today. They must have hunted one another as well. The bird of prey had arrived.

Features: The notable feature about *Boluochia* is the presence of a hooked upper beak, over a toothed lower jaw. The beak is very similar to that of a modern bird of prey and is the earliest example found. It also has strongly curved and sharply pointed toe bones that, in life, must have carried killing talons just like those of a modern hawk. A strong pygostyle indicates a powerful fan of tail feathers, evidently used for manoeuvring while hunting.

Distribution: Liaoning, China.
Classification: Bird, order uncertain.
Meaning of name: From Bolouchi.
Named by: Zhou, 1995.
Time: Barremian stage of the early Cretaceous.
Size: 13cm (5in) – shrike sized.
Lifestyle: Hunter.
Species: B. zhengi.

Left: Boluochia was probably the top of the avian food pyramid at the time, hunting and eating the many smaller birds.

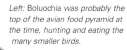

OTHER KNOWN BIRDS

Although our knowledge of early Cretaceous birds comes largely from the marvellous lake deposits of Liaoning, this is not the only area in which birds of this period have been found. Bird bones are very delicate and unfortunately that means that they rarely fossilize. Most of the fossils that have been found were preserved in lake beds in various parts of the world.

Iberomesornis

Named because it was found in Spain. At the time of finding it was thought to be midway between *Archaeopteryx* and modern birds. This was before the knowledge of the spectacularly abundant and diverse bird fauna of Liaoning showed just how varied these intermediate forms actually were. This fossil was found in a similar lake deposit to that of north-eastern China.

Features: The primitive features of *Iberomesornis* include the presence of a toothed bill, the structure of the hip in which the hip bones consist of three separate bones at each side, rather than fused into a unit as in modern birds, and the possession of clawed fingers on the wings. The modern features include the plumage, shoulder bones, the wishbone and the pygostyle that held the tail feathers. It would have been capable of sophisticated flying with its wings, that had a span of about 20cm (8in) or so.

Distribution: Cuenca Province, Spain.
Classification: Bird, Enantiornithes.
Meaning of name: Middle bird from Spain.
Named by: Sanz and Bonaparte, 1992.
Time: Barremian stage of the early Cretaceous.
Size: 10cm (4in) – finch size.
Lifestyle: Hunter of insects and crustaceans.
Species: *I. romerali*.

Right: Iberomesornis *hunted insects in the woodlands along the lakesides of early Cretaceous Spain.*

Eoalulavis

When found this was the earliest bird known to have possessed an alula – a bastard wing – as do modern birds. More recently discovered birds in China also have this feature, but difficulty in correlating the dates between the deposits in Spain and China make it unclear as to which is now the earliest-known bird to have had one.

Features: The single specimen of *Eoalulavis* found so far lacks the head, and so it is not known whether it had teeth like most of its contemporaries. Its stomach contents consist of little water creatures. The alula, to which the genus name refers, is a tuft of feathers attached to a moveable thumb-like structure on the leading edge of the wing. When this is manipulated in flight, it alters the flow of air over the wing and gives increased manoeuvrability in low speed flight.

Left: Eoalulavis *has been restored here with the head of a modern skimmer. The rest of the body suggests a similar lifestyle.*

Distribution: Cuenca Province, Spain.
Classification: Bird, Enantiornithes.
Meaning of name: Early bird with an alula.
Named by: Sanz, Chiappe, Perez-Moreno, Buscalioni, Moratalla, Ortega and Poyato-Ariza, 1996.
Time: Hauterivian stage of the early Cretaceous.
Size: 10cm (4in) – finch sized.
Lifestyle: Hunter of water creatures.
Species: *E. hoyasi*.

Enaliornis

Distribution: Southern England.
Classification: Bird, Hesperornithiformes.
Meaning of name: Sea bird.
Named by: Seeley, 1876.
Time: Albian stage of the early Cretaceous.
Size: 50cm (20in).
Lifestyle: Fish-hunter.
Species: *E. barretti*, *E. seeleyi*, *E. sedgwicki*.

Although it is known only from scrappy remains in marine sediments, *Enaliornis* is important as the first of the ornithuromorpha – a broad grouping that encompasses the modern birds – to have been found in Mesozoic rocks. It shows that adaptation to marine hunting took place very early among the birds. The foot-propelled hesperornithiformes, an order of birds that is now extinct, with their wings reduced to the role of stabilizers, became the dominant marine birds for the rest of the Cretaceous period.

Right: The hesperornithiformes, like Enaliornis, were the main fishing birds of the Cretaceous.

Features: The shape of the foot bones show that *Enaliornis* had paddle-shaped lobes, like the modern grebes and coots, rather than webbed toes, as in loons and ducks. The legs stick out at the side but it could have waddled on land. The pelvis has features similar to those of modern penguins, suggesting a similar lifestyle of both walking on land and swimming in the sea. Some juvenile bones have been found that have been gnawed by terrestrial mammals, showing that *Enaliornis* nested on land.

THE ENANTIORNITHES

Many of the birds of Liaoning belong to a family called the Enantiornithes. The word means "opposite bird" and refers to the articulation of the forelimb to the shoulder blade – the opposite arrangement to modern birds.

The group was first identified by C. A. Walker in 1981, based on late Cretaceous bird fossils from Argentina. Since then many fossil birds that had already been found have been recognized as members of the group. The earliest known is primitive *Noguernornis* from the early Cretaceous of Spain, and the last-known is *Avisaurus* from the Maastrichtian stage of the late Cretaceous of Montana.

They were a diverse group, ranging in size from that of a sparrow to that of a seagull. Studies of juveniles suggest that they grew quickly on hatching, but then their growth slowed and they did not reach adult size for some years – unlike modern birds that grow to full size in months.

The group existed worldwide and probably evolved sometime in the Jurassic, although all known fossils are from Cretaceous rocks. They became extinct by the end of the Cretaceous period along with all the dinosaurs.

They are regarded as being more advanced than *Archaeopteryx* and the confuciusornithic family, but more primitive than modern birds. They lived side by side with the ancestors of the modern birds – the Euornithes. The precise relationships between these groups has not yet been worked out and remains controversial.

Buitreraptor

Buitreraptor is the only known dromaeosaurid from the southern continents. By Cretaceous times the supercontinent of Pangaea had split, and the individual continents were moving apart. The southern portion, Gondwana, was still largely a single landmass but it was separate from the northern continents where all other dromaeosaurids have been found. Until the discovery of *Buitreraptor* it had been assumed that the dromaeosaurids had not existed on Gondwana.

Below: The name Buitreraptor *actually has two references. Firstly, its remains were found in La Buitrera area of Argentina. Secondly, buitre is Spanish for vulture, which could be an indication of its lifestyle.*

Distribution: Argentina.
Classification: Theropoda, Tetanurae, Deinonychosauria.
Meaning of name: Hunter from the Buitrera region.
Named: Makovicky, 2005.
Time: Cenomanian stage of the late Cretaceous.
Size: 1.5m (5ft).
Lifestyle: Hunter.
Species: *B. gonzalezorum*.

Features: *Buitreraptor* is a long-legged, long-armed, bird-like dromaeosaurid, with a body about the size of a rooster. Its head is furnished with a long stork-like beak. In build it seems to be close to the later Madagascar bird *Rahonavis*. The long arms would have supported feathers, although no impression of feathers were found with the specimen.

FAST HUNTERS

The early Cretaceous saw several lines of active theropods developing. They included the maniraptorans,
the so-called "raptors" – a group that encompasses the Deinonychosauria, the Oviraptorosauria and the
Therizinosauria. It includes the ostrich-mimics – omnivorous, fleet-footed theropods that probably lived
like modern ground-living birds. They would all have been warm-blooded and covered in feathers.

Deinonychus

Known from more than nine skeletons, this is the animal over which the debate about whether dinosaurs were warm- or cold-blooded began. One remarkable deposit has several *Deinonychus* skeletons scattered around the remains of an ornithopod, *Tenontosaurus*, indicating that it was a pack hunter. It was the prototype for the "raptors" in *Jurassic Park*, although modern representations have them covered in feathers.

Below: The tail of Deinonychus was held stiff and straight. Each vertebra had bony tendons growing from it that clasped several of the vertebrae behind, solidifying the whole structure into an inflexible pole with only limited movement at the base for balance.

Distribution: Montana, Oklahoma, Wyoming, Utah and Maryland, USA.
Classification: Theropoda, Tetanurae, Deinonychosauria.
Meaning of name: Terrible claw.
Named by: Ostrom, 1969.
Time: Aptian to Albian stages of the Cretaceous.
Size: 4m (13ft).
Lifestyle: Pack hunter.
Species: *D. antirrhopus*.

Features: This medium-sized member of the theropods has a killer claw on the hind foot. Its light, bird-like body is balanced by a stiff, straight tail, and the brainpower is enough to keep the animal balanced while it slashed away with the killer claw on the second toe. The long, heavily clawed hands are angled so that the palms face inwards, enabling it to clutch firmly at its prey.

Utahraptor

This dinosaur is only known from a single specimen consisting of parts of the skull, the claws of the hand and feet, and some tail vertebrae. It is the biggest of the known deinonychosaurids. It is also the earliest; the individuals of the group seem to have become smaller as time went on. *Utahraptor* would have preyed on the big sauropods of the time.

Below: The species name U. ostrommaysorum (illustrated) honours John Ostrom (who first defined the group) and Chris Mays (from the Dinamation International Corporation that financed the excavation).

Distribution: Utah, USA.
Classification: Theropoda, Tetanurae, Deinonychosauria.
Meaning of name: Hunter from Utah
Named by: Kirkland, Gaston and Burge, 1993.
Time: Barremian stage of the early Cretaceous.
Size: 6m (19½ft).
Lifestyle: Hunter.
Species: *U. spielbergi*, *U. ostrommaysorum*.

Features: This dinosaur is essentially a scaled-up version of *Deinonychus*. The killing claw is about 35cm (14in) long, and would have been held up clear of the ground when the animal was stalking. The hands are proportionally larger than those of *Deinonychus*, and have much more blade-like claws which are just as important as the toe claw when hunting. The leg bones are twice as thick as those of the much larger *Allosaurus*, suggesting that they were built for power not speed.

Pelecanimimus

An important group of thereopods of the late Cretaceous period were the ornithomimids, the bird-mimics. These were generally built like ostriches and had toothless beaks. *Pelecanimimus* was an early toothed form with a long, shallow snout. Its partial skeleton was found in Spain in lake deposits that were fine enough to show skin features. It may have been a fish-eating animal.

Features: The 220 tiny teeth in the jaws of this little dinosaur represent the biggest number of teeth in any known theropod. The fine lake deposits in which the specimen was found have preserved the impression of a pouch of skin beneath the jaw, hence the name, and a so a soft, fleshy crest at the back of the head. The skin impress on shows a wrinkled surface devoid of hair or feathers, unlike other small theropods.

Left: Pelecanimimus was probably similar to the ancestral form from which the later ostrich-mimic dinosaurs evolved. However, the advanced hand shows that it could not have been a direct ancestor.

Distribution: Spain.
Classification: Theropoda, Tetanurae, Ornithomimosauria.
Meaning of name: Pelican-mimic.
Named by: Perez-Moreno, Sanz, Buscalloni, Moratalla, Ortega and Rasskin-Gutman, 1994.
Time: Hauterivian to Barremian stages of the early Cretaceous.
Size: 2m (6½ft).
Lifestyle: Generalized hunter.
Species: *P. polydon*.

ORNITHOMIMID TEETH

The ornithomimids were the toothless, ostrich-like running dinosaurs of late Cretaceous times. However, their ancestry is to be found in the early Cretaceous period. Because of their lack of teeth and their general build, it has always been assumed that the ornithomimids were omnivorous, feeding on a mixture of insects and small reptiles, and fruit and seeds when available. The often repeated statement that the theropods were the meat-eating dinosaurs of the Mesozoic period may be an over-simplification.

There has been a long debate about the cause of the toothlessness. Some palaeontologists thought that the teeth of a more conventional theropod just became increasingly fewer, and finally disappeared altogether. Others maintained that the teeth became smaller and smaller, so that the jaw became more and more saw-like until, finally, the teeth became so small that they faded away. The discovery of the early ornithomimid, *Pelecanimimus*, with its jaws full of tiny teeth, suggests the latter is correct.

Scipionyx

This baby dinosaur was discovered in the late 1980s by amateur collector Giovanni Todesco, but it wasn't until he had seen the film *Jurassic Park* that he took it to professional palaeontologists for examination. They recognized it to be the most perfectly preserved theropod ever discovered, with traces of the intestines, windpipe, liver and even muscle fibres.

Features: The extreme youth of this single, articulated specimen makes it difficult to classify. It appears to be a maniraptoran and resembles *Composognathus*, but the hand is different. Because it is a hatchling, the skull is much bigger in relation to its body than would be the case in the adult. The position of the preserved liver suggests that *Scipionyx* breathed like a crocodile instead of a bird.

Below: The tiny specimen of Scipionyx seems to have been very young, not long out of its egg. The estimated length of an adult is about 2m (about 6ft).

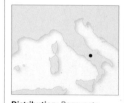

Distribution: Benevento Province, Italy.
Classification: Theropoda, Tetanurae.
Meaning of name: Claw of Scipio (a Roman general) and also Scipione Breislak (a geologist who first studied the formation in which it was found).
Named by: dal Sasso and Signore, 1998.
Time: Aptian stage of the early Cretaceous.
Size: 24cm (9in).
Lifestyle: not known.
Species: *S. samniticus* (after Samnium, the ancient name of the local region).

SPINOSAURIDS

One of the last major dinosaur family to come to the public's attention was the spinosaurids. They were theropod meat-eaters from the early Cretaceous period, with long, narrow snouts armed with many straight, sharp teeth, and they had a distinctive big claw on the thumb. With these adaptations it seems that these dinosaurs were probably specialized fish-eaters.

Baryonyx

Amateur fossil collector William Walker found a huge fossil claw bone in a clay pit in southern England in 1983, and subsequently a team from the (then) British Museum (Natural History), now the Natural History Museum, London, led by Angela Milner and Alan Charig, excavated the almost complete skeleton. This is the first spinosaurid skeleton found, although individual bones were already known.

Features: *Baryonyx* is a large theropod with distinctive narrow jaws and small teeth. The teeth are straighter than those of other meat-eaters, and there is a peculiar rosette of longer teeth at the tip. The forelimb is very strong and armed with a heavy claw. The neck makes less of an S-shape than in other theropods, and the neck and skull form an almost continuous line. The animal's diet is suggested by fish scales and *Iguanodon* bones found in the stomach.

Distribution: Southern England.
Classification: Theropoda, Tetanurae, Spinosauria.
Meaning of name: Heavy claw.
Named by: Charig and Milner, 1986.
Time: Barremian stage of the early Cretaceous.
Size: 10m (33ft).
Lifestyle: Fish-hunter or scavenger.
Species: *B. walkeri*; see also below.

Right: Baryonyx is the biggest meat-eater found in Europe from the early Cretaceous period.

Suchomimus

Suchomimus lived in east Africa a little later than *Baryonyx* lived in Europe. It was big enough to stand in up to 2m (6½ft) of water, chasing fish. However, the east African beds in which it was found contained very few other meat-eating dinosaurs, probably making it the main hunter of the area as well.

Features: When *Suchomimus* was uncovered in the Tenere Desert, Niger, in the late 1990s, scientists were amazed at how similar it looked to *Baryonyx*, with its big claws and long jaws. The only difference is the tall spines along the backbone, which would have supported a low fin in life. However, recent studies of *Baryonyx* suggest that it too might have supported such a fin. It may well be that *Suchomimus* is really a new species of *Baryonyx*.

Distribution: Niger.
Classification: Theropoda, Tetanurae, Spinosauria.
Meaning of name: Crocodile-mimic.
Named by: Sereno, Beck, Dutheil, Gado, Larsson, Lyon, Marcot, Rauhut, Sadleir, Sidor, Varricchio, G. P. Wilson and J. A. Wilson, 1998.
Time: Aptian stage of the early Cretaceous.
Size: 11m (36ft).
Lifestyle: Fish hunter, predator or scavenger.
Species: *S. tenerensis*; see also above.

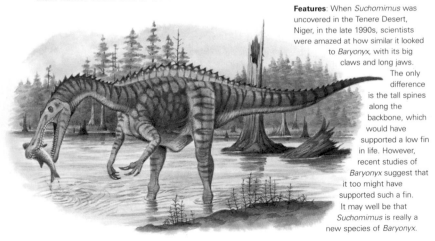

Spinosaurus

The first remains of this remarkable animal were found in Egypt by a German expedition in 1911, and were then lost when the Alte Akademie museum in Munich, Germany, in which it was stored, was destroyed by bombing in 1944. In 1996 Canadian palaeontologist Dale Russell found more remains in Morocco. As the villain of the film *Jurassic Park III*, *Spinosaurus* caught the public's imagination in 2001.

Features: The most significant feature of the skeleton of *Spinosaurus* is the array of spines sticking up from the backbone, reaching heights of almost 2m (6½ft). In life this would have been covered by skin to form a

Right:
In early books
Spinosaurus *is restored with a short deep head, like that of a carnosaur. That was before the discovery of skull material in Morocco, and the realization of how closely related* Spinosaurus *was to the others of the group whose skulls were well known.*

fin or a sail. It may have acted as a heat regulation device, absorbing warmth from the sun or shedding excess body heat into the wind. It may also have been brightly coloured and used for signalling.

Distribution: Egypt and Morocco.
Classification: Theropoda, Tetanurae, Spinosauria.
Meaning of name: Spined lizard.
Named by: Stromer, 1915.
Time: Albian to Cenomanian stages of the Cretaceous.
Size: Maybe up to 17m (56ft).
Lifestyle: Fish hunter, predator or scavenger.
Species: *S. aegyptiacus, S. maroccansus.*

SPINOSAURID JAWS

The jaws and teeth of the spinosaurids are very different from those of any other meat-eating dinosaur. The snout is extremely narrow and very long. The teeth are much straighter than those of other meat-eaters, and on the lower jaw they are very numerous and small. The tip of the upper jaw carries a separate rosette of teeth that corresponds with a hooked structure at the tip of the lower jaw. The nostrils are placed well back on the snout.

Top: Inside jaw view.
Above: The spinosaurid upper jaw with teeth.

These adaptations seem to have been well suited for catching fish. The narrow snout would cleave the water, the small sharp teeth would seize small, slippery prey, and the nostrils would be clear of the surface. In the modern world we see adaptations such as these in the gavial of the Far East. They are also present in river dolphins.

The spinosaurids probably did not rely on a fish diet, and it would be unreasonable for a *Spinosaurus*, as big as a *Tyrannosaurus*, to feed exclusively on such small prey. Stomach contents, and the presence of spinosaurid teeth in the bones of other animals, suggest that they may have fed on land-living animals too. The big claw may have been a killing weapon, but the specialized teeth seem to have been unsuited for hunting. The spinosaurids probably filled out their fish diet with the carrion of dead animals.

Irritator

The name of this dinosaur is derived from the frustration felt by British palaeontologist Dave Martill when faced with the skull. When it was obtained from Brazil it had been doctored by the finder to try to make it look more spectacular and marketable. After it had been prepared properly, it was seen to be the skull of a spinosaurid.

Features: Only the skull of *Irritator* is known, but it obviously came from a spinosaurid. It is the only one known from South America, although another, *Angaturama*, has been described. *Angaturama*, however, is generally regarded as another name for *Irritator*. In 2004, Eric Buffetaut found the tooth of a spinosaurid, probably *Irritator*, embedded in the backbone of a Brazilian pterosaur, suggesting that the diet of these animals was not confined to fish.

Distribution: Brazil.
Classification: Theropoda, Tetanurae, Spinosauria.
Meaning of name: Irritator.
Named by: Martill, Cruikshank, Frey, Small and Clarke, 1996.
Time: Albian stage of the early Cretaceous.
Size: 8m (26ft).
Lifestyle: Fish hunter, predator or scavenger.
Species: *I. challengeri* (also *Angaturama* which is not valid).

Left: The specific name
I. challengeri *(illustrated) refers to Professor Challenger, the hero of Sir Arthur Conan Doyle's novel* The Lost World *in which live dinosaurs are found existing in South America.*

LATE DIPLODOCIDS AND BRACHIOSAURIDS

By Cretaceous times the diplodocids and the brachiosaurids, the main families of sauropods from the Jurassic period, were beginning to die away, but the sauropods themselves were by no means finished. There were still some interesting members, although their line was being taken up by an, until now, lesser known sauropod group.

Amargasaurus

In late Jurassic Argentina *Amargasaurus* was closely related to *Dicraeosaurus*. It too had tall spines jutting up from its backbone. In all diplodocids the spines were split in two, but in *Amargasaurus* this splitting was carried to an extreme. The single, associated skeleton found lacks the tail, and so the total length is uncertain.

Features: Like *Dicraeosaurus*, this diplodocid has a very short neck when compared with the group. The spines sticking up from the backbone between the shoulders and the hips undoubtedly carried a sail, probably used for signalling or for temperature regulation. Those on the neck are paired. They did not carry a sail, as this would have hampered the movement of the neck. They were more likely to have been covered in horn and were possibly used as defensive weapons.

Below: Dicraeosaurus.

Distribution: Argentina.
Classification: Sauropoda, Dicraeosauridae.
Meaning of name: Lizard from Amarga Canyon.
Named by: Salgado and Bonaparte, 1991.
Time: Hauterivian stage of the early Cretaceous.
Size: 12m (39ft).
Lifestyle: Browser.
Species: *A. cazaui, A. groeben.*

Nigersaurus

This dinosaur was found during a very productive field season in Africa by a team from the University of Chicago, USA, in 1997. As well as parts of several adult specimens of *Nigersaurus*, the expedition found the remains of hatchlings. The strange jaw probably signified a change in the vegetation, with low-growing herbs and flowering plants appearing in the landscape for the first time.

Features: The "Mesozoic lawnmower" is the nickname given to this diplodocid on account of the broad, straight front edge to its mouth that projects to each side of its skull, and which is packed with hundreds of needle-like teeth that form a rake-like cutting edge in a single, straight line. The skull is short for a diplodocid. The rest of the body was typical for a diplodocid, although rather smaller than its earlier cousins.

Distribution: Niger.
Classification: Sauropoda, Diplodocidae.
Meaning of name: Lizard from Niger.
Named by: Sereno, Beck, Dutheil, Larsson, Lyon, Mousse, Sadler, Sidor, Varricchio, G. P. Wilson and J. A. Wilson, 1999.
Time: Early Cretaceous.
Size: 15m (49ft).
Lifestyle: Low browser.
Species: *N. taqueti.*

Right: The straight front to the mouth was similar to that of the titanosaurid Bonitasaura. Presumably the two had similar feeding styles.

Cedarosaurus

The details about early Cretaceous sauropods, particularly the brachiosaurids, of North America are not well-known, and have usually been placed in the wastebasket taxon "*Pleurocoelus*". However, more distinct animals, such as *Cedarosaurus*, have been discovered in recent years. Much of one side of the skeleton of a single *Cedarosaurus* individual is known, except for the neck and head.

Features: As with other brachiosaurids, *Cedarosaurus* has front legs that are longer than the hind legs, with long finger bones held vertically, and the humerus and femur the same length. The neck and head are unknown, but they were probably held high as in others of the group. The tail is quite short for a brachiosaurid. Despite the fact that the build is definitely brachiosaurid, the bone structure has certain similarities to that of the later titanosaurids.

Distribution: Utah, USA.
Classification: Sauropoda, Macronaria.
Meaning of name: Lizard from Cedar Mountain.
Named by: Tidwell, Carpenter and Brooks, 1999.
Time: Barremian stage of the early Cretaceous.
Size: 14m (46ft).
Lifestyle: High browser.
Species: *C. weiskopfae*.

Above: With its high shoulders and its presumably long neck, Cedarosaurus would have browsed from high treetops.

Sauroposeidon

An articulated series of four neck vertebrae found in a prison yard in Oklahoma, USA, in 1994 are attributed to *Sauroposeidon*. The vertebrae were so big they were first thought to be fossilized tree trunks. They are very similar to those of *Brachiosaurus*, but about 15–25 per cent bigger, and it must have looked very similar.

Features: The structure of the neck bones suggest that *Sauroposeidon* is proportionally slimmer than *Brachiosaurus*. The neck, when raised, reaches nearly 20m (65½ft) above the ground. As well as being sculpted into thin plates and fine struts, as in other brachiosaurids, the bones have tiny air cells inside them to keep the weight down. It is probably the last of the brachiosaurids of North America, although there is a vertebra known from Mexico that is dated from the Campanian stage of the late Cretaceous.

Distribution: Oklahoma, USA.
Classification: Sauropoda, Macronaria.
Meaning of name: Lizard of Poseidon.
Named by: Wedel, Cifelli and Sanders vide Franklin, 2000.
Time: Albian stage of the early Cretaceous.
Size: 30m (98ft).
Lifestyle: High browser.
Species: *S. proteles*.

Above: Brachiosaurus.

Right: Sauroposeidon is named after Poseidon, the ancient Greek god of earthquakes.

TITANOSAURIDS – THE NEW GIANTS

The titanosaurids were the last family of sauropods to evolve. They may have been related to the macronarians, or even the diplodocids, but their peg-like teeth are the only obvious similarity to the latter. They tended to have broader hips than the other sauropods, resulting in wider trackways and so the group can be identified by their footprints. The genus Titanosaurus *itself is now a wastebasket genus.*

Malawisaurus

This is the earliest known titanosaurid from Africa, found as an 80 per cent complete, associated skeleton. A prepared and mounted skeleton is now on display in the Cultural Museum Centre in Karonga, Malawi. Its remains were originally classed in the wastebasket taxon *Gigantosaurus* – a name that has caused confusion in recent years after the finding of the giant theropod *Giganotosaurus*.

Features: The skulls of all sauropods are flimsy, and rarely preserved. That of *Malawisaurus* is the first known of the titanosaurids, although it is incomplete, consisting of lower and upper jaws and some teeth. The jaw and teeth show that it had a steep face, sloping down from the eyes to the snout, which has become the general model for the restoration of titanosaurid heads. Another general feature of many titanosaurids is the presence of armour on the back, but this is not present in all examples. The hips are more sturdy than those of other sauropods, being fused to six vertebrae rather than the more usual five.

Distribution: Mwakashunguti in the Zambezi valley, Malawi.
Classification: Sauropoda, Titanosauria.
Meaning of name: Lizard from Malawi.
Named by: Jacobs, Winkler, Downs and Gomani, 1993 (but originally named as *Gigantosaurus* by Haughton, 1928).
Time: Aptian stage of the early Cretaceous.
Size: 9m (30ft).
Lifestyle: Browser.
Species: *M. dixeyi*.

Above: Originally Malawisaurus *was thought to have lacked bony armour, but mineral nodules found around the skeleton seem to have been fossils of scutes similar to those of other titanosaurids.*

Agustinia

In the late 1990s an astonishing sauropod was found in Patagonia. If the armour had been found separately it would have been assumed to have come from a stegosaur or an ankylosaur because no other known sauropod has armour like it. It was originally named *Augustia*, but that name was found to have been given already to another animal.

Features: The most astounding feature of this medium-size sauropod is the arrangement of plates on the back. They are like the plates of a stegosaur, but turned sideways. They are rectangular with some drawn out into sideways-pointing spikes. As for the rest of the body, some features define it as a diplodocid, while others are definitely titanosaurid. At the moment the classification is unclear.

Distribution: Argentina.
Classification: Sauropoda.
Meaning of name: Agustin Martinelli's thing (after the discoverer).
Named by: Bonaparte, 1998.
Time: Aptian stage of the early Cretaceous.
Size: 15m (49ft).
Lifestyle: Browser.
Species: *A. ligabuei*.

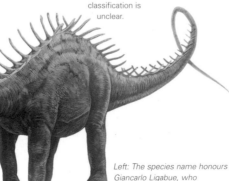

Left: The species name honours Giancarlo Ligabue, who sponsored the excavation of this remarkable animal.

Phuwiangosaurus

Until the discovery of the partly articulated skeleton of *Phuwiangosaurus* in 1992, all known Asian sauropods had been of a very primitive type. Its presence, and the presence of other unique dinosaurs, suggests that this part of South-east Asia was separated from the main Asian landmass of the time and supported a quite different fauna.

Features: *Phuwiangosaurus* differs from other known Asian sauropods because its teeth are narrow rather than spoon-shaped, and its neck vertebrae are broad and flattened from top to bottom, rather than from side to side. The vertebrae have Y-shaped spines. Skull pieces subsequently discovered indicate a skull shape similar to that of the later nemegtosaurids, indicating that they, too, may be part of the great titanosaurid family.

Distribution: Thailand.
Classification: Sauropoda, Titanosauria.
Meaning of name: Phu Wiang county lizard.
Named by: Martin, Buffetaut and Suteethorn, 1994.
Time: Early Cretaceous.
Size: 20m (65½ft).
Lifestyle: Browser.
Species: *P. sirindhornae*.

JOBARIA

Above. Jobaria, a titanosaurid that is a mix of different animals.

A chimera was a monster in ancient Greek mythology, and a mix of several animals. It had the body of a lion, the tail of a dragon and two heads, a lion's and a goat's. In palaeontological parlance a chimera is also a mix of animals, and throughout the literature of palaeontology we find examples, such as the supposed titanosaurid *Jobaria*.

It was found in Tanzania by the German expeditions of the early twentieth century, and named after the great German palaeontologist Werner Janensch. It has been classified as a titanosaurid because of the shape of the legs and feet, but some of the rest of the skeleton appears to come from some sort of diplodocid. It seems that two different animals died and were fossilized close to one another, giving rise to the confusion.

Chubutisaurus

When it was discovered, *Chubutisaurus* was placed in its own family and thought to have been related to the brachiosaurids. However, it did not have the long front legs associated with the brachiosaurids and it is now regarded as a primitive titanosaurid, although this is far from certain. It is known from two partial skeletons but, as usual, the skulls are missing.

Features: This huge sauropod has big air spaces in its vertebrae, the tail is short and the front legs shorter than the hind. Its brachiosaurid features include the articulation of the vertebrae and the great length of the finger bones. It has few resemblances to the titanosaurids, and its classification here is based on the fact that nearly all South American sauropods are members of the titanosaurid group. It is regarded as a very early and unspecialized form.

Distribution: Argentina.
Classification: Sauropoda.
Meaning of name: Lizard from the Chubut Province.
Named by: del Corro, 1974.
Time: Albian stage of the early Cretaceous.
Size: 23m (75½ft).
Lifestyle: Browser.
Species: *C. insignis*.

Left: A number of South American sauropods known only from fragmentary material have been classed as brachiosaurids. It seems more likely that, like Chubutisaurus, they were actually very primitive titanosaurids.

HYPSILOPHODONTS AND IGUANODONTS

After their appearance in the Triassic period and their establishment in the Jurassic, albeit in a rather minor role compared to that of the sauropods, the ornithopods really flourished in the Cretaceous period. Soon they were to become the most diverse and abundant of the plant-eating dinosaurs, and by early Cretaceous times they had already diversified into their major evolutionary lines.

Hypsilophodon

Several skeletons of this dinosaur have been found in the last 150 years, and the original image of this was of a smaller version of *Iguanodon* that could climb trees. Its build and size, similar to the modern tree kangaroo, along with a mistaken observation on the toe bones, led to this idea. It is now known to have been a fast-running ground-dweller.

Below: The common image of Hypsilophodon is of a small animal. However, all specimens so far found are of juveniles, and so it is difficult to estimate the adult size.

Features: *Hypsilophodon* is often regarded as the gazelle of the dinosaur world. Its legs are long and lightweight, with the muscles concentrated around the hips and the thigh bone, a sure sign of a running animal. The toe bones are not evolved for perching as was originally thought, but for speed. The deep skull contains several front teeth behind the beak, as well as chewing teeth at the back.

Distribution: Isle of Wight, England; and Spain.
Classification: Hypsilophodontidae.
Meaning of name: High ridged tooth.
Named by: Huxley, 1869.
Time: Barremian to Aptian stages of the Cretaceous.
Size: 2.3m (7½ft) estimated as an adult size.
Lifestyle: Browser.
Species: *H. foxii, H. wielandi.*

Leaellynasaura

When it was discovered high up in a sea cliff on the south coast of Australia, this little ornithopod was a surprise. In early Cretaceous times this area of Australia was well within the Antarctic Circle. The discovery was the first indication that dinosaurs could cope with the extreme cold and long periods of darkness in high latitudes close to the poles.

Features: The skull of *Leaellynasaura* is distinctive because of the particularly large eye sockets. They, with a big brain cavity, indicate big eyes and possibly the ability to see in the dark. This, in turn, suggests that this dinosaur had a metabolism that allowed it to survive in Antarctic conditions. Classification is a bit uncertain – it would be regarded as a hypsilophodont but for a difference in the shape of the thigh bone and the ridges of the teeth.

Distribution: Victoria, Australia.
Classification: Hypsilophodontidae.
Meaning of name: Leaellyn's (daughter of the discoverers) female lizard.
Named: T. Rich and P. Rich (Leaellyn's parents), 1989.
Time: Aptian to Albian stages of the Cretaceous.
Size: 2m (6½ft).
Lifestyle: Browser.
Species: *L. amicagraphica.*

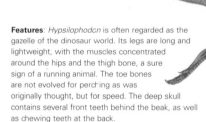

Left: The species name L. amicagraphica honours Friends of the Museum of Victoria, and the National Geographical Society, which funded the research.

Tenontosaurus

Judging by the number of remains that have been found, including 25 skeletons and scattered bones and teeth, *Tenontosaurus* must have been one of the most abundant herbivores in early Cretaceous North America. It was certainly attractive to meat-eaters – one skeleton has been found surrounded by the bodies of several *Deinonychus* that had been killed while attacking it.

Below: Tenontosaurus *was the prey of the North American plains.*

Features: *Tenontosaurus* is like a hypsilophodontid but lacks the teeth on the front part of the jaw. Otherwise it is like an iguanodontid, but the classification is still not clear. Its distinctive feature is its very long tail – longer than the rest of the body – and the network of tendons that supports the spine. Its long forelimbs and strong finger bones suggest that it walked on all fours for most of the time.

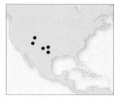

Distribution: Western North America.
Classification: Ornithopoda.
Meaning of name: Tendon lizard.
Named by: Ostrom, 1970.
Time: Aptian to Albian stages of the Cretaceous.

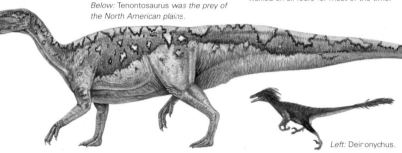

Size: 6.5m (21ft).
Lifestyle: Low browser.
Species: T. tillettorum, T. dossi.

Left: Deinonychus.

EXTREME COLD

The area that is now Victoria in southern Australia was, in early Cretaceous times, deep within the Antarctic Circle. This meant that anything living there would have been subjected to deep cold and a long, dark winter. The vegetation consisted of conifers with thick-skinned needles that were adapted to cold and dryness. There were also ferns, which suggest that the climates were not dry all the time. The mean annual temperature measured by radioactive isotopes in the rocks formed at the time, and by comparing the fossil plants with modern counterparts, was somewhere between 0° and 10°C, like the modern Hudson Bay area, Canada.

The landscape inhabited by the Victoria dinosaurs consisted of a deep rift valley, formed as the continent of Australia was beginning to rip away from that of Antarctica. It could be that such a valley gave shelter from the winter conditions. In any case, it was a hostile world for dinosaurs.
Nevertheless several types inhabited it. As well as the remains of *Leaellynasaura* and other small ornithopods, such as *Atlascopcosaurus*, found in the so-called Dinosaur Cove, were theropods resembling allosaurids and oviraptorosaurids, and even a possible early ceratopsian.

Jinzhousaurus

The Yixian Formation, in Liaoning, China, has yielded little half-bird, half-dinosaur animals. The discovery of *Jinzhousaurus* shows that largish dinosaurs also existed by the Yixian lake. The dinosaur discovery was also important in determining the age of the formation. The presence of such an obviously Cretaceous animal helped to establish that the beds had not been laid down in the earlier Jurassic period, as had previously been surmised.

Features: To look at, *Jinzhousaurus* resembles a small *Iguanodon*. However, its skull shows some advanced features that put the animal somewhere on the evolutionary track towards the later hadrosaurids – the duckbills. It is possible that the iguanodontid line split in three at about this time, one producing the iguanodontids proper, the second producing the hadrosaurids, and the third producing something between, of which *Jinzhousaurus* is the only example so far found.

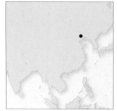

Distribution: Liaoning, China.
Classification: Ornithopoda, Iguanodontidae.
Meaning of name: Jinzhou lizard.
Named by: Wang and Xu, 2001.
Time: Barremian stage of the early Cretaceous.
Size: 7m (23ft).
Lifestyle: Low browser.
Species: J. yangi.

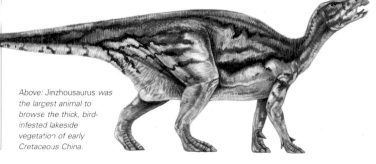

Above: Jinzhousaurus *was the largest animal to browse the thick, bird-infested lakeside vegetation of early Cretaceous China.*

IGUANODONTS

Iguanodontids tended to be bigger than hypsilophodontids, and this used to serve as the difference between the two groups. However, there are a number of differences, especially in the teeth, the forelimbs and the hips. The size of the typical iguanodontids meant that they were too heavy to spend much of their time on hind legs, and so they were basically four-footed animals.

Iguanodon

Famed as being one of the first dinosaurs to be scientifically recognized, *Iguanodon* became something of a wastebasket taxon over the years. It was thought to have been a four-footed, rhinoceros-like animal until complete skeletons were found in a mine, in Belgium, in the 1880s. Thereafter, it was restored in a kangaroo-like pose. Now it is largely regarded as a four-footed animal once more.

Features: *Iguanodon* is the archetypal ornithopod. Its head is narrow and beaked, with tough, grinding teeth. Its hands consist of three weight-bearing fingers with hooves. It has a massive spike on the first finger used for defence or gathering food, and a prehensile fifth finger that works like a thumb. The hind legs are heavy and the three toes are weight-bearing. The long, deep tail balanced the animal as it walked.

Distribution: England; Belgium; Germany; and Spain.
Classification: Iguanodontidae.
Meaning of name: Iguana tooth.
Named by: Boulenger and van Beneden, 1881.
Time: Barremian and Valanginian stages of the early Cretaceous.
Size: 6–10m (19½–33ft).
Lifestyle: Browser.
Species: *I. bernissartensis, I. anglicus, I. atherfieldensis, I. dawsoni, I. fittoni, I. hoggi, I. lakotaensis, I. ottingeri.*

Left: Although Iguanodon *was found and named by Mantell in 1825, the description was based only on teeth. In 2000 the International Commission on Zoological Nomenclature ruled the type species to be* I. bernissartensis *described in 1881, based on complete skeletons from Belgium.*

Altirhinus

Once regarded as a species of *Iguanodon* and called *I. orientalis*, there are enough differences to place *Altirhinus* in a genus of its own. It is known from five partial skeletons and two skulls, which are preserved in enough detail to show the distinctive features. It may represent an intermediate stage between the iguanodontids and the hadrosaurids.

Below: Altirhinus *looked just like* Iguanodon *except for the tall nasal region on the head.*

Features:
Altirhinus, as its name suggests, has a very high nasal region on the skull. This may have been an adaptation to an enhanced sense of smell. It had a greater number of teeth than *Iguanodon*, which provided it with a more efficient food-gathering technique. The beak is wider and flatter than that of *Iguanodon*, rather more like that of one of the later hadrosaurids. It retains the thumb spike, so distinctive of *Iguanodon* and its relatives.

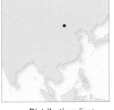

Distribution: East Gobi Province, Mongolia.
Classification: Ornithopoda, Iguanodontidae.
Meaning of name: High nose.
Named by: Norman, 1998.
Time: Aptian and Albian stages of the Cretaceous.
Size: 8m (26ft).
Lifestyle: Browser.
Species: *A. kurzanovi.*

Ouranosaurus

This African genus had the hands of a typical iguanodontid, but the skull had a broad, flat beak like a hadrosaurid. It lived at the same time and in roughly the same area as the sail-backed meat-eater, *Spinosaurus*. Its back structures may have been adaptations to life in the hot dry environment found at the time.

Features: The most obvious feature of *Ouranosaurus* is the huge array of spines jutting up and forming a picket fence along the backbone. The back bone is always shown as supporting a sail, and being used for heat control or signalling, but it is just as likely to have been the basis of a fatty hump that would store nourishment and energy for lean times. Modern hump-backed animals, such as the buffalo and camel, have humps supported by similar skeletal structures. The arid climates of early Cretaceous North Africa may have called for the evolution of specialist food-storage devices.

Distribution: Niger.
Classification: Ornithopoda, Iguanodontidae.
Meaning of name: Brave monitor lizard.
Named by: Taquet, 1976.
Time: Aptian stage of the early Cretaceous.
Size: 7m (23ft).
Lifestyle: Browser.
Species: *O. nigerensis*.

IGUANODON DISCOVERY

Iguanodon was discovered in the 1820s in Sussex, England, by a local country doctor, Gideon Mantell, and his wife Mary. Over several years they unearthed teeth and several bones. There was much discussion among the scientific establishment of the day about what kind of animal the remains, especially the teeth, came from. A fish and a hippopotamus were suggested by the foremost biologists in London and Paris. Eventually Mantell noted the similarity between the teeth and those of a modern iguana lizard – hence the Latin name.

With no other living comparison, Mantell first restored *Iguanodon* as a gigantic lizard, walking on all fours and menaced by a similarly dragon-like, four-footed *Megalosaurus* that had also been found at that time. As such they were restored as full-size statues in the grounds of the Crystal Palace in Sydenham, south London, England, where they stand to this day.

It was only with the discovery of about 40 *Iguanodon* skeletons in a coal mine at Bernissart in Belgium, in the 1880s, that it was obvious what kind of animal *Iguanodon* was. The coal mine was closed for two years while the fossils were excavated. This famous find was studied over the next 40 years by Louis Dollo from the Royal Museum of Natural History, in Brussels.

Muttaburrasaurus

The most complete dinosaur found in Australia so far is *Muttaburrasaurus*, and it is known from two skeletons. The first was found in 1963 by rancher Doug Langdon. The one found at Lightning Ridge in New South Wales had its bones replaced by opal. When alive, this animal may have lived in herds in open woodlands, and fed on ferns, cycads and conifers.

Features: This iguanodontid has a hollow, bony bump on its snout in front of its eyes. This may have had something to do with a sense of smell or an ability to make a noise. The teeth are evolved for slicing rather than for grinding as in the other members of the group. The hands have not been found so we do not know if it had the typical iguanodontid arrangement of fingers with the middle three strong and weight-bearing.

Distribution: Central Queensland, and New South Wales, Australia.
Classification: Ornithopoda, Iguanodontidae.
Meaning of name: Lizard from Muttaburra Station.
Named by: Bartholomai and Molnar, 1981.
Time: Albian stage of the early Cretaceous.
Size: 7m (23ft).
Lifestyle: Browser.
Species: *M. langdoni*.

Left: The first skeleton was kicked to bits by grazing cattle as it lay exposed, and some bones were taken home as souvenirs by locals. When its importance was known, most of the skeleton was subsequently recovered.

EARLY HORNHEADS

As with many of the groups of Cretaceous dinosaurs, the ceratopsians, or the horned dinosaurs, seem to have evolved in central Asia and migrated across to North America, where they later flourished. They appear to have evolved from typical, two-footed plant-eaters from the ornithopod line – the basic, primitive ceratopsian body, showing many similarities to that of the generalized two-footed plant-eater.

Psittacosaurus

Once regarded as an ornithopod, albeit one with a peculiar head, *Psittacosaurus* is now regarded as a transitional form between the primitive ornithopods and the horned dinosaurs – the ceratopsians. *Psittacosaurus* has more species than any other dinosaur, and in times to come it may be split into several genera. Recent studies reveal a series of spines on the tail.

Features: The skull is deep and narrow, and carries a heavy beak. The upper part of the beak is supported by a bone, the rostral, that is only found in the ceratopsians. The back of the skull carries a ridge of bone, probably an anchor for the heavy jaw muscles. This gives the head a square profile which, along with the heavy beak, is rather like that of a parrot. There are no teeth at the front of the mouth, and those at the back are built for chopping. Cheek pouches would have held the food as it was chewed.

Distribution: Thailand; China; Mongolia.
Classification: Ceratopsia.
Meaning of name: Parrot lizard.
Named by: Osborn, 1923.
Time: Aptian stage of the early Cretaceous.
Size: 2m (6½ft).
Lifestyle: Low browser.
Species: *P. mongoliensis*, *P. mazongshanensis*, *P. meileyingensis*, *P. meimongoliensis*, *P. ordosensis*, *P. sattayaraki*, *P. sinensis*, *P. zinjiangensis*.

Yaverlandia

A skull fragment, all that we know of this animal, suggests that it is an early pachycephalosaur. However, this classification is open to dispute. The top of the head is quite flat, and suggests that the heavily domed head that typifies the group evolved very gradually. The other possible pachycephalosaur from Europe is represented by a tooth found in late Cretaceous Portugeuese rocks.

Below: If Yaverlandia proves to be a pachycephalosaur, then the body shape would have been as shown. However, this restoration must be regarded as speculative.

Features: The roof of the skull shows a thickening – the only feature that suggests that this animal is a pachycephalosaur. Yet the traces of the brain shape that are visible in the specimen, particularly the areas that deal with the sense of smell, are quite different from those of known pachycephalosaurs. The way the skull bones knit together is also dissimilar, and so it may not be a pachycephalosaur at all. See the similar confusion that surrounds *Majungatholus* (right).

Distribution: Isle of Wight, England.
Classification: Pachycephalosauria.
Meaning of name: From Yaverland Point.
Named by: Galton, 1971.
Time: Barremian stage of the early Cretaceous.
Size: 2m (6½ft).
Lifestyle: Low browser.
Species: *Y. bitholus*.

Archaeoceratops

This dinosaur is known from two skeletons, one almost complete but lacking the forelimbs, found during the Sino-Japanese Silk Road Dinosaur Expedition in 1992–3. Its discovery seems to suggest that the ceratopsians evolved first in Asia and later evolved into two lines, one of which migrated to North America, where the dinosaurs later flourished and became the great-horned dinosaurs of the late Cretaceous.

Features:
Archaeoceratops is a small, lightweight animal that had the ability to walk on all fours or run on its hind legs. It is one of the most primitive ceratopsians known, with a barely developed neck frill. The head is quite large for the size of the body, and it still retains the three or four teeth in the front of the mouth that are so distinctive of its ornithopod ancestors. There is no sign of any horns.

Left: Archaeoceratops *was a small, fleet-footed rabbit-sized animal, quite unlike its lumbering descendants.*

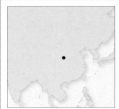

Distribution: Gansu Province, China.
Classification: Marginocephalia, Neoceratopsia.
Meaning of name: Ancient horned face.
Named by: Dong and Azuma, 1997.
Time: Early Cretaceous.
Size: 80cm (31in).
Lifestyle: Low browser.
Species: *A. oshimai.*

THE EARLY CERATOPSIANS

The ceratopsians were the horned dinosaurs. A typical image of such a beast is of a huge rhinoceros-like animal with a solid shield of bone around its neck, and a set of wicked horns ready to inflict deadly damage in defence or offence.

This was true of the later ceratopsians, but the ancestral forms were quite different. The neck shield seems to have evolved before the horns. It probably originated as a supporting shelf that held the powerful jaw muscles that the primitive ceratopsians needed for processing their main food. They seem to have been well adapted to feeding on cycads and cycad relatives. The strong, sharp beak was ideal for selecting the most nutritious part of the plant and ripping it out. The slicing teeth would have chopped up the tough leaves while holding the pulp in the cheek pouches. Very powerful jaw muscles would have been needed for this action which, recent research suggests, involved a forward-and-back as well as an up-and-down action.

As this ridge became bigger it would have functioned as a display structure as well, probably brightly coloured and used in attracting mates or scaring off rivals. The final purpose, that of defence, would have evolved later, once the animals had evolved into big types that would have been too heavy to run away or hide from big meat-eaters.

Liaoceratops

The early Cretaceous lake beds in Liaoning, China, not only produced the fabulous half-bird, half-dinosaur animals that originally made them famous, but also early members of the ankylosaurs, and of the ceratopsians. This fox-sized animal is the earliest-known of the horned dinosaur line. It would have used its shield and horns for display rather than defence, and probably defended itself by running away.

Features: The large head has a pair of horns, pointing sideways, one under each eye. A frill is present and seems to have acted as an attachment for the jaw muscles, judging by the pitted texture that indicates muscle attachment. Its teeth are adapted for slicing rather than grinding. As with other lightweight dinosaurs, it is designed for running on hind legs as well as walking on all fours. It may belong to a line that gave rise to both the psittacosaurids and the ceratopsians proper.

Distribution: China.
Classification: Marginocephalia, Ceratopsia.
Meaning of name: After the Chinese province and village where it was found.
Named by: X. Xu, P. J. Makovicky, X. L. Wang, M. A. Norell, and H. L. You, 2002.
Time: Barremian stage of the early Cretaceous.
Size: 1m (3ft).
Lifestyle: Low browser.
Species: *L. yanzigouensis.*

Left: For all its distinctive ceratopsian features, Liaoceratops *is in some ways even more primitive than* Psittacosaurus, *belonging to the group from which the ceratopsians evolved. Ceratopsian evolution was more complex than first imagined.*

ANKYLOSAURS AND LATE STEGOSAURS

The stegosaurs were the armour-bearers of the Jurassic period. However, with the dawn of the Cretaceous their time had passed, and the ankylosaur family expanded into their niche. The last of the stegosaur family were found in China, and surrounding countries, where they survived well into the age of the ankylosaurs.

Wuerhosaurus

Wuerhosaurus was the last-known of the stegosaurs. Until its discovery in the 1970s, it had been believed that the stegosaurs were restricted to the Jurassic period. *Wuerhosaurus* was found in rocks dating from 10 million years later than any known stegosaur. The main difference between this large stegosaur and *Stegosaurus* itself is in the shape of the plates and also in the relatively shorter body, wider hip bones and head held closer to the ground.

Features: The plates are long and low and, as in *Stegosaurus*, are alternate rather than paired. The body is shorter than *Stegosaurus*, with as few as 11 vertebrae in the back, and the hip bones are broader. The distribution of spikes and plates is uncertain. Some restorations show spikes at the base of the tail, but this may be a mistranslation of a description of "long anterior caudal neural spines" – the projections on top of the tail vertebrae near the hips.

Distribution: Xinjiang Uygur Zizhigu, China.
Classification: Thyreophora, Stegosauria.
Meaning of name: Lizard from Wuerho.
Named by: Dong, 1973.
Time: Valangian to Albian stages of the early Cretaceous.
Size: 6m (20ft).
Lifestyle: Browser.
Species: *W. ordosensis, W. homheni.*

Left: Wuerhosaurus *was squatter than most stegosaurs, evidently an adaptation to feeding on low-growing plants.*

Monkonosaurus

This is the first dinosaur to be found in Tibet but, despite its poor quality, it is quite an important specimen. When it lived, the area of Tibet was at sea level on the edge of the Tethys Ocean. The stratigraphy of the area has still to be worked out, and it is unclear whether this is a Jurassic or a Cretaceous stegosaur.

Features: The fact that we only know of two vertebrae, a part of the hips and three plates means that we cannot tell a great deal about this animal except that it is a medium-sized stegosaur. The plates resemble those of *Stegosaurus*, being big and thin. The spines on top of the vertebrae are quite low, and it does not have such a deep back and tail as other stegosaurs.

Distribution: Eastern Tibet.
Classification: Thyreophora, Stegosauria.
Meaning of name: Lizard from Monko County.
Named by: Zhao, 1983 vide Dong, 1990.
Time: Late Jurassic or early Cretaceous.
Size: 5m (16½ft).
Lifestyle: Browser.
Species: *M. lawulacus.*

Right: Despite its remains being found in the heights of the highest mountain range on Earth, Monkonosaurus *would have lived by the sea.*

Liaoningosaurus

This is a very primitive ankylosaur, hardly more advanced than early Jurassic *Scelidosaurus*. However, since the only skeleton known, although almost complete, is of a juvenile, it is not easy to see what the adult version would have looked like. It was found in the prolific lake beds of the Liaoning Province of north-eastern China.

Features: The obvious feature of this small ankylosaur is the presence of a shield under the hips, the only time such a structure has been found in the dinosaur world. This contrasts with the almost total lack of armour on the back. The only traditional ankylosaurian armour seems to be on the neck, where there are three pairs of spines and several plates. The skull is longer than wide, and the feet bear large claws.

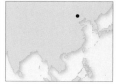

Distribution: China.
Classification: Thyreophora, Ankylosauria.
Meaning of name: Lizard from Liaoning.
Named: Xu, Wang and You, 2001.
Time: Barremian stage of the early Cretaceous.
Size: 34cm (13in), but this is a juvenile.
Lifestyle: Low browser.

Species: *L. paradoxus.*

Above: Skeletons of juvenile dinosaurs are quite rare, and so the one skeleton of Liaoningosaurus *is an important find. However, it is difficult to tell from the skeleton what the adult would have looked like.*

INTERPRETATION AND MISINTERPRETATION

As far as we can tell, all the stegosaurs were extinct by the end of the early Cretaceous period, and were replaced by the ankylosaurs. One exception was *Dravidosaurus*, based on a partial skull, some bones and a tooth found in late Cretaceous rocks in India by Yadagiri and Ayyasami in 1979.

The interpretation is that during Cretaceous times India broke away from the rest of the supercontinent, Gondwana, and was an island separated from all other landmasses of the time. Evolution took its own course there, unaffected by anything that was happening in the rest of the world. There is no reason to doubt that this island continent formed a kind of 'lost world' in which dinosaurs, such as stegosaurs, could survive despite dying out elsewhere. This plausible scenario was refuted in 1996 when Sankhar Chatterjee re-examined the *Dravidosaurus* material and found it to be the remains, not of a stegosaur or even of a dinosaur, but of a plesiosaur, one of the swimming reptiles of the time. The tooth was probably from a small ornithischian dinosaur.

Acanthopholis

This is a wastebasket genus, but as one of the first ankylosaurs to be found, this dinosaur is worth looking at. It is also a *nomen dubium*, since some of the fossils given the name may not belong to this genus. In fact most of the other species consist of parts that belong to ornithopods, sauropods and even turtles.

Features: *Acanthopholis* is known from partial remains found along the beach near Folkestone in Kent, England. It is an armoured, quadrupedal plant-eater. The armour consists of oval, keeled plates set into the skin, and tall conical spikes jutting from the neck and shoulder area. Its classification as a nodosaurid is based on the presence of this armour and on the shape of the teeth. The restoration shown here must be regarded as conjectural. *Acanthopholis* was one of the earliest armoured dinosaurs to have been discovered and so much of the work done on it at the time was speculative.

Distribution: England.
Classification: Thyreophora, Ankylosauria, Nodosauridae.
Meaning of name: Spiny scute.
Named: Huxley, 1867.
Time: Aptian to Cenomanian stages of the early and late Cretaceous.
Size: 5m (16½ft).
Lifestyle: Low browser.
Species: *A. horrida, A. hughesii, A. keepingi, A. macrocercus, A. platypus, A. stereocercus* (all *nomena dubia*).

Above: All the specimens attributed to Acanthopholis *are housed in different museums, and so it is difficult to co-ordinate research on this animal.*

POLACANTHIDS AND OTHER EARLY ANKYLOSAURS

The ankylosaurs had armour that consisted of bony plates that were set into the skin and covered in horn. Their main feature was the armoured pavement that covered the back. Defence consisted of sideways pointing spikes, sharp plates on the tail, or a mace at the tail's end.

Polacanthus

This dinosaur is known from an almost complete skeleton, lacking the skull, and all sorts of isolated bits. Like *Acanthopholis* it was one of the first dinosaurs to be discovered and studied. Another early Cretaceous ankylosaur, *Hylaeosaurus*, also from southern England, was once thought to be the same animal but it differs in the arrangement of the bones in the shoulder.

Features: The armour of *Polacanthus* and its relatives is distinctive. Over the neck, shoulders and back there is a series of spikes that stick up and sideways, with the tallest over the shoulder region. Over the hips is a "buckler" consisting of a compressed mass of bony scutes. Along each side of the tail are thin, sharp plates pointing outwards. We do not know its skull, but it was probably broad at the front, like that of its close relative *Gastonia*.

Below: The pattern of armour on the back of Polacanthus *is well-known.*

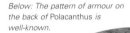

Distribution: England.
Classification: Thyreophora, Ankylosauria, Polacanthidae.
Meaning of name: Many spikes.
Named by: Hulke, 1881.
Time: Barremian stage of the early Cretaceous.
Size: 4m (13ft).
Lifestyle: Low browser.
Species: *P. foxii, P. rudgwickensis.*

Minmi

This dinosaur is known from an almost complete specimen and several other isolated pieces. The stiffened back skeleton suggests that it may have been a fairly fast runner. A cololite, a fossilized lump of stomach contents, shows that it fed on fruit, seed and soft vegetation. It had been so thoroughly chewed that this ankylosaur must have had cheek pouches to hold and process the food.

Features: This primitive ankylosaur has features that are similar to both the ankylosaurids and the nodosaurids. The legs are quite long and the back stiffened by extensions of the vertebrae – the paravertebrae that give it its species name. Unusually for ankylosaurs, the belly is protected by armour. Conical spines stick out in a rim around the hips, presumably to defend against attack from the rear. A supposed club found on the tail was later found to be merely an effect of fossilization.

Distribution: Australia.
Classification: Thyreophora, Ankylosauria.
Meaning of name: From Minmi Crossing, the place where it was found.
Named by: Molnar, 1980.
Time: Aptian stage of the early Cretaceous.
Size: 2m (6½ft).
Lifestyle: Low browser.
Species: *M. paravertebra.*

Left: Minmi *was the first armoured dinosaur to be found in the Southern Hemisphere. Until the discovery of* Mei *its name was the shortest dinosaur name on record.*

Gastonia

Found in the same quarry as *Utahraptor*, *Gastonia* is known from an almost complete skeleton and several individual skulls. It may have fought by head-butting, but its main defence against *Utahraptor* would have been the scything tail with the sideways-pointing blade-like plates, and the long spikes on the shoulders. The plates may have worked like scissors on anything caught between them.

Features: This heavily armoured dinosaur closely related to the European *Polacanthus*, carries several types of armour on its neck, back and tail. Pairs of spines run along the neck and shoulders, and down the sides. Broad plates stick out sideways from the tail. A solid mass of armour of fused ossicles lie across the hips. Smaller ossicles lie in the spaces between the larger armour pieces. The skull itself is broad and solid.

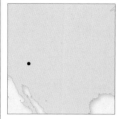

Distribution: Utah, USA.
Classification: Thyreophora, Ankylosauria, Polacanthidae.
Meaning of name: Gaston's (Robert Gaston, the discoverer) thing.
Named by: Kirkland, 1998.
Time: Barremian stage of the early Cretaceous.
Size: 5m (16½ft).
Lifestyle: Low browser.
Species: *G. burgei*.

ANKYLOSAUR CLASSIFICATION
The classification of the ankylosaurs is constantly undergoing revision. For our purposes we can identify several main evolutionary trends in the group.
Below: Scelidosaurus.

The most basal thyreophorans may be ancestral to both the ankylosaurs and the stegosaurs. These primitive forms include animals such as *Scelidosaurus* and *Minmi*. From these, three main ankylosaur branches develop.
Below: Polacanthus.

The first is the polacanthids, including *Polacanthus* and *Gastonia*. They had broad mouths, and armour consisting of spikes on the back and neck, a buckler over the hip and plates on the tail.
Below: Pawpawsaurus.

Next come the nodosaurids, including *Nodosaurus* and *Pawpawsaurus*. Like the polacanthids, they had armour on the neck and shoulders, including sideways-pointing spikes. Unlike the polacanthids, the nodosaurids had a narrow mouth, suggesting more selective feeding.

The final group is the ankylosaurids. This later group is characterized by the club on the end of the tail, used as a weapon. Typical members are the giants *Euoplocephalus* and *Ankylosaurus* from the late Cretaceous. These had the broad mouths of the general-feeding polacanthids.
Above: Euoplocephalus.

Shamosaurus

The earliest-known ankylosaurid is *Shamosaurus*, known from several individuals including a good skull and jaw from Mongolia. It appears that the group evolved in this area and subsequently spread to North America (the two continents were joined at the time). The closely related *Cedarpelta* appeared in North America slightly later.

Features: The beak of the skull is quite narrow – more like that of a nodosaurid than that of an ankylosaurid. There are small horns at each side of the head. As with all ankylosaurids there is a maze of nasal passages inside the skull, probably for increasing the sense of smell, warming the breathed air, or making sounds. The back is well covered by an armour of plates and spines – hence its species name *S. scutatus*, meaning shielded. Otherwise the bones are different enough from the typical ankylosaurids to lead some scientists to put it in its own family.

Distribution: Mongolia.
Classification: Thyreophora, Ankylosauria, Ankylosaurinae.
Meaning of name: Lizard from Shamo (the Gobi Desert).
Named by: Tumanova, 1983.
Time: Barremian to early Aptian stages of the Cretaceous.
Size: 7m (23ft).
Lifestyle: Low browser.
Species: *S. scutatus*.

Left: Shamosaurus had a similar neighbour. Gobisaurus, known since the 1950s but not described until 2001, was another ankylosaurid of a similar appearance to Shamosaurus and from the same area.

MAMMALS

*The mammals evolved at about the same time as the dinosaurs, at the end of the Triassic period.
However, the dinosaurs went on to become the most prominent land animals, and the mammals had a
much more minor role. It was thought that none of the mammals was bigger than a rat, but recent
discoveries are beginning to disprove this.*

Fruitafossor

This creature is actually from the end of the preceding
Jurassic period but important enough to be mentioned here.
Furitafossor shows some specialization in lifestyle. It was a
burrowing squirrel-sized animal, digging into the river banks
of the Morrison Formation streams. It is the earliest-known
digging mammal, predating
any others by 100
million years.

Features: The teeth are quite
specialized for a Mesozoic
mammal, with a single open-
ended root and no enamel,
suggesting that they were
constantly replaced. Similarity to
the teeth of modern armadillos
and aardvarks and other ant-
eating or termite-eating mammals
suggests that *Fruitafossor* ate
termites (ants not having evolved
by this time). The front limbs are
adapted for digging and are very
strong, earning the fossil
the nickname "Popeye",
a reference to that
cartoon character's
outsized arm muscles. It
would have burrowed in
the ground to find food
and to escape from the
local predatory dinosaurs and
may have nested
underground.

Distribution: Colorado, USA.
Classification:
Mammaliformes, order
uncertain.
Meaning of name: Digger
from Fruita fossil site.
Named by: Luo and Wible,
2005.
Time: Kimmeridgian stage of
the late Jurassic.
Size: 15cm (6in).
Lifestyle: Termite-eater.
Species: *F. windsheffeli.*

*Above: Fruitafossor burrowed into
the seasonal river banks hunting
communal insects and possibly
eating plants too.*

Repenomamus

Science had accepted that Mesozoic mammals
were tiny timid animals that scampered about
the dinosaurs' feet and generally kept clear of
them. That was before the discovery of the
perfect articulated skeleton of two species of
badger-sized *Repenomamus* in the lake deposits
of Liaoning, China. One had the bones of a
small *Psittacosaurus* in its stomach area.
These were Mesozoic mammals that
ate dinosaurs!

Features: The teeth of *Repenomamus* are
specialized for meat-eating, with pointy incisors
for holding prey, canines for killing it and
shearing premolars for ripping it up. The deep
muscle attachments on the jaw show that it had
a powerful bite. The molars are small and blunt,
adapted for crushing rather than chewing. The
legs are quite short and stick out at the side like
those of a reptile. It walked on the flats of its
feet like a bear, rather than on its toes like a
running animal. It would not have been able to
run fast, but it would have been very agile.

Distribution: Liaoning, China.
Classification: Theriiformes,
Tricodontia.
Meaning of name: Reptile
mammal.
Named by: Meng, Hu, Wang
and Li, 2005.
Time: Barremian stage of the
early Cretaceous.
Size: 50cm (20in).
Lifestyle: Dinosaur-eater.
Species: *R. robustus,
R. giganticus.*

*Right: Bearing its
meat-eating teeth,
Repenomamus
would have stood
its ground against
bigger meat-
eating dinosaurs.*

Eomaia

Distribution: Liaoning, China.
Classification: Theriiformes, Placentalia.
Meaning of name: Dawn mother.
Named by: Ji, Luo, Yuan, Wible, Zhang and Georgi, 2002.
Time: Barremian stage of the early Cretaceous.
Size: 15cm (6in) – half of which was tail.
Lifestyle: Insectivore.
Species: *E. scansoria*.

Most Mesozoic mammals belong to groups that led to evolutionary dead ends. However, *Eomaia* seems to have been a remote ancestor of most of today's mammals. The subclass that contains the modern placentals is known as the placentalia, and *Eomaia* was probably an early development of this line, predating all others of the group by 40 million years.

Above: Eomaia was probably nocturnal, hunting insects in the branches of trees and bushes around the Liaoning lakes and avoiding the dinosaurs and birds that existed there at the time.

Features: The teeth, with the familiar pattern of incisors, canines, premolars and molars, make *Eomaia* look like the earliest-known placental mammal – the group to which all modern animals except the pouched marsupials and egg-laying monotremes belong. However, the hip bones are similar to those of a marsupial. Its clawed feet suggest that it would have been good at climbing trees. Impressions of hair surround the fossil, and this is sparse on the tail giving it a ratty appearance. As in modern mammals there are two layers of fur in the coat. The long guard hairs throw off water, while the dense layer of underhairs close to the body provides warmth.

OTHER JEHOL MAMMALS

The wide variety of mammals found in the Liaoning lake beds are mostly of groups that have left no descendants. Their classification is based on the teeth – mostly the only fossil remains that had been found before complete articulated skeletons were discovered in Liaoning, China. These were all largely ground-dwelling animals like shrews.

Zhangheotherium – a symmetrodont. The symmetrodonts are characterized by having teeth with three cusps in a line. They are the most primitive of mammals but some scientists do not regard the classification as valid – the tooth shape could have evolved more than once.

Maotherium – another symmetrodont.

Jeholodens – a euticonodont. The eutriconodonts had teeth that had three conical cusps.

Sinobaatar – a multituberculate. The multituberculates had teeth with a large number of cusps.

Yanocodon – a burrowing eurtriconodont with a long cylindrical body, and a very primitive ear structure.

Below: The appearance of Zhangheotherium is based on a complete skeleton.

Sinodelphys

In the Jehol biota – the animal and plant life revealed by the wonderful fossil remains from Liaoning – there are many mammals representing a wide array of evolutionary development. One of these, *Sinodelphys*, is the earliest relative of the marsupials so far found anywhere.

Distribution: Liaoning, China.
Classification: Theriiformes, Marsupialia.
Meaning of name: Chinese opossum.
Named by: Luo, Ji, Wible and Yuan, 2003.
Time: Barremian stage of the early Cretaceous.
Size: 15cm (6in).
Lifestyle: Insectivore.
Species: *S. szalayi*.

Above left: Mammals that looked like modern opossums were common in Cretaceous times.

Features: The marsupial-like characteristics of *Sinodelphys* are the teeth. On each jaw there are four pairs of incisors, the third and fourth of which are diamond-shaped, a pair of canines, and then a forward-pointing premolar. After a space there are the rest of the premolars and molars. This layout of teeth is shared with later primitive marsupials like opossums. In structure these teeth are adapted to an insectivorous diet. The feet and claws show *Sinodelphys* to have been a tree-climber.

CRETACEOUS TURTLES

The turtles represent one of evolution's survivors. Once the basic shape – the heavy shell above and below, the sturdy head that could be pulled in for defence, and the powerful limbs to move the heavy body – was established, it remained largely unchanged for hundreds of millions of years. The representatives that lived in Cretaceous times would not look out of place today.

Archelon

Sharing with elasmosaurs and mosasaurs the late Cretaceous seaway that ran the length of North America, this great turtle – the biggest known – probably fed on slow-moving planktonic prey such as jellyfish. It would have had a similar lifestyle to the modern leatherback turtle of the Caribbean and North Atlantic.

Features: Like that of the modern leatherback turtle, the shell consists merely of an open framework of struts that, in life, would have been covered with a leathery skin. However, these struts stretch out, rib-like, from the backbone rather than running fore and aft as they do in the leatherback. The underside is more robust, consisting of an array of four star-shaped plates. The head is small in proportion, and the toothless jaws carry a sharp beak that would have been covered in horn in life, like that of the modern snapping turtle.

Distribution: Kansas, South Dakota, USA.
Classification: Chelonia.
Meaning of name: Ruling turtle.
Named by: Wieland, 1896.
Time: Campanian to Maastrichtian stages of the late Cretaceous.
Size: 4.6m (15ft).
Lifestyle: Planktonic feeder.
Species: *A. ischyros*.

Left: Archelon, the biggest turtle known, had a very lightweight shell.

Calcarichelys

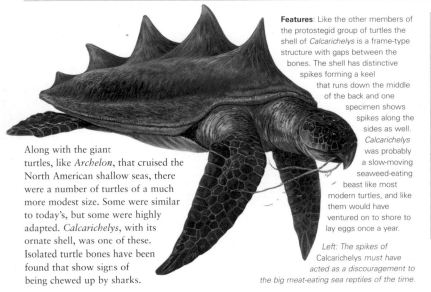

Features: Like the other members of the protostegid group of turtles the shell of *Calcarichelys* is a frame-type structure with gaps between the bones. The shell has distinctive spikes forming a keel that runs down the middle of the back and one specimen shows spikes along the sides as well. *Calcarichelys* was probably a slow-moving seaweed-eating beast like most modern turtles, and like them would have ventured on to shore to lay eggs once a year.

Along with the giant turtles, like *Archelon*, that cruised the North American shallow seas, there were a number of turtles of a much more modest size. Some were similar to today's, but some were highly adapted. *Calcarichelys*, with its ornate shell, was one of these. Isolated turtle bones have been found that show signs of being chewed up by sharks.

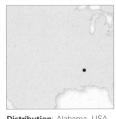

Distribution: Alabama, USA.
Classification: Chelonia.
Meaning of name: Limy turtle.
Named by: Zengerl, 1953.
Time: Campanian stage of the late Cretaceous.
Size: 30cm (1ft).
Lifestyle: Planktonic hunter.
Species: *C. gemma*.

Left: The spikes of Calcarichelys *must have acted as a discouragement to the big meat-eating sea reptiles of the time.*

Araripemys

Distribution: Brazil.
Classification: Chelonia.
Meaning of name: Turtle from Araripe.
Named by: Price, 1973.
Time: Albian stage of the early Cretaceous.
Size: 40cm (16in).
Lifestyle: Omnivorous.
Species: *A. barretoi.*

One of the earliest of the pleurodire turtles – those that withdraw into their shells by swinging the neck to the side, like modern side-necked and snapping turtles – *Araripemys* is known from more than a dozen specimens from the early Cretaceous lake deposits of Araripe in Brazil. This has been a famous fossil site since the 1870s and has yielded the detailed remains of pterosaurs. crocodiles, dinosaurs and fish.

Features: *Araripemys* is a typical pleurodire. It has a long neck which could have been swung around and tucked into the shell. The feet all have five toes, with the shortest on the inside. This indicates that, despite being largely aquatic, it also walked on land with the limbs out to the side in a sprawling gait. The shell is unusual in being very flat, and the neck is very long, even for a pleurodire.

Left: Araripemys *would have resembled many of today's freshwater turtles.*

TURTLE CLASSIFICATION

The turtle line consists of several families, but can be grouped into two major classifications. These are the cryptodires, or the hidden-necked turtles, and the pleurodires, or the side-necked turtles.

The former protect themselves by pulling their heads into their shells by making a vertical S-shaped curve, while the latter do so by swinging their necks sideways under the lip of the upper shell. They are both very old groups. Their ancestors date back to *Proganochelys* in Triassic times. The two groups diverged in the early Cretaceous.

Below: Meiolania, *a cryptodire turtle.*

Below: A pleurodire turtle.

Stygiochelys

The baenids were a very common turtle group from the Cretaceous period and they seem to have been restricted to North America. They were not alone in the streams and swamps of the damp plain on which lived the last of the dinosaurs such as *Tyrannosaurus* and *Triceratops*. Others included early representatives of the modern soft-shelled turtle, *Trionyx*.

Distribution: Montana, USA.
Classification: Chelonia.
Meaning of name: Turtle from the river of Hell.
Named by: Gaffney and Hiatt, 1971.
Time: Maastrichtian stage of the late Cretaceous.
Size: 30cm (1ft).
Lifestyle: Carnivore.
Species: *S. estesi.*

Features: In appearance *Stygiochelys* resembles any modern freshwater turtle belonging to the cryptodire group – the group that protects itself by pulling its head back into the shell. They were stream-living animals that fed on small invertebrates, particularly molluscs. The group survived the late Cretaceous extinction and lived on into the early part of the Tertiary, becoming extinct only in the Eocene.

CHAMPSOSAURS AND CRETACEOUS LIZARDS

The late Cretaceous landscape may have been dominated by the big dinosaurs but, just as today, there were many small animals scampering about in the undergrowth. The lizards, many of them quite similar to modern types, and even related to them, were very numerous.

Champsosaurus

The champsosaurs were a successful, although minor, group of crocodile-like animals. They existed alongside the crocodiles proper in the Cretaceous and, like them, survived the end Cretaceous extinction and went on into the early Tertiary. Their physical similarity is a matter of convergent evolution as the two groups are not closely related. They both had the long snouts and sharp teeth for fishing, the highly placed eyes and nostrils for lying submerged, and the sinuous body and the flattened tail for swimming. It was quite a wide-ranging animal, with remains being found in Europe and North America. Its fossils have been found in deposits laid down in cypress swamps, along with the remains of freshwater bivalves, fish, turtles and true crocodiles.

Features: *Champsosaurus* and its relatives are adapted to the environment and lifestyle of a typical crocodile. The teeth are much smaller and weaker than those of equivalent-sized crocodiles, the tail is a little shorter and there is a single nostril at the end of the snout. The adult females seem to be better adapted for a terrestrial life than the males, probably due to nesting behaviour, but the young of both sexes are quite aquatic.

Below: The long gavial-like snouts and sharp teeth of the champsosaurs mark them out as fish-eaters.

Distribution: Alberta, Canada; Montana, New Mexico, Wyoming, USA; Belgium and France.
Classification: Eusuchia, Choristodera.
Meaning of name: Field lizard.
Named by: Cope, 1876.
Time: Late Cretaceous to early Tertiary.
Size: 1.5m (5ft).
Lifestyle: Fish-hunter.
Species: *C. laramiensis, C. profundus, C. lindoei, C. gigas.*

Sineoamphisbaena

The modern amphisbaenean is a lizard that looks like a worm. When a fossil of a legless lizard was uncovered in China in the early 1990s it was thought to have been an early member of this group. However, it now seems, on the basis of the arrangement and the shapes of the bones of the skull, to be more closely related to the modern skink lizards. Leglessness in lizards has evolved several times, as in glass snakes and slow worms – and indeed in the snakes themselves – and so there is uncertainty about where *Sineoamphisbaena* fits in the family tree.

Features: The tiny snake-like body and bullet head of *Sineoamphisbaena* show it to have been a worm-like burrowing animal. The skull is rounded in front but flattened from top to bottom, suggesting that it burrowed through soil with a shovel action. Like modern worm-lizards it would have fed on worms and other soil-living invertebrates.

Distribution: Nei Mongol, China.
Classification: Lepidosauria, Squamata.
Meaning of name: Chinese amphisbanean.
Named by: Wu, Brinkman, Russell, Dang, Currie, Hou and Cui, 1993.
Time: Late Cretaceous.
Size: 10cm (4in).
Lifestyle: Burrower.
Species: *S. hexatubularis.*

Left: Whatever its relationships, Sineoamphisbaena lived very much like the modern amphisbaeneans.

Primaderma

Distribution: Utah, USA.
Classification: Lepidosauria, Squamata.
Meaning of name: Early skin.
Named by: Nydam, 2000.
Time: Albian to Cenomanian stages of the early and late Cretaceous.
Size: 30cm (1ft).
Lifestyle: Carnivore.
Species: P. nessovi.

The only poisonous lizards that exist today are the Gila monster and the Mexican poisonous lizard, both members of the helodermatidae. A distant relative existed in late Cretaceous times but it does not seem to have been poisonous. Usually we can tell if an animal was poisonous or not by the shape of the teeth – hollow or grooved teeth in a snake indicate where the venom is transmitted from the poison glands into the bite. However, in some reptiles the poison glands are still there but the teeth lack these specializations, especially if the reptile does not rely on the poison to subdue its prey. Hence it can be very difficult to tell whether or not a fossil reptile was poisonous.

Features: *Primaderma* is only known from jaw fragments complete with teeth. This is enough to show that it was a member of the same group as includes the modern Gila monster. However, the teeth do not seem to have the typical poison grooves, but are flat and serrated, and adapted for shearing and tearing flesh. In this they are more like those of a modern monitor lizard to which the group is distantly related.

Below: Although only the jaws and teeth are known, it is thought that the rest of the body of Primaderma *resembled that of the modern Gila monster.*

OLD LIZARDS

Many of the recognizable modern groups of lizards date back to Mesozoic times.

Above: The modern iguana would not have looked out of place in Cretaceous times.

Iguanas – *Pristiguana* from the late Cretaceous of South America.

Geckos – *Ardeosaurus*, known from perfect remains preserved in the late Jurassic lithographic limestone of Solnhofen, Germany. *Bavarisaurus*, found as the stomach contents of the small dinosaur *Compsognathus*.

Monitors – *Chilingosaurus* from the middle Jurassic of China. *Palaeosaniwa*, a big one from the late Cretaceous Hell Creek Formation in Montana, USA.

Tegus and racerunners – *Adamisaurus* from the late Cretaceous of Asia and similar to *Polyglyphanodon* shown here. *Champs,* the biggest of several found in the Hell Creek formation.

Polyglyphanodon

Most of the main lines of lizards were in existence by the end of the Cretaceous period. However, many of the individual genera showed specializations that are unexpected when we consider the modern versions. The early relatives of the modern skinks had at least one vegetarian member, and its teeth were quite highly modified for its way of life, resembling in some ways the teeth of the modern herbivorous iguana lizard.

Features: *Polyglyphanodon* is a big lizard related to the modern skinks. It has unusual teeth. The rearmost teeth are chisel-shaped, with the blades orientated sideways, and when the mouth is closed the top and bottom rows fit neatly between one another. These teeth are similar to those of the Triassic reptile *Trilophosaurus*. They would have been used for wrenching up tough vegetation and then pulping it before swallowing.

Distribution: Utah, USA; Baja California, Mexico.
Classification: Lepidosauria, Squamata.
Meaning of name: Much ornamented teeth.
Named by: Gilmore, 1940
Time: Maastrichtian stage of the late Cretaceous.
Size: 60cm (2ft).
Lifestyle: Browser.
Species: P. sternbergi, P. bajaensis.

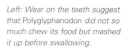

Left: Wear on the teeth suggest that Polyglyphanodon *did not so much chew its food but mashed it up before swallowing.*

SWIMMING LIZARDS

A group of small swimming lizards existed in the Tethys Ocean in late Cretaceous times. Known as the dolichosaurids, they seem to be close to the ancestry of the snakes and of the big swimming lizards, the mosasaurs. They all lived at a time when the shallow shelf seas were particularly extensive and the climate was at its warmest.

Eupodophis

This snake with hind legs is known from a good specimen found in Lebanon and consists of the skull, vertebral column, hip bones and the hind limb. With the hind legs being the only legs present, and these being so small, it is difficult to visualize any land-living activity for this animal. Several hind-legged snakes, including the earlier *Pachyrachis*, are known from this area, suggesting that the shallow seas of the Cretaceous Middle East were where snakes evolved.

Features: The vertebrae and most of the ribs of *Eupodophis* are thickened – a phenomenon known as pachyostosis – adding to the weight of the animal and helped to counteract its natural buoyancy. The tail is tiny, with the limbs positioned very close to the end of the animal. The tail, such as it is, is very primitive and not adapted to swimming. The body is flattened from side to side to aid swimming.The snout is more pointed than that of other contemporary snakes.

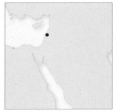

Distribution: Lebanon.
Classification: Lepidosauria, Squamata.
Meaning of name: True footed snake.
Named by: Rage and Escuillié, 2000.
Time: Cenomanian stage of the late Cretaceous.
Size: 85cm (2ft 10in).
Lifestyle: Fish-hunter.
Species: *E. descouensi*.

Right: Eupodophis *and its relatives are sometimes regarded as sea snakes with legs. They swam through the shallow waters with a strong sinuous action.*

Dolichosaurus

Dolichosaurus may have been the Cretaceous equivalent of the Triassic nothosaurs – shallow-water marine hunters. The long neck would have made it easy to hunt into cracks and crevices in rocks and reefs, rather like modern sea snakes. The limbs were really too small to have been much use on land. An occurrence in north-west Germany has the remains of *Dolichosaurus* in a deposit that represents a trough on the sea bed, along with the shells of very large ammonites. The trough was an area of the sea bed that had been scoured out by currents, and the currents had then deposited the animal remains there.

Features: The obvious feature of this swimming lizard is the long neck, produced by an increase in the number of neck vertebrae – 19 to 20 of them – more than double the usual number. The shoulder area and front limbs are reduced, compared with the hip area and hind limbs. This may be significant, as the loss of limbs in snakes started at the front. The jaw joint shows some similarities to that of snakes and of mosasaurs, suggesting a close relationship.

Distribution: Southern England; Germany.
Classification: Lepidosauria, Squamata, Dolichosauridae.
Meaning of name: Thin lizard.
Named by: Owen, 1850.
Time: Cenomanian stage of the late Cretaceous.
Size: 1m (about 3ft).
Lifestyle: Hunter of fish and marine invertebrates.
Species: *D. longicoliis*.

Below: Dolichosaurus *gave its name to the Dolichosauridae family.*

Pontosaurus

This animal is very similar to the others of the dolichosauridae family, but one specimen in particular is unusual – it has traces of the skin preserved. This shows, among other things, that the tail was the main organ of propulsion. It was longer and broader than that of most modern swimming reptiles, and strongly muscled at the base as well as being unusually tall. *Pontosaurus* was closely related to the contemporary mosasaurs, and also to the monitor lizards and the python-like snakes.

Features: The head is covered in scales that are small and polygonal, and form a non-overlapping pavement as do those of a modern monitor lizard. The scales on the neck, body, hind limbs and tail are diamond-shaped and much bigger, arranged in diagonal rows and overlapping like tiles on a roof. These are most like the scales that have been found on contemporary mosasaurs. The skin impression shows that the tail is twice as deep as indicated by the bones.

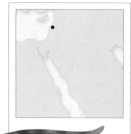

Distribution: Lebanon.
Classification: Lepidosauria, Squamata, Dolichosauridae.
Meaning of name: Sea lizard.
Named by: Kramberger, 1892.
Time: Cennomanian to Toronian stages of the late Cretaceous.
Size: 1m (about 3ft).
Lifestyle: Fish- or invertebrate-hunter.
Species: *P. lesinensis*.

Left: Although it swam with the movements of a sea snake, Pontosaurus had feet with individual toes and claws.

Opetiosaurus

Distribution: Croatia; Germany.
Classification: Lepidosauria, Squamata, Varanoidea.
Meaning of name: Awl toothed lizard.
Named by: Kornbuber, 1901.
Time: Cenomanian stage of the late Cretaceous.
Size: 1.4m (4½ft).
Lifestyle: Fish-hunter.
Species: *O. bucchichi*.

In structure, appearance and lifestyle the aigialosaurs, of which *Opetiosaurus* is a member, represent a transitional form between the terrestrial monitor lizards and the completely aquatic mosasaurs. They evolved in the late Jurassic but it was the late Cretaceous when they became common. A complete skull has been found in Germany. It was an active hunter of fish and other small vertebrates, which it caught in its sharp-toothed jaws.

Features: The head and body are essentially those of a modern monitor lizard, with the same number of vertebrae in the neck. The limbs are a little smaller but do not show much adaptation to a marine way of life. The tail is flattened from side to side, and the tip is turned downwards, suggesting the presence of a fleshy paddle of some sort. In structure it is very close to the land-living monitor lizards, but it is very closely related to the fully aquatic mosasaurs. It must have had a lifestyle that ranged between the two.

Left: Underwater, Opetiosaurus must have swum rather like the modern marine iguana.

MOSASAURS

The ichthyosaurs waned and died out at the beginning of the Cretaceous period. Their place in the Cretaceous period, as the fast-swimming marine predators, was taken by a group of animals called the mosasaurs. These were closely related to the modern monitor lizards, but they had adaptations to marine life that made them the veritable sea serpents of their day.

Clidastes

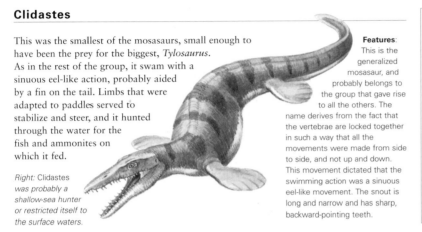

This was the smallest of the mosasaurs, small enough to have been the prey for the biggest, *Tylosaurus*. As in the rest of the group, it swam with a sinuous eel-like action, probably aided by a fin on the tail. Limbs that were adapted to paddles served to stabilize and steer, and it hunted through the water for the fish and ammonites on which it fed.

Right: Clidastes was probably a shallow-sea hunter or restricted itself to the surface waters.

Features: This is the generalized mosasaur, and probably belongs to the group that gave rise to all the others. The name derives from the fact that the vertebrae are locked together in such a way that all the movements were made from side to side, and not up and down. This movement dictated that the swimming action was a sinuous eel-like movement. The snout is long and narrow and has sharp, backward-pointing teeth.

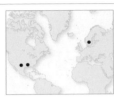

Distribution: USA; Sweden.
Classification: Lepidosauria, Squamata, Varanoidea.
Meaning of name: Locked up.
Named by: Cope, 1868.
Time: Campanian stage of the late Cretaceous.
Size: 4m (13ft).
Lifestyle: Swimming hunter.
Species: *C. propython, C. liodentus, C. moorevillensis.*

Tylosaurus

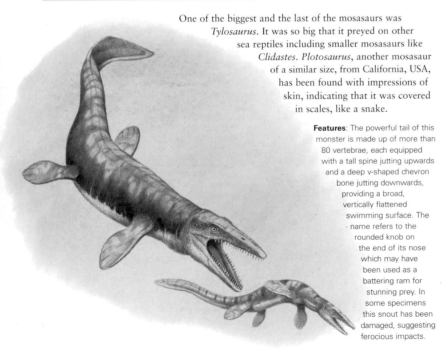

One of the biggest and the last of the mosasaurs was *Tylosaurus*. It was so big that it preyed on other sea reptiles including smaller mosasaurs like *Clidastes*. *Plotosaurus*, another mosasaur of a similar size, from California, USA, has been found with impressions of skin, indicating that it was covered in scales, like a snake.

Features: The powerful tail of this monster is made up of more than 80 vertebrae, each equipped with a tall spine jutting upwards and a deep v-shaped chevron bone jutting downwards, providing a broad, vertically flattened swimming surface. The name refers to the rounded knob on the end of its nose which may have been used as a battering ram for stunning prey. In some specimens this snout has been damaged, suggesting ferocious impacts.

Distribution: Alberta to Texas, USA.
Classification: Lepidosauria, Squamata, Varanoidea.
Meaning of name: Swollen lizard.
Named by: Marsh, 1869.
Time: Maastrichtian stage of the late Cretaceous.
Size: 12m (40ft).
Lifestyle: Swimming hunter.
Species: *T. nepaeolicus, T. proriger, T. kansasensis.*

Left: Tylosaurus was found in Kansas in 1868, and described by Louis Agassiz. Cope described it scientifically in 1869, but it was his great rival Othniel Charles Marsh who named it Tylosaurus.

THE FIRST KNOWN MOSASAUR

In 1780 workers in a sandstone quarry near Maastricht (from which the Maastrichtian stage of the Cretaceous is named) in the Netherlands found a set of fossil jawbones and teeth. They were given to a French army surgeon called Hofmann. Dr Goddin, canon of the local cathedral, successfully sued Hofmann for possession, as the quarry was on his land.

In 1794, in the French Revolutionary wars, the French army bombarded Maastricht but spared the suburb in which the fossil was housed as word had spread to the French authorities about the scientific value of the find. Goddin hid the fossil in a cave, but the occupying French bribed the locals, with 600 bottles of wine, into surrendering it. It was recovered and sent by the army to the Jardin des Plantes in Paris, to be studied by the foremost French anatomist Baron Cuvier, and there it has remained.

The fossil was the jaws of the first mosasaur to have been discovered, and it was soon seen to have been a giant aquatic lizard related to the modern monitors. It was named *Mosasaurus*, meaning "lizard from the Meuse", by British geologist William Conybeare.

Below: The fossil jaw of Mosasaurus.

Platecarpus

We know that mosasaurs ate ammonites because several ammonite shells have been found with tooth marks arranged in the distinctive V-shape of a mosasaur's dentition. The mosasaur seems to have bitten into the shell, turned it and bitten into it again, repeating this several times until the shell collapsed. The soft part would then have been swallowed. No shell fragments have been found in a mosasaur's stomach.

Features: This, the most common mosasaur, has a short body and long tail. Well-preserved jaws show how it swallowed its prey. The teeth are generally straight, but at the back of the roof of the mouth there s a series angled back towards the gullet. A joint in the jaws allows the lower jaws to spread and pull the lower jaw backwards along with the food. These upper back teeth held the food and guided it down the throat.

Distribution: Kansas, USA; also Europe; Africa; and possibly Australia.
Classification: Lepidosauria, Squamata, Varanoidea.
Meaning of name: Flat wrist.
Named by: Cope, 1869.
Time: Turonian to Maastrichtian stage, of the late Cretaceous.
Size: 7.5m (25ft).
Lifestyle: Fish and ammonite hunter.
Species: *P. ictericus, P. tympaniticus, P. bocagei, P. coryphaeus, P. planifrons.*

Above: Mosasaurs are often portrayed with a fin or a crest along the neck and back. This was due to a misidentification of impressions of the throat structure on a specimen of Platecarpus *found in 1899. Williston, who made the discovery, acknowledged his mistake two years later. There is no proof of such a crest.*

Globidens

The strong blunt teeth and powerful jaws show that the mouth was built for crushing. Diet would have consisted of heavy-shelled animals such as turtles or molluscs like ammonites. Although the mosasaurs as a group lived in shallow seas all over the world, they are best known from the chalk deposits laid down in the shallow sea that covered most of North America in late Cretaceous times.

Features: This specialized mosasaur has a short massive head. The short jaws carry bulbous teeth with narrow bases and rounded tops; their upper surfaces are covered in wrinkles. The teeth on the roof of the mouth, typical of other mosasaurs, are absent. Otherwise the skull and backbone show that it is closely related to *Clidastes*. The paddles, like those of other mosasaurs, consist of five toes, showing polyphalangy, (an increase in the number of joints), as in the primitive ichthyosaurs and the plesiosaurs.

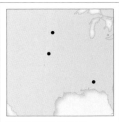

Distribution: Alabama, Kansas, South Dakota, USA.
Classification: Lepidosauria, Squamata, Varanoidea.
Meaning of name: Bulbous teeth.
Named by: Gilmore, 1912.
Time: Campanian to Maastrichtian stage of the late Cretaceous.
Size: 6m (20ft).
Lifestyle: Shellfish-eater.
Species: *G. alabamaensis. G. dakotensis.*

Right: isolated teeth very similar to those of Globidens *have been found in Africa, Europe, the Middle East and South America, suggesting that it may have been quite a wide-ranging genus.*

MORE MOSASAURS

The mosasaurs were the true masters of the late Cretaceous seas. Although their body shape was very conservative they showed a great variation in the head, jaws and teeth indicating specializations into various hunting and eating strategies. Some even advanced into freshwater, hunting in the estuaries and deltas that fringed the shallow seas of the time.

Taniwhasaurus

For some reason New Zealand has always been a good place for mosasaur remains, especially the siltstone deposits at Haumuri Bluff on the northeast coast of South Island. Fossil plesiosaurs and mosasaurs have been extracted there since the 1860s, the work being pioneered by Thomas Cockburn Hood. In late Cretaceous times New Zealand was an extensive area of low-lying land, surrounded by shallow seas that were full of fish and swimming reptiles. Many genera of mosasaur, including *Leidon* and *Moanasaurus* have been excavated there since the later years of the 18th century.

Features: The distinguishing features of *Taniwhasaurus* lie in the arrangement of bones in the skull, which seem to be associated with the structure of the snout, an arrangement that seems to have evolved as the snout lengthened. However, the snout is shorter than in its close relative *Tylosaurus*, and so this genus may have evolved from a long-snouted ancestor. As in *Tylosaurus* there is a strong toothless ramming beak at the front.

Below: The 6m- (20ft-) length quoted here is estimated from the skull of the type specimen. Another skull from the area, referred to as Tylosaurus haumuriensis, *may or may not be of the same species, and this shows an animal reaching 12m (40ft).*

Distribution: New Zealand.
Classification: Lepidosauria, Squamata, Varanoidea.
Meaning of name: Water monster lizard (from Maori mythology).
Named by: Hector, 1874.
Time: Campanian stage of the late Cretaceous.
Size: 6m (20ft).
Lifestyle: Seagoing hunter.
Species: *T. oweni.*

Pluridens

The variety of mosasaurs at the end of the Cretaceous period reflects the range of ichthyosaurs in the preceding Triassic and Jurassic periods. *Pluridens* seems to have had a similar dentition to that of one of the earliest ichthyosaurs *Temnodontosaurus* and like the earlier reptile may also have specialized as an ambush predator, hunting in the shallows seas and lagoons that stretched across West Africa at the time from what is now the Gulf of Guinea to the Mediterranean. Widespread and numerous the mosasaurs may have been, but the heyday of the group was remarkably short-lived. Like so other reptile groups they died out at the end of the Cretaceous.

Features: The "many teeth" of the genus name are on the lower jaw – the only part of the animal to have been discovered so far. There are more than 28 small close-fitting teeth on each side, which is about one and a half times the number found in any other mosasaur. The jaw is long and slender. This combination of features suggests a diet of ammonites and other thin-shelled swimming molluscs.

Distribution: Niger.
Classification: Lepidosauria, Squamata, Varanoidea.
Meaning of name: Many teeth.
Named by: Lingham-Solar, 1998.
Time: Campanian to Maastrichtian stages of the late Cretaceous.
Size: 9m (30ft).
Lifestyle: Mollusc hunter.
Species: *P. walkeri.*

Right: Although only the jaw is known, we can assume that the rest of the body was similar to that of other mosasaurs.

Prognathodon

Distribution: Belgium; South Dakota, Colorado, USA; New Zealand.
Classification: Lepidosauria, Squamata, Varanoidea.
Meaning of name: Forward jaw tooth.
Named by: Dollo, 1889.
Time: Late Cretaceous.
Size: 10m (33ft).
Lifestyle: Fish hunter.
Species: P. solvayi, P. rapax, P. giganeus, P. overtoni, P. stadtmani, P. waiparensis.

The general similarity in the teeth between *Prognathodon* and some of the Triassic placodonts suggest that this had the same lifestyle – feeding on the shellfish of the bed of the sea. The sea in question would have been that which produced the vast deposits of chalk across northern Europe and the midwest of North America at the end of the Cretaceous period. A very large specimen, with a length of 12m (40ft) uncovered in Israel was thought to have been a species of *Prognathodon*. It is now regarded as a separate genus *Oronosaurus*.

Features: Jutting teeth at the front of the upper jaw, the short heavy jaws, the broad blunt snout, the broad powerful conical teeth in the jaws and the large crushing teeth on the palate indicate that this is an animal adapted to picking up and crushing shellfish. In this it would have had a similar diet to *Globidens*, but the latter lacks the well developed teeth on the palate.

Right: The chalk sea would have provided ample shellfish on which specialized reptiles such as Prognathodon *could feed.*

Goronyosaurus

Distribution: Nigeria.
Classification: Lepidosauria, Squamata, Varanoidea.
Meaning of name: Lizard from Goronyo district in Nigeria.
Named by: Assaroli, de Guili, Ficcarelli and Torre, 1972.
Time: Maastrichitan stage of the late Cretaceous.
Size: 7m (24ft).
Lifestyle: Ambush hunter.
Species: G. nigeriensis.

The unique adaptations of the head and jaws of *Goronyosaurus* suggest that it lurked in the obscure waters of muddy estuaries, hunting by smell like modern marine crocodiles. It would have ambushed fast-moving slippery prey, possibly the young of other brackish-water mosasaurs. This is so different from the inferred lifestyle of the other mosasaurs that it indicates the range of types of a single group of successful animals, all specialized for particular lifestyles in particular habitats – even the freshwater areas found in the estuaries where this creature lived.

Features: The very long jaws of *Goronyosaurus* are unlike those of any other mosasaur, having only about ten teeth in each jaw. These teeth are long and pointed, and the tips fit into pits in the opposite jaws when the mouth is closed, holding the closed jaws tightly together. The muscle placement shows that this animal attacked with a snapping bite, like modern crocodiles, and the small eyes and increased sense of smell indicate a dark water habitat.

Left: When Goronyosaurus *was found in 1930 it was regarded as a species of* Mosasaurus.

A VARIED DIET

Above: The pointed jaws of a mosasaur skull.

Mosasaurs were adapted to take all kinds of food in their marine habitat.

Carinodens – a small mosasaur from Belgium with teeth adapted to crush the carapaces of small crustaceans.

Dollosaurus – a little-known type from southern European Russia that was probably a shellfish-eater like *Prognathodon*.

Ectenosaurus – a medium-sized North American mosasaur with a long muzzle that may have been a fish-eater.

Hainosaurus – a Belgian form with a long body and short tail, and with turtle parts found in the stomach contents.

Igdammanosaurus – from Niger, this mosasaur had shell-crushing dentition, not as strong as that of *Globidens*, and so it probably fed on thin-shelled cephalopods rather than tough bivalves.

Plioplatecarpus – a freshwater type from North America that may have hidden in reeds and ambushed soft-bodied cephalopods.

Selmasaurus – from Alabama, USA, with a generalized dentition. It probably fed on anything that moved.

CRETACEOUS PLESIOSAURS

The late Cretaceous seas had as much diversity of plesiosaurs as those of the Jurassic. It seems likely that the different types of plesiosaurs had different hunting areas with, for example, the big whale-like pliosauroids patrolling the open ocean after big prey, and the small-headed, long-necked plesiosauroids preferring the shelf seas and hunting fish and smaller animals.

Hydrotherosaurus

Feeding on fish *Hydrotherosaurus* lived in the shallow waters near the coastlines of continents. We know its diet from the stomach contents of a complete articulated skeleton found in California. The stomach also contained stomach stones, which the animal would have swallowed to help adjust its buoyancy – a technique often used by animals that live and hunt close to the sea floor.

Features: The difference between *Hydrotherosaurus* and the other late-Cretaceous, long-necked plesiosaurs, such as *Elasmosaurus*, lies in the 60 vertebrae in the neck. They are much lower and narrower at the front end, but high towards the shoulders. Early restorations show it with a narrow streamlined body, but this is due to the angle at which the ribs of a complete skeleton were found. It is now thought to have had a squat turtle-like body, with broad shoulder bones and hips like other plesiosauroids.

Distribution: California, USA.
Classification: Plesiosauria.
Meaning of name: Fisherman lizard.
Named by: Welles, 1943.
Time: Maastrichtian stage of the late Cretaceous.
Size: 13m (42ft).
Lifestyle: Fish-hunter.

Species: *H. alexandrae.*

Above: The species was named after the famous fossil collector Annie Montague Alexander (1867–1950), who contributed over 20,000 specimens to the Museum of Palaeontology at the University of California, USA, and was even in the field on her 80th birthday.

Elasmosaurus

The neck of *Elasmosaurus* is so long that it was mistaken for a tail by Edward Drinker Cope when he first investigated it. The species name *E. platyurus* means "big flattened tail" and actually referred to the neck. When his rival Othniel Charles Marsh pointed out the mistake the bad feeling was so intense that it sparked off the so-called "bone wars" of the late nineteenth century in which the two tried to outdo each other in the number of fossil vertebrates discovered.

Features: The enormous neck of *Elasmosaurus* has the greatest number of neck vertebrae in any known animal, 72 in all. The undersides of the shoulder girdle and the hip bones are expanded into broad plate-like structures that anchor the powerful muscles driving the paddles. The head is tiny, but has a very wide gape and carries sharp spike-like teeth. Judging by the stomach contents it was capable of catching the fastest of the fish of the time. As with the other plesiosauroids, the long neck of *Elasmosaurus* would have been used to snatch at fast-moving fish unawares, without having to move the great body too quickly.

Distribution: Kansas, USA.
Classification: Plesiosauria.
Meaning of name: Metal plate lizard (after the flat bones in its pelvis and shoulder girdle).
Named by: Cope, 1868.
Time: Maastrichtian stage of the late Cretaceous.
Size: 12m (45ft).
Lifestyle: Fish-hunter.
Species: *E. platyurus, E. morgani, E. serpentines, E. snowii.*

Above: There is a large number of species of Elasmosaurus. However, many scientists, notably Ken Carpenter of Denver, believe many of them to be separate genera rather than just species.

CRETACEOUS PLESIOSAUR CLASSIFICATION

The Cretaceous seas abounded in plesiosaurs of many different genera, which could be divided into long-necked, small-headed types (plesiosauroids) and short-necked, big-headed types (pliosauroids). It is tempting to think that the two groups were so successful that they survived uninterrupted through both the Jurassic and Cretaceous. However, recent investigations show a different story.

The long-necked plesiosauroids of the Cretaceous have different anatomy from those of the Jurassic – as if the two groups are very distantly related. The Cretaceous long-necked forms seem to be more closely related to the short-necked forms of the Jurassic. The differences lie in the arrangement of bones in the roof of the mouth and the back of the skull, and the way the skull is articulated with the neck.

The implication is that the original long-necked plesiosaurs died out completely at the end of the Jurassic, but then a whole new series of them evolved from the short-necked line and expanded to fill the niches left unoccupied by that extinction. The similarity of the long necks is a matter of convergent evolution rather than of family relationship. As a result the whole classification of Cretaceous plesiosaurs needs to be revised.

Right: A short-necked Jurassic plesiosaur.

Dolichorhynchops

The arrangement of teeth and jaw muscles indicated that *Dolichorhynchops* could make very quick bites at its prey, but that these bites would not have been particularly strong. It seems likely that its prey consisted of the soft-bodied squid that spread throughout the seas at this time. It is known from several complete adult skeletons and partial skeletons of youngsters.

Below: The neck of Dolichorhynchops *was short but a little longer than the head.*

Distribution: Kansas, USA.
Classification: Plesiosauroia, Pliosauridae.
Meaning of name: Long-snouted face.
Named by: Williston, 1902.
Time: Campanian stage of the late Cretaceous.
Size: 5m (17ft).
Lifestyle: Fish-hunter.
Species: *D. osborni.*

Features: The body is short and streamlined, like that of Jurassic *Peloneustes*, but the skull is long and narrow with big eye sockets, like that of an ichthyosaur. The head is much lighter than in the larger pliosauroid genera, and the teeth are small, tightly packed and all of the same size.

Kronosaurus

Although *Kronosaurus* is often illustrated, and its dimensions are often quoted with confidence, its true nature is more problematic. The name was originally applied to skull bones from Queensland, Australia. A mounted skeleton in Harvard University Museum, USA, on which most restorations are based, actually comes from a different horizon and has been inaccurately assembled.

Features: *Kronosaurus* has the biggest skull of any marine reptile known. It is flat-topped and ends in pointed jaws, and is about 2.7m (9ft) long, accounting for about a third of the entire length of the animal. There are probably about 20 vertebrae in the neck, unlike the 30 or so originally attributed to it, hence the revision of the animal's length. It may have lived rather like a sperm whale.

Below: The first specimen of Kronosaurus *found was a piece of jaw with six teeth, found near Hughenden in Queensland, Australia, in 1901. It was thought to have been an ichthyosaur until more complete skull bones were found in 1924. The famous skull was not found until 1931.*

Distribution: Queensland, Australia; and Boyaca, Colombia.
Classification: Plesiosauroia, Pliosauridae.
Meaning of name: Kronos' lizard (after the Greek titan).
Named by: Longman, 1924.
Time: Albian stage of the early Cretaceous (but included here because of its importance).
Size: 9m (30ft), as opposed to the 13m (45ft) often quoted.
Lifestyle: Ocean hunter.
Species: *K. queenslandicus, K. boyacensis.*

THE LAST OF THE PLESIOSAURS

The late Cretaceous seas were dominated by the mosasaurs – the big swimming lizards. However, the plesiosaurs, which had been the most important hunters for most of the Mesozoic since they evolved in Triassic times, were still around. They were divided into two main shapes – those with the small heads and long necks, and those with the long heads and short necks.

Aristonectes

A plesiosaur was found in Argentina in 1941 and named *Aristonectes parvidens*. Another was found in Antarctica in 1989 and named *Morturneria seymourensis*. Work done by Zulma Gasparini from La Plata, Argentina, and colleagues, in 2003 showed that they were the same species at different growth stages. As is customary, the first name given is regarded as the valid one. It was a filter feeder, living rather like the modern crabeater seal, that has similarly fine teeth for sieving krill.

Left: Aristonectes filtered fine crustaceans from the shallow sea waters that surrounded the archipelago formed by South America and the Antarctic Peninsula in the late Cretaceous.

Features: From above, the skull is broad and forms a half-oval shape. From the side, it is low and pointed. The teeth are many, outwardly pointing and needle-like – hence the *parvidens* (meaning small teeth) of the species name – and all the same shape and size. They would have interlocked to form a kind of a sieve. When it was first found, *Aristonectes* was thought to have been the last surviving cryptoclidid (one of the earliest plesiosaur groups), but it is now regarded as having been an elasmosaurid (one of the most advanced groups) and so the cryptoclidid family is now definitely restricted to the Jurassic period.

Distribution: Argentina; Chile; Seymour Island, Antarctica.
Classification: Plesiosauria.
Meaning of name: Best swimmer.
Named by: Cabrera, 1941.
Time: Maastrichtian stage of the late Cretaceous.
Size: 8m (26ft).
Lifestyle: Filter feeder.
Species: *A. parvidens*.

Terminonatator

So far this is the latest plesiosaur of the elasmosaurid family to be found in the Cretaceous deposits of North America, and because of the relative completeness of the skeleton, the one that yields the most information on the bone structure of the whole group. It lived in the shallow seas that were spreading across the Mid-West of North America at the end of the Cretaceous period. It is the smallest adult elamosaurid known. It is evidently an adult as some of the bones are distorted by age, and a healed fracture of the limb shows that the specimen found had an eventful life.

Features: The teeth are large, curved and pointed, striated on the inside edge but smooth on the outside, and becoming smaller towards the back. They interlock to form a fish trap. The details of the skull construction are such as to show that it is a different genus from any other elasmosaurid known. The arrangement of the nasal passages is different from that of other plesiosaurs, in that they run backwards from the external nostrils, not downwards and forwards as in other genera. This may have some bearing on how the animal sensed its environment in life.

Below: Terminonatator *was a fish-hunter in the shallow sea that covered central North America in the late Cretaceous period.*

Distribution: Saskatchewan.
Classification: Plesiosauria.
Meaning of name: Last swimmer.
Named by: Sato, 2003.
Time: Campanian stage of the late Cretaceous.
Size: 7m (23ft).
Lifestyle: Fish-hunter.
Species: *P. ponteixensis*.

Trinacromerum

Distribution: Kansas, Utah, South Dakota, USA.
Classification: Plesiosauria, Pliosauroidea.
Meaning of name: Three-tipped femur.
Named by: Cragin, 1888.
Time: Turonian stage of the late Cretaceous.
Size: 5m (16ft).
Lifestyle: Fish-hunter.
Species: *T. bentonianum*, *T. kirki*.

The skeletons of this plesiosaur occur in the same rocks as those of the earliest mosasaurs. It is possible that they competed for the same food. A skeleton, tentatively identified as *Trinacromerum*, has been found with about 100 stomach stones, presumably swallowed to help in adjusting its buoyancy or to help in digestion. *Trinacromerum*, being one of the long-headed short-necked plesiosaurs, was probably a much more active hunter than the long-necked types, actually swimming after prey rather than just reaching for it.

Right: It has been estimated that Trinacromerum *could swim at speeds of about 30 knots.*

Features: The skull of short-necked *Trinacromerum* is long with elongated narrow jaws containing more than 100 piercing teeth. The front paddle is long and pointed, and the whole animal is evolved for fast swimming and manoeuvring. In appearance it is similar to *Dolichorhynchops* (in fact it has at one time been regarded as a species of this) but it appeared much earlier. Like *Dolichorhynchops* it was an active hunter, chasing after fish like modern penguins or sea lions.

Brachauchenius

PLESIOSAUR STOMACH CONTENTS

In 2001, David Cicimurri, from Clemson University in South Carolina, USA, and Michael Everhart, from the Sternberg Museum in Kansas, USA, published the results of a study of a plesiosaur skeleton with the stomach contents preserved. The animal was a late Cretaceous elasmosaurid called *Styxosaurus*.

Its diet consisted of fish bones that were jumbled together with stomach stones in a lump located just behind the shoulder girdle – where the stomach would have been. The bone fragments were of a small sardine-like fish called *Enchodus*, which must have been the main prey items of this plesiosaur. The fact that stones and bones were mixed up suggests that the stones were swallowed to help digestion rather than for buoyancy. They consisted of rocks of a type that would only have outcropped 600km (373 miles) away at that time, indicating the huge range of travel of this animal.

There were also shark teeth found with the fossil, but these were not thought to have been part of the stomach contents. It seems more likely that they were dropped by sharks that were scavenging the dead body.

This big animal is known from three skulls which are almost complete and even have the lower jaws. There are also two partial skeletons. The first remains were found by Williston in 1903. The biggest and best was found in 1951 by Robert and Frank Jennrich while they were looking for sharks' teeth. It was extracted by G. F. Sternberg.

Features: *Brachauchenius* is one of the last of the big-headed plesiosaurs. It is a much heavier animal than the other plesiosaurs of the time and place, because much of its length is made up of body rather than neck. Its strong conical teeth show that it was an active hunter, probably feeding on other large vertebrates of the inland sea. Despite its short neck, the structure of the skull is similar to that of the long-necked elasmosaurids.

Distribution: Kansas, USA.
Classification: Plesiosauria.
Meaning of name: Short neck.
Named by: Williston, 1903.
Time: Turonian stage of the late Cretaceous.
Size: 11m (36ft).
Lifestyle: Swimming hunter.
Species: *B. lucasi*.

Above: Brachaucherius *was probably related to the huge* Liopleurodon *of the Jurassic.*

LATE CRETACEOUS CROCODILES

Wet lands were extensive in late Cretaceous times. There were many areas of the globe that were occupied by river deltas, estuaries, freshwater lakes and swamps. These were home to all kinds of animals, adapted to an amphibious environment. Chief among these were the crocodiles, many of which would look very familiar to us today.

Deinosuchus

Deinosuchus and the earlier *Sarcosuchus* represent the biggest crocodiles known. Studies show that *Deinosuchus* grew at a rate of about 30cm (1ft) per year, about the same as modern crocodiles, but the prehistoric ones grew for a longer period. *Deinosuchus* prowled the brackish waters of the North American river mouths, ambushing dinosaurs and eating turtles. Until the discovery of the earlier *Sarcosuchus* of Africa, which must have had a similar lifestyle in a similar environment, *Deinosuchus* was regarded as the biggest crocodile that ever lived. Other gigantic Cretaceous crocodiles included *Purussaurus* and *Rhamphosuchus*.

Features: The skull, although gigantic, is similar in proportion to that of a modern crocodile, suggesting a similar lifestyle and feeding habits. Despite its size *Deinosuchus* is very similar to the modern crocodile or alligator, and seems to be close to the ancestry of the two groups. Only skulls are known, and these are about 2m (6½ft) long. Estimates of the length of the whole animal are based on this, and vary considerably.

Below: Deinosuchus *is also known as* Phobosuchus *meaning "horror crocodile".*

Distribution: Texas, North Carolina, Montana, Alabama, Georgia, New Jersey, USA.
Classification: Crocodylomorpha, Alligatoroidea.
Meaning of name: Terrible crocodile.
Named by: Colbert and Bird, 1958.
Time: Campanian stage of the late Cretaceous.
Size: 10m (33ft).
Lifestyle: Hunter or scavenger.
Species: *D. riograndensis, D. rugosus, D. hatchery.*

Stomatosuchus

This massive animal seems to have been a filter feeder. As the biggest animals of today – the baleen whales – thrive on the smallest creatures of their environment, so this enormous crocodile appears to have fed on the tiniest animals of the swamps and lagoons that covered North Africa in Cretaceous times. Unfortunately the only known specimen of *Stomatosuchus* was lost when an allied bombing raid destroyed the Munich Museum in 1944 – the raid that also robbed science of the original remains of the dinosaurs *Spinosaurus* and *Carcharodontosaurus*. As a result it is impossible to subject the fossils to modern scientific analysis to see if this lifestyle was in fact the correct one.

Features: *Stomatosuchus* has a skull that is almost 2m (6½ ft) long, with a long flattened snout. The upper teeth are fine and compact but there appear to be no teeth on the lower jaw. There seems to have been a pelican-like pouch beneath the jaws. Large volumes of water would have been gulped into this pouch and expelled through the fine teeth, extracting the tiny crustaceans on which *Stomatosuchus* fed.

Below: There appear to be very few filter-feeding reptiles in the fossil record, apart from some of the pterosaurs. Stomatosuchus *is a rare exception.*

Distribution: Egypt.
Classification: Crocodylomorpha, Stomatosuchidae.
Meaning of name: Porous crocodile.
Named by: Stromer, 1925.
Time: Cenomanian stage of the late Cretaceous.
Size: 10m (33ft).
Lifestyle: Filter feeder.
Species: *S. inermis.*

TEETH

Crocodiles are often compared to their cousins the theropod dinosaurs – both big reptilian hunters. Indeed many of their adaptations are similar, especially the jaws and teeth. However, it is usually easy to distinguish a fossilized crocodile tooth from a fossilized theropod tooth. In general a theropod tooth is curved and finely serrated like a steak knife, while a crocodile tooth is straight and straight-edged.

However, there was a group of crocodiles that had serrated teeth. Often given the name ziphodont, meaning sword toothed, they existed from Cretaceous to mid-Tertiary times. They have given rise to a deal of confusion. The discovery of their theropod-like teeth in post-Cretaceous deposits has often been wrongly cited as proof that the dinosaurs, particularly the theropod dinosaurs, survived the end-Cretaceous extinction and existed well into the Tertiary.

Brachychampsa

With so many freshwater turtles inhabiting the shallow swamp waters and rivers of late Cretaceous North America, it is hardly surprising that hunting animals evolved to prey on them. The short jaws and broad crushing teeth of *Brachychampsa* and other contemporary crocodiles like *Stangerochampsa*, *Allognathosuchus* and *Procaimanoidea* may be an adaptation to this diet.

Features: The skull of *Brachychampsa* is short and broad, with a rounded snout. The teeth, especially at the back, are broad and bulbous. This suggests that it ate things that were big and hard, like turtles. However, it is likely that turtles did not make up its entire diet, but, like most others of the group, such as the modern alligator, it was an opportunistic feeder, eating anything that came its way.

Right: Brachychampsa *inhabited the swamps and streams of the dinosaur-rich plains of North America at the very end of dinosaur times.*

Distribution: Montana, New Mexico, USA.
Classification: Crocodylomorpha, Mesoeucrocodylia.
Meaning of name: Short alligator.
Named by: Gilmore, 1911.
Time: Maastrichtian stage of the late Cretaceous.
Size: 2m (6½ft).
Lifestyle: Turtle-eater or opportunistic feeder.
Species: *B. montana*, *B. sealeyi*.

Dyrosaurus

Distribution: Montana, USA; Morocco.
Classification: Crocodylomorpha, Dyrosauridae.
Meaning of name: Evil lizard.
Named by: Thomas, 1893.
Time: Maastrichtian stage of the late Cretaceous.
Size: 4m (13ft).
Lifestyle: Fish-hunter.
Species: *D. phosphaticus*.

The dyrosaurids represent a family of crocodiles existing from the late Cretaceous to the early Tertiary, which were outwardly similar to the modern gavials with their long, narrow, toothy snouts. They were strong swimmers and hunted fish in the shallow coastal waters or brackish river mouths.

Features: The long narrow jaws of *Dryosaurus* are full of sharp teeth and form an ideal fish trap. The tail is a broad swimming organ and the limbs are quite long, indicating that it spent as much time on land as in the water. The typical crocodile armour is present but it is reduced compared with others of the crocodile group. The long skull is quite lightweight, with large gaps between the bones.

Below: Dyrosaurus *was the terror of the coastal and estuarine fish of late Cretaceous times.*

TERRESTRIAL AND VEGETARIAN CROCODILES

Although most of the crocodiles were developing into recognizable modern shapes, there were still many very specialized forms. There was evidently still a variety of ecological niches into which the crocodile family could expand, in order to exploit food sources that were not being taken by others.

Libycosuchus

This was the animal that took the place of the hyena in North Africa in Cretaceous times. The big meat-eating dinosaurs of the time, such as *Spinosaurus* or *Carcharodontosaurus*, would inevitably have left large parts of their prey uneaten, and these would have been an adequate food source to support a wide range of scavenging animals. *Libycosuchus* was able to break the bones of the corpses with its strong teeth and jaws.

Features: The skull is very short and hyena-like, with very powerful jaws and strong conical teeth. The whole head is adapted to produce an extremely strong and fast bite. As with the other terrestrial crocodiles the legs are quite long and held in an upright stance. Its small size would have made it quite nimble, necessary for avoiding all the other animals that would be attracted to a large carcass.

Distribution: Niger; Morocco; Egypt.
Classification: Crocodylomorpha, Libycosuchidae.
Meaning of name: Crocodile from Libya.
Named by: Stromer, 1914.
Time: Barremian to Santonian stage of the late Cretaceous
Size: 1m (about 3ft).
Lifestyle: Scavenger.
Species: *L. brevirostris*.

Right: The strong jaws and thick teeth of Libycosuchus *would have been ideal for crushing bones.*

Simosuchus

Vegetarian crocodiles are unusual but not unknown in the fossil record. *Simosuchus* probably adapted to a diet of plants in the face of strong competition among all the crocodiles of the area, going for a food source that was not available to the others. A similar crocodile, *Uruguayasuchus* is known from South America, suggesting a land link between this continent and Madagascar at that time.

Features: The short snout of *Simosuchus* is its notable feature. Its multi-cusped teeth are similar to those of vegetarian dinosaurs such as the ankylosaurs or stegosaurs. The position of its eyes and nostrils shows that it spent all of its time on land rather than in the water. The shovel-shaped snout and the muscular neck suggest that *Simosuchus* may have dug for roots and buried plant material. When it was discovered it was only the second vegetarian crocodile known.

Distribution: Madagascar.
Classification: Crocodylomorpha, Mesoeucrocodylia.
Meaning of name: Pug-nosed crocodile.
Named by: Buckley, 2000.
Time: Barremian stage of the early Cretaceous.
Size: 1.2m (4ft).
Lifestyle: Browser.
Species: *S. clarki*.

Right: Simosuchus *did not have the jaw articulation to produce the strong bite typical of carnivorous crocodiles.*

Baurusuchus

Distribution: Brazil.
Classification:
Crocodylomorpha,
Baurusuchidae.
Meaning of name:
Crocodile from Bauru.
Named by: Prince,
1945.
Time: Turonian to
Santonian stages
of the late
Cretaceous.
Size: 3.5m (12ft).
Lifestyle: Hunter.
Species: *B. pachecoi,
B. salgadoensis.*

Approximately 11 skeletons of a new species of this terrestrial animal were presented to the public in 2005. It was very similar to another terrestrial crocodile, *Pabwehshi*, from Pakistan, suggesting that South America was linked to the Indian sub-continent in Cretaceous times. At that time the Indian subcontinent was part of the southern supercontinent of Gondwana.

Features: The skull is compact and built to absorb the strong forces of biting. The teeth are pointed and conical, some with serrations, and were evidently used for penetrating and ripping up flesh. The nostrils are at the front of the skull, rather than on top, indicating that it was not a water-living animal. This was a fully terrestrial crocodile living and hunting in a particularly hot and arid area of the South American continent.

Left: Baurusuchus *must have resembled the crocodile-like rauisuchians that were the main land predators in previous Triassic times.*

CROCODILE WALK

When comparing the locomotion of reptiles with that of dinosaurs we often state that the former have a sprawling stance while the latter have an upright stance. The former have the weight of the body slung between the legs that stick out to the side, while the latter have the weight at the top of vertical legs that form columns, like those of mammals. When the comparison is made, the stance of crocodiles is often shown as "semi-upright" – a halfway house between the two. The stance is better described as "varied" since modern crocodiles can choose to walk either sprawling or semi-upright depending on whether it is walking slowly or running.

The assumption has always been that the semi-upright stance evolved directly from the sprawling stance of more primitive reptiles, as if on its way to becoming fully upright, as in the dinosaur relatives.

Recent researches, by Stephen Reilly and Jason Elias at Ohio University, USA, show that this is not the case. The knee and ankle joints of modern crocodiles could not have evolved directly from the joints used in a sprawling stance. What's more, when a crocodile runs it straightens its legs as if an upright stance were natural to it. The inference is that ancestral crocodiles had an upright stance, and that the semi-upright stance of all living and most fossil crocodiles evolved from this.

The presence of many fossil crocodiles with upright stances, including many of those featured here, seems to confirm this.

Sphagesaurus

When this genus was first named the description was based on only two teeth found in the late 1940s. It was not until a complete skull was found and published by Diego Pol, in 2003, that it was realized just what a strange crocodile this was, with teeth of different shapes spread out all along its short jaws.

Features: The skull has a short high snout, narrowing towards the tip. There are two pairs of canine-like teeth. The back teeth are grinding teeth. The chewing mechanism could be worked either by an up and down action or by a fore and aft movement of the jaws. There is a palate in the roof of the mouth, separating the air passage from the feeding area, allowing the animal to breathe and chew at the same time.

Below: The skull and the teeth are the only things known of Sphagesaurus, *but it is assumed that the rest of the body is similar to that of the other terrestrial crocodiles of the time.*

Distribution: Southern Brazil.
Classification:
Crocodylomorpha,
Sebecosuchia.
Meaning of name: Bog
lizard.
Named by: Price, 1950.
Time: Cenomanian to
Campanian stages of the late
Cretaceous.
Size: 1m (about 3ft).
Lifestyle: Hunter.
Species: *S. huenei.*

GIANT PTEROSAURS

The last of the pterosaurs were veritable monsters, some with wingspans greater than those of hang-gliders or small aircraft. The biggest modern flying birds, such as the albatross and the Andean condor, would have been puny beside the biggest of the pterosaurs that ruled the skies at the end of the Age of Dinosaurs. Soon the birds would take over the skies.

Pteranodon

The first specimens to be found were in the chalk of Kansas, USA, by O. C. Marsh, in 1870. The specimens were wing-bone fragments and identified as a species of *Pterodactylus*. In 1876 the first skulls were found and the name was changed to *Pteranodon* because of the lack of teeth. Until the 1970s *Pteranodon* was regarded as the biggest animal that was capable of flight.

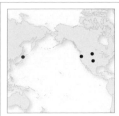

Features: *Pteranodon* has a long head with deep toothless jaws and a long crest that extends backwards, giving the skull a kind of a hammerhead appearance. In the largest species, *P. sternbergi*, the crest projects upwards giving a totally different profile. The head is much bigger than the body. The bones are hollow to keep down the weight, and the dorsal vertebrae are fused together with the ribs to form a solid support for the flight muscles. It probably spent more time soaring than in flapping flight.

Distribution: South Dakota, Kansas, Oregon, USA; and Japan.
Classification: Pterosauria, Pterodactyloidea.
Meaning of name: Wing without teeth.
Named by: Marsh 1876.
Time: Santonian to Campanian stages of the late Cretaceous.
Size: 9m (30ft) wingspan.
Lifestyle: Fish-hunter.
Species: *P. longiceps, P. ingens, P. eatoni, P. marshi, P. walkeri, P. oregonensis, P. sternbergi.*

Nyctosaurus

The size of the crest of *Nyctosaurus* is comparable with the area of the wing. Some palaeontologists argue that in life this may have carried a sail of skin, like that of *Tapejara*, used for display or for aerodynamics. There is no direct evidence that such a sail existed, and many palaeontologists think that a sail is impossible for mechanical reasons.

Right: The restoration given here is rendered without the suggested membrane on the head crest.

Right: Tapejara for comparison.

Features: Since its discovery in the 1870s, *Nyctosaurus* has always been regarded as a small version of *Pteranodon*, but without the crest. It also differs from *Pteranodon* by the shape of the upper arm bone and shoulder joint. Then in 2003 two new skulls were found in the Smoky Hill Chalk at western Kansas, USA, and these bear the most remarkable crests, made up of fine struts of bone like a single deer antler, about three times the length of the skulls themselves.

Distribution: Kansas, USA; and Brazil.
Classification: Pterosauria, Pterodactyloidea.
Meaning of name: Night lizard.
Named by: Marsh, 1876.
Time: Santonian to Maastrichtian stages of the late Cretaceous.
Size: 2.9m (9½ft) wingspan.
Lifestyle: Fish-hunter.
Species: *N. lamegoi, N. gracilis.*

FOOTPRINTS

Pterosaurs were flying reptiles: we have known that for more than 150 years. However it is only in the last few decades that we have determined that they flew by active flapping as birds do, rather than by gliding like flying squirrels. What they did while on the ground is a mystery.

There were two schools of thought. One was that they sprawled on all fours, rather like frogs. The other was that they were fully bipedal like birds. In the 1980s certain fossil trackways became recognized for what they were – marks made by pterosaurs while they were on the ground. They showed four-toed footprints made by the hind feet as quite a wide trackway. The hands were represented by three-fingered marks, with one of the fingers sweeping away to the side, positioned quite close to the hind prints. The analysis is that the animal had been walking bow-legged on its hind feet, with its body more or less upright, supported by the wings in an action that resembled someone walking on crutches.

A more detailed trackway found in France, in 2004, showed that a pterosaur coming in to land slowed its flight to stalling speed close to the ground, and let down its feet gently. It then dragged its toes for a moment, made a short hop and put down its front legs to walk away on all fours, a very precise landing manoeuvre.

Below: Pterosaur tracks.

Zhejiangopterus

Four articulated specimens of this genus are known. It was regarded as a relative of *Pteranodon* and *Nyctosaurus* when it was discovered, but is now thought to be closer to *Quetzalcoatlus*, largely because of the very long neck vertebrae. Complete pterosaur skulls like these are rare, due to the fact that they are made of very porous lightweight bones with the consistency of expanded polystyrene.

Features: As in other members of the group, *Zhejiangopterus* has short wings, long legs (about half as long again as the arm) and a very big head that is extremely narrow. The only crest is a long one beneath the lower jaw, which may have had a structural significance in supporting the very thin bones of the skull. The jaws are toothless and the eye socket is tiny. It has a very long neck made up of elongated vertebrae.

Left: The reason that the skulls of these big-headed pterosaurs rarely fossilize is that they were made of very delicate material, which disintegrated rapidly after death.

Distribution: Zhejiang Province, China.
Classification: Pterosauria, Pterodactyloidea.
Meaning of name: Wing from Zhejiang Province.
Named by: Cai and Feng, 1994.
Time: Santonian stage of the late Cretaceous.
Size: 5m (16ft) wingspan.
Lifestyle: Fish hunter.
Species: *Z. linhaiensis*.

Quetzalcoatlus

The discovery of *Quetzalcoatlus* in the 1970s gave the world what was thought to have been the biggest flying animal that could possibly have existed. It may have lived by fishing in inland waters or by scavenging from the corpses of dead dinosaurs spotted while soaring on rising air over the arid landscape. Now, however, there is evidence for even bigger genera of pterosaurs emerging.

Below: Despite its great size the skeleton was lightly built and the whole animal may have weighed no more than 100kg (220lbs).

Features: *Quetzalcoatlus* has a very big head with a bony crest along the back portion. Like *Pteranodon* and the other big pterosaurs it is toothless. The weight of the skull is kept to a minimum by the fusion of the nostril and the gap that usually exists between the nostril and the eye socket. Since its discovery the wingspan has been revised from 15m (52ft) to 11m (37ft), still very respectable.

Distribution: Texas, USA.
Classification: Pterosauria, Pterodactyloidea.
Meaning of name: From Quetzalcoatl, the plumed serpent of Aztec mythology.
Named by: Lawson, 1975.

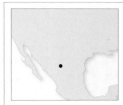

Time: Maastrichtian stage of the late Cretaceous.
Size: 11m (37ft) wingspan.
Lifestyle: Fish- or carrion-eater.
Species: *Q. northropi*. One other unnamed.

BASIC ABELISAURIDS

The abelisaurids were a distinctive family of theropods from the late Cretaceous period. They evolved from the same line as the ceratosaurids, quite distinct from the tetanurans. The abelisaurids were originally thought to have been restricted to South America, but they now seem to have been much more widely distributed across the southern continents.

Masiakasaurus

Distribution: Madagascar.
Classification: Theropoda, Neoceratosauria, Abelisauria.
Meaning of name: Vicious lizard.
Named by: Sampson Carrano and Forster, 2001.
Time: Maastrichtian stage of the late Cretaceous.
Size: 1.8m (6ft).
Lifestyle: Fisher.
Species: *M. knopfleri*.

This unusual abelisaurid is known from the single, incomplete and disarticulated skeleton, including parts of the jaws, the hind limbs and some of the vertebrae. The lower jaw shows a strange arrangement of teeth, with the front teeth pointing forward and hooked upwards. Those at the rear of the jaw are standard.

Features: The strange forward-pointing teeth on the lower jaw indicate that this was an unusual dinosaur. They are similar to the teeth of some pterosaurs that we know were fishing animals, and by analogy *Masiakasaurus* was probably a fishing animal too. Its long neck also suggests this animal hunted fish. Unfortunately, we do not have the front of the upper jaw, and so it is unclear how the mouth actually worked.

Right: This particular species, M. knopfleri, is named after Mark Knopfler, whose guitar music the team were listening to when they made the discovery.

Noasaurus

The killing claw on the hind foot of *Noasaurus* is a textbook example of convergent evolution. Superficially it is similar to the killing claw of the deinonychosaurs, but the two animals are only distantly related. Such a claw evolved independently in response to similar needs for similar hunting styles. Possibly they hunted large animals and inflicted deep bleeding wounds, rather than going for a quick kill.

Distribution: Argentina.
Classification: Theropoda, Neoceratosauria, Abelisauria.

Features: This is a small, active, hunting abelisaurid dinosaur with a deep head, and with a huge killing claw on the second toe of the foot. The claw differs from that of the unrelated deinonychosaurs by the way the muscles are attached, and by being more sharply curved and more manoeuvrable. Recent research suggests that the big claw could not be used for slashing.

Right: More recent research suggests that this restoration is inaccurate, and that the killing claw belongs not on the foot but on the hand. This would result in a far more conventional-looking animal.

Meaning of name: Northwest Argentine lizard.
Named by: Bonaparte and J. E. Powell, 1980.
Time: Maastrichtian stage of the late Cretaceous.
Size: 3m (10ft).
Lifestyle: Hunter.
Species: *N. leali*.

ABELISAURIDS

The evolution of the abelisaurids has, for two decades, been held up as one of the proofs of the shifting palaeogeography of the Cretaceous world.

From the observable evidence, the group evolved in the area that is current-day South America, and then spread across the southern continents. The remains have been found in late Cretaceous rocks of South America, Madagascar and India. As the supercontinent of Pangaea split up in Mesozoic times, the southern section, known as Gondwana, remained as a whole for much longer than the northern part, Laurasia. Animals were able to migrate freely across this area, but by that time there was very little connection between Gondwana and Laurasia.

However, the lack of abelisaurids in Africa has always suggested that the African continent broke away from the rest of Gondwana quite early, before the abelisaurids became established. These big animals could not cross the widening ocean areas between Africa and the other southern continents.

Discoveries in the early part of the twenty-first century have shown that the situation is not as simple as this. A large abelisaurid, *Rugops*, has been found in North Africa, and the discovery of *Tarascosaurus*, in France, showed that somehow the group was able to cross the Tethys Ocean that separated Gondwana from Laurasia.

Abelisaurus

This is the dinosaur that gave its name to the whole abelisaurid group – based originally on *Abelisaurus* and *Carnotaurus* – and typifies the late Cretaceous meat-eaters of the Southern Hemisphere. It is only known from a partial skull, but since its discovery there have been more complete specimens of closely related animals, and so we can quite confidently tell what the living animal looked like.

Features: The deep skull and sharp teeth of the basic abelisaurid shows that it was a powerful hunting dinosaur, analogous to the carnosaurs or the tyrannosaurids of the Northern Hemisphere. The skull is distinctly different from either because it has a particularly big gap in front of the eyes.

Distribution: Argentina.
Classification: Theropoda, Neoceratosauria, Abelisauria.
Meaning of name: Abel's (Roberto Abel, the director of the Argentinian Museum of Natural Science) lizard.
Named by: Bonaparte and Novas, 1985.

Time: Maastrichtian stage of the late Cretaceous.
Size: 6.5m (21ft).
Lifestyle: Hunter.
Species: *A. comahuensis*.

Left: A. comahuensis *is named after the Comahue formation, from where it was excavated.*

Tarascosaurus

It was thought that the abelisaurids were confined to the southern continents, evolving there in isolation, until this animal was excavated in France in the 1990s. Although it is only known from a femur and two vertebrae, these are distinctive enough to put it in the group. Its exact provenance is unclear, although it is certainly from a limited sequence of grey limestone of Campanian age.

Features: Since this animal is only known from a few scraps of bone, it is difficult to give a full account of it. It is thought to have had a big head with a blunt snout, and long dagger-like teeth. The body is long and heavy. The arms are small and three-fingered, and the feet have big claws. Much of this description is, of course, based on our knowledge of other large abelisaurids.

Right: The tarasque, after which the dinosaur was named, was a legendary dragon in Provençal legend. We do not know who discovered the dinosaur or who dug it up.

Distribution: France.
Classification: Theropoda, Neoceratosauria, Abelisauria.
Meaning of name: Dragon lizard.
Named by: LeLoeff and Buffetaut, 1991.
Time: Campanian stage.
Size: 10m (33ft).
Lifestyle: Hunter.
Species: *T. salluvicus*.

ADVANCED ABELISAURIDS

Until the 1980s the abelisaurids were not well understood. Since then the burgeoning study of dinosaurs in South America and Madagascar has shown just how wide-ranging and diverse this group actually was. Well-preserved skeletons give us a clear idea of what these animals looked like – mostly big heavy-bodied meat-eaters.

Aucasaurus

Known from an almost complete skeleton, lacking only the end of the tail, *Aucasaurus* was found in lake sediments in Patagonia in 1999. This makes it the best-known abelisaurid skeleton, and it is used as the basis for several other reconstructions. There was damage to the skull of the skeleton found, suggesting that this individual had been involved in a fight shortly before its death.

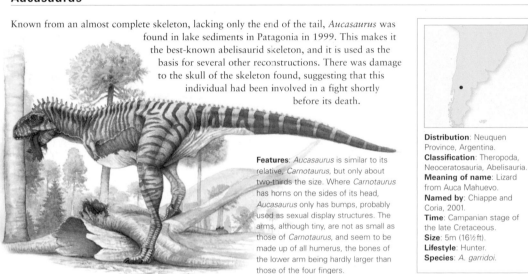

Features: *Aucasaurus* is similar to its relative, *Carnotaurus*, but only about two-thirds the size. Where *Carnotaurus* has horns on the sides of its head, *Aucasaurus* only has bumps, probably used as sexual display structures. The arms, although tiny, are not as small as those of *Carnotaurus*, and seem to be made up of all humerus, the bones of the lower arm being hardly larger than those of the four fingers.

Distribution: Neuquen Province, Argentina.
Classification: Theropoda, Neoceratosauria, Abelisauria.
Meaning of name: Lizard from Auca Mahuevo.
Named by: Chiappe and Coria, 2001.
Time: Campanian stage of the late Cretaceous.
Size: 5m (16½ft).
Lifestyle: Hunter.
Species: *A. garridoi*.

Carnotaurus

An almost complete skeleton of *Carnotaurus* was extracted with difficulty from the hard mineral nodule in which it was preserved in Argentina. The deep skull suggests that it may have had an acute sense of smell, but the strength of the jaws and neck implied by the muscle attachments seem at odds with the weakness of the lower jaw and the teeth.

Features: The head is very short and squashed-looking with a shallow, hooked lower jaw. Two horns stick out sideways from above the eyes, probably being used for sparring with rivals. The arms are

Distribution: Argentina.
Classification: Theropoda, Neoceratosauria, Abelisauria.
Meaning of name: Flesh-eating bull.
Named by: Bonaparte, 1985.
Time: Campanian to Maastrichtian stage of the late Cretaceous.
Size: 7.5m (25ft).
Lifestyle: Hunter.
Species: *C. sastrei*.

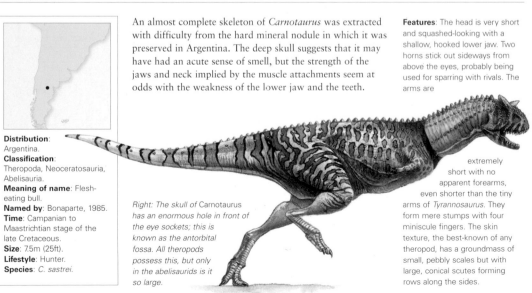

Right: The skull of Carnotaurus *has an enormous hole in front of the eye sockets; this is known as the antorbital fossa. All theropods possess this, but only in the abelisaurids is it so large.*

extremely short with no apparent forearms, even shorter than the tiny arms of *Tyrannosaurus*. They form mere stumps with four miniscule fingers. The skin texture, the best-known of any theropod, has a groundmass of small, pebbly scales but with large, conical scutes forming rows along the sides.

Rugops

The fossil skull of *Rugops* was found in 2000 by a team from *National Geographic* magazine, led by the Chicago Field Museum's Paul Sereno. Several big sauropods were found in the same area, so there was no shortage of food. It was the first abelisaurid to be found in Africa, all others having been found in South America, Madagascar and India. Evidently there was some land connection between Africa and the rest of the continents at the time.

Distribution: Niger.
Classification: Theropoda, Neoceratosauria, Abelisauria.
Meaning of name: Wrinkled face.
Named by: Sereno, Wilson and Conrad, 2004.
Time: Cenomanian stage of the late Cretaceous.
Size: 9m (30ft).
Lifestyle: Scavenger.
Species: R. primus.

Features: The wrinkled face of *Rugops* was the result of the bones being riddled with arteries and veins, leaving grooves etched across the skull. This implies that the head was covered in skin or armour. Holes along the snout suggest the presence of some kind of fleshy display structure. The skull is short and has a rounded snout. The teeth are those of a meat-eater, but are small and weak, suggesting that *Rugops* was not a hunter but a scavenger of dead animals. The skull bones were found lying on the surface of the rock where they had been weathered by the wind and flying sand.

Above: The presence of Rugops *in north Africa suggests that there was a connection between South America and the northern part of Africa as late as 100 million years ago – a good 20 million years later than originally thought.*

HUNTER OR SCAVENGER?

Cannibalism, as observed in *Majungatholus*, is not unusual in the animal world. Nowadays there are at least 14 species of mammal, and many species of reptiles and birds, that are known to kill and eat members of their own kind when conditions become harsh. In Cretaceous times, the environment of the area that is now northern Madagascar had a seasonal climate with no steady supply of food or water, and times of extreme dryness. Conditions such as this are conducive to cannibalistic behaviour among animals.

We do not know whether *Majungatholus* killed its own kind or scavenged from the corpses of those that had already succumbed to starvation and drought. The *Majungatholus* bones in question show sets of tooth marks that match the size and spacing of the teeth in a *Majungatholus* jaw, and also the pattern of serrations on the individual teeth. The evidence is from two different sites, and so the fossils do not represent a one-off incident.

The other suggestion of dinosaur cannibalism is the bone-bed of *Coelophysis* that seemed to suggest that this animal ate its own young in times of stress. This has now been discredited.

Left: Cannibalism.

Majungatholus

The first specimen of *Majungatholus* to be discovered was a small skull fragment. The thickening on the top of this led early researchers to classify this animal as a pachycephalosaurid, hence the name. The term *tholus* means "dome", and its presence in a dinosaur name suggests that the animal was a pachycephalosaurid; and *Majunga* is the province in which it was found. Since the first specimen, several better fossils, including an almost complete skull, have been found, proving it to be an abelosaurid. It seems to have been a cannibal, judging by the toothmarks on some of the bones.

Distribution: Madagascar and possibly India.
Classification: Theropoda, Neoceratosauria, Abelisauria.
Meaning of name: Dome from Majunga (the district in which it was found).
Named by: Sues and Taquet, 1979.

Features: The head is short and broad, broader than in most other theropods, and the snout deep and blunt with thickened bones around the nostrils. The dome-like bulges on top of the skull, the cause of confusion in the original specimen, may have been the bases of horns. The head ornamentation is a single spike rising from above the eyes, and was used for display.

Time: Campanian stage of the late Cretaceous.
Size: 7–9m (23–30ft).
Lifestyle: Hunter and scavenger.
Species: M. atopus.

Left: Majungatholus *is known to have been a cannibal, attacking its own species.*

SUNDRY THEROPODS

Almost weekly the list of meat-eating dinosaurs becomes longer, as new specimens representing completely new species come to light. The 1990s was a particularly fruitful decade in the discovery of new theropods, especially in South America and Africa. In some instances the new discoveries represented animals that had been found but whose original specimens had been lost or forgotten.

Deltadromeus

This theropod resembled the late Jurassic *Ornitholestes* in its anatomy, but it was very much bigger. It was a late-surviving member of this primitive group of meat-eating dinosaurs that are more usually associated with North America. Of the dinosaurs excavated from the Cretaceous beds of North Africa in the 1990s, this would have been the fiercest hunter.

Features: A surprising feature of this animal is the extraordinarily long and delicate limbs. The leg bones are half the thickness of those of a similar-size *Allosaurus*, and the lower limbs approach the proportions of one of the fleet-footed ostrich mimics. The arms are very long for a theropod, and the slim legs make it a very fast runner. The teeth are thin and adapted for stripping flesh rather than for crushing bone. They have very fine serrations but this feature does not seem to be significant in the classification.

Distribution: Kem Kem region, Morocco.
Classification: Theropoda, Tetanurae, Coelurosauria.
Meaning of name: Delta runner.
Named by: Sereno, Duthiel, Iarochene, Larsson, Lyon, Magwene, Sidor, Variccio and J. A. Wilson, 1996.
Time: Late Cretaceous.
Size: 8m (26ft).
Lifestyle: Hunter.
Species: *D. agilis*.

Above: As in many other dinosaurs, the best skeleton found is incomplete. However the cast of a complete skeleton appears in several museums, based largely on guesswork. The head, significantly, is totally speculative.

Carcharodontosaurus

Distribution: Morocco; Tunisia; Algeria; Libya; Niger.

Classification: Theropoda, Tetanurae, Carnosauria.
Meaning of name: Great white shark lizard.
Named by: Stromer, 1931.
Time: Albian to Cenomanian stages of the Cretaceous.
Size: 14m (46ft).
Lifestyle: Hunter.
Species: *C. saharicus*.

Bones of *Carcharodontosaurus* found in Egypt were described by Stromer in the 1920s, but were lost when an air raid destroyed the Bavarian State Museum, Munich, Germany, in which they and other valuable palaeontological specimens were housed. At the time it had not been clear just how big an animal *Carcharodontosaurus* had been. Half a century later, new specimens were found in Morocco by a team from the Field Museum, Chicago, USA. The skull found was one of the biggest known, and the animal must have been one of the biggest meat-eaters that ever lived.

Features: The skull is missing the lower jaw and the snout, but the full length has been estimated at 1.53m (5ft), which is longer than any *Tyrannosaurus* skull yet found. The brain cavity, however, is tiny, much smaller even than that of *Tyrannosaurus*. The upper jaw has cavities for 14 blade-like teeth at each side, and they are curved and finely serrated at the edges, with grooves across the sides to allow blood to flow away.

Right: Carcharodontosaurus was orignally named in 1925 by Deperet and Savornin, who described it as a megalosaur.

Giganotosaurus

For a century we have regarded *Tyrannosaurus* as the biggest meat-eater that ever lived. That was before the discovery of *Giganotosaurus* in Patagonia in 1993, by amateur collector Ruben Carolini. The subsequent excavation unearthed a skull that was bigger than the skull of any *Tyrannosaurus* known. Then they found the jawbone of another individual that was even bigger. Unlike its close relative, *Carcharodontosaurus*, from Africa, whose remains consisted only of the skull and a few scraps of bone, the skeleton of *Giganotosaurus* is 70 per cent complete. It lived at the time that the big titanosaurs were flourishing on the South American plains, and it probably found its prey among them.

A mounted skeleton of *Giganotosaurus* stands in the entrance hall of the Academy of Natural Sciences in Philadelphia, USA.

Features: In the legs, the tibia and the femur are about the same length. This indicates that *Giganotosaurus* was not a running animal. With a principal prey of big sauropods, such as titanosaurs, it would not need speed to help it hunt. It weighed between four and eight tonnes.

Below: Giganotosaurus, Carcharodontosaurus and Tyrannosaurus *were the biggest known meat-eating dinosaurs. Studies carried out in 1999 at the North Carolina State Univeristy, USA, showed that Giganotosaurus and* Tyrannosaurus *had some form of warm-blooded metabolism.*

Distribution: Neuquén Province, Argentina.
Classification: Theropoda, Tetanurae, Carnosauria.
Meaning of name: Giant southern lizard.
Named by: Coria and Salgado, 1995.
Time: Albian stage of the lower Cretaceous.
Size: 15m (49½ ft).
Lifestyle: Hunter.
Species: *G. carolini.*

THE DINOSAURS OF BAHARIYA OASIS

Early in the twentieth century, pieces of bone and tooth were discovered in Egypt. They were collected and described by E. S. von Richenbach, of the Bavarian State Collection of Palaeontology and Historical Geology, in Munich. He named them as *Carcharodontosaurus*, seeing a resemblance to the lifestyle of a killer shark with such teeth. The remains of some of the biggest meat-eating dinosaurs ever were uncovered by German palaeontologists in the early years of the century and housed in the museum, including the famous sail-backed *Spinosaurus*.

Unfortunately, this bone collection, which included the original material of *Spinosaurus*, was destroyed when the museum was caught in a World War II bombing raid. Then, in the 1980s, an expedition from the Field Museum, in Chicago, USA, found a skull in the deserts of Morocco that matched the published descriptions of the *Carcharodontosaurus* remains.

Stromer's original site in Egypt has been pinpointed, and has become the focus of a great deal of palaeontological activity, including the discovery of the enormous sauropod *Paralititan*.

TROODONTS

The troodonts were a family of active little theropods that seem to fall somewhere between the long-legged ostrich mimics, the ornithomimids, and the killer-clawed deinonychosaurs. The skeletons were bird-like and all had big eyes and large brains. They were probably warm-blooded and covered in feathers but, as yet, there is no direct evidence for this.

Troodon

The teeth of *Troodon* were the first part of this animal to be found, and they were thought to have come from a lizard or pachycephalosaur, or even a carnivorous ornithopod, something of an absurdity. They were then seen to have come from a subsequently discovered dinosaur that had been called *Stenonychosaurus*. As the name *Troodon* was invented first, *Stenonychosaurus* had to be dropped.

Features: The long head contains the biggest brain for its body size of any dinosaur, being comparable to that of a modern emu. The hands are long and slim, with three-clawed fingers, and could grasp objects palm-to-palm. The legs are particularly long and each foot has a big killing claw, like that of a *Velociraptor*, on its second toe. Its big eyes, arranged stereoscopically, suggest that it was a hunter of small prey during darkness or at dusk.

Left: The big brain of Troodon *led Canadian palaeontologist Dale Russell to suggest that, had the dinosaurs not become extinct,* Troodon *would have evolved into an intelligent humanoid form by today.*

Distribution: Alberta, Canada; Montana, Wyoming and perhaps Alaska, USA.
Classification: Theropoda, Tetanurae, Troodontidae.
Meaning of name: Tearing tooth.
Named by: Leidy, 1856.
Time: Campanian stage of the late Cretaceous.
Size: 2m (6½ft).
Lifestyle: Stealthy crepuscular hunter.
Species: *T. formosus*.

Saurornithoides

When the scattered remains of *Saurornithoides* were discovered on the Central Asiatic Expedition by the American Museum of Natural History, they were thought to have been the remains of a toothed bird (the skeleton of the whole troodontid group is particularly bird-like). *Saurornithoides* hunted by stealth and ambush rather than by pursuit. The teeth, like those of other troodonts, were small and sharp, adapted for gripping small prey-like lizards or mammals, rather than shearing flesh.

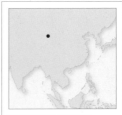

Features: In general, the skull is shorter than that of *Troodon* and similar to that of *Velociraptor*. However, it has many more teeth, with 38 in each upper jaw compared with 30 in *Velociraptor*. The big brain seems to be enlarged in the auditory region, suggesting that it had a very good sense of hearing, and the eyes face forwards giving good stereoscopic vision, a valuable aid to hunting small prey. The middle foot bone is a mere splint, a feature that the troodonts share with the ostrich mimics and even the tyrannosaurs. This may have lessened the stress on the foot during running.

Above: The brain of Saurornithoides *was one of the largest known among dinosaurs. It was six times larger than that of a crocodile of a similar weight.*

Distribution: Mongolia.
Classification: Theropoda, Tetanurae, Troodontidae.
Meaning of name: Bird-shaped lizard.
Named by: Osborn, 1924.
Time: Campanian to Maastrichtian stages of the late Cretaceous.
Size: 2m (6½ft).
Lifestyle: Stealthy hunter.
Species: *S. mongoliensis, S. junior, S. asiamericanus, S. isfarensis*.

Borogovia

This troodontid is known only from the hind limbs, which were at first attributed to *Saurornithoides*. There are several other small theropods known from the area, including *Saurornithoides*, and there must have been enough food types to enable them all to survive. The name derives from a fictitious animal, the borogove, in Lewis Carroll's poem *Jabberwocky*.

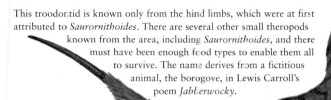

Features: *Borogovia* was a much slimmer animal than *Saurornithoides*, as suggested by the very delicate toe bones. The killing claw on the second toe is much straighter than that found on any of the other troodontids, and is quite small. These killing claws seem to have become smaller as the group evolved. Otherwise there is no reason to doubt that the rest of the skeleton is typically troodontid. The leg bones are so much like those of *Saurornithoides* that some palaeontologists regarded it as a specimen of *S. junior*.

Above: Borogovia, Saurornithoides *and* Tochisaurus *have been found at the same site, leading to the suggestion that they are all the same animal.*

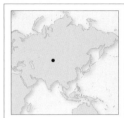

Distribution: Bayankhongor, Mongolia.
Classification: Theropoda, Tetanurae, Troodontidae.
Meaning of name: From borogove.
Named by: Osmólska, 1987.
Time: Campanian to Maastrichtian stages of the late Cretaceous.
Size: 2m (6½ft).
Lifestyle: Hunter.
Species: *B. gracilicrus*.

EGG MOUNTAIN

In 1979 the famous hadrosaur nesting sites in Montana were being excavated and studied by a team from Princeton University, USA, when a seismic company began a survey of the area on behalf of an oil company. As this involved drilling and explosions, the team did a rapid hunt for fossils across the whole area before any damage was done. Right beside one of the boreholes they discovered a whole nesting site and a total of 52 eggs from a small dinosaur. These were totally different from the hadrosaur sites they had been studying. Along with the nests there were the remains of lizards, small mammals and dinosaur bones.

The bones belonged to the small hypsilophodont *Orodromeus*, and it was assumed that the site represented an *Orodromeus* nesting colony. However, when the remains were studied properly, it was found that the eggs contained *Troodon* babies and it was actually a *Troodon* nesting site. The original site had been on an island in an alkaline lake, and a parent *Troodon* had been bringing *Orodromeus* remains back to feed its babies. The seismologists appreciated the importance of the site and diverted the line of the survey in order to leave it alone. As the site was located on the top of a knoll, it eventually came to be known as Egg Mountain.
Left: Troodon *at Egg Mountain.*

Byronosaurus

The smallest of the troodontids was *Byronosaurus*. It is known from one of the best-preserved troodontid skulls ever found. It was discovered in 1994, and from bits of bone found by expeditions during the two subsequent years in the very rich fossil locality of Ukhaa Tolgod, in the Gobi Desert. All troodontids, except for *Troodon* itself, have been found in Asia.

Features: What distinguishes *Byronosaurus* from all other troodontids is the unserrated teeth. As a rule the theropods have teeth that were serrated like steak knives, with other rare exceptions being among the spinosaurids and primitive ostrich mimics. Primitive toothed birds also have teeth that are not serrated. The mouth has a palate separating it from the nasal passages, and the structure of the snout suggests that *Byronosaurus* had a very sensitive nose.

Distribution: Ukhaa Tolgod, Mongolia.
Classification: Theropoda, Tetanurae, Troodontidae.
Meaning of name: Byron lizard (after a sponsor of the American Museum of Natural History's Palaeontological Expeditions).
Named by: Norell, Mackovicky and Clark, 2000.
Time: Campanian stage of the late Cretaceous.
Size: 1.5m (5ft).
Lifestyle: Hunter.

Species: *B. jaffei*.

Left: By the time Byronosaurus *was found there were eight troodonts known, seven of which were found in Asia. The group probably evolved in Asia.*

ORNITHOMIMIDS

The ornithimimids have become famous as the ostrich mimics. In general structure they were similar to modern, ground-living birds with a lightly built skeleton, compact body, long neck and small skull. Although all the ostrich mimics have a very similar body plan, they are distinguished by details of the beak, the hands and the body proportions.

Archaeornithomimus

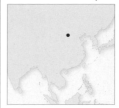

Distribution: Erenhot City, Inner Mongolia.

Originally described by Gilmore in 1933 as a species of *Ornithomimus*, *Archaeornithomimus* is known from limb bones and vertebrae. Many scientists regard it as a *nomen dubium* since there is so little to study, and it is possible that the remains represent not an ornithomimid but quite a different kind of theropod dinosaur.

Features:
Archaeornithomimus appears to be very similar to both *Struthiomimus* and *Gallimimus*, but it lived about 30 million years before either of them and is slightly smaller. Its fingers are much smaller than those of the other ornithomimids, and the third finger is particularly short. It has straight claws on its fingers, a rather primitive feature. Like its bigger relatives, it used speed as a means of escaping from predators.

Left: Whatever the true classification of Archaeornithomimus, it was evidently a fast runner, suggesting that it was an active predator, hunting small reptiles and mammals.

Classification: Theropoda, Tetanurae, Ornithomimosauria.
Meaning of name: Early Ornithomimus.
Named by: Russell, 1972.
Time: Early to late Cretaceous.
Size: 3.5m (11½ft).
Lifestyle: Omnivore or hunter.
Species: *A. asiaticus*, *A. bissektensis*.

Garudimimus

Distribution: Bayshin Tsav, Mongolia.
Classification: Theropoda, Tetanurae, Ornithomimosauria.

Meaning of name: Garuda (an Indian deity) mimic.
Named by: Barsbold, 1981.
Time: Coniacian to Santonian stages of the late Cretaceous.
Size: 4m (13ft).
Lifestyle: Omnivore.
Species: *G. brevipes*.

Although all the ornithomimids were built for speed, *Garudimimus*, a fairly early form, was not in the same league. The relatively short leg and heavy foot show that it was not as speedy as the more advanced forms. The foot had the vestige of the first toe, whereas all other ornithomimids were purely three-toed with the first and fifth toes lost.

Right: The crest on the skull of Garudimimus was tiny. However, in life it may have been covered in horn and would have appeared much larger. If so, it would have been used for display and communication.

Features: The skull has a more rounded snout than others of the group, and larger eyes, and the side of the skull is more like that of a primitive theropod. It has a small crest in front of its eyes, something no other known ornithomimid possesses. The leg is different too, having a much shorter lower section and foot bones than the other ornithomimids, and four toes rather than three. The short ilium bone in the hip suggests that the musculature of the legs was quite weak, an argument against its being a powerful runner.

Gallimimus

The ornithomimid featured in the film *Jurassic Park* was *Gallimimus*. It was a fairly good representation, except that it is now thought that these animals were covered in feathers, which would make sense if they were to be as active as they were portrayed in the film. There are skeletons of juveniles that have allowed scientists to study the growth pattern of ornithomimids in general.

Distribution: Bayshin Tsav, Mongolia.

Classification: Theropoda, Tetanurae, Ornithomimosauria.
Meaning of name: Chicken mimic.
Named by: Osmólska, Roniewicz, Barsbold, 1972.
Time: Maastrichtian stage of the late Cretaceous.
Size: 6m (19½ft).
Lifestyle: Omnivore.
Species: *G. bullatus*, *G. mongoliensis*.

Right: Like the other ostrich mimics and modern birds Gallimimus *had hollow bones. This device allowed for a reduction of weight in the body, without reducing the strength, and enabled the animal to move quickly.*

The main difference between the two known Gallimimus *species is the shape of the fingers. G. mongoliensis had shorter hands and would not have grasped as well.*

Features: *Gallimimus* is the largest known type of ornithomimid, but it has shorter arms in proportion to the other species. The hands, too, are quite small and the fingers are not very flexible. The head is quite long and graceful and, as in nearly all ornithomimids, the jaws have no teeth. The beak of the lower jaw is shovel-shaped, and the big eyes are situated on the sides of the head, so it did not have binocular vision.

HERD OR SOLITARY ANIMAL?

The ornithomimids were made famous by their appearance in the film *Jurassic Park* when a whole herd of *Gallimimus* weaved across the countryside, acting as a unified mass. At the time and for a long while after, this behaviour was regarded as good cinema but poor science because there was no evidence that *Gallimimus*, or any other ornithomimid, lived in a herd.

Then, in 2003, a paper was published by Yoshitsugu Kobayashi and Jun-Chang Lu that described the discovery of a bed of bones in Inner Mongolia consisting of 14 ornithomimid individuals, 11 of them youngsters. A new genus, *Sinornithomimus*, was established based on these specimens, and lay somewhere between *Archaeornithomimus* and *Anserimimus* on the evolutionary scale. The important aspect of the find was that it seemed to show evidence that ornithomimids lived in large groups, with adults protecting the young. The bone bed may have been the result of either an accident befalling a big herd in which a high proportion of the youngsters died, or a catastrophic event that killed the whole herd, which consisted of many youngsters and a few adults. Studies of the juvenile bones suggested that the adults could run faster than the young.

Anserimimus

Anserimimus is known from a partial skeleton that includes an incomplete forelimb. The strong arms of this ornithomimid suggest that it may have dug in the ground for food such as roots, insects or dinosaur eggs. It would have been a fast runner, giving the lie to its name of "goose mimic".

Features: What set this species apart from other ornithomimids is the size of the muscle attachments of the upper arm. This must have meant that the forelimbs were particularly strong. The bones of the hand are bound closely together to give a rigid structure, and claws on the hand are flat and hoof-like. Otherwise the skeleton is very much like that of the other ornithomimids.

Distribution: Mongolia.
Classification: Theropoda, Tetanurae, Ornithomimosauria.
Meaning of name: Goose mimic.
Named by: Barsbold, 1988.
Time: Campanian to Maastrichtian stages of the late Cretaceous.
Size: 3m (10ft).
Lifestyle: Omnivore.
Species: *A. planinychus*.

Left: "Goose mimic", "chicken mimic", "ostrich mimic" and "emu mimic" – all evocative names that emphasize these animals' resemblance to modern ground-dwelling birds. However, their skulls were more like those of the extinct ground-dwelling birds from New Zealand, the moas, in being sturdily constructed and strong.

ADVANCED ORNITHOMIMIDS

The late ornithomimids were the fastest dinosaurs known. In fact, the known fossils of this group of advanced ostrich-mimics seem to reflect the known fossils of horse ancestors, a sequence leading from small generalized beasts to large, elegant, long-legged creatures adapted for speed. The presence of feathers is suggested by pits in the arm bones of a specimen in Tyrrell Museum, Alberta, Canada.

Struthiomimus

This is the animal that gave rise to the term ostrich mimic, which is often used instead of the slightly more formal ornithomimid. It was the first complete ornithomimid skeleton to be found. It was a fast runner on the late Cretaceous plains of North America. Its main predators would have been sickle-clawed deinonychosaurs and the tyrannosaurid

Albertosaurus, from which it would have escaped by a sudden turn of speed.

Features: The small head, lack of teeth, long neck, compact body and long legs are the ostrich-like features of this dinosaur. Its non-ostrich-like features are its long arms with three-fingered hands, and the long tail. It is very similar to its close relative *Ornithomimus,* the main differences being its slightly smaller size and longer tail. Nevertheless many scientists regard it, along with many other genera of ornithomimids, as merely a species of *Ornithomimus.* Stomach stones have been associated with the skeleton. Usually only plant-eating animals have them, so this find indicates that *Struthiomimus* was partly vegetarian.

Left: Although the specimen of Struthiomimus *was fairly complete, it was quite badly damaged, resulting in the ongoing confusion as to whether or not it is really a specimen of* Ornithomimus.

Distribution: Alberta, Canada.
Classification: Theropoda, Tetanurae, Ornithomimosauria.
Meaning of name: Ostrich-mimic.
Named by: Osborn, 1917.
Time: Campanian stage of the late Cretaceous.
Size: 3–4.3m (10–14ft).
Lifestyle: Omnivore.
Species: *S. sedens.*

Ornithomimus

The image of *Ornithomimus* took many decades to compile after the first very fragmentary specimen was found in the 1880s. It was not until 1917 that a good skeleton of *O. edmontonicus,* complete, but lacking the skull, was discovered in Canada, and the true bird-like nature was appreciated. Three species of *Ornithomimus* existed together, their slight differences in beak shapes suggesting that they ate different foods, some preferring insects, with others opting for small reptiles or plants.

Features: The head is small and carries a fluted toothless beak. The neck and tail are long, and the body is more compact than that of other ornithomimids. The legs are very long for the size of the body, although not quite as long as those of other ornithomimids, and show that it was a running animal. The arms are quite long and slender, and carry three-clawed fingers, the first longer than the others.

Above: Ornithomimus *is the best-known of the ornithomimids, and the remains are quite widely dispersed. Several other ornithomimid genera are regarded by some scientists as species of* Ornithomimus.

Distribution: Alberta, Canada; to Texas, USA.
Classification: Theropoda, Tetanurae, Ornithomimosauria.
Meaning of name: Bird mimic.
Named by: Marsh, 1890.
Time: Campanian to Maastrichtian stages of the late Cretaceous.
Size: 4.5m (15ft).
Lifestyle: Omnivore.
Species: *O. antiquus, O. edmontonicus, O. velox, O. lonzeensis* and *O. sedens.*

Dromiceiomimus

The big eyes and the shape of the beak and hands suggest that *Dromiceiomimus* specialized in hunting small prey in the twilight hours. The discovery of an adult skeleton with two young indicates that there was some form of family structure. There is a suggestion, that because of the wide hips. *Dromiceiomimus* did not lay eggs but gave birth to live young.

Features: The shins are long, longer in proportion than those of any other ornithomimid, and this indicates a very fast runner. It was probably the fastest dinosaur known, with the ability to reach speeds of 73kph (45mph). The eyes are bigger than those of other ornithomimids. The shape of the muzzle and the weak jaw muscles suggest a diet of insects, and the hands seem more adapted to scraping in the ground than clutching big prey.

Above: Dromiceiomimus was originally described by William A. Parks, a Canadian palaeontologist, in 1926, as a species of Struthiomimus.

Distribution: Alberta, Canada.
Classification: Theropoda, Tetanurae, Ornithomimosauria.
Meaning of name: Emu mimic.
Named by: Russell, 1972.
Time: Campanian to Maastrichtian stages of the late Cretaceous.
Size: 3.5m (12ft).
Lifestyle: Omnivore or hunter of small animals.
Species: *D. brevitertius*, *D. samuelli*.

VEGETARIAN OR CARNIVORE?

The way of life of the ornithomimids seems to have been different from that of all other theropods. In general, the typical theropod was the hunter of the time, whether it was a small animal preying on mammals and small dinosaurs, or a gigantic, dragon-like monster preying on the biggest plant-eaters of the day.

In contrast, the lack of teeth in the jaw precluded the average ornithomimid from such a lifestyle. Instead of teeth it had a beak, like that of a bird. This has led palaeontologists to assume that they were omnivorous animals, feeding on both plant and animal food. There is, however, a strong argument against the vegetable part of this diet; there seems to have been no means of processing such food. Modern plant-eating birds, lacking teeth for chewing, swallow grit and stones to help them grind up the food. This is true of some of the sauropod plant-eaters that lacked chewing teeth. Gastroliths have been found in the stomach areas of skeletons of some of them. However, no gastroliths have ever been found associated with an ornithomimid skeleton, with the exception of *Struthiomimus*.

The fluting on the beak has led to a suggestion that the beak was used for straining shrimps and other tiny animals from lake water, but this is not a generally accepted view. The consensus is that a typical ornithomimid fed on small animals, such as insects or lizards.

Deinocheirus

The only thing that is known of this supposed ornithomimid is a pair of enormous arms, with huge, clawed hands. Since the description in 1970, scientists have speculated on whether the arms were unusually large for the size of animal, or whether they were in the same proportions as in the other ornithomimids, which would have meant an animal as big as *Tyrannosaurus*.

Features: Each arm is 2.6m (8½ft) long. The lower arm is about two-thirds the length of the upper arm, and the hand with the three equal-length fingers is about the same length as the lower arm.

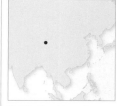

Distribution: Mongolia.
Classification: Theropoda, Tetanurae, Ornithomimosauria.
Meaning of name: Terrible hand.
Named by: Osmólska and Roniewicz, 1970.
Time: Maastrichtian stage of the late Cretaceous.
Size: 7–12m? (23–39ft?).
Lifestyle: Unclear.
Species: *D. mirificus*. One other, unnamed.

Left: The fingers have strongly curved claws that are 25cm (10in) long. In life these were covered in horn. We know nothing more about this mysterious animal.

OVIRAPTORIDS

Oviraptorids were a family of extremely bird-like dinosaurs, not only in their general build but also in the presence of the beak and the fact that the shoulders were strengthened by a collarbone. It has even been suggested that the accepted classification of oviraptorids is wrong, and that they should be classed as birds instead of dinosaurs.

Oviraptor

Distribution: Mongolia.
Classification: Theropoda, Tetanurae, Orviraptorosauria.
Meaning of name Egg stealer.
Named by: Ostrom, 1924.
Time: Campanian stage of the late Cretaceous.
Size: 1.8m (6ft).
Lifestyle: Specialist feeder.
Species: *O. philoceratops, O. mongoliensis.*

The genus name *Oviraptor* derives from the belief that the first *Oviraptor* found had been eating the eggs of a ceratopsian dinosaur. The mouth seems to have evolved for a crushing action, and the current diet suggestions point to shellfish or nuts. There seems to be a variation in size and shape of the skull crest, maybe a sign of different stages of growth and maturity, or different species. The skull, on which most restorations, including this one, are based, is thought to have belonged to the related oviraptorid *Citipati*.

Features: As with all other oviraptorids the head is short and carries a heavy toothless beak at the end of its well-muscled jaws. A hollow crest, like that of a cassowary, which sticks up on the head was probably used for display and intimidation. There is a pair of teeth on the palate. The skull is extremely lightweight and has very large eye sockets.

Right: Oviraptor hands are very long. Its eggs are about the size of a hot-dog bun.

Khaan

The three almost complete articulated skeletons that have been excavated give a good idea of what this oviraptorid looked like. It was first thought to have been a specimen of the related oviraptorid, *Ingenia*, showing just how little variation there was between animals of this group. Several oviraptorids lived in the same area at the same time. The specific name of the type species *K. mckennai* honours American palaeontologist Malcolm McKenna.

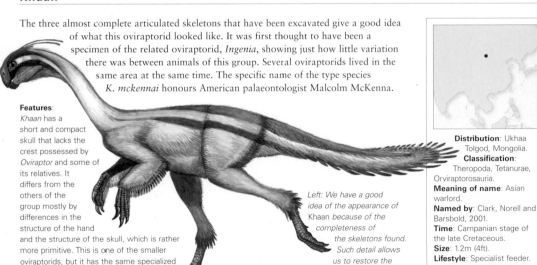

Features:
Khaan has a short and compact skull that lacks the crest possessed by *Oviraptor* and some of its relatives. It differs from the others of the group mostly by differences in the structure of the hand and the structure of the skull, which is rather more primitive. This is one of the smaller oviraptorids, but it has the same specialized head, long neck, huge hands, big feet and short tail.

Left: We have a good idea of the appearance of Khaan because of the completeness of the skeletons found. Such detail allows us to restore the appearance of others of the group that are not so complete.

Distribution: Ukhaa Tolgod, Mongolia.
Classification: Theropoda, Tetanurae, Orviraptorosauria.
Meaning of name: Asian warlord.
Named by: Clark, Norell and Barsbold, 2001.
Time: Campanian stage of the late Cretaceous.
Size: 1.2m (4ft).
Lifestyle: Specialist feeder.
Species: *K. mckennai.*

OVIRAPTOR DIET

With the oviraptorids' absurdly short heads, huge, toothless beaks and pair of little teeth on the palate, their diet was clearly understood when specimens were first discovered but has since become a mystery. The American Museum of Natural History expeditions to the Gobi Desert in the 1920s produced a whole host of new dinosaurs. The most abundant were the little ceratopsian, *Protoceratops*, found with nesting sites full of eggs. The first *Oviraptor* was found close to one of these nests, and it was believed to have been raiding the *Protoceratops* nest when it was overcome and killed by a sandstorm. This made perfect sense, since the mouth of *Oviraptor* was ideal for breaking into hard-shelled eggs, and its hands were perfectly shaped to clutch eggs of that size.

Then, in the 1990s, another American Museum of Natural History expedition found the remains of an identical nest, but this time with an oviraptorid sitting on it brooding the eggs. So the original *Oviraptor* had not been raiding a *Protoceratops* nest at all, it had been on its own nest. But we still do not know what *Oviraptor* could have eaten with its very specialized mouth parts.

Nomingia

The remarkable pygostyle on the end of the tail of *Nomingia* must have supported a fan of feathers. Since this was manifestly not a flying animal, the feathers must have been used for display purposes. The fan's effect was probably reinforced by display feathers on the arms, but the forelimbs and the head are missing from the only known specimen.

Features: This is the only dinosaur known with a pygostyle, a structure formed of the fusion of tail vertebrae. In birds, the entire tail is formed of a pygostyle and it is used as a mounting platform for the fan of tail feathers. In *Nomingia* it is on the end of a short length of tail formed of stubby vertebrae, but it probably represents the base of a tail fan as well. The legs are proportionally longer than in other oviraptorids. It is possible that the second toe bears a killing claw.

Distribution: Bugin Tsav, Mongolia.
Classification: Theropoda, Tetanurae, Orviraptorosauria.
Meaning of name: From the Nomingiin Gobi, part of the Gobi Desert.
Named by: Barsbold, Osmólska, Watabe, Currie and Tsogtbataar, 2000.
Time: Maastrichtian stage of the late Cretaceous.
Size: 1.8m (6ft), but this takes into account a shorter tail than is usual in the oviraptorids.
Lifestyle: Specialist feeder.
Species: *N. gobiensis*.

Left: Despite the specialization of the tail, Nomingia *seems to be quite a primitive oviraptorid in other respects.*

Chirostenotes

This animal is assembled from bits of skeleton excavated over a period of time. The hand was found in 1924, the feet (named *Macrophalangia*) in 1932, and the jaws (named *Caenagnathus*) in 1936. It was not realized that they all came from the same genus of animal until, in 1988, an unprepared skeleton was found in a museum collection where it had lain unstudied for 60 years.

Features: Each hand has three narrow, clawed fingers with the middle finger bigger than the others. *Chirostenotes* belongs to a side branch of the oviraptorid family known as the caenagnathids. They have toothless jaws, but the head is not as extremely specialized as we see in the oviraptorids proper, although it does have a spectacular crest.

Right: Chirostenotes *was closely related to another caenagnathid, called* Elmisaurus, *from Mongolia, showing that this was quite a wide-ranging group.*

Distribution: Alberta, Canada.
Classification: Theropoda, Tetanurae, Orviraptorosauria, Caenignathidae.
Meaning of name: Slim hand.
Named by: Gilmore, 1924.
Time: Campanian stage of the late Cretaceous.
Size: 2.9m (9½ft).
Lifestyle: Specialist feeder.
Species: *C. sternbergi*, *C. pergracilis*.

THERIZINOSAURIDS

The therizinosaurids represent a rare, exclusively Cretaceous family of dinosaurs, which is still undergoing revision. They may not represent a coherent group. The biggest is Therizinosaurus, which could be so different from the rest that the others have to be put into a different family altogether, and called the segnosaurids. They seem to show a mixture of plant-eating and meat-eating features.

Therizinosaurus

Distribution: Mongolia, Kazakhstan and Transbaykalia.
Classification: Theropoda, Tetanurae, Therizinosauria.
Meaning of name: Scythe lizard.
Named by: Maleev, 1954.
Time: Campanian stage of the late Cretaceous.
Size: 8–11m (26–36ft).
Lifestyle: Unclear.
Species: *T. cheloniformis*.

Therizinosaurus was first thought to have been a turtle-like animal (hence the species name). That was the best that could be made of the original remains, which consisted of some flattened ribs and a set of enormous arms and claws. Over the last half century more pieces have come to light and these give a better picture of the whole animal.

Features: The claws of this animal are its most remarkable feature. The hand has three of them, roughly the same length, with the longest measuring 71cm (28in). In life they would have been covered with a horny sheath perhaps half as long again. These massive hands are supported by arms that are 2.1m (7ft) long. Most of the rest of the animal has been restored on the basis of other better-known relatives, but this is the giant of the group.

Segnosaurus

Distribution: Mongolia.
Classification: Theropoda, Tetanurae, Therizinosauria.
Meaning of name: Slow lizard.
Named by: Perle, 1979.
Time: Cenomanian to Turonian stages of the late Cretaceous.
Size: 4–9m (13–30ft).
Lifestyle: Unclear.
Species: *S. galbinensis*.

We know of only a few remains of *Segnosaurus*, but the whole animal can be restored by comparison with its close relatives. The head is based on the skull of the closely related *Erlikosaurus*, and the feathery covering comes from an early Cretaceous form, *Beipiaosaurus*, found perfectly preserved in the Liaoning sediments, in China.

Features: The fragmentary remains of this genus show the typical down-turned jaw with the leaf-shaped teeth, and the hip bones with the swept-back pubis that give the impression of an ornithischian dinosaur. These are important details in establishing the make-up of this whole line of dinosaur. It is such an important animal that the name Segnosauria has been proposed as an alternative name for the group. Fossil eggs about the size of duck eggs have been attributed to these animals.

Right: The restoration is largely based on the partial remains of three skeletons. It differs from other therizinosaurids by the arrangement of teeth in the jaw.

Nothronychus

The first therizinosaurid to be found outside Asia is the most complete. It lived at the edge of the shallow sea that covered most of central North America at the time, in swampy deltas in the area of the Arizona/New Mexico borderlands. It is one of the few dinosaurs to be found from the early part of the late Cretaceous. Its name comes from the similarity between it and the giant ground sloths that existed until a few million years ago.

Features: The almost complete skeleton of this animal forms the basis for most modern restorations of therizinosaurids. It has a small head on a long neck, leaf-shaped teeth (suggesting a herbivorous habit), a heavy body with broad hips (also suggesting a partial diet of plant material), heavy hands and a short, stumpy tail. It carried itself with a more upright stance than other theropods, and its hind legs are relatively short. The hip girdle has the swept-back bird-like ischium bone that is usually only seen in plant-eating dinosaurs.

Distribution: New Mexico, USA.
Classification: Theropoda, Tetanurae, Therizinosauria.
Meaning of name: Sloth claw.
Named by: Kirkland and Wolfe vide Stanley, 2001.
Time: Turonian stage of the late Cretaceous.
Size: 4.5–6m (15–18½ ft).
Lifestyle: Unclear.
Species: *N. mckinleyi*.

Right: The resemblance to a ground sloth lies in its upright stance and the enormous claws on its hands. The claws were probably used for pulling down vegetation in the swampy forests where it lived.

THERIZINOSAUR DIET

With heads containing leaf-shaped, plant-shredding teeth, cheek pouches (indicated by the depressions at the side of the head), sharp cutting beaks, huge ripping claws and heavy, pot-bellied bodies, the therizinosaurids' lifestyle has always been a mystery. The hip bones, although in most respects saurischian, have the swept-back pubis typical of the ornithischians. This could be an adaptation to a plant-eating diet, to accommodate the big plant-processing digestive system. However, we see this feature also in the deinonychosaurs – as unambiguous a family of meat-eaters as we can find.

One intriguing interpretation of the huge claws is a comparison with the big claws of modern ant-eating animals, such as ant-eaters, aardvarks, armadillos and the like. In these animals the claws are an adaptation to ripping into nests and logs to reach tiny insects. However, it seems unlikely that such a diet would support an animal as big as *Therizinosaurus*.

The most likely interpretation is that the therizinosaurs were plant-eaters, the huge claws being used for ripping branches down from trees. What environmental pressures induced a family of meat-eating theropods to evolve into such a lifestyle is still a mystery.

Neimongosaurus

Neimongosaurus is known from two partial skeletons, one of which has most of the backbone and nearly all the limb bones. We know of only part of the skull, including the brain box. Mysterious arm and claw bones found in the same area in 1920, and named *Alectrosaurus*, may belong to *Neimongosaurus* and shed light on the missing hand bones.

Features: The long neck and the short tail, as well as the air spaces in the vertebrae and the arrangement of the shoulder muscles, suggest that these animals are closely related to the oviraptorids. Unfortunately, the hand bones are absent and we cannot compare them with the rest of the group. The jaw is deep and markedly down-turned, and bears a broad beak. The teeth are very similar to those of some ornithischians, suggesting a plant diet.

Distribution: Nei Mongol, China.
Classification: Theropoda, Tetanurae, Therizinosauria.
Meaning of name: Lizard from Nei Mongol.
Named by: Xu, Sereno, Kuang and Tan, 2001.
Time: Cenomanian to Campanian stages of the late Cretaceous.
Size: 2.3m (7½ ft).
Lifestyle: Unclear.
Species: *N. yangi*.

Left: The shoulder girdle is very much like that of an oviraptorid, as are the vertebrae, which are full of air spaces.

ALVAREZSAURIDS

Bird or dinosaur? The fleet-footed, compact-bodied, pointy-snouted alvarezsaurids represent one of those enigmatic groups that could be classed as either. Their bird-like features include specialized forelimbs with the breast bone, fused ankle bones and narrow skull. The non-bird-like features include the huge claw and long tail.

Alvarezsaurus

The skull and forelimbs of *Alvarezsaurus* (the most important features of the group) are missing from the only skeleton found, and it was not until other members of the group were discovered that scientists could appreciate just how unusual this animal was. It was originally restored with the three-fingered hands of a typical coelurosaurid: this restoration still appears today.

Features: The lack of spines on the back vertebrae, resulting in a compact body with no ridge down its back, show it to have been very bird-like. The tail is flattened from side to side and is very long, about twice the length of the body and neck. The neck is long and flexible. It had the long, lightweight feet of a running animal. *Alvarezsaurus* appears to be the most primitive of the Alvarezsaurid group.

Distribution: Neuquen, Argentina.
Classification: Theropoda, Tetanurae, Alvarezsauria.
Meaning of name: Alvarez's (historian Don Gregorio Alvarez) lizard.
Named by: Bonaparte, 1991.
Time: Coniacian to Santonian stages of the late Cretaceous.
Size: 2m (6½ft).
Lifestyle: Insectivore.
Species: *A. calvoi*.

Right: The marked contrast between this alvarezsaurid and the bird Archaeopteryx *has been noted.* Archaeopteryx *was obviously evolved to fly, but it had very few bird-like features. Alvarezsaurus, on the other hand, had many bird-like features but was certainly not evolved to fly. Evidently the various bird-like features and the various features adapted for flight evolved over and over again.*

Patagonykus

Distribution: Neuquen Province, Argentina.
Classification: Theropoda, Tetanurae, Alvarezsauria.
Meaning of name: Patagonian claw.
Named by: Novas, 1996.
Time: Turonian stage of the late Cretaceous.
Size: 2m (6½ft).
Lifestyle: Insectivore.
Species: *P. puertai*.

The discovery of *Patagonykus* in western Argentina in 1996 was important in establishing the alvarezsaurids as a group. The fragmentary skeleton was similar to *Alvarezsaurus*, but it had the distinctive short, powerful forelimb with the single claw that was such an important feature of the more complete Asian relative, *Mononykus*. *Patagonykus* and *Alvarezsaurus* both lived in South America, about as far away as it was possible to get in the late Cretaceous world from Mongolia where the rest of the group have been found. The alvarezsaurids represented a very widespread group.

Features: The arm is very distinctive, with quite a short, thin humerus, but a massive ulna that projects well back from the elbow joint, suggesting a very powerful leverage. The claw bone is almost as big as this ulna. The joint between the first tail vertebra and the hips is very flexible, indicating a great deal of movement here. This may show an ability to sit down on the heavy pubis bone, with the tail out of the way.

Right: Patagonykus seems to have been midway between the primitive members of the group, such as Alvarezsaurus, and the more advanced types, such as Mononykus.

Mononykus

The alvarezsaurid *Mononykus* was found in 1923 by an expedition from the American Museum of Natural History, but its significance was not realized at the time. It was referred to as an "unidentified bird-like dinosaur". Another expedition from the Museum found better examples in the 1990s. It was originally named *Mononychus*, but unfortunately that name had already been given to a kind of beetle.

Features: The upper and lower arms and the single claw are of equal length, and support a very strong musculature, attached to a keeled breast bone. This is interpreted as a digging adaptation. The long legs are very bird-like, with the fibula reduced to a vestige, and this has led some palaeontologists to think that *Mononykus*, and the rest of the alvarezsaurids, should be regarded as birds.

Distribution: Bugin Tsav, Mongolia.
Classification: Theropoda, Tetanurae, Alvarezsauria.
Meaning of name: Single claw.

Named by: Perle, Norrell, Chiappe and Clark, 1993.
Time: Campanian stage of the late Cretaceous.
Size: 0.9m (3ft).
Lifestyle: Insectivore.
Species: *M. olecranus*.

ALVAREZSAURID LIFESTYLE

The alvarezsaurids have always been a mystery. In their adaptations they were quite unlike any other dinosaur. The strange, stubby arms and the single massive claw were obviously an adaptation to some very specialized lifestyle. The current thinking is that these forelimbs were adapted for digging. The bird-like breast bone suggests powerful arm muscles, but they were certainly not used for flying. They were probably used for digging into the mounds of termites and other colonial insects, breaking up the concrete-like walls so that the living chambers could be reached by the long, narrow prehensile jaws and, perhaps, a long tongue. A similar lifestyle has been suggested for the therizinosaurids, but the large size of these animals seems to rule this out. It is a much more realistic prospect for the alvarezsaurids, which were considerably smaller animals and could possibly have subsisted on such a diet. Termite mounds are known from as far back as the Triassic period.

The particularly long legs would have been used for escape from predators, since they lacked any other form of defence. Indeed an alvarezsaurid ankle bone found by Othniel Charles Marsh in the 1880s was thought to have come from one of the fleet-footed ostrich mimics.

Right: Mononykus arm bones.

Shuvuuia

This dinosaur is known from two very well-preserved skulls. In fact it is the only member of the alvarezsaurids for which the skull is known. The appearance of the head of all the other alvarezsaurids shown in these restorations is based on these specimens. Some fossils originally attributed to *Mononykus* are thought to have come from *Shuvuuia*, the two are so similar.

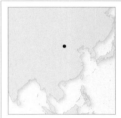

Features: The notable feature of the skull is the joint between the long, pointed snout and the bone in front of the eye. This means that the mouth could open very widely, hinged upwards in front of the brain case. The teeth, set in a continuous groove, are numerous and small, very similar to those of primitive birds. Analysis of the fossils shows traces of a chemical beta keratin only found in feathers, indicating that it had a feathered covering of some sort.

Right: Shuvuuia and Mononykus were so similar that many fossil bones previously attributed to Mononykus are now regarded as having belonged to Shuvuuia. However, Shuvuuia had two additional tiny claws on the hand that Mononykus lacked.

Distribution: Ukhaa Tolgod, Mongolia.
Classification: Theropoda, Tetanurae, Alvarezsauria.
Meaning of name: Bird.
Named by: Chiappe, Norell and Clark, 1998.
Time: Campanian stage of the late Cretaceous.
Size: 1m (3ft).

Lifestyle: Insectivore.
Species: *S. deserti*.

CRETACEOUS BIRDS

After the main spectacular early diversification of the early birds in early Cretaceous times, as seen in the remains in the lake deposits of China, many of the late Cretaceous birds began to look distinctly modern. However, there were still quite a few types that were unlike anything we have today. The bird group was still evolving and had not yet settled down to the families that have survived until today.

Rahonavis

Below: Rahonavis *combines the features of a bird and of a hunting dinosaur.*

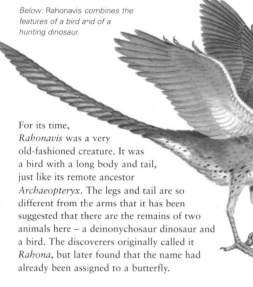

For its time, *Rahonavis* was a very old-fashioned creature. It was a bird with a long body and tail, just like its remote ancestor *Archaeopteryx*. The legs and tail are so different from the arms that it has been suggested that there are the remains of two animals here – a deinonychosaur dinosaur and a bird. The discoverers originally called it *Rahona*, but later found that the name had already been assigned to a butterfly.

Features: If it were not for the long arm bones *Rahonavis* would have been identified as a deinonychosaur such as *Velociraptor*, with its long bony tail, and the killing claw on its second toe. However, the arms are evidently adapted to be *Archaeopteryx*-type wings and have marks along the bone indicating where flight feathers were attached. It is a relict type that survived in Madagascar long after its relatives died out elsewhere. Alternatively it may indicate that the dromaeosaurs evolved flight, and then lost it again in their evolution, and that *Rahonavis* is actually a dromaeosaur.

Distribution: Madagascar.
Classification: Bird, order uncertain.
Meaning of name: Cloud menace bird.
Named by: Forster, Sampson, Chiappe and Krause, 1998.
Time: Campanian stage of the late Cretaceous.
Size: 50cm (20in) including tail – crow-sized.
Lifestyle: Hunter.
Species: *R. ostromi.*

Ichthyornis

One of the first Cretaceous birds to be found was *Ichthyornis*. Its name refers to the fact that the vertebrae are shaped like those of a fish. The ichthyornids appear to have been a side branch of bird evolution, that died out leaving no descendants. The original restorations of *Ichthyornis*, made in the 1880s show a bird with an overly large head. This is because the restorations were based on specimens of several individuals, all of different sizes.

Features: *Ichthyornis* has the build of a typical sea bird. However, its jaws are big, heavy and full of teeth. The toothed jaw is so reptile-like that when he described it Marsh thought at first that it belonged to a swimming lizard. He later realized his mistake. Then in the 1950s the jaw was re-identified as that of a mosasaur, but subsequent *Ichthyornis* discoveries in the 1970s showed that *Ichthyornis* really did have these big toothy jaws.

Distribution: Kansas, USA.
Classification: Bird, Ichthyornithiformes.
Meaning of name: Fish bird.
Named by: Marsh, 1883.
Time: Barremian stage of the early Cretaceous.
Size: 20cm (8in) – gull sized.
Lifestyle: Fish hunter.
Species: *I. dispar.*

Left: Ichthyornis *hunted for fish in the shallow Niobrara Sea that covered much of central North America in late Cretaceous times.*

Hesperornis

Below: Hesperornis *hunted for fish among the mosasaurs and plesiosaurs of the shallow seas at the end of the Cretaceous period.*

Distribution: Kansas, USA.
Classification: Bird, Hesperornith formes.
Meaning of name: Western bird.
Named: Marsh, 1872.
Time: Maastrichtian stage of the late Cretaceous.
Size: 2m (6½ft).
Lifestyle: Fish hunter.
Species: *H. regalis*.

The hesperornithiformes were man-sized flightless sea birds that preyed on small fish. They lived in the northern reaches of the Niobrara Sea – the coolest part – where they swam and fed like modern penguins. *Hesperornis* was typical. The largest known was *Canadaga*, from Canada, but a smaller genus *Baptornis* was also common. They were not purely marine, as their remains have been found in river deposits that were formed well inland.

Features: *Hesperornis* along with its relatives has lost all trace of its wings or shoulder structure. The feet are big and were obviously used for swimming. The neck is very long and flexible for snapping at fish, and the jaws are long and tooth lined with teeth running the full length of the lower jaw but restricted to the rear in the upper. The feet are broad and carry paddle-shape lobes. The loose joint at the front of the jaw suggest that the jaws were flexible enough to swallow very large prey, like the modern pelican. Its legs stick out at the side to help it to swim.

BIRD COLOURS

We cannot tell what colour fossil birds were, any more than we can tell the colour of any other long-extinct animal. We can assume, however, that they were as varied and brightly-coloured as today's birds.

A bird's coloration comes from its feathers. The colour of a feather can be due to the pigments within the feather structure, or it can be due to the effect of the light waves bouncing off the microscopic surface contours and interfering with one another. The latter effect gives the impression of changing colour patterns and is called iridescence.

Bird colours and patterns have generally evolved to assist their survival in some way. It may be that for a ground-bird some kind of cryptic coloration would be advisable – mottled browns and dark colours would camouflage a ground-living bird and hide it from predators. On the other hand, breeding males are usually brightly coloured, to make themselves attractive to females, often outdoing one another in the brightness of the display. The modern birds of paradise and peacock take this adaptation to an extreme. Scientists are unsure why this trait evolved, but it is possible that it arose because a very obvious colour display would make a bird an appealing target to a predator. The female would conclude that if a male could survive with such a disadvantage then it would be worthwhile breeding from it.

Some fossil feathers, mostly from Tertiary rocks, are so finely preserved that some indication of pattern is visible. However, telling the original colour is impossible.

Yandangornis

Across the late Cretaceous plains of China ran this ground-dwelling bird. It may have been flightless through having lost its powers of flight, or it may never have had flying ancestors in the first place. In appearance, with its long tail and its long running legs, *Yandangornis* would have resembled a swift-footed dinosaur more than a bird.

Distribution: China.
Classification: Bird, order uncertain.
Meaning of name: Bird from Yandang Mountain.
Named by: Cai and Zhao, 1999
Time: Santonian stage of the late Cretaceous.
Size: 60cm (2ft).

Features: The size of the arm bones indicates that this was a ground-dwelling bird. Its long legs make it a runner. The toes are not adapted for perching. It is evidently a very primitive bird judging by the long tail. The jaws supported a beak in life and lack teeth. Some scientists suggest that it may be an oviraptorid dinosaur, rather than a bird, but the wide breast bone and the structure of the hind limbs are definitely bird-like.

Lifestyle: Insect hunter.
Species: *Y. longicaudatus*.

Left: Yandangornis *must have been a fast hunter of ground-living animals.*

MORE BIRDS

Not only are there new dinosaurs being discovered on an almost weekly basis, but new Mesozoic birds are also continually coming to light. It used to be acknowledged that bird fossils had been found on every continent except Antarctica, but now even that remote landmass is revealing an ancient bird fauna dating from dinosaur times.

Patagopteryx

This flightless bird had a robust body and small wings. The original investigators decided that because it must have looked and behaved much like the ratites – the modern flightless birds – it was their distant relative. However it does not seem to have been related, and would have acquired its flightlessness totally independently.

Features: The powerful running feet have fused bones, as do modern birds. The wings are only half the length of the legs, the wishbone is absent and there is no keel on the breastbone, so it could not have had the muscles that would have made it a flying creature, giving a lie to the literal meaning of the name. The long hind leg has a very short femur and long lightweight foot bones. These features are usually the signs of a running animal. The second toe has a strongly curved claw, like the dromaeosaurids, but this does not seem to have been used as a weapon.

Distribution: Argentina.
Classification: Bird, order uncertain.
Meaning of name: Patagonian wing.
Named by: Alvarenga and Bonaparte, 1992.
Time: Santonian stage of the late Cretaceous.
Size: 70cm (2¼ft) – hen sized.
Lifestyle: Omnivore.
Species: *P. deferrariasi.*

Left: Patagopteryx probably ran in flocks across the plains of Cretaceous South America.

Gargantuavis

It is surprising to find ostrich-sized flightless birds existing alongside dinosaurs, where the two groups would be competing directly. It has therefore been suggested that *Gargantuavis* actually lived on islands along the northern edge of the Tethys Ocean at the time. Eggs that have been attributed to the sauropod *Hypselosaurus* may actually have been laid by *Gargantuavis.*

Features: *Gargantuavis* is known only from the pelvis and part of the femur. These bones definitely come from a bird and not a dinosaur because there are as many as ten vertebrae fused to the hip bones and this fusion is particularly solid – both bird traits. Its presence in late Cretaceous times dismisses an old idea that the big flightless birds did not evolve until the dinosaurs became extinct and left their ecological niches vacant.

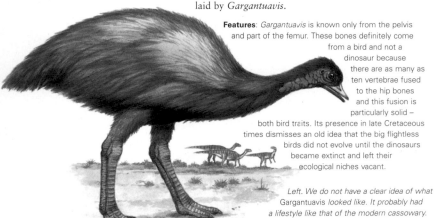

Distribution: Southern France.
Classification: Aves, order uncertain.
Meaning of name: Giant bird.
Named by: Buffetaut and Le Loeuff, 1998.
Time: Barremian stage of the early Cretaceous.
Size: 2m (6½ft) – ostrich sized.
Lifestyle: Unknown.
Species: *G. philoinos.*

Left. We do not have a clear idea of what Gargantuavis looked like. It probably had a lifestyle like that of the modern cassowary.

There have been long arguments about whether the various orders and families of modern birds existed before the end of the Cretaceous period and so shared the world with their dinosaur relatives. Scientific orthodoxy has always maintained that the modern groups evolved with a "big bang" immediately after the extinction of the dinosaurs. The fossil record always seemed to indicate this.

However, the distribution of the modern groups across the various continents, and the biological history as deduced from genetic studies, has indicated that their origin lies back in the Cretaceous period. Now fossils are coming to light to show that much of the radiation did indeed take place in Cretaceous times. It appears that the anseriformes (the ducks and geese), the galliformes (the hens) and the ratites (the big flightless birds) were in existence by the end of the Cretaceous and survived the extinction event.

Below: The last dinosaurs shared their world with modern birds.

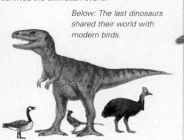

Teviornis

The fossil bird family known as the presbyornithids may have belonged to the order anseriformes, the group to which the modern ducks and geese belong, but may have been closer to the flamingos. It was a widespread family, with Tertiary specimens found from Maryland, North America, through southern England to Mongolia. However, in 2002, a late Cretaceous example was found in the Gobi Desert, pushing back the origin of these birds into the Mesozoic.

Below: If Teviornis had the same proportions as the later, better-known presbyornithids, it would have looked rather like a long-legged goose.

Distribution: Southern Mongolia.
Classification: Bird, Neognathae, Presbyornithidae.
Meaning of name: Bird of Victor Treschenko, the discoverer.
Named by: Kurochkin, Dyke and Karhu, 2002.
Time: Maastrichtian stage of the late Cretaceous.
Size: 15cm (6in) – duck sized.
Lifestyle: Water feeder.
Species: *T. gobiensis*.

Features: All that is known of this important genus is an arm bone, but enough of it is present to identify it as a member of the presbyornithids. Other, even less clear, specimens have been tentatively identified as members of the group, and so it may be that they were abundant around the late Cretaceous lakes of Asia, perhaps living and flying in flocks as modern geese and ducks do.

Vegavis

Distribution: Vega Island, Antarctica.
Classification: Bird, Neognathidae. Anseriformes.
Meaning of name: Bird from Vega Island.
Named by: Clarke, 2005.
Time: Maastrichtian stage of the late Cretaceous.
Size: 60cm (2ft) – goose sized.
Lifestyle: Water feeder.
Species: *V. iaai*.

The Anseriformes, the ducks and the geese, were around in the late stage of dinosaur times. A specimen collected in Antarctica in 1992 but not studied properly for about 12 years was found to be the complete skeleton of a duck-like bird from the end of the Cretaceous period.

Features: A major idea in the evolution of birds was that the modern groupings did not evolve until after the dinosaurs and pterosaurs had become extinct, leaving space for them to develop and radiate into the forms that we know today. The features of the partial skeleton of *Vegavis* place it firmly within the Anseriformes – the modern ducks and geese. This is the first time that a truly modern bird group has been identified in Mesozoic rocks.

Left: The ducks lived with the dinosaurs and survived the event that drove their neighbours to extinction.

DROMAEOSAURIDS

Dromaeosaurids, a taxonomic sub-group of the Deinonychosaurs, were found in the Gobi Desert in the 1920s. Several types have been found since in Asia and North America. Debate continues as to whether they evolved in Asia and migrated to North America, or vice versa. The discovery of remains in Europe in the 1990s suggests that the evolution took place in North America, since Europe was closer to it at that time.

Bambiraptor

This small, fast dromaeosaurid, about the size of a goose, was discovered by 14-year-old Wes Linster in 1994. The skeleton was about 95 per cent complete, and included such delicate elements as throat and ear bones. Its name gives the totally wrong idea of the vicious little predator with its killer claws, its cunning hunting skills and binocular vision. The species name, *B. feinbergorum*, honours the sponsors.

Features: The complete skeleton shows all the dromaeosaurid features, the killing claw on the second toe, the clawed hands held in a wing-like position, the wishbone, the deep, narrow head with serrated teeth, and the tail vertebrae lashed together as a stiff rod. This has the largest brain-to-body weight ratio of any dinosaur. The very long arms resemble clawed wings, and there is evidence of thin, hair-like feathers. *Bambiraptor* is the most bird-like dromaeosaurid so far discovered.

Distribution: Montana, USA.
Classification: Theropoda, Tetanurae, Deinonychosauria.
Meaning of name: Baby hunter.
Named by: Burnham, Derstler, Currie, Shou and Ostrom, 2000, amended Olshevsky, 2000.
Time: Campanian stage of the late Cretaceous.
Size: 1m (3ft).
Lifestyle: Hunter.
Species: *B. feinbergorum*

Adasaurus

Known from a partial skeleton, including an incomplete skull, the *Adasaurus* fossil found shows an elderly individual. The original paper was not very detailed, and the skeleton itself was in rather poor shape, being aged and diseased, and so it is difficult to tell quite what this animal looked like. It seems to belong to the dromaeosaurid family.

Features: This small dromaeosaurid has the typical, deep narrow skull of the group, but it has a slender hand, and a smaller killing claw on the second toe of its foot than most of its relatives. The many bird-like features, including the bird-like hip with the strongly swept-back pubis bone, have led some scientists to think that this should be classed as a large flightless killer bird rather than a dromaeosaurid dinosaur.

Distribution: Mongolia.
Classification: Theropoda, Tetanurae, Deinonychosauria.
Meaning of name: Ada's lizard.
Named by: Barsbold, 1983.
Time: Campanian to Maastrichtian stages of the late Cretaceous.
Size: 2m (6½ft).
Lifestyle: Hunter.
Species: *A. mongoliensis*.

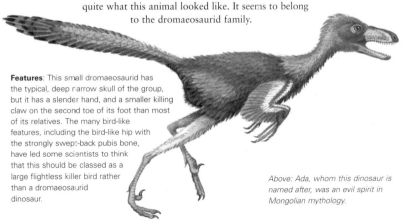

Above: Ada, whom this dinosaur is named after, was an evil spirit in Mongolian mythology.

GENUS SIZE

Dromaeosaurids seem to have come in a wide range of sizes. The smallest was more or less fixed at goose-size *Bambiraptor*, but new discoveries keep pushing the upper limit larger. Until the 1990s tiger-sized *Deinonychus* was regarded as the giant of the family. The group was generally thought of as having small, intelligent fast-moving predators that hunted by stealth and cunning, and which killed the big animals of the time by inflicting deep wounds, and waiting for them to bleed to death.

Then, in 1993, *Utahraptor* was discovered. At 7m (23ft) long it became the largest. This was topped by the discovery of the hand and foot bones, in South America, of *Megaraptor*, an animal that was about 8m (26ft) long, and which had a killing claw measuring 35cm (14in) in circumference, without the horny sheath. Explanations of this remarkable size abounded. The Cretaceous South American continent was famed for its giant fauna, and the giant deinonychosaurs were thought to have been part of this phenomenon. However, new specimens of this animal suggest that the big foot claw was really a hand claw, and that *Megaraptor* was not a deinonychosaur at all.

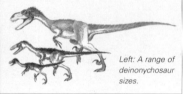

Left: A range of deinonychosaur sizes.

Variraptor

The first member of the group to have been found in Europe turned up in southern France, near Marseilles, in 1992. It existed late in the Mesozoic period, and there is speculation as to whether it existed right to the end of the Cretaceous period. Some scientists regard it as having been a scavenger, rather than a hunter, unlike the rest of the group who were hunters.

Distribution: France.
Classification: Theropoda, Tetanurae, Deinonychosauria.
Meaning of name: Hunter from the Var Department.
Named by: LeLoeuff and Buffetaut, 1998.

Features: The skull is long and lightly built with very sharp teeth in the jaws. The forearms are very powerful. However, there is doubt as to just how valid these fossil bones are, and they might belong to more than one animal. If so, this may well be a *nomen dubium*.

Time: Campanian to Maastrichtian stages of the late Cretaceous.
Size: 3m (10ft).
Lifestyle: Hunter.
Species: *V. mechinorum*.

Pyroraptor

Pyroraptor's name is derived from the fact that its remains were found in an area where the rocks had been cleared of vegetation after a forest fire. The specimen consists of the limb bones, a few vertebrae and some teeth. The material is more clearly dromaeosaurid than the remains of *Variraptor*, and this may well be the first definite member of the group to be found in Europe.

Features: The few bones found, including the diagnostic killing claw on the second toe, suggest that this may resemble some of the half-bird, half-dinosaur animals from Liaoning in China. It is a fast-moving, intelligent hunter. Although it is difficult to know, the fossils of *Variraptor* may be part of a *Pyroraptor*. If this proves to be the case, the name *Variraptor* will take precedence.

Left: Although Pyroraptor was similar to dinosaurs that lived in North America, the other dinosaurs that were found with it were most similar to South American forms, suggesting a land connection to both continents at that time.

Distribution: Provence, France.
Classification: Theropoda, Tetanurae, Deinonychosauria.
Meaning of name: Forest fire hunter.
Named by: Allain and Taquet, 2000.
Time: Campanian to Maastrichtian stages of the late Cretaceous.
Size: 2.5m (8ft).
Lifestyle: Hunter.
Species: *P. olympius*.

MORE DROMAEOSAURIDS

Without a doubt the dromaeosaurids were the most important active predators of late Cretaceous times. With their clawed hands that could grasp prey between their palms, their killing claw on the hind foot, and their mental agility that would have enabled them to balance and slash at the same time, they represented a fearsome group of animals.

Dromaeosaurus

Distribution: Alberta, Canada, and Montana, USA.
Classification: Theropoda, Tetanurae, Deinonychosauria.
Meaning of name: Running lizard.
Named by: Matthew and Brown, 1922.
Time: Campanian stage.
Size: 1.8m (6ft).
Lifestyle: Hunter.
Species: *D. albertensis, D. cristatus, D. gracilis, D. explanatus.*

Dromaeosaurus was the first of the group to have been discovered, and led to the establishment of the family. It is surprisingly poorly known, although a cast of a complete mounted skeleton, prepared by the Tyrrell Museum in Alberta, Canada, appears in several museums throughout the world. Its construction was made possible by knowledge of others of the group that have been discovered more recently.

Left: The famous dinosaur hunter Barnum Brown found the first and best of the Dromaeosaurus remains on the banks of the Red Deer River in Alberta, Canada, in 1914, naming it eight years later.

Features: The jaws are long and heavily built, and the neck is curved and flexible. The snout is deep and rounded. The tail is stiff and straight, articulated only at the base, stiffened by bony rods growing backwards from above and below the individual vertebrae. This would have helped the animal to balance while hunting prey. Its large eyes gave it excellent vision, and the size of the nasal cavities suggest that it could hunt by smell as well. The killing claw on the second toe is smaller than that of others of the group, but still efficient.

Saurornitholestes

This hunting dinosaur is known from the remains of three individuals. One remarkable occurrence is of a *Saurornitholestes'* tooth embedded in a pterosaur bone. It is not impossible to imagine an active predator like this snatching a pterosaur from the sky, but it is more likely that it scavenged the carcass of a pterosaur that had already died.

Features: The shape of the skull suggests a bigger brain than many of its relatives, but a poorer sense of smell. The teeth are also different from those of *Dromaeosaurus*, but otherwise the two animals are very similar, having grasping hands with sharp claws and a killing claw on the second toe. It was originally classed among the troodontids, but now some palaeontologists regard it as a species of *Velociraptor*.

Right: Saurornitholestes seems to be an amalgam of different animals. The head is very much like that of Velociraptor, while the rest of the skeleton (what has been found of it) is more like that of Deinonychus.

Distribution: Alberta, Canada.
Classification: Theropoda, Tetanurae, Deinonychosauria.
Meaning of name: Lizard bird thief.
Named by: Sues, 1978.
Time: Campanian stage of the late Cretaceous.
Size: 2m (6½ft).
Lifestyle: Hunter.
Species: *S. langstoni.*

DROMAEOSAUR ANCESTRY

A theory put forward by American palaeontologist Greg Paul, in 2000, suggests that the dromaeosaurids actually descended from a flying ancestor. Their ancestor would have been *Archaeopteryx*, or some relative of *Archaeopteryx*, back in late Jurassic times. It would have possessed the *Archaeopteryx* jaws and teeth, the clawed wings and the long, bony tail. At some subsequent time the descendants would have lost their powers of flight, and taken up a ground-dwelling existence. The wings would have atrophied, but the jaws and teeth, the claws, and the long tail would have been retained. They would have kept their plumage and their warm-blooded metabolism, but not their flight feathers. Like modern flightless birds, such as

Unenlagia

This dinosaur caused some confusion when it was discovered in the 1990s. It was so bird-like that it was initially regarded as a kind of bird. At one time it was thought to have been a juvenile specimen of the contemporary *Megaraptor*, but that seems unlikely as *Megaraptor* now appears to have been a totally different type of dinosaur.

Features: The shoulder joint of this dinosaur would allow for flapping, as though it had wings rather than the more normal dinosaur arms. It was too big for a flying animal, and it is possible that wing-like arms could have been used for stabilization and steering while the dinosaur was running at speed. This

Distribution: Argentina.
Classification: Theropoda, Tetanurae, Deinonychosauria.
Meaning of name: Half bird.
Named by: Novas, 1997.
Time: Turonian to Coniacian stages of

ostriches and emus, they would have become bigger and heavier, and eventually they would have looked as though they had never flown, so well would they have adapted to ground-living. This would account for the bird-like skeletons of these animals, and the suggestion of bird-like musculature in animals such as *Unenlagia*.

appears to add weight to the theory that dromaeosaurids evolved from flying ancestors. Certainly it shows how closely these animals are related to birds.

the late Cretaceous.
Size: 2–3m (6½–10ft).
Lifestyle: Hunter.
Species: *U. comahuensis*.

Left: Archaeopteryx (bottom) and Velociraptor (top).

Left: The only specimen found consists of 20 bones found in river-laid sedimentary rocks in Argentina by Fernando Novas of the Museum of Natural History in Buenos Aires. Its name is a mixture of Latin and the local Mapache language.

Velociraptor

Perhaps the best-known of the dromaeosaurids, *Velociraptor* is known from several specimens, the first found by the American Museum of Natural History expedition to Mongolia in the 1920s. Found in 1971, a famous fossil consisted of a complete *Velociraptor* skeleton wrapped around that of a *Protoceratops*. The two had been preserved in the middle of a fight, possibly engulfed in a sandstorm.

Features: The 80 very sharp curved teeth in a long snout, flattened from side to side, the three-fingered hands, each finger equipped with eagle-like talons, and the curved killing claw, 9cm (3½in) long, on the second toe of each foot show this to have been a ferocious hunter. Its long, stiff tail functioned as a balance while running and making sharp turns. A covering of feathers would help to keep the animal insulated, a necessity for its active, warm-blooded lifestyle.

Distribution: Mongolia; China; and Russia.
Classification: Theropoda, Tetanurae, Deinonychosauria.
Meaning of name: Fast hunter.
Named by: Osborn, 1924.
Time: Campanian stage of the late Cretaceous.
Size: 2m (6½ft).
Lifestyle: Hunter.
Species: *V. mongoliensis*. Several other genera, such as *Deinonychus*, *Saurornitholestes* and *Bambiraptor*, have been regarded as species of *Velociraptor* in the past.

TYRANNOSAURIDS

If the dromaeosaurids represented the ultimate small and agile killers, then the tyrannosaurids were undoubtedly the typical killer giants, especially of Asia and North America. They were among the biggest meat-eaters that ever walked the Earth. The tyrannosaurids are classed with the dromaeosaurids in the Coelurosauria, a division that formerly encompassed only the smallest of the meat-eating dinosaurs.

Albertosaurus

The earliest dinosaur remains to be found in Alberta, Canada, were *Albertosaurus* bones. They were found in 1884 by J. B. Tyrrell, after whom the world-famous dinosaur museum in Drumheller was named. It was one of the most abundant predators of the North American Cretaceous plains, and despite its great weight it was probably a fast runner, running down its prey, which would have consisted of duckbills.

Features: *Albertosaurus* is very similar to its later cousin *Tyrannosaurus*, but is only about half the size. It is much better known as its remains are more numerous. Its skull is heavier, with smaller gaps in it surrounded by thicker struts of bone, the muzzle is longer and lower and also much wider, and the jaw is considerably shallower. The arms, although small, are a little larger than those of *Tyrannosaurus*.

Distribution: Alberta, Canada; Alaska, Montana, Wyoming, USA.
Classification: Theropoda, Tetanurae, Tyrannosauroidea.
Meaning of name: Lizard from Alberta.
Named by: Osborn, 1905.
Time: Campanian to Maastrichtian stages of the late Cretaceous.
Size: 8.5m (28ft).
Lifestyle: Hunter.
Species: *A. sarcophagus*, *A. grandis*.

Nanotyrannus

This is significant in being the smallest tyrannosaurid yet recovered from the late Cretaceous. It is known from a skull discovered in 1942 and studied in the 1980s, when its significance was realized. Modern analytical techniques such as CAT scanning, a medical technique that produces 3D images of body interiors, were brought to it, revealing features inside that had not been noted in any other tyrannosaurid skull.

Distribution: Montana, USA.
Classification: Theropoda, Tetanurae, Tyrannosauroidea.
Meaning of name: Little tyrant.
Named by: Bakker, Currie and Williams, 1988.
Time: Maastrichtian stage of the late Cretaceous.
Size: 5m (16½ft).
Lifestyle: Hunter.
Species: *N. lancensis*.

Features: The skull is 57.2cm (22½in) long. It is long and low, with a narrow snout, broadening towards the rear. The eye sockets point forward, giving stereoscopic vision. Turbinals (scrolls of bone inside the nose) are present, showing an enhanced sense of smell or a cooling device. Some palaeontologists regard *Nanotyrannus* as a juvenile specimen of something better known, or even a dwarf species of *Albertosaurus* or *Gorgosaurus*.

Left: A new skeleton found in Montana in 2000 by the Burpee Museum in Illinois, USA, was thought to have been a second specimen of Nanotyrannus. It was later established that it was a juvenile Tyrannosaurus.

Alioramus

The Asian tyrannosaurid *Alioramus* is poorly known, and only from the jaw bones, part of the skull and some foot bones. Its name derives from the fact that this dinosaur represents a distinct branch of tyrannosaurid evolution that left the main evolutionary line early in the late Cretaceous. The animal was lightly built, and well adapted to chasing prey over open landscapes.

Features: The long, rugged snout of *Alioramus* is distinctive. A ridge down the nose shows six prominent bumps that may be horn cores, two side by side, with four in a single row in front of them. The long jaws have many more teeth – 18 in the front of each lower jaw alone – than any other tyrannosaurid, suggesting that this is a very primitive member of the group. The eyes are large and well adapted for hunting.

Right: The only specimen of Alioramus found consisted of a partial skull and some foot bones found in Ingenii Höövör Valley, in Mongolia. The name, meaning "different branch," refers to the fact that it diverged early from the main tyrannosaurid line.

Distribution: Mongolia.
Classification: Theropoda, Tetanurae, Tyrannosauroidea.
Meaning of name: Different branch.
Named by: Kurzanov, 1976.
Time: Maastrichtian stage of the late Cretaceous.
Size: 6m (19½ ft).
Lifestyle: Hunter.
Species: *A. remotus*.

TYRANNOSAURID ARMS

The small arms of a tyrannosaurid have always been a puzzle. Of what use are a pair of arms that are too small to reach the mouth, or anything else? The arm bones of a 12m (39ft) tyrannosaurid are about the same length as those of a human, but about three times as thick. The muscle scars show that the upper arm was very heavily muscled. There are only two fingers on each hand. The second finger is much larger than the first, and the third has disappeared altogether. The elbow joint was not very flexible. The hands were positioned so that the palms faced towards each other.

There are three main theories about the purpose of these arms. First, that they were used for clutching prey to the chest so that the jaws and teeth could reach it. Second, that they were used for grasping the female while mating. And third, that they were used to stabilize the body as the great animal stood from a lying position. We don't know for sure.

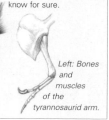

Left: Bones and muscles of the tyrannosaurid arm.

Appalachiosaurus

The finding of a tyrannosaurid in eastern North America in the early 1980s, and its scientific publication in 2005, was a surprise. Until then all members of the group had been found in western North America and Asia – separated from eastern North America by the Cretaceous inland sea. The ancestors must have spread across the North American continent before the inland sea cut the landmass in two.

Features: The single skeleton of *Appalachiosaurus* consists of most of the skull, the hind legs and parts of the tail and hips. This is enough to show that it is a typical primitive medium-size tyrannosaurid. The skull is similar to that of *Albertosaurus* and the ornamentation on it is small – unlike the bumps and knobs of *Alioramus* and other primitive tyrannosaurids. The most remarkable feature of this animal is the fact that it was found outside the expected geographic range of the group.

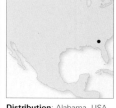

Distribution: Alabama, USA.
Classification: Theropoda, Tetanurae, Tyrannosauroidea.
Meaning of name: Lizard of the Appalachians.
Named by: Carr, Williamson and Schwimmer, 2005
Time: Campanian stage of the late Cretaceous.
Size: 7m (23ft).
Lifestyle: Hunter.
Species: *A. montgomeriensis*.

Left: The skeleton of Appalachiosaurus was found in marine mudstones. The body had been washed out to sea by river currents and deposited on the sea bed.

THE LAST TYRANNOSAURIDS

At the end of the Age of Dinosaurs, the biggest of the meat-eaters of the Northern Hemisphere were those in the tyrannosaurid family. By then, their body shape had settled into a consistent design, with each of the late examples becoming almost indistinguishable from the others. Most of the descriptions here concentrate on the small departures from the basic Tyrannosaurus *anatomy.*

Tyrannosaurus

Perhaps the best-known of all dinosaurs, *Tyrannosaurus* held the record for the biggest and most powerful land-living predator of all time for a century, until the discovery of the big carnosaurs, such as *Carcharodontosaurus* and *Giganotosaurus*, in the 1990s. About 20 skeletons of *Tyrannosaurus* are known, some articulated and some scattered, and so the appearance of this dinosaur is known with confidence.

Features: The skull is short and deep, and solid compared with that of other big meat-eaters. The teeth are 8–16cm (3–6in) long and about 2.5cm (1in) wide. Those at the front are D-shaped, built for gripping, while the back teeth are thin blades, evolved for shearing meat. The eyes are positioned so that they give a stereoscopic view forward. The ear structure is like that of crocodiles, which have good hearing.

Distribution: Alberta to Texas, USA.
Classification: Theropoda, Tetanurae, Tyrannosauroidea.
Meaning of name: Tyrant lizard.
Named by: Osborn, 1905.
Time: Maastrichtian stage of the late Cretaceous.
Size: 12m (39ft).
Lifestyle: Hunter or scavenger.
Species: *T. rex*, although *Daspletosaurus*, *Gorgosaurus* and *Tarbosaurus* are sometimes regarded as species of *Tyrannosaurus*.

Tarbosaurus

Tyrannosaurid *Tarbosaurus* is the largest Asian predator known, a close cousin of *Tyrannosaurus*. Indeed some regard it as a species of *Tyrannosaurus*, named *T. bataar*. Three skeletons were found by a Russian expedition to the Nemegt Formation, in the Gobi Desert, in the 1940s. Since then there have been almost as many *Tarbosaurus* skeletons as *Tyrannosaurus* skeletons found.

Features: *Tarbosaurus* is very similar to *Tyrannosaurus*, but it is less heavily built. It has a larger head with a shallower snout and lower jaw, and slightly smaller teeth. The other differences are in minor points of the shape of the individual skull bones. These features are slightly more primitive in *Tarbosaurus*, and so the early evolution may have taken place in Asia. Had it been found in North America, *Tarbosaurus* would have been regarded as a species of *Tyrannosaurus*.

Right: The two Tarbosaurus *skeletons found in Mongolia in the 1940s are currently mounted in the Palaeontological Institute of the Russian Academy of Sciences in Moscow, Russia.*

Distribution: China and Mongolia.
Classification: Theropoda, Tetanurae, Tyrannosauroidea.
Meaning of name: Alarming lizard.
Named by: Maleev, 1955.
Time: Maastrichtian stage of the late Cretaceous.
Size: 12m (39ft).
Lifestyle: Hunter or scavenger.
Species: *T. efremovi*, *T. bataar*.

Daspletosaurus

When C. M. Sternberg found the first skeleton of *Daspletosaurus* in 1921 he regarded it as a species of *Gorgosaurus*. However, it was revealed to be a much heavier animal. It is so similar to the slightly later *Tyrannosaurus* that it is sometimes regarded as its ancestor. Bonebeds of *Daspletosaurus* found in Montana, USA, suggest that it may have hunted in packs.

Features: The difference between *Daspletosaurus* and *Tyrannosaurus* is in the teeth. *Daspletosaurus* has even larger teeth than *Tyrannosaurus*. The neck and back are stockier, and the foot slightly shorter and heavier. Although slightly smaller, *Daspletosaurus* has a stronger build than *Tyrannosaurus*, and was much

Distribution: Alberta, Canada; and Montana, USA.
Classification: Theropoda, Tetanurae, Tyrannosauroidea.
Meaning of name: Frightful lizard.
Named by: D. A. Russell, 1970.
Time: Campanian stage of the late Cretaceous.
Size: 9m (30ft).
Lifestyle: Hunter or scavenger.
Species: *D. torosus* and one other, unnamed.

Right: There have been approximately six good specimens of Daspletosaurus *found, as well as quite a few scattered remains. The best mounted skeleton is in the Royal Tyrrell Museum, in Alberta, Canada.*

stronger than its contemporary *Albertosaurus*. It possibly fought the big, slow-moving, horned dinosaurs, while its more agile relatives preyed on the swift-footed duckbills.

HUNTER OR SCAVENGER?

A long-running debate among scientists concerns the lifestyle of the tyrannosaurids, and *Tyrannosaurus* in particular. Was it the fearsome hunter, the terror of the Cretaceous plains and forests, as it has always been portrayed? Or was it a slow-moving animal, eking out a living as a scavenger, subsisting only on the corpses of animals that had already died or been killed by more active hunters?

Evidence for the former includes the position of the eyes. These pointed forward giving stereoscopic vision, essential for a hunter of swift prey. The nostrils contained turbinal bones, thin sheets of bone that would have carried moist, sensitive tissue. This would have enhanced the sense of smell, which would be useful for either a hunter or a scavenger.

On the other hand, a *Tyrannosaurus* was a very big animal, perhaps too big to be capable of much sustained speed or activity. Once up to speed, the slightest stumble would have resulted in a crash that might have been fatal. The *Tyrannosaurus* tooth marks found on ceratopsian bones show signs of flesh being scraped off a dead animal, rather than chunks bitten off a live one, but this is not to say that the ceratopsian had not been killed by the *Tyrannosaurus* in the first place. A bite mark in the backbone of a duckbill was made by a *Tyrannosaurus* on a living animal. It is, in fact, very likely that both lines of evidence are valid, and that *Tyrannosaurus* and its relatives were active hunters, but did not pass up the chance of devouring any corpse that they came across.

Gorgosaurus

Gorgosaurus is known from more than 20 skeletons, since the first was found and named by Lambe in 1914. In the 1970s, a study of the tyrannosaurids indicated that *Gorgosaurus* and *Albertosaurus* were the same, and the name *Gorgosaurus* was dropped. Another study in 1981 showed that they were different, however, and so the name was reinstated.

Features: The several skulls found have different numbers of teeth, but are thought to belong to animals of different growth stages. There is a pair of short horns above the eyes which appears in two types, one in which the horns point up and forwards, and another in which they are longer and much more horizontal. It may be that this indicates two sexes or more than one species.

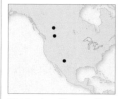

Distribution: Alberta to New Mexico, USA.
Classification: Theropoda, Tetanurae, Tyrannosauroidea.
Meaning of name: Gorgon lizard.
Named by: Lambe, 1914.
Time: Campanian stage of the late Cretaceous.
Size: 9m (30ft).
Lifestyle: Hunter or scavenger.
Species: *G. libratus*.

Left: The original specimen of Gorgosaurus *consisted of a complete skeleton with a crushed skull. Subsequent finds have shown different numbers of teeth in the jaws, but this is probably due to the ages of the different individuals.*

BIG TITANOSAURIDS

As the late Cretaceous period dawned, the sauropods appeared to have had their day, their role as the principal herbivores having been taken over by the ornithopods. However, the surviving titanosaurid family, a branch of the sauropod group, flourished, spreading out mostly over the southern continents, with some examples on the northern lands. They developed into the biggest land animals known.

Andesaurus

This very big animal is known from only a handful of vertebrae and some leg bones. It lived in an area of river plains and enclosed lagoons, a landscape clothed by a thick covering of conifers and ferns, in what is now the Comahue region of Patagonia, in Argentina. The remains were found by Alejandro Delgado while swimming in a lake.

Right: The beds in which the only specimen was found contain many iguanodont footprints. It seems that the ornithopods were the main herbivores of the area at that time, and the titanosaurids did not become really important until later.

Features: *Andesaurus* must be one of the biggest sauropods that ever lived. The few tail bones that have been found have ball-and-socket joints, rather than the usual flat faces like other sauropods. This suggests strength and flexibility in the tail. It comes from the same period of history as another titanosaurid giant, *Argentinosaurus*, but there are enough differences to show that they are separate genera. The back vertebrae have high spines, giving the animal a tall ridge down its back. The specimen is incomplete, and consists of some vertebrae, the hind legs and a few ribs.

Distribution: Argentina.
Classification: Sauropoda, Macronaria, Titanosauria.
Meaning of name: Lizard of the Andes.
Named by: Calvo and Bonaparte, 1991.
Time: Albian stage of the Cretaceous.
Size: 40m (131ft).
Lifestyle: Browser.
Species: *A. delgadoi*.

Paralititan

When *Paralititan* was discovered in 2000, it was seen to be the biggest dinosaur ever to have been found in Africa. Its discovery came as a surprise, as its presence in North Africa suggests that even in the late Cretaceous period there was some land link to the continent of South America, the place where the gigantic titanosaurids then flourished. It was found in rocks that were deposited in vegetated tidal flats and channels.

Features: *Paralititan* is very similar in its build to the huge titanosaurids of South America, Although the only specimen known consists of about 100 fragments of 16 different bones, we can judge the size of the beast by the fact that the upper arm bone, the humerus, is 1.69m (5½ft) long, the longest complete humerus known. The animal's weight has been estimated at a massive 70–80 tonnes. The bones are titanosaurid, and so its size can be calculated by comparing it with other titanosaurids.

Left: Paralititan's specific name honours Ernst Stromer, a German palaeontologist who worked in this area of Africa a century before Paralititan was found.

Distribution: Egypt.
Classification: Sauropoda, Macronaria, Titanosauria.
Meaning of name: Giant of the beaches.
Named by: Lamanna, Lacovara, Dodson, Smith, Poole, Giegengack and Attia, 2001.
Time: Albian or Cenomanian stage of the Cretaceous.
Size: 24–30m (79–98ft).
Lifestyle: Browser.
Species: *P. stromeri*.

THE TITANOSAURID SKELETON

Titanosaurids are known from very incomplete remains. Few titanosaurid skulls have been found but the skull fragments that we know show a wide, steeply sloping head and thin, peg-like teeth with tapering crowns. As a rule the neck is relatively short, the front legs are about three-quarters of the length of the back legs, and there is a shortish tail.

For all that the titanosaurids are the last group of sauropods to appear, the backbones are very primitive in aspect, lacking the deep hollows and the weight-bearing flanges and plates of the earlier groups. Likewise, the pelvis seems to have been different. It is fused to the backbone by six vertebrae rather than five as in the diplodocids. It is less robust than that of the earlier relatives and is much wider. This suggests that the living animals walked with a different gait from that of the diplodocids or the brachiosaurids. Titanosaurid footprints are easily recognized by their wide gait, as though they are spreading their legs further to the side than other sauropods. This contrasts with the narrow trackways of ciplodocids and brachiosaurids that seem to show the animals putting the left foot just in front of the right foot in a rather mincing manner. Since titanosaurid trackways are known from middle Jurassic times, the group is older than it appears from the body fossils that have been preserved.

Left: Prints of a diplodocid (above), titanosaurid (below).

"Bruhathkayosaurus"

When the remains of this dinosaur were first discovered in the 1980s it was thought they belonged to some kind of monstrous theropod. However, Chatterjee reclassified it as a titanosaurid in 1995. All in all, it is something of a *nomen dubium*, but the remains are important in belonging to the biggest animal found in India so far.

Features Despite the difficulties in actually identifying this huge animal, it seems that the tibia is about 25 per cent longer than that of South American *Argentinosaurus*. This would make it far bigger than *Argentinosaurus*, currently accepted as the biggest dinosaur. However, the uncertainty about the identity of "*Bruhathkayosaurus*" keeps it out of the record books. Its classification as a sauropod is based on the fact that nothing else is as big.

Below: Argentinosaurus.

Distribution: Tamilnadu, India.
Classification: Sauropoda, Macronaria, Titanosauria.
Meaning of name: Huge body lizard.
Named by: Yadagiri and Ayyasami, 1989.
Time: Maastrichtian stage of the late Cretaceous.
Size: 40m (131ft).
Lifestyle: Browser.
Species: "*B.*" *matleyi*.

Antarctosaurus

The original *Antarctosaurus* skeleton found is very complete for such a big animal, and consists of the skull and jaws, the shoulders and parts of the legs and hips. There is some doubt as to whether all this belonged to the same animal. Although it was found in South America, questionable material referred to *Antarctosaurus* has also been unearthed in India and Africa.

Features: *Antarctosaurus* has slim legs for the size of animal, and a small head with big eyes and a broad snout with only a few peg-shaped teeth at the front. The jaws are rather squared off at the front, like *Bonitasaura*, suggesting a low browsing or grazing feeding habit. Although it is one of the better-known southern sauropods, there is still a great deal of confusion about whether all the pieces belong to the same beast. One specimen has a thigh bone that measures 2.3m (7ft 8in), one of the biggest dinosaur bones ever found.

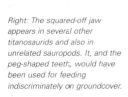

Right: The squared-off jaw appears in several other titanosaurids and also in unrelated sauropods. It, and the peg-shaped teeth, would have been used for feeding indiscriminately on groundcover.

Distribution: South America.
Classification: Sauropoda, Macronaria, Titanosauria.
Meaning of name: Southern lizard.
Named by: von Huene, 1929.
Time: Campanian to Maastrichtian stages of the late Cretaceous.
Size: 40m (131ft).
Lifestyle: Browser.
Species: *A. wichmannianus; A. jaxarticus, A. brasiliensis*.

SOME ARMOURED TITANOSAURIDS

An unusual feature about titanosaurids, and one that has come to light during the past decade or two, is the fact that at least some of them carried armour on their backs. This seemed just so unlikely for such large animals that the evidence was ignored for years, and researchers preferred to think that armour found with some titanosaurid remains actually belonged to ankylosaurs that had died nearby.

Saltasaurus

The first indication that some of the titanosaurids may have been covered in armour came with the discovery of *Saltasaurus*, and the re-examination of its close relative, *Laplatasaurus*, in the 1970s. Armour plates had been discovered long before this in South America, and von Heune attributed them to ankylosaurs in 1929. Since this discovery, other armoured titanosaurids have been excavated all over the world.

Features: The armour plates for which this dinosaur is famous consist of rough-surface discs 10–12cm (4–4¼in) in diameter, which probably supported conical horns. They are embedded in a tightly packed mass of tiny nodular bones covering the back and sides. The back and the base of the tail is as broad as in other titanosaurids. It has a rather short neck for a sauropod, and a long, whip-like tail.

Distribution: Argentina to Uruguay.
Classification: Sauropoda, Macronaria, Titanosauria.
Meaning of name: Lizard of Salta Province.
Named by: Bonaparte and Powell, 1980.
Time: Campanian to Maastrichtian stages of the late Cretaceous.
Size: 12m (39ft).
Lifestyle: Browser.
Species: *S. loricatus*.

Ampelosaurus

There are plenty of remains of *Ampelosaurus*, including several specimens that together show us what the whole skeleton looks like. The remains were found in a late Cretaceous river bed, and its ancestors probably reached France from South America via Africa. It is different from the other European titanosaurids, suggesting that the group diversified after reaching Europe. It is Europe's best-known sauropod.

Features: This is one of the titanosaurids that shows unmistakeable signs of being covered in armour, the first to have been found in the Northern Hemisphere. Heavy conical spines set in a matrix of tiny plates cover the shoulders. Behind them the plates become bigger and bigger until the hips are plated with a whole pavement of them. Stomach stones show that it swallowed stones to aid digestion.

Distribution: France.
Classification: Sauropoda, Macronaria, Titanosauria.

Meaning of name: Vineyard lizard (after the site of its discovery, the vineyard of Blanquette de Limoux).
Named by: Le Loeuff, 1995.
Time: Maastrichtian stage of the late Cretaceous.
Size: 15m (49ft).
Lifestyle: Browser.
Species: *A. atacis*.

Above: The late Cretaceous rocks in the valley of the river Aude, in France, in which Ampelosaurus *remains were found, are the richest in Europe for late Cretaceous dinosaur remains.*

Laplatasaurus

This is known from a large number of fossil bones, mostly neck and limb bones, that may or may not be from the same animal. It was found in three separate quarries by the famous German palaeontologist Friedrich von Huene, in 1927. Several individuals were represented, including the remains of juveniles. It is an important animal because it was the first sauropod to have been found that had a back covered in armour.

Features: *Laplatasaurus* is very similar to the better-known *Saltasaurus*, but is about half as big again. Like *Saltasaurus*, its back is covered in an armour of round plates set in a matrix of fine bony studs. It is a slimly built animal. There are no skull bones available to allow us to reconstruct the head. The range of remains suggests the animal lived in herds or family groups.

Above: Laplatasaurus *was originally described as* Titanosaurus australis *by Lydekker in 1893. So many sauropods were assigned to the genus* Titanosaurus *that it is no longer used.*

With no head or neck remains it is impossible to judge the full length of the animal.

Distribution: Argentina, Uruguay, Madagascar.
Classification: Sauropoda, Macronaria, Titanosauria.
Meaning of name: Lizard from La Plata.
Named by: von Heune, 1929.
Time: Campanian to Maastrichtian stages of the late Cretaceous.
Size: 18m (59ft).
Lifestyle: Browser.
Species: *L. araukanicus, L. madagascariensis.*

SAUROPOD ARMOUR

There were theories around as long ago as 1896 that sauropods were armoured, but it was not until the discovery of six partial skeletons of *Saltasaurus* in the 1970s that the theories were proved right. Since then there has been speculation as to the purpose of the armour. The obvious answer is defence, and with such big meat-eaters as the abelisaurids inhabiting South America at the time, it may be a reasonable idea.

Another possible explanation is a structural one. Members of earlier sauropod groups, such as the diplodocids and the brachiosaurids, supported their great weights with backbones that were extremely lightweight and carved out into hollows and stress-bearing struts and flanges. The vertebrae of the titanosaurids were much more conventional and did not have these strange structures. This meant that they were comparatively weak. It has been proposed by Jean Le Loeuff of the dinosaur museum at Esperaza that the armour of the titanosaurids served to stiffen the back, rather like the carapace of a crab, so that there was not too much strain placed on the backbone.

Neuquensaurus

This is one of those dinosaurs originally thought to have been a species of *Titanosaurus*. The remains of this animal consist of the scattered bones of several individuals. When they were found the armour pieces were at first thought to have belonged to an ankylosaur, a group unknown in South America. An excellent mounted skeleton of *Neuquensaurus* stands in the Museo de La Plata, in Argentina.

Features: *Neuquenosaurus* appears to be a medium-sized, advanced, titanosaurid like *Saltasaurus*, which occurred in the same area but in slightly earlier strata. The front feet are supported by columnar toes, held vertically, like others in the group. The thick body armour consists of bony plates and conical bumps across its back. It is possible that it was a species of *Saltasaurus*, despite their difference in ages, *Neuquensaurus* appearing somewhat later than *Saltasaurus*.

Distribution: Rio Negro Province, Argentina.
Classification: Sauropoda, Macronaria, Titanosauria.
Meaning of name: Lizard from the Nuequén Group.
Named by: Powell, 1992.
Time: Maastrichtian stage of the late Cretaceous.

Size: 15m (49ft).
Lifestyle: Browser.
Species: *N. australis.*

Left: The titanosaurids, like Neuquensaurus, *were probably herd-living animals, moving about the South American plains in large numbers to protect themselves from the big meat-eating dinosaurs of the time – the abelisaurids.*

EASTERN TITANOSAURIDS

The titanosaurid family is most famously known from South America, where they flourished as the main plant-eaters in the relative absence of the ornithopods. However, they were dispersed worldwide. The ornithopods seem to have predominated over the rest of the continents at that time, but several important titanosaurid genera are known from all continents, except Australia and Antarctica so far.

Nemegtosaurus

Our understanding of *Nemegtosaurus* is based on a partial skull that has features similar to *Brachiosaurus* and *Diplodocus*. It was classed as a diplodocid-like *Dicraeosaurus* for decades despite the fact that the diplodocids were unknown in rocks younger than the earliest Cretaceous. It was discovered by a Polish/Mongolian expedition in the 1960s.

Features: The skull is long, lightly constructed and has a drawn-out snout. The snout is slightly bent downwards. The eye sockets are large and contain a sclerotic ring (a ring of fine bones that supports the eyeball).

Distribution: Mongolia.
Classification: Sauropoda, Macronaria, Titanosauria.
Meaning of name: Lizard from the Nemegt Formation.
Named by: Nowinski, 1971.
Time: Campanian or Maastrichtian stage of the late Cretaceous.
Size: 12m (39ft).
Lifestyle: Browser.
Species: *N. pachi*, *N. mongoliensis*.

Above: Discovery of the more complete Rapetosaurus *(so complete as tc become a benchmark when comparing titanosaurids) showed that* Nemegtosaurus *was actually a titanosaurid.*

The lower jaw is lightly built, and the teeth long and pointed. It is thought by some to be the same as *Opisthocoelicaudia*, a titanosaurid from which, coincidentally, the skull and neck are missing.

Aegyptosaurus

Distribution: Egypt.
Classification: Sauropoda, Macronaria, Titanosauria.
Meaning of name: Egyptian lizard.
Named by: Stromer, 1932.
Time: Cenoman an stage of the late Cretaceous.
Size: 15m (49ft).
Lifestyle: Browser.
Species: *A. baharijensis.*

When found in 1932 this was the most complete titanosaurid skeleton known. It is one of the Egyptian dinosaurs destroyed during World War II, but there is plenty of published research based on it. It was probably the food source for the huge meat-eaters, *Carcharodontosaurus* and *Spinosaurus*, found in that area, and was a contemporary of the even bigger *Paralititan*. Both would have fed on the abundant vegetation.

Features: The single, partial skeleton known consists of only a few vertebrae, but it has an almost complete set of limbs (or had until it was destroyed), and this gives us an idea of the limb proportions of the whole group. However, the points of muscle attachment are different enough to distinguish *Aegyptosaurus* from other well-known titanosaurids such as *Saltasaurus* and *Argentinosaurus*. Despite this detailed knowledge of the leg bones it is not clear how closely related it is to other titanosaurids.

Above: Aegyptosaurus *was one of the many spectacular dinosaurs found by famed German palaeontologist Ernst Stromer von Reichenbach in the early part of the 20th century.*

Rapetosaurus

Rapetosaurus is the most complete titanosaurid to be discovered. The skeleton found was a juvenile, but adult skulls were discovered. Its discovery brings together strands of the group's classification. It shows that the sauropods most closely related to the titanosaurids were brachiosaurids.

Features: The neck is quite long, consisting of 16 vertebrae. The skull is brachiosaurid-like in the number of teeth and the articulation of bones around the big gaps in the skull. However, it is more like a diplodocid's in the length and overall shape. Other titanosaurid skull fragments from

Distribution: Madagascar.
Classification: Sauropoda, Macronaria, Titanosauria.
Meaning of name: Mischievous giant in local legend, lizard.
Named by: Rogers and Forster, 2001.
Time: Maastrichtian stage of the late Cretaceous.
Size: 15m (49ft).
Lifestyle: Browser.
Species: *R. krausei*.

Above: The species,
Rapetosaurus krausei, *honours American palaeontologist Dave Krause from the Field Museum, Chicago, who was the leader of the expedition that found it in a period covering five field seasons.*

other areas show different shapes and sizes of bones, suggesting that there was a great variation of head shapes within the titanosaurid group.

'TITANOSAURUS'

Although the titanosaurid group contains the biggest land animals that ever existed, the name is misleading since the majority were smaller than members of the other sauropod groups. The genus that gave its name to the group, *Titanosaurus*, is not very well known. The name was applied in 1877 by Richard Lydekker, a British palaeontologist, on the basis of two vertebrae and part of a leg bone. Since then *Titanosaurus* has become a wastebasket taxon, with about 14 species applied to it. This gives the genus an unlikely range that stretches from Argentina to southern Europe, Madagascar, India and Laos. These species cover a period of 60 million years which is also unlikely for a single genus.

Since the features that Lydekker used to define *Titanosaurus* are now thought to have no scientific significance, *Titanosaurus* has become a *nomen dubium*.

Below: The red marks indicated the distribution of 'Titanosaurus' *remains.*

Opisthocoelicaudia

Known from a complete skeleton but lacking the head and neck, found by a Polish-Mongolian expedition to the Gobi Desert in 1965, this may be the body of *Nemegtosaurus*, which is known from the skull found just a few kilometres away. The Gobi Desert at the time of the dinosaurs was covered in subtropical forests, with moist but seasonal climates.

Features: The feature that gives this animal its name is the shape of the tail vertebrae. In most sauropod groups the tail vertebrae were articulated on the ball and socket principle, with the socket facing forward. In *Opisthocoelicaudia*, the socket faces backward. This has since been found in other titanosaurids. Its back is almost level, and the short tail is held stiff and straight and off the ground.

Distribution: Mongolia.
Classification: Sauropoda, Macronaria, Titanosauria.
Meaning of name: Tail vertebrae with the socket to the rear.
Named by: Borsuk-Bialynicka, 1977.
Time: Campanian to Maastrichtian stages of the late Cretaceous.
Size: 12m (39ft).
Lifestyle: High browser.
Species: *O. skarzynskii*.

MISCELLANEOUS TITANOSAURIDS

As with other widespread animal groups, there were many different types of titanosaurid in different parts of the world. The members of the family ranged from the true titanosaurids of South America to the dwarf species of the European islands. Although a complete skull was not identified until recently, the variation of shapes in isolated skull bones suggests that titanosaurids had varied head shapes.

Hypselosaurus

The dinosaur *Hypselosaurus* is known from the scattered remains of at least ten individuals. Fossilized eggs, approximately 30cm (12in) in diameter and lying in groups of five, found near Aix in southern France (eggs-en-Provence, as some wit put it), have been attributed to *Hypselosaurus*, although this has not been scientifically confirmed. Another theory is that the eggs actually belong to a contemporary flightless bird, *Gargantuavis*.

Distribution: France and Spain.
Classification: Sauropoda, Macronaria, Titanosauria.
Meaning of name: High ridge lizard.
Named by: Matheron, 1869.
Time: Maastrichtian stage of the late Cretaceous.
Size: 12m (39ft).
Lifestyle: Browser.
Species: *H. priscus*.

Right: Hypselosaurus' eggs are spherical, more than twice the size of ostrich eggs and have a volume of 2 litres (½ gallon).

Features: It is difficult to restore *Hypselosaurus*. It is a large, four-footed, long-necked plant-eater, of a typical sauropod shape, but is known only from disarticulated remains. Comparing it with other titanosaurids, it seems to have had more robust limbs than its relatives. The teeth are weak and peg-shaped. We do not know whether it was covered in armour like some titanosaurids.

Magyarosaurus

The smallest known adult sauropod, *Magyarosaurus*, was found in Romania and Hungary. In late Cretaceous times this area of Europe was an island chain, and it seems likely that dwarf dinosaurs evolved on these islands to make the best use of limited food supplies. Other contemporary dwarf dinosaurs from this area include the duckbill, *Telmatosaurus*, and the ankylosaur, *Struthiosaurus*.

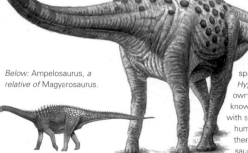

Below: Ampelosaurus, a relative of Magyarosaurus.

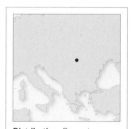

Features: *Magyarosaurus* is probably related to *Ampelosaurus*, and is probably one of the armoured forms. Its unusual stature makes it difficult to classify. Some scientists believe it to be a small species of "*Titanosaurus*", or even of *Hypselosaurus*, rather than a genus in its own right. In fact, the several specimens known are not consistent, and there are those with slender humeri and those with robust humeri. This may be a sexual difference, or there may be more than one genus of dwarf sauropod represented.

Distribution: Romania; Hungary.
Classification: Sauropoda, Macronaria, Titanosauria.
Meaning of name: Lizard of the Magyars, the main ethnic group living in Hungary.
Named by: von Huene, 1932.
Time: Maastrichtian stage of the late Cretaceous.
Size: 6m (19½ft).
Lifestyle: Low browser.
Species: *M. dacus*, *M. transylvanicus*.

Gondwanatitan

The single, partial skeleton known of *Gondwanatitan* consists of bones from all over the body. It was discovered in the mid-1980s by Fausto L. de Souza Cunha, palaeontologist of the National Museum of Rio De Janeiro, and researcher Jose Suarez. It lived by the sides of lakes, marshes and rivers that were common in central Brazil in late Cretaceous times.

Features: *Gondwanatitan* is a relatively small and lightly built titanosaurid. The skeletal features, particularly the articulation of the tail vertebrae, show it to have been quite different from, and more highly evolved than, any of the better known of the titanosaurids. The tibia is straight whereas in other titanosaurids it is curved. The spines on the tail bones point forward, suggesting that it is most closely related to *Aeolosaurus*.

Left: Gondwanatitan is named after the supercontinent Gondwana, which encompassed the whole landmass of the Southern Hemisphere at the time that this dinosaur lived.

Distribution: São Paulo state, Brazil.
Classification: Sauropoda, Macronaria, Titanosauria.
Meaning of name: Giant from Gondwana.
Named by: Kellner and de Azevedo, 1999.
Time: Maastrichtian stage of the late Cretaceous.
Size: 8m (26ft).
Lifestyle: Browser.
Species: *G. faustoi*.

DWARF SPECIES

The flora and fauna of islands have always evolved differently from those of mainland areas in response to local conditions. One of the results of this is dwarfism, the development of small versions of animals found elsewhere. During the last Ice Age, elephants evolved into dwarf forms that roamed the rugged outcrops of Malta. Miniature mammoths existed on islands off the coast of California, USA. On Caribbean islands the 4m (13ft)-high South American giant ground sloth evolved into a form that was no bigger than a domestic cat. Modern equivalents include the Shetland pony of the Scottish islands. Until a few thousand years ago a dwarf human, *Homo floresiensis*, inhabited the island of Flores in Indonesia.

In these circumstances there may be no predators. As there is only a limited amount of food in such areas, large populations of small individuals make the best use of the resources. Miniature dinosaurs roamed the island chains that lay across the shallow seas of late Cretaceous Europe.

Below: Size comparison of a dwarf species.

Aeolosaurus

The dinosaur *Aeolosaurus* was discovered in 1987 when much of the skeleton, including the backbone and leg bones, was found. However, an additional find in 1993 by Salgado and Coria produced evidence of back armour. It seems likely that *Aeolosaurus* inhabited the swampy lowlands and coastal plains of late Cretaceous Argentina, while other sauropods fed on the vegetation of the surrounding uplands. Eggs were also found associated with some of the bones.

Distribution: Patagonia.
Classification: Sauropoda, Macronaria, Titanosauria.
Meaning of name: Windy lizard (from the windswept plains of Patagonia).
Named by: Powell, 1987.
Time: Campanian or Maastrichtian stage of the late Cretaceous.
Size: 15m (49ft).
Lifestyle: High browser.
Species: *A. rionegrinus*.

Features: *Aeolosaurus* is known from many skeletal parts, including pieces of armour about 15cm (6in) in diameter. The forward-pointing spines on the tail vertebrae close to the hips are regarded as proof of its ability to rise on its hind legs, propped by its tail, to feed from high conifer branches. Some other sauropods would have been able to do this.

MORE MISCELLANEOUS TITANOSAURIDS

Titanosaurids belonged to the sauropod group. In general, although sauropods were more typical of Jurassic and early Cretaceous landscapes, the dinosaur Alamosaurus, *a titanosaurid, was among the last of the dinosaurs to have existed at the end of the Age of Dinosaurs. In all their era spanned 150 million years from the end of the Triassic to the end of the Cretaceous.*

Bonitasaura

When found, the remains of *Bonitasaura* were thought to have been a mixture of a titanosaurid skeleton and that of a diplodocid. The shape of the skull, with its broad-fronted jaws, is extremely similar to that of the diplodocid, *Nigersaurus*. This indicates a rapid evolutionary expansion of the titanosaurids once the diplodocids, their evolutionary cousins, had died out, with the former taking over all the niches of the latter.

Features: The amazing feature of this titanosaurid is the skull. The front of the jaw is broad and square, evidently adapted to cropping plants near the ground. The front teeth are short and pencil-like, rather like those of a diplodocid. Directly behind them, where the skull narrows, the jaws are equipped with horny blades evolved for slicing vegetation. Apart from this, the rest of the animal is similar to its relative, *Antarctosaurus*. The species is named to honour Leonardo Salgado, the famous Argentinian sauropod specialist.

Distribution: Patagonia.
Classification: Sauropoda, Macronaria, Titanosauria.
Meaning of name: Female lizard from La Bonita Hill, where it was discovered.
Named by: Apesteguia, 2004.
Time: Maastrichtian stage of the late Cretaceous.
Size: 9m (30ft).
Lifestyle: Browser.
Species: *B. salgadoi*.

OTHER TITANOSAURIDS

The titanosaurid *Pellegrinisaurus*, and other titanosaurids that were once assigned to the now defunct genus "*Titanosaurus*", probably inhabited the hilly areas of the South American island continent. They fed on the upland flora of conifers and ferns, and were menaced by the theropod *Abelisaurus*, while other titanosaurids such as *Aeolosaurus* and the relatively few ornithopods present lived on the nearby coastal lowland plains.

Since *Pellegrinisaurus* is known from a single specimen consisting of part of a backbone and a femur, little can be said about its appearance. It was first thought to have been a species of *Epachthosaurus* when it was discovered in 1975, but work 20 years later showed that it was sufficiently different to have been a separate genus.

NESTING IN PATAGONIA

One of the most spectacular dinosaur nesting sites in the world was discovered, and studied, in the late 1990s in Campanian rocks in Patagonia. Called Auca Mahueveo, the site stretches for several square kilometres. It was used regularly by hundreds or thousands of dinosaurs that returned to the site each year.

The site represents the floodplain of a river that periodically overflowed its banks. Mud deposited each time it flooded buried and suffocated the eggs that were exposed there, and preserved them. Eggs discovered in subsequent layers showed that dinosaurs were not discouraged by random disasters.

Each egg is almost spherical, measuring 13–11.5cm (5–4½in), and the best examples preserve the structure of the eggshell, the imprint of the internal membrane and even the baby complete with lizard-like skin. The embryos have an "egg tooth" on the front of the upper jaw, as birds do. This was used to help in breaking through the egg shell. It is clear that the eggs were laid by titanosaurids, but the exact genus has not been identified. The embryonic skin lacks armour, which suggests that it developed after the animal hatched.

Left: Bonitasaurus.

Alamosaurus

After a gap of 35–40 million years when there were no sauropods in North America, the genus returned with *Alamosaurus*. Until the discovery of an as yet unnamed titanosaurid in the late Cretaceous period of Chihuahua Province in Mexico in 2002, this was the sole representative of the group in North America. It migrated from South America across the land-bridge of Central America established in late Cretaceous times.

Features: *Alamosaurus* is larger, but more lightly built than its close relative *Saltasaurus*, and does not seem to have possessed armour. Several partial skeletons are known from various parts of North America. A site in Texas has produced the remains of an adult and two well-grown juveniles, suggesting that it lived in a family structure. A study has suggested that the total population of *Alamosaurus* in Texas at any one time would have been about 350,000. That would represent a density of one per 2km².

Distribution: New Mexico, Utah and Texas, USA.
Classification: Sauropoda, Macronaria, Titanosauria.
Meaning of name: Lizard from the Ojo Alamo Sandstone (alamo being the local Spanish name for the cottonwood tree).
Named by: Gilmore, 1922.
Time: Maastrichtian stage of the late Cretaceous.
Size: 21m (69ft).
Lifestyle: Browser.
Species: *A. sanjuanensis.*

Left: Remains of Alamosaurus *have been found in the latest Cretaceous rocks from New Mexico to Utah, USA. However none has been found in Alaska, where similar aged deposits yield other dinosaurs. Perhaps climate was a factor in this distribution.*

Epachthosaurus

When it was found, *Epachthosaurus* was thought to have been buried in Cenomanian deposits, from the base of the late Cretaceous. This fitted well with the primitive aspect of the bones. However, later studies showed it to have come from the end of the period, making it a primitive throwback to an earlier type. Primitive forms must have existed alongside the more advanced.

Features: This primitive titanosaurid shares its features with many earlier types. It differs from other titanosaurids in the structure of the back, the shape of the articulations between the vertebrae, and the presence of a very strong linkage between the hip and the backbone giving it a very strong back structure. The single skeleton known is almost complete except for the neck and head, and the tip of the tail, but shows no sign of armour plates. The join between the hips and the backbone consists of six vertebrae that are completely fused.

Right: An articulated skeleton of Epachthosaurus *has been found, with its hind legs crumpled beneath it and its front legs splayed out. It lay on its stomach in death.*

Distribution: Argentina.
Classification: Sauropoda, Macronaria, Titanosauria.
Meaning of name: Heavy lizard.
Named by: J. E. Powell, 1990.
Time: Maastrichtian stage of the late Cretaceous.
Size: 15–20m (49–65½ft).
Lifestyle: Browser.
Species: *E. sciuttoi.*

SMALL ORNITHOPODS

The late Cretaceous period was the time of the big ornithopods. However, there were still many smaller types that scampered around the feet of the giants. Some retained their primitive features from earlier times, but some were quite advanced. Their narrow beaks were able to select food from the newly evolved flowering plants on the ground.

Thescelosaurus

The classification of *Thescelosaurus*, known from complete and articulated remains, has been moved between the hypsilophodontids and the iguanodontids for most of the last century, and the authorities are still not sure where it sits. One specimen, nicknamed Willo, has the remains of the internal organs, including the heart, preserved. Its rather primitive structure is surprising considering it is among the last of the dinosaurs to have lived.

Above: The specific name, Thescelosaurus neglectus, refers to the fact that the skeleton was not studied for 22 years after it was discovered and collected.

Features: *Thescelosaurus* is more heavily built than other medium-sized ornithopods, and the legs are relatively shorter. This is not an animal built for speed. As with its relatives, it has a beak at the front of the mouth but an unusual arrangement of three different kinds of teeth – sharp in the front, canine-like at the side and molar-like grinding teeth at the back. The four-toed foot is one of the primitive features of *Thescelosaurus*.

Distribution: Alberta, Canada, Saskatchewan, Colorado, Montana, South Dakota and Wyoming, USA.
Classification: Ornithopoda.
Meaning of name: Wonderful lizard.
Named by: Gilmore, 1913.
Time: Campanian to Maastrichtian stages of the late Cretaceous.
Size: 4m (13ft).
Lifestyle: Browser.
Species: *T. neglectus*. Several others, all unnamed.

Orodromeus

Skeletons of this small ornithopod were found in Montana with the remains of dinosaur eggs and nests at a site known as Egg Mountain. For some time it was believed that *Orodromeus* was the builder of the nests, but later it was found that they actually belonged to the meat-eater *Troodon* and that *Orodromeus* represented the prey.

Features: This is a fairly primitive small ornithopod. Its legs show it to have been a speedy animal, hence its name. The head is small, with a beak and cheek pouches. The neck is long and flexible, and the tail is stiffened by bony rods that helped to keep it straight for balance while running. There are three toes on the hind foot, and four fingers on the hand.

Distribution: Montana, USA.
Classification: Ornithopoda, Hypsilophodontidae.
Meaning of name: Mountain runner.
Named by: Horner and Weishampel, 1988.
Time: Campanian stage of the late Cretaceous.
Size: 2.5m (8ft).
Lifestyle: Low browser.
Species: *O. makelai*.

Left: Fossil finds of Orodromeus suggest that it lived in small groups.

Oryctodromeus

Although the appearance of *Oryctodromeus* is very similar to that of the other hypsilophodonts, the manner of its discovery is very unusual. Its remains were found in the structure of an underground burrow about 2m (6ft) long, and consisted of an adult, and two juveniles about half the size of the adult. It probably had a lifestyle similar to that of modern bears – living on the surface but digging dens in which to breed.

Features:
Oryctodromeus's only physical adaptations to a burrowing habit lie in its slightly stronger shoulders, its broad shovel-like beak that would have helped it to dig, and its tail, which is lacking the usual stiffening rods. This last feature would make the tail more flexible for the restricted space of the burrow. The burrow is the same size as the adult, suggesting that the dinosaur actually made it, and did not just occupy some other animal's lair.

Left: It is possible that other hypsilophodonts, such as Orodromeus, were able to dig burrows and raise their families there as well. The burrow was filled with sand, and had turned to sandstone.

Distribution: Montana, USA.
Classification: Ornithopoda, Hypsilophodontidae.
Meaning of name: Digging runner.
Named by: Varricchio, Martin and Katsura, 2007.
Time: Cenomanian stage of the late Cretaceous.
Size: 2.1m (6⅔ft).
Lifestyle: Low browser, sometimes burrower.
Species: *O. cubicularis.*

DISCOVERY OF THE HEART

The internal organs of the *Thescelosaurus* specimen nicknamed Willo, named after the wife of the rancher on whose property it was found, consist of cartilaginous ribs and plates, tendons attached to the vertebrae, and most importantly, the heart.

The grapefruit-size lump in the chest was suspected of being the heart but there was a great deal of doubt. Finally it was scanned using a medical CAT scanning device, which is used in hospitals to build up a three-dimensional image of a patient's internal organs, and this proved without a doubt that it was a heart. The heart has four chambers, a double pump and a single aorta, making it more like that of a bird or mammal than that of a reptile, and suggests a very high metabolic rate. The work was done at North Carolina State University and the North Carolina Museum of Natural Sciences.

Below: The heart on the left is that of a crocodile, while that on the right is that of a dinosaur.

Paired systemic aorta

Single systemic aorta

Valve

Right Ventricle

Left Ventricle

Gasparinisaura

Ornithopods are rare in South America when compared with the sauropods, and so the discovery of *Gasparinisaura* was quite important. This small plant-eater is known from a complete juvenile specimen and about 15 other incomplete skeletons. It is classed as a small iguanodontid, but recent studies show that it may have been closer to the hypsilophodontids.

Features: As in the hypsilophodontids, *Gasparinisaura* has very narrow front bones in the hips, which suggest that it may be related to this group, but the feature may have developed independently. It is a rather dainty-looking dinosaur, but with quite sturdy legs for the size of animal and very small arms. The head is short, ending in a beak, and the mouth contains diamond-shape chopping teeth. The eyes are large, but this may be because the specimen is a juvenile.

Distribution: Neuquen Province, Argentina.
Classification: Ornithopoda.
Meaning of name: Dr Zulma Gasparini's lizard.
Named by: Coria and Saldago, 1996.
Time: Coniacian to Santonian stages of the late Cretaceous.
Size: 0.8m (2½ft), but this may be a juvenile.
Lifestyle: Low browser.

Species: *G. cincosaltensis.*

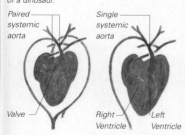

Left: The discovery of parts of several individuals in a small area suggests that Gasparinisaura may have been gregarious.

THE IGUANODONTID-HADROSAURID CONNECTION

The most important plant-eaters in the Northern Hemisphere at the end of the dinosaur era were the hadrosaurids, or duck-billed dinosaurs. They evolved from iguanodontids halfway through the Cretaceous. At the time many ornithopods had features from both groups, suggesting a transition stage.

Telmatosaurus

We know *Telmatosaurus* from several fragmentary skulls from individuals of different ages, as well as pieces of the rest of the skeleton. There are also clutches of two to four eggs that have been attributed to *Telmatosaurus*. It is one of the few hadrosaurids to have been found in Europe. Its relative smallness may be due to the fact that it lived on islands.

Features: What we know of *Telmatosaurus* suggests that it was very similar to *Iguanodon*, despite the fact that it occured at the end of the Cretaceous period. Like *Equijubus*, the appearance of its skeleton supports the idea that the hadrosaurids evolved from the iguanodontid line. It has a deep skull like *Gryposaurus*, but lacks the duck-like bill possessed by other members of the group; instead it has a long, drawn-out snout.

Distribution: Romania; France; and Spain.
Classification: Ornithopoda, Iguanodontidae.
Meaning of name: Marsh lizard.
Named by: Nopcsa, 1899.
Time: Maastrichtian stage of the late Cretaceous.
Size: 5m (16½ft).
Lifestyle: Browser.
Species: *T. transylvanicus*, *T. cantabrigiensis*.

Right: Telmatosaurus is very primitive, despite its late appearance, which suggests that its island habitat acted as a refugium – an isolated area where animals continued to exist despite dying out elsewhere.

Gilmoreosaurus

One of the earliest Asian hadrosaurids, *Gilmoreosaurus* was originally thought to have been a species of *Mandschurosaurus*. The confusion is typical of its history. Fossils that were originally thought to be from *Bactrosaurus* now seem to be *Gilmoreosaurus*. Another genus, *Cionodon*, from Asia, is now regarded as *Gilmoreosaurus* as well.

Features: *Gilmoreosaurus* is quite a lightweight animal for its size, but has particularly strong legs. Its feet are rather like those of *Iguanodon*, emphasizing the close relationship between the two groups. As with other medium-size ornithopods it had the ability to walk either on its hind legs, balanced by the heavy tail, or on all fours, taking the weight of its forequarters on its strengthened fingertips.

Distribution: Mongolia.
Classification: Ornithopoda, Hadrosauridae.
Meaning of name: Charles Whitney Gilmore's lizard.
Named by: Brett-Surman, 1979.
Time: Cenomanian to Maastrichtian stages of the late Cretaceous.
Size: 8m (26ft).
Lifestyle: Browser.
Species: *G. mongoliensis*, *G. atavus*, '*G. arkhangelskyi*'.

Left: Although it was found in 1923, and partially studied in 1933, Gilmoreosaurus was not properly studied until 1979. One species, G. arkhangelskyi, is a nomen dubium, as the partial skeleton seems to be made of different animals.

Equijubus

Although *Equijubus* was found in late early Cretaceous beds, it is included here because of its significance in hadrosaurid evolution. It consists of a complete skull with an articulated jaw, combining features found in the iguanodontids and the hadrosaurids. Its presence suggests that the hadrosaurids evolved from the iguanodontids somewhere in Asia at the end of the early Cretaceous, or the beginning of the late Cretaceous.

Features: The significance of *Equijubus* is that it had the long jaws of the iguanodontids, but the elaborate tooth arrangement consisting of a large number of small, tightly packed grinding teeth in very mobile jaws seen in the hadrosaurids. This shows a transition between the two groups, but *Equijubus* is regarded as the earliest and most primitive known of the hadrosaurids. The rest of the skeleton could have come from the group of ornithopod dinosaurs.

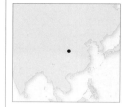

Distribution: Gansu Province, China.
Classification: Ornithopoda, Hadrosauridae.
Meaning of name: Horse's mane (local name of the site of discovery).
Named by: You, Luo, Shubin, Witmer, Zhi-lu Tang and Feng Tang, 2003.
Time: Albian stage.
Size: 5m (16½ft).
Lifestyle: Browser.
Species: *E. normanii*.

Right: The genus name Equijubus *is derived from the Chinese translation of Ma Zong, the name of the mountain range where this dinosaur was found. The specific name comes from David Norman (the British expert on iguanodontids).*

EVOLUTIONARY PROGRESS

The hadrosaurids were the most abundant of the plant-eaters at the end of the age of dinosaurs. They evolved with the development of flowering plants, and may have diversified in response to the spread of this new food source.

There is a theory, proposed by Bob Bakker in the 1980s, that flowering plants evolved in response to heavy grazing by hordes of herbivorous dinosaurs. If low-growing plants were being eaten at a great rate, then evolution would favour the ability of a plant to regenerate itself quickly, and hence the seed of a flowering plant, with its self-contained food supply feeding its fertilized embryo, became more efficient than the spore of the fern which is cast off and fertilized by a hit-and-miss lottery.

The shape of a hadrosaurid neck is consistent with a low feeder. The distinctive S-shape is very similar to that of a modern horse or a buffalo, bringing the snout and the mouth down to where the low plants are growing. There is a suggestion that the traditional restoration, with the snaky neck, is wrong. Perhaps the curve of the neck was filled with muscle, like that of a horse. The two types of hadrosaurid, the lambeosaurines and the hadrosaurines, have differently-shaped duckbills that must reflect a different feeding strategy. The narrow-beaked lambeosaurines would be selective about what plants they ate, while the broad-beaked hadrosaurines would have taken great mouthfuls indiscriminately. They would both also have been able to rear up on their hind legs to scrape leaves from trees. Neither type would have eaten grass, which did not evolve significantly until after the dinosaurs were extinct.

Protohadros

This dinosaur is based on a skull and some bones from the rest of the body. When it was found in Texas, USA, by Gary Bird, it was hailed as the most primitive hadrosaurid yet discovered, even though it was not the earliest, hence its name. Scientific opinion has changed and it is now regarded as a very specialized member of the iguanodontid group.

Features: The lower jaw of *Protohadros* is very large. The snout is turned down at the front, which suggests a habit of grazing low-growing vegetation, rather than browsing from bushes or overhanging branches. It does not have the flexible mouth that produced the food-grinding action seen in either iguanodontids or hadrosaurids. Its food would have been the aquatic plants of the delta streams in its habitat, scooped up by the broad, down-turned mouth.

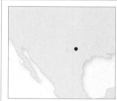

Distribution: Texas, USA.
Classification: Ornithopoda, Iguanodontidae.
Meaning of name: First hadrosaurid.
Named by: Head, 1998.
Time: Cenomanian stage of the late Cretaceous.
Size: 6m (19½ft).
Lifestyle: Low browser.
Species: *P. byrdi*.

Right: The location of the Protohadros *discovery in Texas, USA, confounded the accepted idea that the hadrosaurids evolved in Asia. However, its reclassification as an iguanodontid has removed this ambiguity.*

PRIMITIVE HADROSAURIDS

The dinosaurs of the hadrosaurid family differed from each other in the shape of the head. They were bulky animals with heavy tails. The hind legs were longer than the front, and their feet had three broad, hoofed toes. The fingers were fused together by a mitten of skin that encased a weight-bearing pad. Often portrayed as bi-pedal animals, they would have spent most of their time on all fours.

Secernosaurus

The first hadrosaurid to be described from South America, a continent not known for its advanced ornithopod fauna was *Secernosaurus*. It is based on a partial skull and some pelvic and tail bones found in 1923, but it was not described until 1979. The translation of the name refers to the fact that it was found a long way away from any other duckbilled dinosaur.

Features: The remains are so scrappy and so isolated from any other duckbill that it is difficult to compare *Secernosaurus* with any others in the group. There is not enough of the skull to determine whether it had a crest. In fact it is only the hip bones, similar to those of

Right: Until the discovery of Secernosaurus, it was assumed that the only big plant-eaters in Cretaceous South America were the titanosaurid sauropods. It had been thought that the hadrosaurids had never crossed from North America.

Edmontosaurus and *Shantungosaurus*, that reveal this animal to be a duckbill. The similarity to those two suggests that it is a member of the flat-headed group. Until more are found, *Secernosaurus* will remain a mystery.

Distribution: Rio Negro, Argentina.
Classification: Ornithopoda, Hadrosauridae.
Meaning of name: Separate lizard.
Named by: Brett-Surman, 1979.
Time: Maastrichtian stage of the late Cretaceous.
Size: 3m (10ft).
Lifestyle: Browser.
Species: *S. koerneri*.

Mandschurosaurus

This was the first dinosaur to have been unearthed in China. It was dug up by a Russian expedition in 1914 but wasn't scientifically described for another 16 years. The skeleton was very fragmentary and, although it is mounted in the Central Geological and Prospecting Museum in Leningrad, this mount is mostly recreated from plaster.

Features: As with *Secernosaurus* there really is not enough of the original skeleton to tell us much about the appearance of this animal. What we can tell is that it was a very large member of the duckbills, and was probably one of the flat-headed kind, without a crest; there are no skull remains that would enable

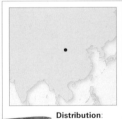

Distribution: Mei Mongol Zizhiqu, China.
Classification: Ornithopoda, Hadrosauridae.
Meaning of name: Manchurian lizard, from the Amur River.
Named by: Riabinin, 1930.
Time: Campanian to Maastrichtian stages of the late Cretaceous.
Size: 8m (26ft).
Lifestyle: Browser.
Species: *M. amurensis*.

us to verify this. It had the typical weight-bearing hand of the hadrosaurids. It is quite possible that it is an iguanodontid rather than a hadrosaurid – or something else altogether.

Left: Not until the discovery of Mandschurosaurus in 1914 were dinosaurs known from China. Since then, however, China has become an important locality in the hunt for dinosaurs of all periods.

HADROSAURID TEETH

The hadrosaurid chewing mechanism was quite complex. As in iguanodontids, the upper teeth were mounted on jaw bones that were loosely articulated to the skull. When the lower jaw was brought upwards, the upper jaw bones moved outwards slightly, allowing for the crowns of the teeth to slide past one another at an angle, and for a strong grinding action. The grinding surface of the teeth was formed of ridges of hard enamel. Hard though they were, the teeth constantly wore away and had to be replaced.

A solid mass of tightly interlocked teeth, packed like dates in a box, grew up from the lower jawbone and down from the upper jawbone to present a fresh grinding surface. There may have been 2,000 teeth growing in a hadrosaurid mouth, but only those at the surface were used at any one time. This was an ideal arrangement for grinding down tough plant material, such as conifer leaves and cones. They were scraped off the branches of the trees by the broad, toothless, horny bill, transferred to the back of the mouth where this grinding battery was situated, and held in cheek pouches while they were processed.

Left: The upper jaw bones were set loosely in the skull, allowing movement while chewing.

Claosaurus

We know *Claosaurus* from an almost complete skeleton lacking the skull. It is the oldest-known of North American hadrosaurids, and it is possible that it is close to the ancestry of the later members of the group, but without the evidence of the skull bones this cannot be proved. It was probably a member of the flat-headed, crestless hadrosaurine group.

Features: *Claosaurus* is a comparatively small hadrosaurid, and is known to be one of the most primitive of the duckbilled group by the presence of a little bone of the first toe, all other hadrosaurids having lost this. The teeth are also very primitive, resembling those of the earlier *Camptosaurus*. The few tail bones preserved suggest that the tail was relatively longer than that of later, more advanced, hadrosaurids.

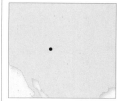

Distribution: Kansas, USA.
Classification: Ornithopoda, Hadrosauridae, Hadrosaurinae.
Meaning of name: Broken lizard.
Named by: Marsh, 1890.
Time: Campanian stage of the late Cretaceous.
Size: 3.5m (11½ft).
Lifestyle: Low browser.
Species: *C. agilis*.

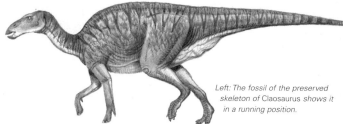

Left: The fossil of the preserved skeleton of Claosaurus *shows it in a running position.*

Bactrosaurus

The oldest-known member of the true lambeosaurines was *Bactrosaurus*, although a recent find of another specimen suggests that it did not possess the crest that was characteristic of the group. It was first found by the American Museum of Natural History's expedition to Mongolia in 1923, and the appearance has been reconstructed from the remains of at least six skeletons.

Features: *Bactrosaurus* looked much more like an iguanodontid than a hadrosaurid, since it was heavy for its size. It has the distinctive tooth batteries of the hadrosaurids, although these batteries contain fewer teeth than is usual for a hadrosaurid. The vertebrae of the back have high club-shaped spines. They give the genus its name. As with the other lambeosaurines, *Bactrosaurus* is more heavily built than its crestless cousins among the hadrosaurines.

Right: Bactrosaurus fed on a variety of plants.

Distribution: China and Mongolia.
Classification: Ornithopoda, Hadrosauridae, Lambeosaurinae.
Meaning of name: Club-spined lizard.
Named by: Gilmore, 1933.
Time: Indeterminate stage of the late Cretaceous.
Size: 6m (20ft).
Lifestyle: Browser.
Species: *B. johnsoni*, *B. kyzylkumensis*.

LAMBEOSAURINE HADROSAURIDS

The lambeosaurine hadrosaurids were those with ornate hollow crests on their heads. Each genus could be distinguished from the next by the shape of this crest, and the crests would have been brightly coloured to enhance this function. Their duck-like bills were narrower than those of the other group of hadrosaurids – the hadrosaurines.

Tsintaosaurus

The characteristic crest of *Tsintaosaurus* has always been a bit of a mystery, being so unlike the crests of other hadrosaurids. At times the crest was thought to have been just the way the skull was preserved, with a sliver of bone out of place, and that it was just a damaged specimen of *Tanius*. Later discoveries of more specimens show that the crest did actually exist.

Distribution: Shandong Province, China.
Classification: Ornithopoda, Hadrosauridae, Lambeosaurinae.
Meaning of name: Lizard from Tsintao, the city close to the place where it was found.
Named by: Young, 1958.
Time: Campanian to Maastrichtian stages of the late Cretaceous.
Size: 10m (36ft).
Lifestyle: Browser.
Species: *T. spinorhinus*.

Features: The obvious feature of *Tsintaosaurus* is the thin, hollow crest jutting directly upward from the middle of the head, like the horn of a unicorn. It may have been a display feature, but it is also possible that the crest supported a flap of skin that could be inflated with air from the nostrils, to make some kind of a sound signal. This would make sense in view of the resonating chambers that have been found inside the skull.

Left: The name Tsintaosaurus *is an anglicized version of Ch'ing-tao, or Qingdao, which means "green island". The species name denotes the fact that it had a spine on its nose.*

Charonosaurus

Charonosaurus comes from late Maastrichtian deposits, the Yuliangze Formation at Jiayin, China, at the end of the Age of Dinosaurs and after most of the other lambeosaurines had died out. The best specimen is a partial skull, but many other pieces of bone were found in a bonebed suggesting that this animal lived in big herds close to a river that flooded frequently.

Distribution: China.
Classification: Ornithopoda, Hadrosauridae, Lambeosaurinae.
Meaning of name: Charon's lizard (after the ferryman in the underworld of ancient Greek legend).
Named by: Godefroit, Zan and Jin, 2000.
Time: Maastrichtian stage of the late Cretaceous.
Size: 13m (42½ft).
Lifestyle: Browser.
Species: *T. jiayinensis*.

Features: The crest of *Charonosaurus* is unclear from the partial remains of the skull that have been found. However, the shape of the rest of the skull suggests that it was long and hollow like that of *Parasaurolophus*. *Charonosaurus* was a very big animal, one of the biggest of the lambeosaurines and half as big again as *Parasaurolophus* to which it seems closely related.

Right: The long forelimbs suggest that Charonosaurus *was habitually a quadruped, but it would have moved about on its hind legs from time to time.*

Parasaurolophus

Probably the most familiar lambeosaurine hadrosaurid, and the one with the most flamboyant crest, was *Parasaurolophus*. Three species are recognized, and each has a different curve of the crest. Different crest sizes may indicate different sexes. Computer studies in the New Mexico Museum of Natural History and Science have produced a deep trombone-like sound that would have been produced by *Parasaurolphus* snorting through its crest.

Distribution: New Mexico, USA; to Alberta, Canada.
Classification: Ornithopoda, Hadrosauridae, Lambeosaurinae.
Meaning of name: Almost *Saurolophus*.
Named by: Parks, 1922.
Time: Campanian to Maastrichtian stages of the late Cretaceous.
Size: 10m (33ft).
Lifestyle: Browser.
Species: *P. walkeri*, *P. cyrtocristatus*, *P. tubicen*.

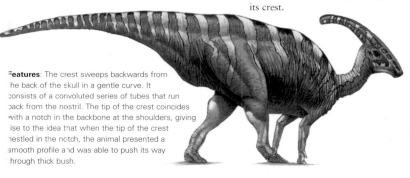

Features: The crest sweeps backwards from the back of the skull in a gentle curve. It consists of a convoluted series of tubes that run back from the nostril. The tip of the crest coincides with a notch in the backbone at the shoulders, giving rise to the idea that when the tip of the crest nestled in the notch, the animal presented a smooth profile and was able to push its way through thick bush.

Lambeosaurus

This is a well-known dinosaur, and gives its name to the whole group. Its remains were discovered in 1889, but it was not recognized as a distinct genus until 1923. More than 20 fossils have been found. The wide geographical range of the finds suggests that it lived all along the western shore of the late Cretaceous inland sea of North America.

Features: The hollow crest on the top of the head is in the shape of an axe, with a squarish blade sticking up and a shaft pointing backwards. The square portion is the hollow part with the convoluted nasal passages, while the spike is solid. The crest of the larger species, *L. magnicristatus*, has a larger hollow portion, bigger than the skull itself, and a very small spike. The skin is thin and covered in small polygonal scales.

Distribution: Alberta, Canada; and Montana to New Mexico, USA.
Classification: Ornithopoda, Hadrosauridae, Lambeosaurinae.
Meaning of name: Lawrence M. Lambe's lizard.
Named by: Parks, 1923.
Time: Campanian stage of the late Cretaceous.
Size: 9–15m (30–49ft).
Lifestyle: Browser.
Species: *L. lambei*, *L. laticaudus*, *L. magnicristatus*.

MORE LAMBEOSAURINE HADROSAURIDS

We know that lambeosaurines spread across the whole of the northern continent, since their remains have been found in Russia, China, Canada and the USA. This shows that for at least some of the late Cretaceous period, the northern continents were continuous across the area where the Bering Strait now separates them.

Hypacrosaurus

Several good specimens of *Hypacrosaurus* have been found, the first by famous American dinosaur hunter Barnum Brown in 1910. Subsequent discoveries of another species of *Hypacrosaurus* included eggs and youngsters in various stages of growth, allowing palaeontologists to chart the growth pattern and family life of the animal.

Features: Tall spines along the backbone gave *Hypacrosaurus* a deep ridge, or a low fin, along its back, which was probably used for display. Alternatively they may have supported a fatty hump used as a food store for lean seasons, as with modern camels. The crest is shorter than that of other lambeosaurines. It is semicircular like that of *Corythosaurus* but smaller and thicker, and it has a short spike jutting backwards.

Distribution: Alberta, Canada; and Montana, USA.
Classification: Ornithopoda, Hadrosauridae, Lambeosaurinae.
Meaning of name: Below the top lizard.
Named by: Brown, 1913.
Time: Maastrichtian stage of the late Cretaceous.
Size: 9m (30ft).
Lifestyle: Browser.
Species: *H. stebingeri*, *H. altispinus*.

Left: Hypacrosaurus is the most primitive known of the lambeosaurine hadrosaurids.

Corythosaurus

One of the best-known hadrosaurids is *Corythosaurus*. The remains of more than 20 individuals are known, many with complete skulls and some with pebbly skin impressions. The abundance of the remains suggest that *Corythosaurus* travelled in groups. As in other hadrosaurids, it would have spent much of its time on all fours, but could also run on its hind legs.

Below: The scaly texture of the skin is known, and is similar to that of other hadrosaurids. However, there is evidence of three rows of broad scales along the belly. This may have protected its underside while it moved through thick undergrowth.

Distribution: Alberta, Canada, and Montana, USA.
Classification: Ornithopoda, Hadrosauridae, Lambeosaurinae..
Meaning of name: Corinthian lizard (referring to the crest, which is shaped like a Corinthian helmet).
Named by: Brown, 1914.
Time: Campanian stage of the late Cretaceous.
Size: 10m (33ft).
Lifestyle: Browser.
Species: *C. casuarius*.

Features: The crest of *Corythosaurus* is distinctive, being half a disc, about the size of a dinner plate. As in other lambeosaurines, the nasal passages within this disc are convoluted and are connected to the nostrils. The part of the brain dealing with the sense of smell is also close to the crest. The crest is in two different sizes, with the smaller crest probably belonging to the female of the species.

LAMBEOSAURINE CRESTS

The flamboyant crests of the lambeosaurines consisted of hollow tubes that were connected to the nasal passages. It is possible that their function was to modify the air that was breathed. If the tubes were lined with moist membrane (something that could never be proved), cold dry air would be warmed and moistened before travelling to the lungs.

In some instances the part of the brain that dealt with the sense of smell was located close to the crest. This suggests that the huge volume of nasal passage would have had something to do with this sense, helping the animal to detect slight variations in scents in the air. A third possibility is that the crest was a communication device. Air snorted through these tubes would have made distinctive honking noises like the sound of a bass trombone. Such distinctive sounds would carry for long distances through forests and across swamps, and would be a valuable means of keeping a herd together. There is no reason why all these functions could not have been combined.

Left: The different lambeosaurine crests.

Olorotitan

Excavated in 1999–2000 on the banks of the Amur River in far eastern Russia, this is the most complete Russian dinosaur skeleton ever found, and the most complete lambeosaurine outside North America. The skeleton, including the head with the spectacular crest, is now on display in the Amur Natural History Museum in Blagoveschensk, Russia. It is very closely related to the North American *Corythosaurus* and *Hypacrosaurus*.

Features: The crest is huge, sweeping up and back, with a hatchet-like blade on its rear edge. This presumably carried some kind of skin ornamentation. The neck is longer than that of other lambeosaurine hadrosaurids, and so is the pelvis. The tail is even more rigid than that of other hadrosaurids, but that may be due to disease affecting the tail vertebrae of the one individual that has been found.

Below: Despite being very closely related to the North American Corythosaurus *and* Hypacrosaurus, Olorotitan *is different in a number of ways. The neck is longer than in the other genera, and the joints of the tail show that the tail was even more rigid.*

Distribution: Kundur, Russia.
Classification: Ornithopoda, Hadrosauridae, Lambeosaurinae.
Meaning of name: Giant swan.
Named by: Godefroit, Bolotsky and Alifanov, 2003.
Time: Maastrichtian stage of the late Cretaceous.
Size: 12m (39ft).
Lifestyle: Browser.
Species: *O. arharensis*.

CRESTED HADROSAURINE HADROSAURIDS

*The hadrosaurine hadrosaurids lacked the flamboyant hollow crests of the lambeosaurines.
Instead they either had flat, crestless heads or heads ornamented with bony bumps or solid crests
formed of spikes of bone. They also tended to have broader jaws, and longer and slimmer limbs
than their hollow-crested relatives.*

Aralosaurus

This Asian hadrosaurine hadrosaurid is known from the back half of an articulated skull
with the distinctive nasal bone from a subadult specimen, along with some limb bones and
vertebrae from several other individuals unearthed from the same deposits.
Despite its current classification, it may have
been a lambeosaurine.

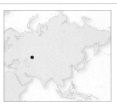

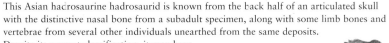

Distribution: Kazakhstan.
Classification: Ornithopoda,
Hadrosauridae,
Hadrosaurinae.
Meaning of name: Lizard of
the Aral Sea, central Asia.
Named by: Rozhdestvensky,
1968.
Time: Turonian to Coniacian
stages of the late
Cretaceous.
Size: 6–8m (19½–26ft).
Lifestyle: Browser.
Species: *A. tuberiferus*.

Features: The skull of *Aralosaurus* has
a peculiar arched area on the nasal region,
swept back to a kind of point just in front of its
eyes, suggesting that it may have actually been
one of the crested lambeosaurines. Otherwise
the skull is similar to that of *Gryposaurus*. It
appears to have had different teeth in the upper
and lower jaws, but it is difficult to see the
significance of this.

*Above: Like other
hadrosaurines
Aralosaurus would
have had a broad beak. The skull
would have been tall at the back
and broad at the front.*

Gryposaurus

Distribution: Alberta,
Canada.
Classification: Ornithopoda,
Hadrosauridae,
Hadrosaurinae.
Meaning of name: Hook-
nosed lizard.
Named by: Lambe, 1914.
Time: Campanian stage of
the late Cretaceous.
Size: 8m (26ft).
Lifestyle: Low browser.
Species: *G. notabilis,
G. incurvimanus, G. laltidens.*

This dinosaur was once thought to have been the same genus as *Kritosaurus*, the original
species being *K. notabilis*. *Gryposaurus* lived much further north than *Kritosaurus*, and
this was probably the only thing that distinguished them (the two were certainly related).
It is known from several specimens, including ten complete skulls and 12 fragmentary skulls,
along with skin impressions.

Features: In appearance *Gryposaurus* is similar
to *Kritosaurus*, with its narrow, deep
skull and highly arched nostrils.
The tooth shape is
slightly different though,
suggesting a slightly
different diet. The skin
is covered in smooth
scales less than
4mm (⅙in) in
diameter, and
there are cone-
shape plates
1.3cm

(½in) in diameter,
spaced 5–7cm
(2–2⅘in) apart on
the tail.

Kritosaurus

The original description of *Kritosaurus* is based on a poorly preserved skull. It is traditionally shown as a flat-headed hadrosaurine hadrosaurid of a subgroup that lacked crests. However, it now seems more likely to have belonged to the subgroup with the small solid crests, such as *Saurolophus*. Other hadrosaurines such as *Anasazisaurus* and *Naashoibitosaurus* may be species of *Kritosaurus*.

Distribution: New Mexico and Texas, USA; and a species in Argentina.
Classification: : Ornithopoda, Hadrosauridae, Hadrosaurinae.
Meaning of name: Separated lizard.
Named by: Brown, 1910.
Time: Campanian to Maastrichtian stages of the late Cretaceous.
Size: 10m (33ft).
Lifestyle: Browser.
Species: *K. australis*, *K. navajovius*.

Features: *Kritosaurus* has a flat skull with a ridge of bone just below the eyes, giving the snout a "Roman nose" appearance. The high ridge on the nose probably supported a flap of skin at each side, and this may have cooled the inhaled air, or helped make noises for signalling to one another. Alternatively, it may have been a horn used for battling rival mates.

HADROSAURINE COMMUNICATION

If the lambeosaurines could signal to one another by using their hollow crests as musical instruments, then how did their close relatives the hadrosaurines manage? The hadrosaurines had much broader beaks than the lambeosaurines, and the tops of their skulls presented a broad platform of bone around the nasal openings. It is quite possible that this platform was covered by a flap of skin that surrounded the nostrils. Such a flap of skin would have been inflated as the animal breathed, and could have been controlled to produce a sound, very much like that produced by the inflatable throat pouch of many modern frogs.

This would explain the solid crests of some of the hadrosaurines, especially among the Maiasaurini and the Saurolophini. The flap of skin could have been supported by this crest, and the various crest shapes in the many species would give a different shape of skin flap, and so a different sound would be created to identify all the herds.

Right: An inflated hadrosaurine nose flap.

Kerberosaurus

The skull of *Kerberosaurus* was found in a river-deposited bonebed along with many indeterminate skeletal parts. The presence of this hadrosaurid, similar to *Prosaurolophus* and *Saurolophus*, the type of animal common in North America but unusual for Russia, shows the exchange of fauna that could take place between Asia and North America in late Cretaceous times.

Features: The 1m- (3ft-) long skull, the only part of the animal to have been found, looks very much like that of *Prosaurolophus* or *Saurolophus*, but it differs in the detail of the individual bones and how they are articulated with one another. The significant thing about this animal is the fact that it lived in eastern Asia at this time. Its ancestors migrated there from North America, where the hadrosaurines had evolved.

Distribution: Blagoveshchensk, Russia.
Classification: Ornithopoda, Hadrosauridae, Hadrosaurinae.
Meaning of name: Lizard of Cerberus (a monster dog in Greek mythology).

Named by: Bolotsky and Godefroit, 2004.
Time: Maastrichtian stage of the late Cretaceous.
Size: 10m (33ft).
Lifestyle: Browser.
Species: *K. manakini*.

Left: The single species is named in honour of Colonel Manakin, a pioneer of dinosaur discovery in the Amur region of Russia.

MORE CRESTED HADROSAURINES

The Maiasaurini and the Saurolophini represent advanced subgroups of the Hadrosaurinae, distinguished by the possession of solid crests, formed by the backward extension of the nasal bones. These crests were quite unlike the hollow crests of the Lambeosaurinae, and may have served as a support for some skin structure, whether a ballooning nose flap or the end of a frill down the neck.

Brachylophosaurus

This was quite a rare hadrosaurine hadrosaurid, but it is known from several remains including one, nicknamed "Leonardo", that preserves the skin, the muscles of the neck and the gastric tract that includes the remains of its last meal. The solid nature of the skull suggests that it may have taken part in head-butting behaviour.

Features: The short crest implied in the name consists of a flat plate on the top of the head with a short spine sticking back from the rear of the skull. The skull itself is quite tall for a hadrosaurine hadrosaurid, and has a steep face. The forelimbs are relatively long. The skin is covered in fine scales, the biggest being on the lower limbs. The two species known are so similar that they may represent a male and female of the same species, the only difference being that the skull of *B. goodwini*, the only part known, has a depression before the crest.

Distribution: Alberta, Canada; to Montana, USA.
Classification: Ornithopoda, Hadrosauridae, Hadrosaurinae.
Meaning of name: Short-crested lizard.
Named by: C. M. Sternberg, 1953.
Time: Santonian to Campanian stages of the late Cretaceous.
Size: 7m (23ft).
Lifestyle: Browser.
Species: *B. canadensis*, *B. goodwini*.

Right: The term duckbill becomes something of a misnomer when dealing with Brachylophosaurus. Rather than being broad and duck-like, the bill is flattened from side to side and downturned.

Maiasaura

This hadrosaurine is known from a nesting colony found in Montana in the 1970s. It lived in big herds and nested in groups, with possibly as many as 10,000 returning to the same area every year. They probably did so for protection. More than 200 skeletons, embryos, hatchlings, immature and mature adults were found.

Features: The crest is a short, broad projection above the eyes, and is solid, distinguishing it from the hollow-crested lambeosaurines. The crest, and a pair of triangular projections on the cheekbones, form the basis for the definition of the Maiasaurini. They have batteries of self-sharpening teeth, and a jaw mechanism that allowed the surfaces to grind past each other to chew up tough vegetation.

Distribution: Montana, USA.
Classification: Ornithopoda, Hadrosauridae, Hadrosaurinae.
Meaning of name: Good mother lizard.
Named by: Horner and Makela, 1979.
Time: Campanian stage of the late Cretaceous.
Size: 9m (30ft).
Lifestyle: Browser.
Species: *M. peeblesorum*.

Saurolophus

Confusingly, *Saurolophus* is only very distantly related to the much more popular *Parasaurolophus*. The type species *S. osborni* is known from the remains of at least three individuals. Another species, *S. angustirostris*, is known from the Gobi desert. Some palaeontologists think it should be the same species as *S. osborni*, but others think it is a different genus altogether.

Distribution: Alberta, Canada; with a species in Mongolia.
Classification: Ornithopoda, Hadrosauridae, Hadrosaurinae.
Meaning of name: Lizard crest.
Named by: Brown, 1912.
Time: Maastrichtian stage of the late Cretaceous.
Size: 9–12m (30–39ft).
Lifestyle: Browser.
Species: *S. angustirostris*, *S. osborni*, "*S. krischtofovici*".

Right: The uniting feature between the species of Saurolophus is the presence of the backward-pointing spike above the eye.

Features: The distinguishing feature of *Saurolophus* is the prominent spine that rises above the eyes and projects backwards. This is formed from the nasal bones that extend backwards and may have been associated with some sound-producing mechanism. The skull is quite narrow for a hadrosaurid, especially across the snout where we would expect to see the duck-like bill. The original species was the most complete to have been found in Canada at the time (1911). The Asian species is much larger.

DINOSAUR NESTS

The nesting site of *Maiasaura* in Montana gives us a unique insight into the family life of hadrosaurids. Each nest was about 2m (6½ft) across and formed of a low mound of mud. The nests were situated at least an adult's length from the next, just the right distance to prevent the adults from pecking at one another while they rested on the eggs. The eggs were laid in a hollow in the top of the mound and covered with a layer of plant material which provided warmth as it rotted, like a compost heap. The remains of the nests lie in successive layers of rock, suggesting that the herds returned year after year to use the same nesting ground. Bones from all age groups were discovered on the site, suggesting that the hatchlings remained in the nest for some time and were looked after by their parents until they were well grown. For the rest of the year the great mass of *Maiasaura* would have split up into smaller bands and migrated to more productive feeding grounds. The presence of the remains of egg-eating lizards and of nest-infecting beetles reveal some of the hazards that the nesting *Maiasaura* had to face.

Prosaurolophus

As its name suggests, this is possibly the ancestor of *Saurolophus* which appeared a little later in the geological succession. It is known from the remains of more than 30 individuals of varying ages. It inhabited the forest plains of late-Cretaceous Canada, and its diet consisted of the conifers, cycads, ginkgoes and flowering plants that grew there.

Features: The crest is a bony lump just in front of the eyes, rising out of a mass of small knobs and developing into a backward-pointing spike. It s not nearly as big as that of *Saurolophus*. The face is sloping and the muzzle broad and flat, but not as widely flared as on other hadrosaurids. The difference in the shape of the skull between the species may be due to crushing during the fossilization process, and it may be that there is only one species known.

Distribution: Alberta, Canada; to Montana, USA.
Classification: Ornithopoda, Hadrosauridae, Hadrosaurinae.
Meaning of name: Before *Saurolophus*.
Named by: Brown, 1916.
Time: Coniacian to Campanian stages of the late Cretaceous.
Size: 8–9m (26–30ft).
Lifestyle: Browser.

Species: *P. maximus*, *P. blackfeetensis*, *P. breviceps*.

Above: Although the official length of Prosaurolophus is given as 8–9m (26–30ft), there is a suggestion that it may actually have reached lengths of 15m (50ft).

FLAT-HEADED HADROSAURINES

The flat-headed hadrosaurines were the Edmontosaurini, and were among the last of the dinosaurs to evolve. It seems likely that they evolved from the solid-crested types. It is possible that they signalled to each other by bellowing, using flaps of skin on their flat heads to amplify the sound, but as yet this has not been proved.

Anatotitan

Often seen in old books as *Anatosaurus* and even as *Trachodon*, *Anatotitan* is known from two good skeletons including the skulls. It rested on all fours, but walked on its powerful back legs with the three hoof-like toes on each foot taking the weight. It was among the last of the dinosaurs to have existed.

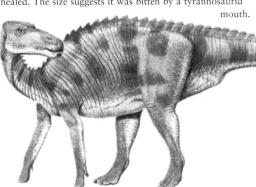

Features: This is the most duck-like of the duckbilled dinosaurs, with a particularly broad and flat beak, and jaws that are toothless for over half their length. However, the hardened horny bill that covered the front of the mouth is actually quite different from the sensitive organ found on a duck. There are slight knobs above the eyes, but apart from that there is no sign of a crest. Although bigger than *Edmontosaurus*, it is more lightly built.

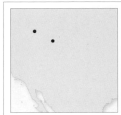

Distribution: Montana to South Dakota, USA.
Classification: Ornithopoda, Hadrosauridae, Hadrosaurinae.
Meaning of name: Giant duck.
Named by: Brett-Surman, 1990.
Time: Maastrichtian stage of the late Cretaceous.
Size: 10–13m (33–43ft).
Lifestyle: Browser.
Species: *A. copei*, *A. longiceps*.

Edmontosaurus

The most abundant of the herbivores at the end of the Cretaceous period, *Edmontosaurus* is known from many skeletons. One skull shows signs of theropod tooth marks suggesting an attack to the neck, while another has a chunk bitten out of the top of the tail. The bite subsequently healed. The size suggests it was bitten by a tyrannosaurid mouth.

Features: The skeleton of *Edmontosaurus* is regarded as the benchmark, the shape to which all other hadrosaurids are compared. The tail is deep and heavy, used as a balance when it walked on hind legs, and the hands have weight-bearing pads on the fingers to support it while it stood on all fours. The neck is flexible, allowing the duckbilled head to reach the food growing all around it.

Left: We know that Edmontosaurus *had leathery skin with non-overlapping scales, as mummified remains of it have been found associated with two skeletons.*

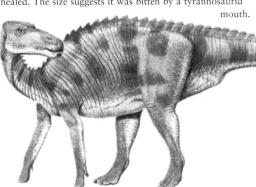

Distribution: Alberta, Canada; to Wyoming and maybe Alaska, USA.
Classification: Ornithopoda, Hadrosauridae, Hadrosaurinae.
Meaning of name: Edmonton lizard.
Named by: Lambe, 1917.
Time: Maastrichtian stage of the late Cretaceous.
Size: 13m (43ft).
Lifestyle: Browser.
Species: *E. annectens, E. regalis, E. saskatchewanensis.*

Shantungosaurus

Shantungosaurus appears to have been the biggest ornithopod known, an adult weighing several tonnes. It was probably the biggest animal ever to walk on two legs, although it would have spent most time on all fours. It is so similar to *Edmontosaurus* that it could be a species of that Canadian hadrosaurid. *Shantungosaurus* is based on an almost complete skeleton in the Beijing Geological Museum, built from the remains of five individuals.

Features: The features that distinguish *Shantungosaurus* from *Edmontosaurus* are those that relate to size – bigger limbs, stronger bones, a more powerful back to take the weight, and so on.

Right: All the specimens that went into the mounted skeleton came from the same quarry. The bones were all disarticulated, but the number of jaw bones indicates that there were at least five individuals present.

Its estimated weight was around 3–3½ tonnes. The head is 1.5m (5ft) long. The fact that the single, complete skeleton is made up of several individuals, picked out of 30 tonnes of fossilized material, suggests that it was a herd-dweller moving about the plains in large numbers to avoid the tyrannosaurs.

Distribution: Shandong Province, China.
Classification: Ornithopoda, Hadrosauridae, Hadrosaurinae.
Meaning of name: Lizard from Shandong.
Named by: Hu, 1973.
Time: Maastrichtian stage of the late Cretaceous.
Size: 12–15m (39–49ft).
Lifestyle: Browser.
Species: *S. giganteus*.

TRACHODON AND HADROSAURUS

Many old books refer to a duckbilled dinosaur called *Trachodon*. The name was first given to some teeth found in Montana in 1855 by Joseph Leidy. The first good dinosaur remains in North America were found two years later in New Jersey. They were of a hadrosaurid that was named *Hadrosaurus* by Leidy. Unfortunately the later find lacked the skull, and was thought to have been an iguanodontid. The first hadrosaurid skull, with beak and teeth, was found by Cope in 1883. The teeth were similar to Leidy's original, and so *Trachodon* became the genus to which all duckbilled dinosaurs belonged – an early example of a "wastebasket taxon".

Soon all sorts of other hadrosaurids came to light, with all shapes of skulls, and many other genera were established. The original *Trachodon* teeth could not confidently be assigned to any of them, and so the name fell into disuse. For a while a hadrosaurine hadrosaurid called *Anatosaurus* became the most likely owner of the *Trachodon* teeth but later, due to the complexity of allocating scientific names, *Anatosaurus* became *Anatotitan*. The name *Trachodon* is no longer used.

Incidentally, the first good North American dinosaur, *Hadrosaurus*, has fallen into disfavour too. The lack of a skull means that it cannot be classified with certainty, which is why, important as it is, it is not featured here.

Tanius

The Chinese geologist H. C. Tan (or Tan Xi-zhou) collected most of the original specimen of *Tanius* in 1923 at Laiyang, in Shandong Province, China. Other specimens that have been assigned to this genus have since been identified as other hadrosaurids.

Features: As the skull is incomplete, it is not entirely certain that the skull found belonged to the Hadrosaurinae. It is possible that it is a *Tsintaosaurus* that has lost its distinctive crest. In either case the body is large, with a deep tail. The specialized, self-sharpening teeth with mobile jaws allowed it to feed on tough vegetable matter. The hind legs are longer than the front, and it could walk either on its hind legs or on all fours. Most of the front of the head is missing, hence the uncertainty about its identity.

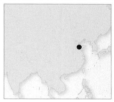

Distribution: China.
Classification: Ornithopoda, Hadrosauridae.
Meaning of name: After Tan.
Named by: Wiman, 1929.
Time: Undetermined stage of the late Cretaceous.
Size: 9m (30ft).
Lifestyle: Browser.
Species: *T. sinensis*.

Right: Whatever it was, Tanius had the appearance of other big ornithopods.

SMALL BONEHEADS

The pachycephalosaurids are the bone-headed dinosaurs. The presence of members of the pachycephalosaurid family in both Asia and North America at the end of the Cretaceous period is yet another indication that the two continents were joined at that time. It seems as if the main radiation of the group took place in Asia, and that it then migrated east to colonize most of North America.

Wannanosaurus

This small animal is known from part of the skull and some of the rest of the skeleton, including the vertebrae and limbs. It probably lived in small groups in coastal or upland regions, running away from predators rather than fighting them with its thick skull. The skull was probably used when the species fought among themselves for mates or for dominance in the herd.

Above: Balanced by the stiff heavy tail, Wannanosaurus would have walked on hind legs between feeding areas, going down on to all fours to eat the soft leaves that seem to have been its main diet.

Features: *Wannanosaurus* is a tiny pachycephalosaurid, and is small for any dinosaur. It walked on its hind legs and had short arms and a stiff tail, the vertebrae of which are lashed together by bony tendons. It has a flat roof to the skull, rather than the characteristic dome of the rest of the group, which gives it the appearance of a small ornithopod. It is distinguished from other pachycephalosaurids by the shape of the teeth – which are low and fan-shaped – and by the upper arm bone which is markedly curved.

Distribution: Shantung Province, China.
Classification: Marginocephalia, Pachycephalosauria.
Meaning of name: Lizard from Wannan.
Named by: Hou, 1977.
Time: Campanian stage of the late Cretaceous.
Size: 60cm (2ft).
Lifestyle: Low browser.
Species: *W. yansiensis*.

Goyocephale

The skull and most of the skeleton is known from this pachycephalosaurid. It is the best-known example of a primitive member of the group, and most of the restorations of the primitive flat-headed pachycephalosaurids – with the strong and heavy backbone, the broad hip bones that may have held the muscles that allowed it to charge enemies with its head, the hind legs longer than the front, and the stiff tail – are based on what we know from the remains of *Goyocephale*.

Features: There are two pairs of sharp incisor teeth in the mouth, those in the lower jaw coinciding with notches in the upper jaw. The heavy top of the skull is rough and covered with protuberances. As in other members of the group, the thickened skull is formed from the frontal and parietal bones. The openings between the skull bones are quite large, indicating the animal's primitive nature. More evolved pachycephalosaurids had progressively smaller openings until they disappeared altogether.

Left: The species name, G. lattimorei, honours the British anthropologist Owen Lattimore, who travelled in Mongolia in the 1950s and 1960s, and published a series of popular books about the area.

Distribution: Mongolia.
Classification: Marginocephalia, Pachycephalosauria.
Meaning of name: Adorned head.
Named by: Perle, Maryańska and Osmólska, 1982.
Time: Santonian to Campanian stages of the late Cretaceous.
Size: 2m (6½ft).
Lifestyle: Low browser.
Species: *G. lattimorei*.

THICK HEADS

Usually a dinosaur skull is very fragile. It is the first part of the skeleton to collapse, since it is made up of lightweight struts of bone. The exceptions are the horned dinosaurs and the pachycephalosaurids or boneheads.

We know surprisingly little about the body of the typical pachycephalosaurid; most restorations are based on the skeletons of only two genera. However, the heads are well known since, unlike other dinosaurs, the skulls are made up of very thick pieces of bone, firmly sutured together, and they tend to fossilize easily. The name of the group reflects this (pachycephalosaurid means thick-headed lizard).

Sometimes the skulls are found in areas known to have been river or marine sediments and show a great deal of wear. The contemporary interpretation for the location of such fossil finds and the damage to the skull is that the animals lived in mountain areas, and the skulls were washed down to the plains, or the seas, by rivers. The thickness of the bones meant that the skulls survived the rough passage. If this interpretation is accurate, it is almost the only evidence that there is of the existence of mountain-living dinosaurs. The bones of other mountain-dwellers would probably have been destroyed long before they reached any area of deposition where they could be preserved.

Micropachycephalosaurus

This small dinosaur has the distinction of having the longest dinosaur name on record, despite being among the smallest dinosaurs known. It is only known from its lower jaw and some bone fragments. It is one of few types of pachycephalosaurid found in China; nearly all others come from North America.

Features: The jawbone of *Micropachycephalosaurus* looks like that of a pachycephalosaurid. Some scientists think that the associated skeletal material does not belong to the same animal, as the bones do not seem to be the right size. However, going by what is known of other pachycephalosaurids, *Micropachycephalosaurus* was a bipedal animal, with a big head and a stiff tail.

Distribution: China.
Classification: Marginocephalia, Pachycephalosauria.
Meaning of name: Tiny pachycephalosaurid.
Named by: Dong, 1978.

Time: Campanian stage of the late Cretaceous.
Size: 50cm (1½ft).
Lifestyle: Low browser.
Species: *M. hongtuyanensis.*

Left: Micropachycephalosaurus was a veritable miniature compared with the other dinosaurs that were around at the time.

Homalocephale

The vertebrae of the back and tail, hips and hind legs, together with the skull, are the only parts of the pachycephalosaurid skeleton known. The top of the skull is very thick, but not particularly dome-shaped, and is made of porous, non-rigid bone full of blood vessels. It forms more of an even table surface than a raised dome, hence the meaning of the name.

Below: The thickness of the skulls of the pachycephalosaurids has been taken as an indication that they butted one another. They may have butted each other on the flanks as often as head-on.

Features: The partial skeleton shows *Homalocephale* to have had wide hip bones and a very broad base to the tail, possibly allowing the internal organs to lie quite far back. It may also show the ability of this animal to sit back on its tail like a kangaroo. The tail was made solid by a rigid basketwork of tendons, suggesting its use as a strong support like one leg of a tripod.

Distribution: South-western Mongolia.
Classification: Marginocephalia, Pachycephalosauria.
Meaning of name: Even head.
Named by: Maryańska and Osmólska, 1974.
Time: Maastrichtian stage of the late Cretaceous.
Size: 3m (10ft).
Lifestyle: Low browser.
Species: *H. calathoceros.*

BONEHEADS

Most of what we know about the pachycephalosaurids is based on the range of skulls that have been found; there are very few skeletons. For a time most members of the group were regarded as species of Stegoceras, *for want of any evidence to the contrary. Current knowledge suggests the subtle differences in skull pattern indicate that there are more genera of pachycephalosaurid than was first appreciated.*

Stegoceras

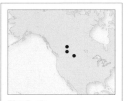

Distribution:
Alberta, Canada;
and Montana, USA.
Classification:
Marginocephalia,
Pachycephalosauria.
Meaning of name: Roof
horn.
Named by: Lambe, 1902.
Time: Campanian stage of
the late Cretaceous.
Size: 3m (10ft).
Lifestyle: Low browser.
Species: *S. validum*,
S. browni (formerly
Ornatotholus browni).

This is the best known of the pachycephalosaurids, with dozens of skull fragments known and also a partial skeleton. Most of the reconstructions of other pachycephalosaurid genera are based on this skeleton. The structure of the bone in the head dome was such that the bone fibres aligned to absorb impact from the top. The vertebrae of the neck and back were very strong, lashed together with strong tendons that prevented twisting, and aligned to absorb shocks emanating from the head end. The hips were particularly wide and solid. All this is consistent with the idea that the dome was used as a weapon, like a battering ram.

Below: The muzzle is wide and the teeth set quite far apart, suggesting that Stegoceras *had a different diet from other boneheads.*

Features: The dome on the head of *Stegoceras* is high, but not as high as that of others in the group, and is surrounded by a frill of little horns and knobs. The teeth at the front of the jaw are very widely set and the muzzle is broad compared with other pachycephalosaurids. This may indicate a less selective feeding strategy. The very broad hips suggest that the pachycephalosaurids gave birth to live young, (this is not widely accepted).

Colepiocephale

This dinosaur was discovered by fossil-hunter L. M. Sternberg in 1945 and regarded as a species of *Stegoceras*, but it was studied again by palaeontologist Robert Sullivan in 2003 who found enough features of the dome to distinguish it, and put it in a genus of its own. It is the oldest definite pachycephalosaurid found in North America (although another has been found in slightly earlier strata but has not yet been studied).

Features: The dome has quite an oblique slope to it, producing a flattened aspect to the front of the head, and is somewhat triangular in top view. The bones of the side of the skull are not as complex as those of other pachycephalosaurids. The range in dome shapes among the different genera of pachycephalosaurids suggests that they functioned as display

devices, enabling individuals to distinguish one herd from another. In this way they would be similar to modern birds that have big spectacular beaks or flamboyant feather crests to distinguish one species from another.

Distribution: Alberta, Canada.
Classification:
Marginocephalia,
Pachycephalosauria.
Meaning of name:
Knucklehead.
Named by: Sullivan, 2003.
Time: Campanian stage of
the late Cretaceous.
Size: 1m (3ft).
Lifestyle: Low browser.
Species: *C. lambei*.

Above: The "knucklehead" of the name refers to the shape of the dome head that has the appearance of a finger joint.

The popular image of a pachycephalosaurid is of a pair of rival males, excited by the prospect of leading the herd and mating with the females, head-butting one another furiously. The

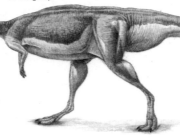

Above: Pachycephalosaurids probably butted one another on the flanks.

vision of mountain goats doing the same thing is difficult to dispel. The idea of this behaviour was put forward by American palaeontologist Ed Colbert in 1955.

However, there are flaws to this evocative scene, as pointed out by Mark Goodwin of the University of California, at Berkeley. For one thing, the shapes are wrong. Two dome-heads on a collision course need to have pinpoint accuracy to be effective, otherwise they would just glance off one another. Also, of all the fossils of bonehead skulls found so far, none of them shows the kind of damage that would have been sustained in such an engagement. It now seems more likely that the dome-head of the pachycephalosaurid was used as a battering ram, but not for head-on attacks.

Hanssuesia

This dinosaur was identified as *Stegoceras sternbergi* by Brown and Schlaikjer in 1943. They gave it a new species name because they found it to be different from *S. validus*. In the study and reclassification of the pachycephalosaurids, conducted by Robert Sullivan of the State Museum of Pennsylvania in 2003, enough differences were found to put it in a brand new genus of its own.

Features: This pachycephalosaurid has a very low and round dome. The joins between the bones that make up the dome are different from those in other pachycephalosaurids. The bones at the side of the skull are smaller too. These features are significant enough to determine that this is a new genus of dinosaur. The most obvious difference is that there is no shelf around the back of the dome as with other members of the group.

Distribution: Alberta, Canada.
Classification: Marginocephalia, Pachycephalosauria.
Meaning of name: From Hans-Dieter Sues, the Canadian pachycephalosaurid expert.
Named by: Sullivan, 2003.
Time: Cenomanian stage of the late Cretaceous.
Size: 2.5m (8ft).
Lifestyle: Low browser.
Species: *H. sternbergi*.

Left: When the bones were found they were thought to have belonged to Troodon, *but that was before* Troodon *was known to have been a meat-eater. Later they were reassigned to the bonehead group.*

Tylocephale

Knowledge of *Tylocephale* is based on a single damaged skull. It is closely related to *Prenocephale* and may well have been a new species of this genus. The distribution suggests that pachycephalosaurids evolved in Asia and migrated to North America, and then as a group, represented by *Tylocephale*, migrated back again.

Features: *Tylocephale* has the tallest dome of any pachycephalosaurid. The highest portion of it is further back than in others of the group, and it is quite narrow from side to side. There are small spikes around the back of the skull, similar to those of *Stegoceras*. The teeth are quite large for a pachycephalosaurid.

Distribution: Mongolia.
Classification: Marginocephalia, Pachycephalosauria.
Meaning of name: Swollen head.
Named by: Maryańska and Osmólska, 1974.
Time: Campanian stage of the late Cretaceous.
Size: 2.5m (8ft).
Lifestyle: Low browser.
Species: *T. gilmorei*, *T. bexelli*.

Left: Tylocephale and the other boneheads may have lived in inland or even mountainous areas.

LATE BONEHEADS

The pachycephalosaurids were among the last of the dinosaurs to evolve and flourish. Representatives from this group existed until the end of the Age of Dinosaurs. They began as small rabbit-sized animals, some with a slight thickening to the skull, but towards the end of the Cretaceous they became quite large and the skull ornamentation developed to spectacular proportions.

Pachycephalosaurus

Distribution: Montana, South Dakota and Wyoming, USA.
Classification: Marginocephalia, Pachycephalosauria.
Meaning of name: Thick-headed lizard.
Named by: Brown and Schlaikjer, 1943 based on Gilmore, 1931.
Time: Maastrichtian stage of the late Cretaceous.
Size: 8m (26ft).
Lifestyle: Low browser.
Species: *P. wyomingensis*, plus one other unnamed species.

Although this animal is often shown in restorations, little is known about it except for the skull. It was the biggest pachycephalosaurid known, hence its fame. Its three kinds of teeth suggest that it fed on a mixed diet of leaves, seed, fruit and insects. *Pachycephalosaurus*, *Stygimoloch* and a new species described in 2006 *Dracorex hogwartsia* ("king dragon from Harry Potter's Hogwarts Academy") are a family characterized by their horns.

Features: The muzzle of *Pachycephalosaurus* is quite long and narrow, and carries a number of tall spikes. Around the rear of the dome is a complex array of nodules and lumps. The length given is based on the assumption that the rest of the body is in the same proportion to the skull as in the others of the group, and the restoration is based on the bodies of other pachycephalosaurids.

Below: The dome on the head of Pachycephalosaurus *was 20cm (8in) thick. Estimates of its total length vary from 4.5–8m (15–26ft).*

AN ALTERNATIVE EXPLANATION

The head-butting activities of the pachycephalosaurids came under scrutiny with research published in 2004 by Mark Goodwin of the University of California, at Berkeley, and Jack Horner of Montana State University. They found that the radiating bone structure that was thought to have given strength to the head dome and which made it a powerful battering ram was only present in juvenile specimens, and not in adults, when it was assumed the head-butting would most likely have taken place. The head dome continued to grow as the animal grew, with the bone structure altering all the time. In addition, the bone carried blood vessels that suggest that the dome was covered with a horny cap while the dinosaur was alive. This would have been used for species recognition. As a result we cannot put together an accurate restoration of a pachycephalosaurid without knowing what this horn ornamentation was like.

Right: The dome may have been the base of a tall horn.

Stygimoloch

The name of this pachycephalosaurid derives from its frightful appearance. Moloch was a horned devil in Hebrew mythology, and in Greek legends the river Styx was the river that the dead had to cross to reach the Underworld. The fossils were found in the Hell Creek formation in Montana, and this was a further inspiration for the name. The first *Stygimoloch* horn core was found in 1896 and regarded as part of a *Triceratops* skull. In the 1940s, when pachycephalosaurids were recognized, it was classed as a species of *Pachycephalosaurus*.

Features: The most obvious feature of *Stygimoloch* is the array of horns projecting from the rim of the dome. The head is quite long and the dome is high, narrow and thin. From the front this presents a startling apparition of ornamentation, with long horns surrounded by clusters of more stubby spikes, that would evidently have been very effective as a threat or defence display, very much like those of some of the horned ceratopsians.

Distribution: Montana to Wyoming, USA.
Classification: Marginocephalia, Pachycephalosauria.
Meaning of name: Horned devil from the river of death.
Named by: Galton and Sues, 1983.
Time: Maastrichtian stage of the late Cretaceous.
Size: 3m (10ft).
Lifestyle: Low browser.
Species: *S. spirifer*.

Above:
Stygimoloch is known mostly from the skull. There have been five partial skulls found, but there have been other parts of the skeleton found in remains from North and South Dakota, USA.

Sphaerotholus

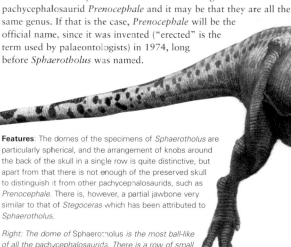

Distribution: Montana and New Mexico, USA.
Classification: Marginocephalia, Pachycephalosauria.
Meaning of name: Spherical dome.
Named by: Williamson and Carr, 2003.
Time: Maastrichtian stage of the late Cretaceous.
Size: 2m (6ft).
Lifestyle: Low browser.
Species: *S. goodwini*, *S. edmontonense*.

Sphaerotholus is known from two skulls, one of which is one of the most complete pachycephalosaurid skulls to have been found. Despite this, the animal's actual identity is rather unclear. The remains are very similar to those of the Mongolian pachycephalosaurid *Prenocephale* and it may be that they are all the same genus. If that is the case, *Prenocephale* will be the official name, since it was invented ("erected" is the term used by palaeontologists) in 1974, long before *Sphaerotholus* was named.

Features: The domes of the specimens of *Sphaerotholus* are particularly spherical, and the arrangement of knobs around the back of the skull in a single row is quite distinctive, but apart from that there is not enough of the preserved skull to distinguish it from other pachycephalosaurids, such as *Prenocephale*. There is, however, a partial jawbone very similar to that of *Stegoceras* which has been attributed to *Sphaerotholus*.

Right: The dome of Sphaerotholus *is the most ball-like of all the pachycephalosaurids. There is a row of small knobs around the back of the skull.*

PRIMITIVE ASIAN CERATOPSIANS

The main division of the Marginocephalian group is represented by the ceratopsians, or ceratopians.
These are the horned dinosaurs, of which there are a small number of families. Their origin can be traced
back into early Cretaceous times, but it was in the late Cretaceous period that they really came into their
own. The early forms were quite graceful little animals but they soon evolved into pig-sized beasts.

Graciliceratops

When *Graciliceratops* was discovered in 1975, the skeleton was referred to as a specimen of *Microceratops gobiensis*, but Paul Sereno of Chicago subsequently identified it as the juvenile of something quite different. The name derives from its small size and light build, and its bipedal stance shows that the group had its origins in the two-footed plant-eaters.

Distribution: Omnogov, Mongolia.
Classification: Marginocephalia, Ceratopsia, Neoceratopsia.
Meaning of name: Graceful horned face.
Named by: Sereno, 2000.
Time: Santonian to Campanian stages of the late Cretaceous.
Size: 0.9m (3ft), but this is immature. The adult was probably 2m (6½ft).
Lifestyle: Low browser.
Species: *G. mongoliensis*.

Above: This primitive ceratopsian was evidently able to move swiftly on its hind legs, unlike its heavy successors.

Features: Although this dinosaur is only known from a juvenile skeleton, there is enough to show that it is basically a bipedal animal with a front limb that is smaller than the hind limb. The hind limbs show that it was capable of running swiftly. As with all primitive ceratopsians, it has a beak at the front of its mouth and a ridge of bone, not quite a shield, around the back of the skull.

Protoceratops

There have been dozens of skeletons of *Protoceratops* found, both adult and juvenile, and so the whole growth pattern is known. It was found by the expeditions to the Gobi Desert undertaken by the American Museum of Natural History in the 1920s. It seems to have lived in herds, and its remains are so abundant that it has been termed the "sheep of the Cretaceous".

Features: *Protoceratops* is a heavy animal with short legs, a deep tail and a heavy head. Although a member of the horned dinosaurs, it does not have true horns. Two forms of adult are known, a lightweight form with a low frill, and a heavier form with a big frill and a bump on the snout where a horn would have been. These probably represent the two sexes, with the males having the heavier head.

Distribution: China and Mongolia.
Classification: Marginocephalia, Ceratopsia, Neoceratopsia.
Meaning of name: Before the horned heads.
Named by: Granger and Gregory, 1923.
Time: Santonian and Campanian stages of the late Cretaceous.
Size: 2.5m (8ft).
Lifestyle: Low browser.
Species: *P. andrewsi*.

CERATOPSIAN DEVELOPMENT AND DISTRIBUTION

The basal ceratopsians used to consist merely of *Protoceratops*, found in the 1920s. Then in the 1960s and 1970s a Polish-Mongolian expedition turned up all kinds of other small primitive ceratopsians at various places in the Gobi Desert.

It now seems likely that the ceratopsians evolved in Asia, and developed into fairly small, compact animals. Some time later they migrated eastwards, crossing the land bridge that is now the Bering Strait, entering North America where they evolved into the huge shield-necked, multi-horned, rhinoceros-like animals that have been known for more than a century. Strangely enough, the primitive forms also existed in North America, and continued largely unchanged until the end of the Age of Dinosaurs.

Below: Asia was once joined to America.

Asia

North America

Bagaceratops

Bagaceratops is known from about two dozen skulls, five of them complete, and some bits of the rest of the skeleton from several juveniles and adults. It is closely related to *Protoceratops*, but is smaller, hence the name. Like *Protoceratops* this is a desert-dweller, and is often found preserved in desert sandstone, having been overwhelmed by sandstorms or collapsing dunes.

Features: Unlike its hornless relatives, *Bagaceratops* sports a small horn on the snout. It also has a distinctive triangular frill around the neck forming a shield. Another distinction is that it lacks the pointed teeth at the front of the upper jaw, relying instead on its beak to gather food. Despite these features, and the fact that it came later, *Bagaceratops* is considered to be a more primitive animal than *Protoceratops*.

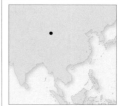

Distribution: Mongolia.
Classification: Marginocephalia, Ceratopsia, Neoceratopsia.
Meaning of name: Small horned head.
Named by: Maryańska and Osmólska, 1975.
Time: Campanian stage of the late Cretaceous.
Size: 1m (3ft).
Lifestyle: Low browser.
Species: *B. rozhdestvenskyi*.

Left: So many specimens of Bagaceratops, *young and old, are known that scientists have a good idea of how the individuals grew and developed.*

Breviceratops

Some scientists think that *Breviceratops* is a synonym for *Bagaceratops*. Indeed the two must have looked very much like one another. They lived at about the same time and had the same lifestyle, grazing the low-growing scrubby plants on a bleak desert landscape. In size, *Breviceratops* was intermediate between *Bagaceratops* and *Protoceratops*.

Features: Although some scientists think *Breviceratops* was the same as *Bagaceratops*, or even *Protoceratops* (it was at first thought to have been a species of *Protoceratops*), there are a number of important differences. Unlike *Bagaceratops*, *Breviceratops* has no sign of a horn and has two upper front teeth. It has a straight, lower jaw, unlike the curved jaw of *Protoceratops*, and the bones that form the frill do not flare outwards as widely as in *Protoceratops*. Notable though these differences are, there is still a possibility that *Breviceratops* may be a juvenile *Protoceratops*, and that the differences just represent different growth stages.

Distribution: Khulsan, Mongolia.
Classification: Marginocephalia, Ceratopsia, Neoceratopsia.
Meaning of name: Short horned face.
Named by: Maryańska and Osmólska, 1990.
Time: Santonian to Campanian stages of the late Cretaceous.
Size: 2m (6½ft).
Lifestyle: Low browser.
Species: *B. kozlowskii*.

Left: Classification of Breviceratops *is not helped by the fact that the most important specimen was stolen from the Palaeontological Institute of the Russian Academy of Sciences, in Moscow, in 1996.*

NEW WORLD PRIMITIVE CERATOPSIANS

The basal ceratopsians were mostly found in Asia, but several were discovered in North America too. Those in North America were mostly from a slightly later date, in fact one was from the end of the Age of Dinosaurs, suggesting that they evolved in Asia and later migrated eastwards. There some evolved into the big horned dinosaurs, while others remained in a small and primitive state.

Montanoceratops

Montanoceratops is the state fossil of Montana. It is known from two partial skeletons, the first studied by Brown and Schlaikjer in 1942, and named *Leptoceratops cerorhynchus*. C. M. Sternberg later realized that it was different enough to be a separate genus. There are two known specimens, the first found in 1916 and the other 80 years later in 1996, just a few metres away from the first.

Features: As noted by Sternberg in the 1950s, in almost every respect *Montanoceratops* is more advanced than *Protoceratops*, and much more so than *Leptoceratops*, both of which it superficially resembles. The first three vertebrae of the neck are fused into a solid lump, presumably as an adaptation to carrying the heavy skull. The long spines on the vertebrae of the tail give it a deep profile until mid-length, after which it tapers rapidly to a point.

Distribution: Montana, USA.
Classification: Marginocephalia, Ceratopsia, Neoceratopsia.
Meaning of name: Horned head from Montana.
Named by: Brown and Schlaikjer, 1942.
Time: Maastrichtian stage of the late Cretaceous.
Size: 3m (10ft).
Lifestyle: Low browser.
Species: *M. cerorhynchus*.

Zuniceratops

The earliest known ceratopsian to have carried a pair of brow horns is *Zuniceratops*. It is the oldest-known ceratopsian from North America. Its presence in New Mexico at such an early stage of the late Cretaceous suggests that the ceratopsians with the brow horns evolved in North America rather than Asia. A bonebed shows that it may have lived in herds.

Features: This is quite a lightweight horned dinosaur, rather like *Protoceratops* or *Montanoceratops* but less heavily built, with a well-developed neck shield containing large openings, and a pair of horns over the eyes. There is no horn on the snout.

Distribution: New Mexico, USA.
Classification: Marginocephalia, Ceratopsia, Neoceratopsia.
Meaning of name: Horned face of the Zuni tribe.
Named by: Wolfe and Kirkland, 1998.
Time: Turonian stage of the late Cretaceous.
Size: 3.5m (11½ft).
Lifestyle: Low browser.
Species: *Z. christopheri*.

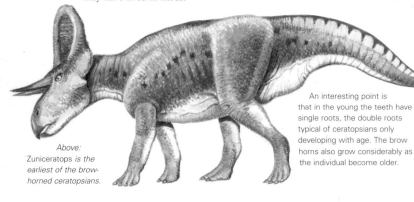

An interesting point is that in the young the teeth have single roots, the double roots typical of ceratopsians only developing with age. The brow horns also grow considerably as the individual become older.

Above:
Zuniceratops *is the earliest of the brow-horned ceratopsians.*

Turanoceratops

Despite the fact that all its relatives lived in North America *Turanoceratops* is included here. So far this is the only brow-horned ceratopsian to have been found in Asia. However, the material is very fragmentary and it is difficult to make any definitive statement about it. It does, however, appear to have a pair of horns above the eyes, and the double-rooted teeth of the more advanced ceratopsians.

Features: The double-rooted teeth so distinctive of the later North American ceratopsians are present here, long before they appeared in North American forms. This interesting fact is about the only thing we can say about *Turanoceratops*. A fragmentary skull with evidence of a pair of horns above the eyes, and bits of vertebrae and shoulder bone are all that are known, unfortunately not enough to tell us anything more.

Distribution: Kazakhstan.
Classification: Marginocephalia, Ceratopsia, Neoceratopsia.
Meaning of name: Turanian horned head.
Named by: Nessov and Kaznyshkina, 1989.
Time: Cenomanian to Turonian stages of the late Cretaceous.
Size: 2m (6½ft).
Lifestyle: Low browser.
Species: *T. tardabilis*.

Right: Apart from Turanoceratops, brow-horned ceratopsians are unknown from Asia. It is possible that it returned to Asia after evolving in North America, and that the migration was two-way.

CERATOPSIAN DIET

The food of ceratopsians has always been a puzzle. They were herbivores, we know that, but what exactly were the plants that they ate?

The basal ceratopsians had relatively simple teeth which worked by sliding past one another in a kind of a scissor movement. Their muzzles were quite delicate, suggesting that they were quite selective about what they ate. The heavy ridge at the back of the skull probably evolved first as a base for strong jaw muscles. (The later development into a neck shield was for quite another purpose, such as defence or display.) Their food of choice was probably the shoots and leaves of cycads. These would have been pecked off with the beak and then chopped up in the mouth, the cheeks holding it while the powerful jaws chopped it into bits. Later ceratopsians used the same action but with whole batteries of slicing teeth, with four or five growing upwards to replace each one that was wearing away.

Below: Ceratopsians ate shoots and leaves.

Leptoceratops

Leptoceratops may have been bipedal like the early forms, or it may have walked on all fours. The surprising feature of this animal is its extreme primitiveness, despite which it is among the last of the dinosaurs to have existed, at the end of the Cretaceous period, sharing the North American landscape with the biggest and most advanced ceratopsian, *Triceratops*.

Features: *Leptoceratops* has a slender body with short forelimbs. The skull is deep and the jaws carry primitive, single-rooted teeth. The teeth are adapted for crushing rather than for slicing as in other ceratopsians. There are no horns on the head. The neck shield is flattened from side to side, and carries a tall central ridge and smooth rear border. The front foot has five toes, with claws rather than hooves. All in all, it is a very primitive animal.

Below: Leptoceratops *is known from five skulls and parts of skeletons from North America. However, an almost identical limb bone has been found in Australia.*

Distribution: Alberta, Canada; to Wyoming, USA.
Classification: Marginocephalia, Ceratopsia, Neoceratopsia.
Meaning of name: Slender horned face.
Named by: Brown, 1914.
Time: Maastrichtian stage of the late Cretaceous.
Size: 3m (10ft).
Lifestyle: Low browser.
Species: *L. gracilis*.

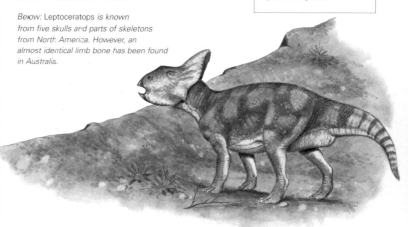

SHORT-FRILLED CERATOPSIDS

*The big horned dinosaurs of the late Cretaceous period belonged to the group called the Ceratopsidae.
They were only found in North America, with the possible exception of* Turanosaurus *from Kazakhstan.
They nearly all had massive rhinoceros-like bodies and heavy, armoured heads. They can be divided into
two subgroups, the short-frilled Centrosaurinae and the long-frilled Chamosaurinae.*

Avaceratops

Although the ceratopsids are generally
big animals, *Avaceratops* is quite small.
It is known from an almost complete
skeleton missing only the hip bones,
much of the tail and, frustratingly, the
roof of the skull including the horn
cores. The skeleton found is not an
adult, since most of the skull came
apart before it fossilized, but it
was almost fully grown
when it died.

Features: This small ceratopsid has a short frill that is
quite thick. Like other centrosaurines it has a short, deep
snout, a powerful lower jaw with batteries of double-
rooted shearing teeth, and a beak like that of a parrot.
It is assumed that like other centrosaurines, it has a
bigger horn on the nose than above the eyes. It may be
a juvenile or subadult of some
other genus such as
Monoclonius.

Distribution: Montana, USA.
Classification:
Marginocephalia, Ceratopsia,
Centrosaurinae.
Meaning of name: Ava's
horned face (from Ava Cole,
the wife of the discoverer).
Named by: Dodson, 1986.
Time: Campanian stage of
the late Cretaceous.
Size: 2.5m (8ft), but this was
a juvenile. The grown animal
was probably 4m (13ft).
Lifestyle: Low browser.
Species: *A. lammersi.*

Right: Avaceratops *looked like
a diminutive version of its
giant contemporaries.*

Centrosaurus

Distribution: Alberta,
Canada.
Classification:
Marginocephalia, Ceratopsia,
Centrosaurinae.
Meaning of name: Pointed
lizard.
Named by: Lambe, 1904.
Time: Campanian stage of
the late Cretaceous.
Size: 6m (19½ft).
Lifestyle: Low browser.
Species: *C. cutleri,
C. apertus.*

The animal to which the short-frilled group
owes its name is known from at least
15 skulls and pieces of bone from animals
of all stages of growth. The first part of the
skeleton to be found was the back of the
neck shield, with its
hook-shaped horns
which give the
animal its
name (not
the single
nose horn).

Features: *Centrosaurus* is noted for
the big, single horn on its snout, as well as
smaller horns over the eyes and others like
hooks on the neck shield. The edge of the
shield has bony growths. The big horn
curves forward in some individuals, leans
back in others, and yet sticks straight up in
others, a variation that palaeontologists do
not seem to think significant. A pair of
openings, or *fenestrae*, on the
neck shield keeps the weight down.

SKIN TEXTURE

It is not often that skin impressions are found from dinosaurs. However, there are skin impressions known from ceratopsians. These formed as the dead animal sank into river mud, and the impressions left in the mud solidified as the skin itself rotted away. The skin impressions known come from the area of the hips. These show plates of about 5cm (2in) in diameter, set in irregular rows 5cm (2in) apart, separated by a mass of 1cm (½in) scales. The plates may have been covered in horn, like those on the back of a crocodile. It is possible that the skin had a totally different texture on different parts of the body, especially the underside.

Unfortunately there is no evidence of the kind of skin that covered the face or shield, or even of the shapes of the keratinous parts of the horns, something that would be so valuable to artists of dinosaur restorations.

Below: A ceratopsian skin impression.

Monoclonius

For a long time there has been confusion over the name *Monoclonius*. Many specimens that had been attributed to it have now been reclassified as *Centrosaurus*, the uniting feature being the single nose horn that curved backwards. The spectacular *Centrosaurus* mount in the American Museum of Natural History was labelled *Monoclonius* until 1992, long after the palaeontologists knew better.

Features: *Monoclonius* is an average-size centrosaurine, but the skull is particularly long, more like that of a long-frilled ceratopsine. The single horn on the nose curves backwards, and there are no signs of horns above the eyes or across the shield. The edge of the shield is scalloped. The shield is rather thin compared with others of the group, suggesting it was used for display.

Distribution: Alberta, Canada; to Montana, USA.
Classification: Marginocephalia, Ceratopsia, Centrosaurinae.
Meaning of name: Single stem, referring to a single-rooted tooth found with the original specimen but which turned out to be something else.
Named by: Cope, 1876.
Time: Campanian stage of the late Cretaceous.

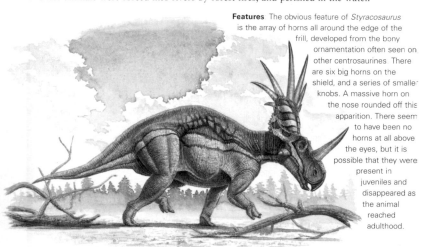

Size: 6m (19½ft).
Lifestyle: Low browser.
Species: *M. crassus, M. fissus, M. recurvicornis.*

Styracosaurus

Bone beds of several thousand *Styracosaurus* individuals are known, but there are few undamaged skulls, the only good one being the one on which Lawrence Lambe based the original description in 1913. Bone beds incorporating masses of charcoal suggest that herds of these animals were forced into rivers by forest fires, and perished in the water.

Features: The obvious feature of *Styracosaurus* is the array of horns all around the edge of the frill, developed from the bony ornamentation often seen on other centrosaurines. There are six big horns on the shield, and a series of smaller knobs. A massive horn on the nose rounded off this apparition. There seem to have been no horns at all above the eyes, but it is possible that they were present in juveniles and disappeared as the animal reached adulthood.

Distribution: Alberta, Canada; to Montana, USA.
Classification: Marginocephalia, Ceratopsia, Centrosaurinae.
Meaning of name: Spike lizard.
Named by: Lambe, 1913.
Time: Campanian to Maastrichtian stages of the late Cretaceous.
Size: 5.5m (18ft).
Lifestyle: Low browser.
Species: *S. albertensis, S. ovatus, S. sphenocerus.*

CENTROSAURINES

Although the Centrosaurinae, a family of the Ceratopsidae, are collectively known as the short-frilled ceratopsians, it does not mean that the frills were small. Most frills were ornamented with spikes and horns that made them look very imposing, and give the impression of belonging to a much larger animal. They tended to have small brow horns, the main horn being a big one on the nose.

Pachyrhinosaurus

The massive lump of bone on the nose of *Pachyrhinosaurus* gave the impression of its being a horned dinosaur with no significant horn. This bony mass may have been used as a battering ram when sparring, or as the base of a massive horn built of keratin that has not fossilized.

Features: The distinctive feature of *Pachyrhinosaurus*, and the one that gives it its name, is the massive shelf of bone on the nose where, in other centrosaurines, there is a horn. It also has a small horn above the eyes and along the centre line of the neck shield, and hook-shaped horns at the shield's top edge. The rest of the body is exactly the same shape as in that of other centrosaurines.

Distribution: Alberta, Canada, to Alaska.
Classification: Marginocephalia, Ceratopsia, Centrosaurinae.
Meaning of name: Thick-nosed lizard.
Named by: C. M. Sternberg, 1950.
Time: Maastrichtian stage of the late Cretaceous.
Size: 7m (23ft).
Lifestyle: Low browser.
Species: *P. canadensis*.

Left: The broad bony lump on the snout of Pachyrhinosaurus *is similar to that of the modern rhinoceros – and that supports a keratinous horn.*

Achelousaurus

A horned dinosaur with bony knobs on its head instead of horns, *Achelousaurus* is known from a huge bone bed in the Two Medicine Formation in Montana, USA. As in *Pachyrhinosaurus*, there is a possibility that these knobs were the bases of keratinous horns.

Features: The distinctive horn arrangement of *Achelousaurus* consists of a bony lump over the nose, smaller than that of *Pachyrhinosaurus* but deeply wrinkled. It also has a pair of smaller lumps above the eyes. As the animal became older, the lump of bone on the nose grew taller and began to point forwards, while those above the eyes became pitted. A pair of horn cores, flattened in cross-section, project from the back of the frill and splay outward.

Distribution: Montana, USA.
Classification: Marginocephalia, Ceratopsia, Centrosaurinae..
Meaning of name: Lizard of Acheloo (a Greek god whose horns were snapped off by Hercules).
Named by: Sampson, 1995.
Time: Campanian to Maastrichtian stages of the late Cretaceous.
Size: 6m (19½ft).
Lifestyle: Low browser.
Species: *A. horneri*.

Right: It is possible that Achelousaurus *is a species of* Pachyrhinosaurus; *the two are definitely closely related.*

Einiosaurus

This dinosaur *Einiosaurus*, together with *Achelousaurus* was established in 1995 when American palaeontologist Scott Sampson revised the Centrosaurinae. He established that the centrosaurines consisted of an evolutionary line with *Pachyrhinosaurus* at the basal end, and *Centrosaurus* and *Styracosaurus* at the other. *Einiosaurus* and *Achelousaurus* occupied a position between the two. The classification of these animals is determined by the ornamentation of the head.

Features: The nasal horn of this centrosaurine is large, compressed from side to side, and curved forwards and downwards. Two blades, rather than horns, protruded above the eyes. At the back of the frill two horns round in cross-section, projected straight backwards, adding to the

length of the skull and making it a spectacular apparition when the animal had its head down and was seen from the front.

Right: Einiosaurus was known before Sampson's study, but it was thought to have been a species of Styracosaurus.

Distribution: Montana, USA.
Classification:
Marginocephalia, Ceratopsia, Centrosaurinae.
 Meaning of name:
 Bison lizard (in the local Blackfoot language).
 Named by:
 Samson, 1995.
 Time: Campanian stage of the late Cretaceous.
Size: 6m (19½ft).
Lifestyle: Low browser.
Species: *E. procurvicornis.*

HEADS FOR SHOW

The only thing that distinguishes one genus of centrosaurine from another is the arrangement of features on the skull, the layout of horns and the shape of the neck shield. The bodies of the animals are, to all intents and purposes, identical.

The family was very restricted in number when it existed at the end of the Cretaceous period in its native North America. The geology of the area shows that each genus evolved very quickly, taking between half a million and one million years. In herds they roamed the plains between the modern Rocky Mountains, USA, and the shallow continental sea, with different ceratopsians keeping to their own group. As among the herds of grass-eaters on the African plains today, each type of animal kept to its own herd, avoiding the others. The distinctive frills and horns of ceratopsians probably evolved as a means of recognition, so that each member of a herd could identify its own kind.

The fact that the horns and shields were tough and heavy suggests that they were also used in combat. This was likely to have been combat within the species, with big males struggling for dominance in the herd, rather than against predators.

Brachyceratops

This dinosaur is known only from six subadult skeletons. Current understanding of the growth and development of the horn and shield arrangement in centrosaurines suggests that they may represent immature specimens of established genera such as *Styracosaurus* or *Monoclonius*.

Features: *Brachyceratops* has a small nose horn. The immature nature of even the largest of the specimens is shown by the fact that the horn is not fully fused to the skull. The shield is thin but broad in relation to the rest of the skull and lacks the holes seen in adults of the group. The tooth row is short, and all the skeletal remains are smaller than would be expected.

Distribution: Montana, USA.
Classification:
Marginocephalia, Ceratopsia, Centrosaurinae.
Meaning of name: Short-horned head.
Named by: Gilmore, 1914.
Time: Campanian to Maastrichtian stage of the late Cretaceous.
Size: 1.8m (6ft) as found, but perhaps 4m (13ft) as adult.
Lifestyle: Low browser.
Species: *B. montanaensis.*

Left: With only juvenile remains to work with, it is impossible to determine the exact classification of a Brachyceratops.

CHASMOSAURINES

The chasmosaurines represent the second of the two families of advanced ceratopsians. They were also rhinoceros-sized beasts, but with long neck shields. They generally had long snouts, and the brow horns were usually bigger than the nose horn. Like the Centrosaurinae, they were confined to the North American continent where they lived in herds migrating across the open plains.

Chasmosaurus

Distribution: Alberta, Canada; to Texas, USA.
Classification: Marginocephalia, Ceratopsia, Chasmosaurini.
Meaning of name: Chasm reptile.
Named by: Lambe, 1904.
Time: Maastrichtian stage of the late Cretaceous.
Size: 6m (19½ft).
Lifestyle: Low browser.
Species: *C. belli*, *C. russelli*.

In the 1880s the dinosaur-bearing beds of Canada were being opened up. Many spectacular horned dinosaurs were found, and by the turn of the century about half-a-dozen species of the spectacularly frilled *Chasmosaurus* had been identified. More detailed study by Canadians Stephen Godfrey and Robert Holmes in 1995 whittled these species down to two.

Features: The obvious feature of *Chasmosaurus* was the vast frill, like a huge triangular sail, around the back of its head. The weight of this structure was kept to a minimum by the holes (or chasms, hence the name) that reduced it essentially to a framework of struts. In life this shield would have been covered in skin, and was probably brightly coloured to act as a display organ.

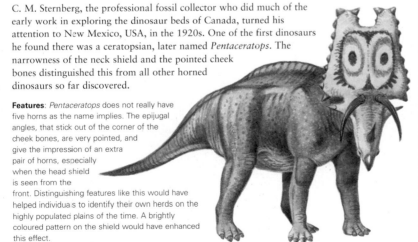

Pentaceratops

C. M. Sternberg, the professional fossil collector who did much of the early work in exploring the dinosaur beds of Canada, turned his attention to New Mexico, USA, in the 1920s. One of the first dinosaurs he found there was a ceratopsian, later named *Pentaceratops*. The narrowness of the neck shield and the pointed cheek bones distinguished this from all other horned dinosaurs so far discovered.

Features: *Pentaceratops* does not really have five horns as the name implies. The epijugal angles, that stick out of the corner of the cheek bones, are very pointed, and give the impression of an extra pair of horns, especially when the head shield is seen from the front. Distinguishing features like this would have helped individuals to identify their own herds on the highly populated plains of the time. A brightly coloured pattern on the shield would have enhanced this effect.

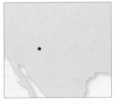

Distribution: New Mexico and Colorado, USA.
Classification: Marginocephalia, Ceratopsia, Chasmosaurini.
Meaning of name: Five-horned face.
Named by: Osborn, 1923.
Time: Campanian to Maastrichtian stages of the late Cretaceous.
Size: 6m (19½ft).
Lifestyle: Low browser.
Species: *P. sternbergi*.

Anchiceratops

The famous fossil collector Barnum Brown found the first remains of *Anchiceratops* in the Red Deer River Valley, Canada, in 1912. Charles Sternberg found another in 1924. This second one had a longer snout, smaller horns and a much thinner shield. They were then thought to have been two species, *A. ornatus* and *A. longisostris*, but are now regarded as a male and female of *A. ornatus*.

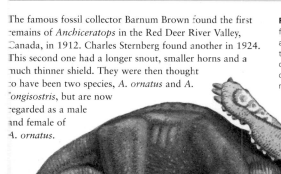

Features: The distinguishing feature of *Anchiceratops* is the array of points around the rim of the shield. The brow horns are quite short for a ceratopsine ceratopsian, and curve forwards rather elegantly. The snout is long and narrow. We do not know the body skeleton of *Anchiceratops* since only skulls have been found, but it seems that the body would have resembled that of its close relatives as all the ceratopsians had the same body shape and were distinguished by their heads.

Distribution: Alberta, Canada.
Classification: Marginocephalia, Ceratopsia, Chasmosaurini.
Meaning of name: Almost-horned head.
Named by: Brown, 1914.
Time: Campanian to Maastrichtian stages of the late Cretaceous.
Size: 6m (19½ ft).
Lifestyle Low browser.
Species: *A. ornatus*.

CERATOPSIAN SKULL

Ceratopsians are unusual among dinosaurs in that many skulls have been discovered. Usually the skull is the least likely part of the dinosaur skeleton to be preserved, being made up of a flimsy framework of bony struts, but ceratopsian skulls are very solid and heavy, and do not fall apart as easily as those of other dinosaurs.

Many ceratopsian genera are known only from the skull. Despite the great solidity, the ceratopsian skull is still made up of the elements found in every other dinosaur skull. Well-preserved skulls show the sutures, the lines along which each bone is joined to the next.

The neck shield, the most prominent part of a ceratopsian skull, is made up of the parietal and the squamosal bones. The *fenestrae*, or the holes in the shield, are gaps between these bones. In some ceratopsians, such as *Arrhinoceratops*, the *fenestrae* are quite small (and in *Triceratops* they are missing altogether) but in others, such as *Chasmosaurus*, they are so big that they make up the greater part of the area of the shield. In life the *fenestrae* would have been covered by skin, and may even have been filled with the muscles that powered the jaws.

Below: The bones that make up the skull of a ceratopsian.

Arrhinoceratops

The last of the long-frilled ceratopsians was *Arrhinoceratops*. Since its neck shield was quite short, it was originally misassigned as a member of the short-frilled Centrosaurinae by its describer W. A. Parks of Toronto. In the 1970s the skull was studied again by Helen Tyson of Alberta, Canada, who proved that it was unambiguously a member of the long-frilled Chasmosaurinae.

Features: *Arrhinoceratops* does not really lack a horn on the nose, as the name indicates. The horn core is there on the skull, although it grew from a slightly different position compared with the chasmosaurines, and is considerably smaller than that of most other ceratopsians. The face is rather short. The neck shield is quite thick and carries only small holes. Only a single skull has been found, and the rest of the skeleton is unknown.

Distribution: Alberta, Canada.
Classification: Marginocephalia, Ceratopsia, Chasmosaurini.
Meaning of name: Lacking a nose horn.
Named by: Parks, 1925.
Time: Maastrichtian stage of the late Cretaceous.
Size: 6m (19½ ft).
Lifestyle: Low browser.
Species: *A. brachyops*.

Left: Only one fossil of Arrhinoceratops has been found. It was probably one of the rarest of the ceratopsians on the North American plains.

MORE CHASMOSAURINES

The last of the ceratopsians were the long-frilled types. They included the biggest and best-known of them all, Triceratops. Some reached almost the size of modern elephants, and they lived right at the end of the Age of Dinosaurs. They ranged across North America and Canada, from Colorado in the south to Alberta and Saskatchewan in the north.

Triceratops

Although *Triceratops* was the biggest of the long-frilled chasmosaurine ceratopsians (the living animal weighed something like 4.5 tonnes), its frill wasn't as long as that of its relatives. It was more in the proportion of its short-frilled cousins, the centrosaurines.

When it was discovered, it was only known from a pair of horn cores. However, the whole skulls were so solid that they began to turn up quite regularly as complete fossils. Over the years so many different skulls of *Triceratops* have been unearthed that at one time there were 16 species attributed to the genus. These have now been combined, so that only the two given here are acknowledged, the common *T. horridus* and the bigger, but rarer, *T. prorsus*. Some authorities regard these as male and female *T. horridus*.

Features: This is the biggest and best-known of all ceratopsians. Its three magnificent horns give it its name. The horns on the fossilised skulls are only cores – they would have been covered in horny sheaths that made them much bigger. The neck shield is massive, with no holes in it, and is bordered by little knobs of bone. The teeth are arranged to work as shears, and powered by strong jaw muscles.

Distribution: Wyoming, Montana, South Dakota, and Colorado, USA; and Alberta and Saskatchewan, Canada.
Classification: Marginocephalia, Ceratopsia, Chasmosaurini.
Meaning of name: Three-horned face.
Named by: Marsh, 1889.
Time: Maastrichtian stage of the late Cretaceous.
Size: 9m (30ft).
Lifestyle: Low browser.
Species: *T. horridus, T. prorsus.*

Diceratops

Diceratops was for a time thought to be a species of *Triceratops*. It was classified as a separate genus when it was found by R. S. Lull in 1905, but later became *Triceratops hatcheri* when the distinctive skull features were put down to disease. Work by Catherine Forster of the University of Pennsylvania on the fossils in 1990 reinstated it as a genus in its own right.

Features: *Diceratops* is a very large example of a chasmosaurine. As the meaning of the name suggests, it has two horns on its skull, above the eyes. There is only a trace of a nose horn. Another feature is the presence of holes, or *fenestrae*, in the neck shield, showing that it is a different animal from *Triceratops*. The shape and arrangement of the individual bones that make up the neck shield are also different from *Triceratops*, but these would not be visible on the living animal.

Distribution: Wyoming, USA.
Classification: Marginocephalia, Ceratopsia, Chasmosaurini.
Meaning of name: Two-horned head.
Named by: Lull, 1907.
Time: Maastrichtian stage of the late Cretaceous.
Size: 9m (30ft).
Lifestyle: Low browser.
Species: *D. hatcheri*.

CERATOPSIAN POSTURE

We speak of the ceratopsians as rhinoceros-like, a reference to the horns on the head. However, for a long time the rest of the body was depicted as being very unrhinoceros-like. The standard depiction of *Triceratops*, and of any other big ceratopsian – for they all had almost identical bodies and limbs – has the heavy body supported mostly by straight hind legs. The legs were held, pillar-like, under the body like those of heavy mammals such as elephants and, indeed, rhinoceri. This is the standard modern view of dinosaur leg deployment. The front legs, however, were always shown splayed, with the upper leg more or less horizontal and the elbow bent at right angles. This seemed to make a lot of sense, since such a position would give flexibility, allowing the front part of the body to turn quickly, being pivoted at the hips with strong hind legs, so that the shield and horns could be presented to the enemy whatever its tactics. However, the modern view is that the front legs were held straight like the hind legs, as in all other big dinosaurs. This would give a body that did look like a rhinoceros.

Footprints have been found in the foothills of the Rocky Mountains near Denver, USA, that have been identified as ceratopsian trackways. The front prints fall very slightly outside the track of the back prints. If we reconstruct the body of a ceratopsian as having narrower shoulders than hips, then it does seem as if the front legs were at least slightly splayed.

Torosaurus

This dinosaur had the distinction of having the longest skull of any land-living animal. Much of the length is due to the enormous sweep of the neck shield (although a recent *Pentaceratops* skull may be longer). One specimen shows evidence of cancerous growths forming lesions within the bone of the neck shield.

Features: The skull is the only part of *Torosaurus* that is known. It is very much like that of *Triceratops* to which it is closely related, having three stout, forward-pointing horns, two long horns over the eyes and another shorter one on the snout. The shield, however, is quite different, having a pair of huge *fenestrae* that kept the weight down. The vast area presented by the frill was undoubtedly brightly coloured and used for display.

Distribution: Wyoming, Montana, South Dakota, Colorado, Utah, New Mexico, and Texas, USA; and Saskatchewan, Canada.
Classification: Marginocephalia, Ceratopsia, Chasmosaurini.
Meaning of name: Punctured lizard.
Named by: Marsh, 1891.
Time: Maastrichtian stage of the late Cretaceous.
Size: 6m (19½ft).
Lifestyle: Low browser.
Species: *T. latus*.

Above: Torosaurus had the longest skull of any dinosaur.

INDETERMINATE ANKLYOSAURS
AND STEGOPELTINAE

*Not every dinosaur can be neatly slotted into a taxonomic classification. In every group, including the
ankylosaurs, there are some that we cannot quite classify. Also in this group there is a subclassification,
the Stegopeltinae, which again, the palaeontologists cannot quite place.*

"Heishansaurus"

This is an animal that is known only from bits of bone and
some armour found in the Heishan, or Black Mountain,
area of Gansu Province in central China. When it was found
it was thought to have been a pachycephalosaurid, so
undiagnostic were the remains. It has even been classed as a
hadrosaurid.

Features: Whatever this animal
is, it seems to have had a heavy
head, the thickness of the partial
skull suggesting an armoured
animal, which led to its original
classification as a
pachycephalosaurid. It is more
likely to have been an
ankylosaurid, since the armoured
dinosaurs were quite
common in central Asia
at that time. The only
other parts of the
skeleton are
fragmentary
vertebrae, some ribs
and plates of armour.
There is so little to
say about it that it
must be regarded
as a *nomen
dubium.*

Distribution: China.
Classification: Thyreophora,
Ankylosauria.
Meaning of name: Lizard
from Heishan.
Named by: Bohlin, 1953.
Time: Campanian to
Maastrichtian stages of the
late Cretaceous.
Size: Not known.
Lifestyle: Not known.
Species: "*H.*" *pachycephalus.*

Pawpawsaurus

Known from a well-preserved adult skull and
some juvenile material *Pawpawsaurus* is the
earliest nodosaurid that has been found so far.
The main specimen is a complete skull found in
marine sediments – the remains of an animal
that had been washed out to sea. As in all
nodosaurids, most of the defensive armour
would have been on the neck and shoulders.
The typically narrow mouth suggests that it
was a more particular eater than either the
ankylosaurids or the polacanthids.

Features: This early nodosaurid has the typical
nodosaurid pattern of armour plates fixed to the
skull instead of being embedded in the skin. As
in later examples of nodosaurids, the mouth is
narrow and carries a beak. Unlike the later forms,
however, it does not have a palate separating the
mouth from the nasal chambers. It is the first of
the group to be found with the armoured eyelids.

Distribution: Texas, USA.
Classification: Thyreophora,
Ankylosauria, Nodosauridae.
Meaning of name: Lizard
from the Paw Paw Formation,
in north-central Texas.
Named by: Lee, 1996.
Time: Albian to Cenomanian
stages of the early and
late Cretaceous.
Size: 4.5m (15ft).
Lifestyle: Low
browser.
Species: *P. campbelli.*

Glyptodontopelta

Distribution: New Mexico.
Classification: Thyreophora, Ankylosauria, Stegopeltinae.
Meaning of name: Mimic of the shield of *Glyptodon* (an armadillo-like mammal from the Pleistocene).
Named by: Ford, 2000.
Time: Maastrichtian stage of the late Cretaceous.
Size: 5m (16½ ft).
Lifestyle: Low browser.
Species: *G. mimus.*

Glyptodontopelta is known from the armour of the hip regions and other pieces. Gilmore found it in 1919, but could not place it with any accuracy although he acknowledged that it was similar to that of *Polacanthus* and *Stegopelta*. Tracy Ford created the new classification Stegopeltinae in 2000 to accommodate it, along with *Stegopelta* and one other as yet unnamed genus.

Features: The armour consists of large, irregular hexagonal or pentagonal scutes with flat peaks. They are set in a mosaic that is packed so closely that, unlike other ankylosaurids, there is no space for smaller scales between them. It therefore resembles the armour of the Pleistocene mammal *Glyptodon*. This shield covered the hips, as with *Polacanthus* or *Stegopelta*, but it is not clear whether or not it extended to the rest of the body. The remainder of the restoration here is based on the general body shape of these animals.

Left: Dating from the end of the Cretaceous period, Glyptodontopelta must have been one of the last of the nodosaurid group of ankylosaurs.

ANKYLOSAUR CLASSIFICATION

The main distinction between the two groups of the ankylosaurs, the nodosaurids and the ankylosaurids, used to be the presence or the absence of the tail club. The ankylosaurids had the heavy club at the end of the tail, while the nodosaurids did not and were armed with spikes along their shoulders and sides instead.

This distinction is no longer valid. The early Cretaceous polacanthids, which lack the tail club, have proved to be more closely related to the clubbed ankylosaurids than the clubless nodosaurids. At one time it was thought that polacanthids did have a small club on the end of the tail, but this was found to be a mistake.

The distinction is now based on more subtle differences in the skeleton, and the fact that nodosaurids tended to have narrow mouths while the mouths of ankylosaurids were broader. This difference suggests a difference in diet, with the narrow-mouthed nodosaurids being more selective in their grazing than the broad-mouthed ankylosaurids.

Stegopelta

This is known from a fragmentary skeleton, consisting of a jawbone and bones from parts of the whole skeleton. Since its discovery by Williston in 1905, it has been thought of as a specimen of *Nodosaurus*, and was an invalid junior synonym of this genus. Now it has been placed in the new grouping Stegopeltinae by Tracy Ford in his reassessment of the group.

Features: *Stegopelta* has quite long foot bones for an ankylosaur. As in *Glyptodontopelta* the armour consists of broad, low scutes across the hip area in a pattern that is so tightly packed that there is no room for finer armour between. The rest of the skeleton appears to be that of a typical nodosaurid, suggesting that apart from the distinctive armour the Stegopeltinae did not differ greatly from the rest of the group.

Distribution: Wyoming, USA.
Classification: Thyreophora, Ankylosauria, Stegopeltinae.
Meaning of name: Roof shield.
Named by: Williston, 1905.
Time: Albian to Cenomanian stages of the early and late Cretaceous.
Size: 6m (19½ ft).
Lifestyle: Low browser.
Species: *S. landerensis.*

Left: A single shoulder spine was found among the scrappy remains of Stegopelta, suggesting that it had spiny armour in the neck and shoulder region.

NODOSAURIDS

The nodosaurids are (generally) the ankylosaurs without the tail clubs. They were more primitive than the ankylosaurids and came slightly earlier in the Cretaceous period. They tended to have armour that consisted of spines and spikes, and narrow mouths indicating that they had a more specialized diet and were more selective about their food than their club-tailed cousins.

Edmontonia

One of the best-known of the nodosaurids is *Edmontonia*. The original specimen was found in 1924 in Alberta, Canada, by collector George Paterson, but others, some almost complete, have since been found all across North America. It was once regarded as the same animal as *Panoplosaurus*. One species, *E. schlessmani*, is often given its own genus, *Denversaurus*, and it may be the same as *E. rugosidens*.

Features: This is a classic nodosaurid, with a broad back covered in armour and a wicked array of huge spikes sticking outwards, forwards and slightly downwards from each shoulder. The largest spike is split, giving it two points. The tail is very long. The skull is long and narrow, and angled downwards, adapted for grazing low vegetation and being selective about what it ate. Acoustical studies suggest it had the ability to make honking sounds.

Distribution: Alberta, Canada; Montana, South Dakota, Texas and Alaska, USA.
Classification: Thyreophora, Ankylosauria, Nodosauridae.
Meaning of name: From Edmonton.
Named by: C. M. Sternberg, 1928.
Time: Campanian to Maastrichtian stage of the late Cretaceous.
Size: 7m (23ft).
Lifestyle: Low browser.
Species: *E. longiceps, E. australis, E. rugosidens, E. schlessmani.*

Right: The two tines on the end of the main shoulder spike may have interlocked with those of rivals in mating contests, resulting in a trial of strength between big males.

Niobrarasaurus

The skeleton of *Niobrarasaurus* was found in 1930 by Virgil Cole, prospecting for oil-bearing rocks in Kansas, USA. He thought it was a plesiosaur, but had it shipped to his old college at the University of Missouri. There it was identified as a dinosaur and named *Hierosaurus* by Dr. M. G. Mehl. It was renamed as *Niobrarasaurus* by Ken Carpenter and his team in 1995.

Features: *Niobrarasaurus* is a typical nodosaurid, with a broad, armoured back and a slim tail. The armour consists of broad plates along the back and short spines along the sides. The foot bones are very short. As the original description was not very scientific, it was studied again in the 1990s and renamed. In 2003 a piece of the leg left behind by Cole 70 years earlier was discovered at the same Kansas site.

Distribution: Kansas, USA.
Classification: Thyreophora, Ankylosauria, Nodosauridae.
Meaning of name: Lizard from the Niobrara chalk.
Named by: Carpenter, Delkes and Weishampel, 1995 (but originally as *Heirosaurus* by Mehl, 1935).
Time: Campanian stage of the late Cretaceous.
Size: 5m (16½ft).
Lifestyle: Low browser.
Species: *N. coleii.*

Animantarx

This dinosaur has the distinction of being the first to be found by radiometric survey. Lamal Jones, a technician at the University of Utah, USA, knowing fossil bones to be slightly radioactive, surveyed a likely fossil site in Utah and persuaded the university to excavate the spot where low-level radiation seemed strongest.

Features: *Animantarx* is known from remains that consist of a partial skull with its jawbone, and a partial skeleton consisting of backbones, ribs, shoulder structure and parts of front and rear legs. It is a medium-size nodosaurid that resembles *Pawpawsaurus* with armour plates like upturned boats. The skull has a very high cranium and two pairs of short horns, one pair behind the eyes and another on the cheeks.

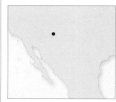

Distribution: Utah, USA.
Classification: Thyreophora, Ankylosauria, Nodosauridae.
Meaning of name: Animated living fortress.
Named by: Carpenter, Kirkland, Burge and Bird, 1999.
Time: Cenomanian to Turonian stages of the late Cretaceous.
Size: 3m (10ft).
Lifestyle: Low browser.
Species: *A. ramaljonesi.*

Top: Pawpawsaurus.

Left: "A 12ft-long dinosaur, looking like an armadillo but bigger than a cow," was how Don Burge, one of the team that studied Animantarx, *described it.*

BROAD AND NARROW MOUTHS

The main difference between the head of a nodosaurid and that of an ankylosaurid is the breadth of the mouth. A nodosaurid tended to have a pear-shaped skull that narrowed towards the jaws, unlike the hourglass-shaped jaws of an ankylosaurid with its broad beak. At the time these dinosaurs thrived, flowering plants had evolved, and there would have been plenty of leafy and seed-bearing undergrowth to be grazed. The nodosaurids, with their narrower mouths, must have been selective in their food choices, unlike the broad-mouthed ankylosaurids.

Common to both groups is the presence of a palate in the roof of the mouth (typical in mammals but rare in dinosaurs). The palate is a shelf that separates the airways of the nostrils from the foodways of the mouth, allowing that animal to eat and breathe at the same time. This would have speeded up the eating process.

The teeth of both groups were quite small, shaped like little hands and designed for chopping. Primitive types had pointed teeth at the front as well.

Below: The difference in the shape of the head between a nodosaurid (left) and an ankylosaurid (right) is obvious in top view.

Struthiosaurus

This is the smallest known of the nodosaurids. Its remains have been found all across Europe, in areas that are known to have been parts of an island chain during late Cretaceous times. Its small stature is taken to be proof that animals on islands tend to develop dwarf forms to make the best use of limited resources.

Features: *Struthiosaurus* resembles its larger relatives, but is more lightly built. The armour consists of three pairs of sideways-projecting spikes on the neck, at least one pair of tall spines over the shoulders and a double row of triangular plates sticking up along the tail. The back is covered in keeled scutes separated by a groundmass of ossicles, and there seems to be a well-marked boundary between the armoured back and the skin of the underside.

Distribution: Austria; France; and Hungary.
Classification: Thyreophora, Ankylosauria, Nodosauridae.
Meaning of name: Ostrich lizard.
Named by: Bunzel, 1871.
Time: Campanian stage of the late Cretaceous.
Size: 2m (6½ft).
Lifestyle: Low browser.
Species: *S. austriacus, S. ludgunensis, S. transylvanicus.*

MORE NODOSAURIDS

The nodosaurids spread across the Northern Hemisphere. Their remains have been found in Europe,
Asia and North America. They do not seem to have penetrated to Gondwana, however, as hardly any
ankylosaurs have been found in the southern continents. The Australian primitive ankylosaur, Minmi,
and the scattered remains in Argentina are the only exceptions.

Anoplosaurus

When *Anoplosaurus* was found,
early in the history of dinosaur
discovery, it was thought to be a
relative of *Iguanodon*, one of
the few dinosaurs known at that
time. It is now known to be a
nodosaurid, closely related to
the American genera
Silvisaurus and *Texasestes*.
It was once put into the
wastebasket taxon
Acanthophoplis.

Features: All that is really known of the first species of this animal,
A. curtonotus, is a few vertebrae from the neck, but some of the
remains are mixed with the bones of an iguanodont, hence the
confusion. It seems to be a primitive nodosaurid. *A. major* is better
known, with parts of the jawbone, vertebrae of the neck and back,
pieces of rib and bones of the legs and toes
found.

Distribution:
Cambridgeshire,
England.
Classification:
Thyreophora,
Ankylosauria, Nodosauridae.
Meaning of name: Lizard
without a weapon.
Named by: Seeley, 1878.
Time: Cenomanian stage of
the late Cretaceous.
Size: 5m (16ft).
Lifestyle: Low browser.
Species: *A. curtonotus,
A. major, A. tanyspondulus*.

Panoplosaurus

Panoplosaurus was one of the last of the
nodosaurids before they died out and the
armoured dinosaur niche was filled by the
club-tailed ankylosaurids. It is one of the
best-known, with two partial skeletons and
three skulls discovered. Until recently it was
thought to have been a species of
Edmontonia, but now it is
regarded as a genus
in its own
right.

Features: *Panoplosaurus* has a wide, pear-shaped
skull, toothless at the front and with broad nostrils. It
is similar to *Edmontonia*, but differs in having no long
spines projecting from the shoulders. Instead, the
armour is restricted to a series of thick, bony plates,
each with a prominent keel along the centre. The
plates are broadest over the shoulder and neck.
Short spikes run along the sides of the body and tail.

Distribution: Alberta,
Canada; to Montana, USA.
Classification: Thyreophora,
Ankylosauria, Nodosauridae.
Meaning of name: Totally
armoured reptile.
Named by: Lambe, 1919.
Time: Campanian stage of
the late Cretaceous.
Size: 7m (23ft).
Lifestyle: Low

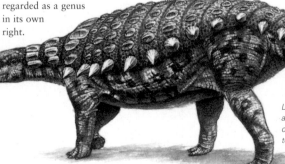

*Left: As with most
ankylosaurs, the armour plates
of* Panoplosaurus *are fused
to the skull, rather than being
merely embedded
in the skin.*

browser.
Species: *P. mirus*.

Sauropelta

The best-known of the nodosaurids is *Sauropelta*, and it is known from several almost complete skeletons. It appeared quite early in the geological succession, and had a number of primitive and unspecialized features. It is also one of the largest of the group, with a long tail that accounted for much of its length.

Features: *Sauropelta* is a nodosaurid with a very long tail, which is more highly armoured than was originally thought. The notable feature about its armour is the array of four pairs of spines projecting up from the neck. The armour is arranged in transverse rows of big, bony studs interspersed with smaller, pebbly armour. Its primitive features consist of the lack of a palate in the mouth and unfused neck bones, unlike its relatives.

Distribution: Wyoming, Montana and Utah, USA.
Classification: Thyreophora, Ankylosauria, Nodosauridae.
Meaning of name: Lizard shield.
Named by: Ostrom, 1970.
Time: Aptian to Cenomanian stages of the early and late Cretaceous.
Size: 8m (26ft).
Lifestyle: Low browser.
Species: *S. edwardsorum*.

Left: Sauropelta lived in the same area and at the same time as many of the sickle-clawed dromaeosaurids. It would have needed its armour to guard against them.

WHO WAS PALAEOSCINCUS?

In old dinosaur books, up to the late 1970s, an ankylosaur referred to as *Palaeoscincus* may appear. This dinosaur genus was based on a single tooth found in 1855 in Montana, USA, by fossil collector F. V. Hayden and named by the pioneer American palaeontologist Joseph Leidy in Philadelphia, USA, the following year. The name means "ancient skink" and refers to the similarity between this tooth and that of the modern skink lizard. Apart from that there were no clues as to what this animal was, since dinosaur studies were in their infancy at that time.

As the fossil discoveries of the late nineteenth century gained momentum, armoured dinosaurs of various kinds started coming to light, all with this distinctive kind of tooth. Soon *Palaeoscincus* began to be depicted as a generalized ankylosaur, complete with nodosaurid side spines and an ankylosaurid tail club. It was invariably shown in a sprawling posture, with the legs held out to the side like those of a lizard. This chimerical animal slipped into the public consciousness and continued to turn up in books for over a century until the demand for more accurate and scientific popular literature put it to rest relatively recently.

Silvisaurus

We know *Silvisaurus* from the front end of the skeleton including the skull. It was found in a stream bed where it had lain exposed and damaged by being trampled underfoot by drinking cattle. It was partly protected by being embedded in an iron nodule, but this also meant a great deal of effort in extracting and preparing it.

Features: *Silvisaurus* has a short beak at the front of the mouth with small, pointed teeth. It has large cheekbones and quite a long neck. The air-passages through the skull are cavernous, and suggest the ability to make loud honking noises for signalling. The armour consists of rows of thick, rounded plates over the back and sharp spines on its shoulders. It is possible that there were also spines down each side of its tail.

Distribution: Kansas, USA.
Classification: Thyreophora, Ankylosauria, Nodosauridae.
Meaning of name: Forest lizard.
Named by: Eaton, 1960.
Time: Aptian to Cenomanian stages of the early and late Cretaceous.
Size: 4m (13ft).
Lifestyle: Low browser.
Species: *S. condrayi*.

Right: In reality we know nothing about the appearance of Silvisaurus behind the shoulder region It is reasonable to surmise that it would be similar to other nodosaurids.

ANKYLOSAURIDS

The later and more advanced family of the ankylosaurs was the Ankylosauridae. They tended to have fewer spectacular spines than the nodosaurids, broader mouths and the shorter, stiffer tail with the distinctive club on the end. They were generally more lightly built, using the club for active defence rather than the more passive reliance on spikes and spines.

Aletopelta

The distribution of *Aletopelta* is interesting since it was found in California, in a rock sequence that had travelled north from Mexico since the end of the Cretaceous. It is the first time that ankylosaurids are known to have lived in Mexico. The name reflects this change of locale. The species name honours Walter P. Coombs, the ankylosaur specialist. The only specimen consists of the limbs, teeth and armour.

Features: Despite the fact that it is an ankylosaurid, *Aletopelta*'s back carried plates and spines as though it were a nodosaurid. A mass of armour over the hips is similar to that found in *Stegopelta*. There are heavy plates on the neck, sideways-pointing spines running along the body and tail, and a pair of tall spines, probably over the shoulders. The femur was considerably longer than the bones of the lower leg, showing that this animal did not run.

Distribution: California, USA.
Classification: Thyreophora, Ankylosauria, Ankylosauridae.
Meaning of name: Wandering shield.
Named by: Ford and Kirkland, 2001.
Time: Campanian stage of the late Cretaceous.
Size: 6m (20ft).
Lifestyle: Low browser.
Species: *A. coombsi*.

Cedarpelta

We know of parts of the skeletons of several *Cedarpelta* individuals including juveniles and adults from the same location. One skull has come apart completely allowing the individual bones of an ankylosaur head to be described in detail for the first time. These show that the decoration on the head developed from the bones of the skull rather than from the armour of the skin. As with other members of the ankylosaurid family, this dinosaur had eight vertebrae solidly fused to the hip bones, making for an exceptionally strong link between backbone and hind legs.

Features: This is a large and quite primitive ankylosaurid. The partial skull found has a strangely narrow snout with teeth at the front. This probably shows that it adapted to being selective about the plants that it ate. The skull is very highly ornamented. This ankylosaurid is probably related to *Shamosaurus* from the early Cretaceous of Mongolia, suggesting that there was migration between the two areas at that time.

Distribution: Utah, USA.
Classification: Thyreophora, Ankylosauria.
Meaning of name: Cedar Mountain formation shield.
Named by: Carpenter, Kirkland, Burge and Bird, 2001.
Time: Albian stage of the early or late Cretaceous.
Size: 10m (33ft).
Lifestyle: Low browser.
Species: *C. bilbeyhallorum*.

Right: It is not clear whether Cedarpelta was a nodosaurid or an ankylosaurid. The narrow mouth would suggest the former.

Gobisaurus

Closely related to *Shamosaurus*, the almost complete skeleton of *Gobisaurus* was found in 1959 or 1960 by the Sino-Soviet expeditions of the time, but was not studied for another 40 years. Hence the 'overlooked' inference in the species name. It was eventually studied by the Royal Tyrrell Museum in Drumheller, and the Canadian Museum of Nature in Ottawa, Canada, when it was brought to the West as part of a travelling exhibition.

Distribution: Alshan, China.
Classification: Thyreophora, Ankylosauria, Ankylosauridae.
Meaning of name: Lizard from the Gobi Desert.
Named by: Vickaryous, Zhao, Russell and Currie, 2001.
Time: Maastrichtian stage of the late Cretaceous.
Size: 7m (23ft).
Lifestyle: Low browser.
Species: *G. domoculus*.

Features: *Gobisaurus* is almost identical to *Shamosaurus*, except in the articulation of the bones of the jaw. The jawbones are a little longer as well. The eye sockets face the side, as is to be expected from a grazing animal, but face slightly more forwards than in most other ankylosaurids. The armour consists of bony plates, horns on the skull and a tail club. The name is a direct reference to *Shamosaurus*, since *Shamo* is the local name for Gobi.

Left: The species name is from the Latin domo *and* oculus *meaning "defective eye," referring to the fact that it was overlooked for decades.*

TAIL CLUB

The ankylosaurid tail club consists of bones embedded in the skin, which became enlarged and fused to each other in two distinctive lobes. At least half of the rest of the tail consisted of vertebrae fused to one another, and lashed together by tendons that became bone. The result is like a medieval mace, with a heavy head at the end of a stiff shaft. The muscles at the base of the tail were very powerful, and it seems obvious that this structure was used as a weapon. It would have been carried clear of the ground and could be swung sideways with considerable force. The legs of a big theropod, such as a tyrannosaurid, would have been vulnerable to this weapon.

A more modern equivalent can be found in the Ice Age mammal *Doedicurus*. This was one of the glyptodonts, big armoured animals that resembled giant armadillos. *Doedicurus* had a spiked club at the end of an armoured tail, which was probably used for the same purpose.

Another intriguing theory is that the tail club was a decoy. At first glance it would have resembled the long neck and small head of a sauropod, and a hungry theropod may have wasted its effort launching an attack at the wrong end of the animal.

Crichtonsaurus

The single specimen of *Crichtonsaurus* that has been found consists of the skull, parts of the lower jaw, vertebrae from the neck, back and tail, and the forelimbs and shoulder structure. It is now housed and displayed in the Palaeontology Museum of Liaoning. Its discovery helped to fix the date of the beds in which it was found.

Features: *Crichtonsaurus* is a fairly small ankylosaurid. The skull is broad at the back and narrows to the beak at the front. The jawbone is thin and unornamented. The teeth are symmetrical and small, like other ankylosaurid teeth. The armour consists of plates, some of which have been reconstructed as sticking upwards in a double row like that of a stegosaurid. There is a bony club on the end of the tail.

Distribution: Liaoning Province, China.
Classification: Thyreophora, Ankylosauria, Ankylosauridae.
Meaning of name: Michael Crichton's (the author of Jurassic Park) lizard.
Named by: Dong, 2002.
Time: Cenomanian to Toronian stages of the late Cretaceous.
Size: 3m (10ft).
Lifestyle: Low browser.
Species: *C. bohlini*.

BIG ANKYLOSAURIDS

The proliferation of ankylosaurids across both Asia and North America suggests that these animals lived in an era when other big meat-eaters lived and against which the ankylosaurids needed to be protected. Indeed this was the time and the stalking ground of the great, formidable tyrannosaurids. Slow-moving beasts such as the ankylosaurids would need heavy armour to protect themselves against such hunters.

Shanxia

Shanxia is known from a single partial skeleton found in 1993 by the Hebei Geological Survey. It consists of a partial skull, some of the backbone and limb bones and, sadly, only one piece of armour. It was found in river deposits with few other fossils, and has been impossible to date accurately.

Features: What distinguishes *Shanxia* from the other ankylosaurids is the shape of the horns on the rear of the skull. There are two pairs of horns and they are flattened, pointed and swept sideways and backwards rather than pointing straight out at the side as in other ankylosaurids. However, the way the skull articulates to the neck is more like that of a nodosaurid, but palaeontologists do not think that significant in the classification.

Left: Shanxia was probably similar in appearance to the other club-tailed ankylosaurs of the time.

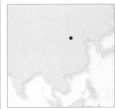

Distribution: China.
Classification: Thyreophora, Ankylosauria, Ankylosauridae.
Meaning of name: From Shanxi Province.
Named by: Barrett, You, Upchurch and Burton, 1998.
Time: Late Cretaceous.
Size: 3.5m (11½ft).
Lifestyle: Low browser.
Species: *S. tianzhenensis*.

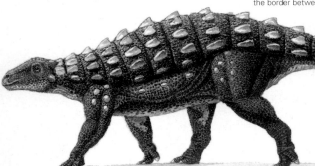

Tsagantegia

Tsagantegia is known from a single skull indicating that it was a medium-size ankylosaurid that must have been very similar to *Shamosaurus* and *Talarurus*. The skull is distinctive enough to show that it is a separate genus, but the lack of any fossils from the rest of the skeleton is very frustrating.

Features: The skull, the only known part of this animal, is long and flat for an ankylosaur. There is a prominent ring of bone around the eye socket. The eye sockets are situated just behind the midpoint of the skull. There is armour on the roof of the skull, formed from masses of small, bony knobs, but this is quite low and not very prominent. The snout is wider than that of *Shamosaurus* or *Cedarpelta*. *Tsagantegia* was found in the Gobi Desert, close to the border between Mongolia and China.

Left: The tail of an ankylosaurid was held stiff and straight, like the wooden shaft of a club. All flexibility was at the base, where powerful muscles could swing the weapon from side to side.

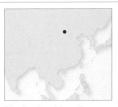

Distribution: Mongolia.
Classification: Thyreophora, Ankylosauria, Ankylosauridae.
Meaning of name: From Tsagan Teg.
Named by: Tumanova, 1993.
Time: Cenomanian stage of the late Cretaceous.
Size: 6m (19½ft).
Lifestyle: Low browser.
Species: *T. longicranialis*.

ANKYLOSAURID BRAIN

The brain cavity of an ankylosaur is quite well-known, since the armour-covered heads were well protected and often fossilized. The part of the brain that controlled movement and general activity is quite small compared with that of other dinosaurs, such as the ornithopods. This indicates that they were quite slow in moving about. The shapes of the legs, with a long thigh bone compared to the lower leg, is that of a slow-moving animal too.

The most highly developed part of the brain is that which deals with the sense of smell. This, combined with the complex maze of nasal passages that we find in many ankylosaurid skulls, suggests that this may have been the primary sense on which they relied. The nasal passages were probably also lined with membranes to moisten and warm the air as it passed down to the lungs. Most ankylosaurids, certainly the Asian forms, lived in very dry environments and would benefit from this. It is also possible that the nasal passages were for generating sounds for communication. The nodosaurids, on the other hand, did not have such complex nasal passages, merely paired tubes that passed from the nostrils straight back to the throat.

Ankylosaurus

The ankylosaur dynasty, both nodosaurid and ankylosaurid, reached its climax in *Ankylosaurus* itself. This is the most familiar of the ankylosaurids, but is only known from a skull, some vertebrae and pieces of armour and a few teeth. Until the ankylosaurs were re-examined in the 1970s, *Ankylosaurus* was portrayed as a mixture of ankylosaurid types (with the tail club), and the nodosaurids (with the side spikes). It was the largest and last of the ankylosaurs.

Features: The armour consists of ovals of embedded bone, each one supporting a horny covering. They are quite smooth compared with those of other ankylosaurids. The tail, stiffened by having the vertebrae lashed together by fused, bony tendons, carries a bony club at the end. The skull is broad, with two pairs of sideways-pointing horns at the rear corners. There are no teeth at the front of the mouth, just the broad beak.

Distribution: Alberta, Canada; and Wyoming, USA.
Classification: Thyreophora, Ankylosauria, Ankylosauridae.
Meaning of name: Fused lizard.
Named by: Brown, 1908.
Time: Maastrichtian stage of the late Cretaceous.
Size: 11m (36ft).
Lifestyle: Low browser.
Species: *A. magniventris*.

Below: Large plates on the neck and shoulders, and smaller plates in rows along the sides, characterize the armour of Ankylosaurus *and the other ankylosaurids.*

MORE BIG ANKYLOSAURIDS

As with the horned ceratopsians, many ankylosaurids are known only from their skulls – they were heavily armoured and could fossilize well. This poses no problem in the ceratopsids, whose bodies were all more or less the same. With the ankylosaurids, however, the variation in the arrangement of armour on the back means that there must be some speculation about the appearance of the whole animal.

Tianzhenosaurus

The almost complete skeleton and a skull that have been found are very much like *Saichania*; however, the skull was relatively smaller. *Shanxia* may even be the same animal as *Saichania*, which is known from a partial skull and skeleton. This was described by Barrett, You, Upchurch and Burton, also in 1998, and if the two are the same it will cause problems in naming.

Features: *Tianzhenosaurus* is similar to *Saichania* and may even be the same thing. However, the skull is slightly smaller and more triangular, and the skull roof is covered in big, irregular tubercles. Two blunt horns project outwards and backwards from the rear of the skull. The front bone of the lower jaw is quite long, as is the snout, and the tooth rows are parallel. There is a big club on the end of the tail.

Distribution: Sichuan Province, China.
Classification: Thyreophora, Ankylosauria, Ankylosauridae.
Meaning of name: Lizard from Tianzhen.
Named by: Pang and Zheng, 1998.
Time: Maastrichtian stage of the late Cretaceous.
Size: 4m (13ft).
Lifestyle: Low browser.
Species: *T. youngi*.

Above: The species name of Tianzhenosaurus youngi *honoured Chung Chien Young (or Yang Zhongquian) the founder of Chinese vertebrate palaeontology on his 100th birthday.*

Nodocephalosaurus

This seems to be closely related to the Asian *Saichania* and *Tarchia* from about the same time. However, it was found a long way away, and we only know it from a partial skull. Other pieces of skeleton found in the area were once regarded as parts of *Panoplosaurus* or *Euoplocephalus*, but it is possible that they belong to *Nodocephalosaurus*.

Features: The skull of *Nodocephalosaurus* is high and arched, and the snout turned downwards as if for low grazing. The armour on the head is distinctive, and consists of hollow, bulbous, bony lumps arranged in a symmetrical pattern, quite unlike the low-profile, flat-head armour of the other North American ankylosaurids, such as *Euoplocephalus*. Its Asian relatives possess this feature, showing the closeness of their kinship.

Left: The only specimen known of Nodocephalosaurus *is a single partial skull. Although this tells us how it is related to other ankylosaurids, it does not give us a clear idea of what the rest of the animal looked like.*

Distribution: New Mexico.
Classification: Thyreophora, Ankylosauria, Ankylosauridae.
Meaning of name: Node-headed lizard.
Named by: Sullivan, 1999.
Time: Campanian stage of the late Cretaceous.
Size: 6m (20ft).
Lifestyle: Low browser.
Species: *N. kirtlandensis*.

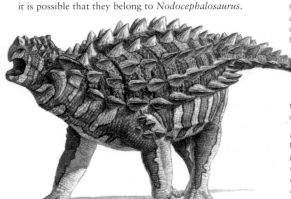

Saichania

Two complete skulls and an almost complete skeleton with armour still in position, are the fossils found of *Saichania*, and make it the best-preserved Asian ankylosaurid. The Asian ankylosaurids have mostly been found in rocks deposited under desert conditions. It seems that they were more suited to dry landscapes than the North American ankylosaurids that have been found in more varied environments.

Features: The armour around the neck is distinctive in having crescent shapes. It also has plates on the underside. There are tall spikes on the back and also along the sides. The forelimbs are short and massive, and consequently the animal is very low-slung. The skull has a complex set of nasal passages, probably as a device for cooling and campening the inhaled air. This is in keeping with the arid environment that it inhabited.

Distribution: Mongolia.
Classification: Thyreophora, Ankylosauria, Ankylosauridae.
Meaning of name: Beautiful (referring to the state of preservation).
Named by: Maryańska, 1977.
Time: Campanian stage of the late Cretaceous.
Size: 7m (23ft).
Lifestyle: Low browser.
Species: *S. chulsanensis*.

Tarchia

This is the largest ankylosaurid known from Asia and also the last surviving. It is known from the remains of at least seven individuals, and appears to have been quite common. The original skull carries injuries that had been inflicted in life and then healed. The damage is consistent with an attack by a big meat-eater, such as a tyrannosaurid.

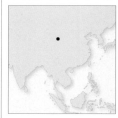

Features: The head is big and quite tall, with horns at the rear corners. The snout is broad and blunt, and the teeth small and weak. The brain case is considerably larger than that of its relative, *Saichania*, but it is still quite small for the size of animal. The size of the head supports the idea that ankylosaur heads became more massive as the group evolved. The tail club is quite large.

Left: The massive musculature of the hips of Tarchia and the other ankylosaurs, and the specialized bone structure needed to anchor it, meant that there is no trace of the bird-like structure that is implied in the fact that they belonged to the "bird-hipped" dinosaur group.

Distribution: Mongolia.
Classification: Thyreophora, Ankylosauria, Ankylosauridae.
Meaning of name: Brainy.
Named by Maryańska, 1977.
Time: Campanian to Maastrichtian stages of the late Cretaceous.
Size: 8.5m (28ft).
Lifestyle: Low browser.
Species: *T. gigantea*, *T. kielanae*.

THE LAST ANKYLOSAURIDS

Some members of the ankylosaurid group, as well as being the last dinosaurs to evolve, and taking over from their cousins the nodosaurids, existed right up to the end of the Age of Dinosaurs. They could be regarded as the climax of the dinosaur dynasty that lasted 160 million years. Whatever the disaster that wiped dinosaurs out, it was observed by the ankylosaurs, tyrannosaurs, ceratopsians and hadrosaurids.

Maleevus

All that is known of *Maleevus* is the jawbone and part of the skull, found by E. A. Maleev in 1952. These are so similar to *Talarurus* that many palaeontologists regard it as a species of this genus. Some even think that there is so little information about the fragments that it cannot be identified as anything with any certainty.

Features: *Maleevus* is regarded as being almost identical to its relative *Talarurus*, except for details of the hind portion of the skull. Like *Talarurus* it is a large ankylosaurid. It has armour in transverse bands, and a club formed by three masses of fused bones in two lobes at the end of a long tail which is stiffened by solidified tendons. Its broad mouth has a cutting beak adapted for biting off large mouthfuls of food indiscriminately.

Distribution: Mongolia.
Classification: Thyreophora, Ankylosauria, Ankylosauridae.
Meaning of name: From E. A. Maleev (a Russian palaeontologist).
Named by: Turmanova, 1987.
Time: Cenomanian to Turonian stages of the late Cretaceous.
Size: 6m (19½ ft).
Lifestyle: Low browser.
Species: *M. disparoserratus*.

Left: The genus name honours the Russian E. A. Maleev who did much to open Mongolia to palaeontologists in the 1950s.

Talarurus

Talarurus is known from the remains of at least five individuals. At the time Maleev, who discovered it, also found the skull that is now regarded as *Maleevus disparoserratus*, naming it *Syrmosaurus disparoserratus*. The name *Talarurus* means "wicker tail" and refers to the interwoven tendons that kept the tail straight and stiff, forming a rigid shaft for the club.

Features: *Talarurus* has a smaller tail club than *Euoplocephalus*, to which it is closely related. The completeness of the remains shows that the hind foot has four toes. This is possibly the condition that existed among the primitive ankylosaurids, while the more advanced forms, such as *Euoplocephalus*, had three. The armour is arranged in transverse bands, with no sign of the upward-pointing spines of *Euoplocephalus*. As in other ankylosaurids, half of the tail consisted of fused vertebrae.

Distribution: Mongolia.
Classification: Thyreophora, Ankylosauria, Ankylosauridae.
Meaning of name: Wicker tail.
Named by: Maleev, 1952.
Time: Cenomanian to Turonian stages of the late Cretaceous.
Size: 6m (19½ ft).
Lifestyle: Low browser.
Species: *T. plicatospineus*.

Right: The remains of Talarurus are so similar to those of Maleevus, that some scientists think that the two are the same genus. If this is so, then Talarurus must take priority, having been named first.

Euoplocephalus

This is without doubt the ankylosaurid that is best known to science. There are more than 40 specimens known, including 15 skulls, suggesting that this was the most common ankylosaurid in North America at the time. They have never been found in groups, and were probably solitary foragers. The forelegs were quite supple, suggesting that this animal could dig for roots and buried stems.

Below: Different species appear to have differently shaped clubs. E. tutus has a broad heavy club, while E. acutosquameus has a smaller pointed club.

Features: The eyelids are armoured with movable slabs of bone, the first time this is seen in an ankylosaur. The skull is quite light with tortuous air passages, probably to warm or moisten the air before it reached the lungs. The back is armoured with heavy, bony nodules set into leathery skin. Spines, up to 15cm (6in) long, stick up from the neck and shoulders.

Distribution: Alberta, Canada; to Montana, USA.
Classification: Thyreophora, **Classification**: Thyreophora, Ankylosauria, Ankylosauridae.
Meaning of name: Completely well-armoured head.
Named by: Lambe, 1910.
Time: Campanian to Maastrichtian stages of the late Cretaceous.
Size: 6m (19½ft).
Lifestyle: Low browser.
Species: *E. tutus, E. acutosquameus.*

GEOGRAPHICAL LOCATION OF ANKYLOSAURS

As we have noted, the ankylosaurs were confined almost entirely to the northern continents. However, there have been isolated bones found in late Cretaceous rocks of Argentina that seem to belong to ankylosaurs – the muscle scars and the shape of the articulations show this. There are also isolated armour plates.

The remains are too scrappy for any identification, but they seem to come from nodosaurids rather than ankylosaurids. They occur in the same rock sequences as duckbills, another rare group for South America. It appears that about this time there was some land connection with the continent of North America and, for a brief period, there was an exchange of animals between the two. However, this invasion does not seem to have been much of a success, since ankylosaur remains appear to be very limited in the Southern Hemisphere. As the world passed out of the Age of Dinosaurs, the continents were beginning to separate and diverge, and were developing their distinctive suites of animal types.

Pinacosaurus

More than 15 specimens are known of this genus, a good reflection of how well the armoured backs fossilize. It is one of the few ankylosaurs for which the juveniles are known, after two finds of several of them huddled together, killed by a sandstorm. Such finds give us a good idea of the relation of the armour to the skeleton, something that is difficult to understand from an adult specimen.

Features: The skeleton of this ankylosaurid is relatively light, with more slender limb bones and smaller feet than others of the group. The front foot has five toes while the rear foot has four. The shoulder blades are much more slender than in any other ankylosaurid, but we cannot determine the significance of this. The skull has more openings than is usual in an ankylosaurid. The armour consists of deeply keeled plates. One species, *P. mephistocephalus*, has a pair of devil-like horns, hence the species name.

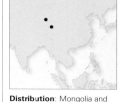

Distribution: Mongolia and China.
Classification: Thyreophora, Ankylosauria, Ankylosauridae.
Meaning of name: Plank lizard.
Named by: Gilmore, 1933.
Time: Santonian to Campanian stages of the late Cretaceous.
Size: 5.5m (18ft).
Lifestyle: Low browser.
Species: *P. grangeri, P. mephistocephalus.*

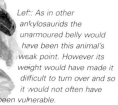

Left: As in other ankylosaurids the unarmoured belly would have been this animal's weak point. However its weight would have made it difficult to turn over and so it would not often have been vulnerable.

THE MAMMALS, POISED FOR TAKEOVER

As the dinosaurs reached the pinnacle of their existence, the mammals continued to expand and diversify as they had been doing for the previous 150 million years. However, they were still, for the most part, little insignificant scurrying beasts that ate insects or seeds. Most of the modern groups known today had evolved, but their time was still to come.

Zalambdalestes

There is a possibility that *Zalambdalestes* is a holotherian – the group to which most modern mammals belong – close to the ancestry of the rodents and rabbits, but this is not universally accepted. The teeth seem to be intermediate between these and the teeth of earlier mammals. It looked very much like a modern elephant shrew.

Below: Zalambdalestes *hunted insects through the undergrowth.*

Features: *Zalambdalestes* is one of the few late Cretaceous mammals for which the skeleton is well known. The snout is long and needle-like, like that of a shrew, and there is a long gap between the front teeth and those at the back. It probably found its food in the dark by smell, or even by echolocation. The tail is long and the rear legs are quite strong, with long foot bones, suggesting that it may have moved about in a jumping action like modern kangaroo rats.

Below: Modern jumping rodent resembling Zalambdalestes.

Distribution: Mongolia.
Classification: Theriiformes, Holotheria.
Meaning of name: V-shaped thief.
Named by: Gregory and Simpson, 1926.
Time: Campanian stage of the late Cretaceous.
Size: 20cm (8in).
Lifestyle: Insectivore.
Species: *Z. lechei.*

Cimolestes

This widespread genus may have been the ancestor of the carnivores – the modern cats, dogs and weasels – even though at this stage it would have had an insectivorous diet. Some of the larger species may have been big enough to take vertebrate prey. It was successful enough to have survived from the late Cretaceous through to the Tertiary. The remains of *Cimolestes* consist mostly of teeth from widely separated parts of the world. The teeth are tiny and have been found only by the fine sieving of sediments. A large number of species have been named, but the validity of most of these is open to question.

Features: The teeth of *Cimolestes* worked with a scissor-like shearing action that was very similar to that of the carnassial, or meat-shearing, teeth of the modern carnivores. There is some difference between the species of *Cimolestes*, with the smaller forms being adapted to eating insects and the larger species tending towards a more omnivorous diet. They would have been predominantly tree-dwellers keeping out the way of the bigger animals of the time.

Below: As we know little more than the teeth of Cimolestes, *this restoration must be regarded as rather conjectural.*

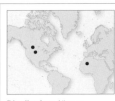

Distribution: Alberta, Saskatchewan, Canada; Montana, Wyoming, New Mexico, USA; Morocco.
Classification: Theriiformes, Holotheria.
Meaning of name: Insect thief.
Named by: Marsh, 1889.
Time: Maastrichitan stage of the late Cretaceous to Palaeocene.
Size: 20cm (8in).
Lifestyle: Insectivore.
Species: *C. incisus, C. cerberoides, C. cuspulus, C. lucasi, C. magnus, C. propalaeoryctes, C. simpsoni, C. stirtoni.*

Cimexomys

Distribution: Montana, USA.
Classification: Theriiformes, Allotheria.
Meaning of name: Insect mouse.
Named by: Sloan and van Valen, 1965.
Time: Campanian stage of the late Cretaceous.
Size: 30cm (1ft).
Lifestyle: Insectivore or egg-stealer.
Species: *C. minor*, *C. antiquus*, *C. magnus*, *C. judithae*, *C. gratus*, *C. hausoi*, *C. gregoryi*.

One of the late representatives of the multituberculate group of primitive mammals, *Cimexomys* lived in the late Cretaceous along with the dinosaurs and outlived them into Palaeocene times. Their presence near the *Troodon* nests in Montana, USA, suggests that they may have preyed on insects that are always found in nests, or may have fed on the eggs themselves. The "insect" part of the name is a reference to the Bug Creek area in Montana where it was first found.

Features: Usually the multituberculates are known only from isolated teeth, but one occurrence consists of jawbones as well as teeth from the region of the *Troodon* nests at Egg Mountain in Montana. The upper dentition consists of the many-cusped teeth typical of the multituberculate group, while the lower jaw carries rodent-like incisors at the front and a huge shearing tooth at the back.

Left: With only the teeth and jaws to work with, it is almost impossible to produce an accurate restoration of Cimexomys.

MORE LATE CRETACEOUS MAMMALS

Alymlestes – like *Zalambdalestes* from Kazahkstan.
Barunlestes – like *Zalambdalestes* but with a shorter skull, from Mongolia.
Kulbeckia – the oldest-known of the Zalambdalestes group, from Uzbekistan.
Paranyctoides – perhaps an ancestor of the insectivores, from Uzbeckistan.
Schowalteria – a monster as far as Mesozoic mammals is concerned, fox-sized from Canada, but only known from its snout.
Stygimys – known from the Hell Creek formation in Montana, USA.
Bryceomys – tiny, about half the size of a house mouse, from Utah, USA.
Dakotamys – like *Cimexomys*, from Utah, USA.
Uzbekbaatar – like *Cimexomys*, from Usbekistan.
Essonodon – another monster, weighing 2.5 kilos (5½lb), but little is known about it, from Canada and Montana, USA.
Barbatodon – a multituberculate from Romania.
Kogaionon – a multituberculate from Belgium.
Deltatheridium – a marsupial from Mongolia
Asioryctes – an early placental from Mongolia, known from a skull and preserved brain cavity.
Catopsbaatar – a relatively large multituberculate from Mongolia.
Kryptobataar – from Mongolia, a multituberculate with limb bones preserved, showing it to have had a sprawling lizard-like posture.
Ausktribosphenos – a controversial find, a supposed placental from the middle Cretaceous of Australia.

Ukhaatherium

Fossil mammals can be found anywhere. Unfortunately they usually only occur as individual, almost microscopic, teeth – teeth being harder and more easily fossilized than bones. However, occasionally a complete skeleton is found. Skeletons of many animals, including dinosaurs, but also small things like lizards, birds and mammals, have been found in the fossil site of Ukhaa Tolgod in Mongolia, which was discovered in 1993. Among them was the primitive mammal known as *Ukhaatherium*.

Features: The skeleton is similar to that of the primitive insect-eaters like the modern tenrecs of Madagascar, but is much more primitive. The hips are quite narrow, like those of a modern marsupial, and this suggests that it may have given birth to poorly developed young after a short period of gestation. The feet, with their long claws, appear modern compared with the rest of the skeleton.

Below: Ukhaatherium *hunted insects beneath the feet of the great dinosaurs in the area that is now the Gobi Desert.*

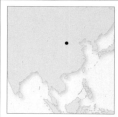

Distribution: Mongolia.
Classification: Theriiformes, Placentalia.
Meaning of name: Mammal from Ukhaa Tolgod.
Named by: Novacek, 1997.
Time: Campanian stage of the late Cretaceous.
Size: 25cm (10in) including the tail.
Lifestyle: Insectivore.
Species: *U. nessovi*.

THE TERTIARY PERIOD

The beginning of the Tertiary period, after things had settled from the chaos that was the end-Cretaceous mass extinction, was a time of tropical climates and environments. Mammals and birds flourished to take the place of the great reptiles that had become extinct. As the period progressed the climates cooled and dried. Grasses became important and for the first time open grassy prairies spread over the continents. As a result we can see the Tertiary as two distinct periods – the early Tertiary or Palaeogene, in which the forest conditions prevailed, and the late Tertiary or Neogene, characterized by the grasslands.

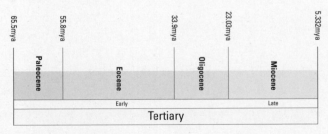

Above: The timeline shows the epochs that make up the Tertiary period of Earth's history.

1 Embolothorium
2 Eosimias
3 Andrewsarchus
4 Sarkastodon
5 Harpagolestes

BIRDS

With the passing of the pterosaurs at the end of the Cretaceous period, the birds established themselves as the true masters of the air. By the early Tertiary all the bird groups that we would recognize today were in existence. However, there were also a number of very specialized forms that have since become extinct, and it is those on which we concentrate here.

Copepteryx

By late Cretaceous times there were birds that had given up the powers of flight and had taken to a purely aquatic way of life. This trend continued throughout the Tertiary period. Many big swimming birds existed in the early Tertiary hunting the shoals of fish and squid that survived the end-Cretaceous extinction and they may have waned as the seals and sealions developed later.

Left: The larger species of Copepteryx must have stood taller than a man while on land.

Features: The skull of a typical bird from the plotopterid family is similar to that of a modern gannet, while the body is more like that of a darter. It lived like a penguin, relying on its powerful paddle-like wings to propel it through the water. As in the modern penguin the swimming motion was a kind of an underwater flight, with the paddle-wings acting just as conventional bird's wings do in air. It is not clear whether they are actually related to penguins or the similarities came about by convergent evolution. Other genera of the group are known, including *Plotopterium* itself and *Tonsala*, both from the Pacific coast of North America.

Distribution: Japan.
Classification: Bird, Neornithes.
Meaning of name: Oar wing.
Named by: Olson and Hasegawa, 1996.
Time: Oligocene epoch of the early Tertiary.
Size: 2m (6½ft).
Lifestyle: Fish-hunter.
Species: *C. hexeris*, *C. titan*.

Diatryma

The shape of the meat-eating dinosaur was such a good one that Nature seemed reluctant to let it go. The gastrornithiformes were ground-living predators with tiny forelimbs and huge beaks. It has always been believed that the gastrornithiform group was related to the cranes – hence their popular name "terror crane". Modern theories suggest that they were closer to the ducks.

Right: Diatryma preyed upon the small mammals of the early Tertiary.

Features: *Diatryma* is a giant flightless bird with massive running legs. Its wings are tiny and the breastbone shows no sign of any flying structure. The bones are hollow, like those of other birds, indicating a light build suitable for fast running and leaping. The huge bill has been taken as a sign of a flesh-eating creature, or alternatively as a plant-eater like a parrot. The general consensus, though, is that *Diatryma* was a hunter.

Distribution: Wyoming, USA; also (as *Gastornis*) France; Germany; England.
Classification: Bird, Neornithes, probably Anseriformes.
Meaning of name: Across Trym (a giant in Norse mythology).
Named by: Cope, 1876.
Time: Paleocene and Eocene epochs of the early Tertiary.
Size: 2m (6½ft).
Lifestyle: Hunter.
Species: *D. gigantica*.

A WIDESPREAD GROUP

In 1855 the remains of a giant flightless bird were found in Tertiary deposits near Paris. Others were found elsewhere in France, in Germany and in England. The bird was named *Gastornis* by its finder and a tentative reconstruction of the skeleton was attempted in 1881. This showed a 3m- (10ft-) high bird with a very long neck and relatively small lizard-like head.

The reconstruction was regarded as accurate, and so when a giant bird with a huge head was found later in Wyoming, USA, and named *Diatryma* nobody thought that the two creatures could have been the same. It was not until the original *Gastornis* material was re-examined in the 1990s that it was found to have been closely related to *Diatryma*, and may even have been the same genus. If this proves to be the case, then the name *Gastornis* will have precedence.

The gastornithiformes have a representative in China.

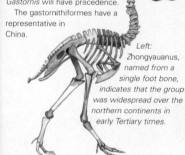

Left: Zhongyauanus, named from a single foot bone, indicates that the group was widespread over the northern continents in early Tertiary times.

Presbyornis

This must have been one of the commonest water birds of early Tertiary times, judging by its abundance and the frequent occurrence of its footprints in lake muds. It nested in vast colonies by the lake sides and fed by filtering tiny organisms from the waters with its broad shallow bill. Nesting sites, containing both bones and eggshells, are common in the Green River Formation in Wyoming, USA.

Left: Presbyornis lived in huge flocks around the lakes of North America in the early Tertiary.

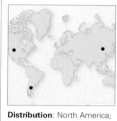

Distribution: North America; Mongolia; Patagonia.
Classification: Bird, Neornithes, Anseriformes.
Meaning of name: Elderly bird.
Named by: Wetmore, 1926.
Time: Eocene epoch of the early Tertiary.
Size: 1m (about 3ft).
Lifestyle: Filter feeder.
Species: *P. isoni*, *P. pervetus*.

Features: At first *Presbyornis* was taken as a relative of the flamingo. In appearance it looks rather like a duck but with extremely long legs. In its anatomy it combines features of both the duck family, especially in the head, and the family of waders, indicating that these two modern groups may have shared an ancestor. The family evolved in late Cretaceous times and survived into the Tertiary.

Neocathartes

Distribution: North America.
Classification: Bird, Neornithes, Gruiformes.
Meaning of name: New vulture.
Named by: Wetmore, 1944.
Time: Eocene to Oligocene epochs of the early to late Tertiary.
Size: 45cm (18in).
Lifestyle: Hunter or scavenger.
Species: *N. grallator*.

When first discovered the describer of *Neocathartes* gave it the genus name *Eocathartes*. This had already been used for an extinct vulture and so it was changed. Despite its resemblance to a vulture, *Neocathartes* belonged to the bird group that includes the cranes, the bustards and the rails. Although it was a flying bird, it was evidently adept at hunting on the ground, and probably only took to the sky in an emergency – rather like the modern secretary bird.

Features: *Neocathartes* has the appearance of a vulture with very long legs. The resemblance to a vulture would have come about by convergent evolution – similar body shapes developing in response to the requirements of similar lifestyles in similar environments. The sharp beak and the clawed feet show it to have been an active hunter of small prey or a carrion-eater.

Left: Neocathartes was one of the many flesh-eating birds that evolved early in the Tertiary.

PRIMITIVE MAMMALS

With the dinosaurs out of the way, the mammals took over. The beginning of the Tertiary saw a blossoming of all different mammal types in many different family lines, and all adapted to varied environments. Some flourished but then died away, leaving no modern descendants, or descendants that are so few they are regarded as living fossils.

Obdurodon

The monotremes are the modern egg-laying mammals. These consist of a single species of platypus and two species of echidna living in Australia. At one time the group was more widespread. *Obdurodon* was a platypus, whose remains have been found in Australia, with a close relative, *Monotrematum*, which may be a species of the same genus, found in South America. The difference in shape of the skull and bill between *Obdurodon* and the modern platypus suggests that the former fed by digging into the underwater banks of streams rather than on the stream bed like its modern relative.

Features: The main difference between *Obdurodon* and the modern platypus is the presence of teeth. A modern platypus has teeth but only as a juvenile – *Obdurodon* retained its teeth throughout life. Apart from that, little is known about this beast. It is only known from the skull of one species and parts of a jaw and pelvis of the other.

Distribution: Australia; Argentina.
Classification: Mammaliformes, Monotremata.
Meaning of name: Tough tooth.
Named by: Archer and Tedford, 1975.
Time: Paleocene to Miocene of the early and late Tertiary
Size: 30cm (1ft).
Lifestyle: Bottom feeder.
Species: *O. dicksoni*, *O. insignis* also *Monotrematum sudamericanum*.

Right: It is assumed that, like the modern platypus, Obdurodon was an aquatic-bottom feeder.

Taeniolabis

The multituberculates were one of the orders of primitive mammal that lived alongside the dinosaurs during the Mesozoic. Some survived the extinction of the dinosaurs and continued into the early Tertiary. Beaver-sized *Taeniolabis* was the biggest of these. In fact it is the biggest mammal known from the beginning of the Tertiary.

Features: Although unrelated, *Taeniolabis* has a tooth structure similar to that of modern rodents – with gnawing incisors that kept themselves sharp through constant use, and small premolars and molars with broad complex grinding surfaces. The heavy square-shaped skull with its short, blunt snout shows that this was a vegetarian animal, unlike most of the mammals from the earliest Tertiary.

Distribution: New Mexico, Montana, USA.
Classification: Allotheria, Multituberculata.
Meaning of name: Ribbon lips.
Named by: Cope, 1882.
Time: Paleocene epoch of the early Tertiary.
Size: 60cm (2ft).
Lifestyle: Gnawer and rooter.
Species: *T. taoensis*, *T. sulcatus*, *T. lamberti*.

Right: As with most primitive mammals, Taeniolabis is known mostly from teeth, and so the restoration is speculative.

Ptilodus

Distribution: Wyoming, North Dakota, Montana, New Mexico, USA; Alberta, Saskatchewan.
Classification: Allotheria, Multituberculata.
Meaning of name: Soft-haired one.
Named by: Cope, 1881.
Time: Palaeocene epoch.
Size: 50cm (20in).
Lifestyle: Arboreal omnivore.
Species: *P. fractus, P. gnomus, P. kummae, P. mediaevus, P. montanus, P. tspsiensis, P. wyomingensis.*

Most of the multituberculates that survived into the Tertiary were small unspecialized animals, as they were in Mesozoic times. Many of them took to living in trees, where their rodent-like dentition could be used for stripping the husks off nuts and seeds.

Below: Ptilodus and the other small multituberculates were the early Tertiary equivalent of the squirrels.

Features: The skeleton of *Ptilodus* is about the size of that of a squirre . Like a squirrel it is adapted to life in the trees, with grasping toes, sharp claws, feet that could be reversed for going head-first down tree trunks and a long tail. Unlike that of a squirrel, the tail is prehensile and would have been used as an extra limb. As in other multituberculates, the dentition is like that of a rodent, with gnawing teeth at the front, but there is an enormous blade-like premolar on the lower jaw.

MULTITUBERCULATES

The name multituberculates refers to the shape of the teeth that were crowned with many cusps. For all that they were the only major group of mammals to become completely extinct, the multituberculates had a long and distinguished history, evolving in Jurassic times and not dying out until the Oligocene. During their 100-million-year lifespan they developed into all shapes and lifestyles.
Lambdopsalis – a prairie-dog-like burrowing multituberculate from China.

GONDWANATHERES

The gondwanatheria represent a puzzling group of mammals from the late Cretaceous and early Tertiary of the southern continents – South America, Madagascar and India. It is unclear where they fit into the general evolution of the mammals but there is a possibility that they were an offshoot of the multituberculates. They are known only from their teeth and so a restoration is impossible. The important feature is that the teeth are high crowned, similar to those that are found later in the grass-eating mammals. The puzzle is that there were no grasses in late Cretaceous and early Tertiary times, and the gondwana-theres that have been found seemed to have lived in mangrove swamps.
Sudamerica – a typical gondwanathere based on part of a lower jaw from Argentina.

Psittacotherium

The taeniodonts were a short-lived group, flourishing very early in the Tertiary and dying out by the end of the Eocene. Their early members were small unspecialized opossum-like animals, like *Onychodectes*, but they established themselves as a group of heavy digging animals adapted to scraping in the ground for tubers and roots.

Features: *Psittacotherium* is one of the heaviest of the taeniodonts. It has a large square head, a stout neck and powerful forelimbs with five fingers – the middle three being huge and adapted for digging. The strong hind legs and heavy tail would have braced the animal while it was scraping in the ground. The jaws are massive and the front teeth huge, for ripping up tough underground stems. The back teeth are simple crushing and grinding pegs.

Distribution: New Mexico, Wyoming, Montana, USA.
Classification: Taeniodonta.
Meaning of name: Parrot beast.
Named by: Schoch, 1986.
Time: Palaeocene epoch of the early Tertiary.
Size: 60cm (2ft).
Lifestyle: Rooter.
Species: *P. multifragrum.*

Right: The underground stems and roots of the Palaeocene forests would have provided plenty of food to be exploited by specialist rooters like Psittacotherium.

XENARTHRA

The Xenarthra, formerly known as the Edentates, is the order to which the anteaters, the sloths and the armadillos belong. The name refers to the multiple joints between the vertebrae, and this eventually led to a backbone structure that could support the heavy weight of armour or enormous bodies found in later members of the group. The early members, however, were quite small and unspectacular.

Metacheiromys

The palaeanodonts – the ancient rootless ones – may not have been true xenarthrans but were closely related. They were a family of armadillo-like creatures that scraped in the soil after small insects. They probably lived in burrows in the thick soils of the early Tertiary tropical forests. They were mostly restricted to North America but one genus, *Palaeanodon*, is known from both North America and Europe, showing a connection between the continents in late Palaeocene and early Eocene times. *Metacheiromys* is the best-known of the group.

Features: *Metacheiromys* is mongoose-like in build but has a head like that of an armadillo, with long narrow jaws. The teeth are highly modified – the cheek teeth have vanished and have been replaced by horny pads for crushing insect carapaces. The claws on the forefeet are much larger than those on the hind, and the elbow joint is very powerful indicating a digging mode of life. The jaw structure

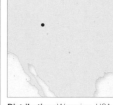

Distribution: Wyoming, USA.
Classification: Placentalia, Xenarthra.
Meaning of name: Almost a mouse with a hand.
Named by: Wortman, 1903.
Time: Middle Eocene epoch of the early Tertiary
Size: 45cm (18in).
Lifestyle: Insectivore.
Species: *M. marshi, M. tatusia, M. dasybus, M. osborni.*

Right: Metacheiromys *scraped for ants in the peaty floor of the Eocene forests.*

suggests that the group is ancestral to the pangolins.

Eomanis

The pangolins today consist of a single genus, but they were more widespread in early Tertiary times. *Patriomanis* is known from the Oligocene of North America, *Necromanis* from the late Eocene of France, and *Eomanis*, which was essentially a modern-looking pangolin, is known from the middle Eocene of Germany where perfect specimens were found in the oil shale quarries at Messel. Two species are known, one about a third larger than the other.

Features: *Eomanis* is built just like a modern pangolin, down to the scales on the back formed from fused hairs. These scales overlap one another as on a shingle roof, allowing the animal to curl up into an armoured ball when danger threatens. It is not known if in life it possessed the deeply rooted tongue – all the way back to the hip bones – of the modern pangolin, but the stomach contents reveal vegetable matter as well as the expected ants. It was apparently more omnivorous than the modern representative.

Distribution: Germany.
Classification: Placentalia, Xenarthra.
Meaning of name: Dawn pangolin.
Named by: Storch, 1978.
Time: Middle Eocene epoch of the early Tertiary.
Size: 50cm (20in).
Lifestyle: Anteater and omnivore.
Species: *E. waldi, E. krebsi.*

Below: If we saw Eomanis *in life, we would find it difficult to distinguish it from a modern pangolin.*

Eurotamandua

Distribution: Germany.
Classification: Placentalia, Xenarthra.
Meaning of name: European Tamandua (the modern collared anteater).
Named by: Storch, 1981.
Time: Middle Eocene epoch of the Tertiary.
Size: 90cm (3ft).
Lifestyle: Anteater.
Species: E. joresi.

Before the discovery of the perfect skeleton of *Eurotamandua* in the Messel oil shale pits in Germany, scientists believed that the anteater family had always been restricted to South America. Indeed the earliest-known true anteater, *Prototamandua*, had been found in the Miocene deposits of Argentina. Despite its similarities to modern anteaters, there is a possibility that *Eurotamandua* is closely related to *Eomanis* and the pangolins, with the anteater resemblance being a matter of convergence.

Features *Eurotamandua* has the build and appearance of the modern collared anteater, *Tamandua*. It has a long tubular snout with no teeth. The front legs are strong with huge claws for digging at termite mounds and ant nests. There is a long and powerful tail that would have been used for support as the animal tore at insect colonies with its claws. The Messel oil shale fossils are so detailed that even the ants are preserved.

Right: It has been suggested that Eomanis krebsi *is actually a juvenile specimen of* Euroatamandua.

EDENTATE

The name "edentate" or "toothless" was first given by Cuvier to the animals that had few or no teeth and subsisted on insects. These included the anteaters, the sloths and the armadillos. However, the name was not totally correct in that, of this group, only the eleven species of living anteaters were actually toothless. The others have teeth, albeit reduced, rootless and sometimes without enamel. The giant armadillo has more than 100 little teeth, more than almost any other mammal.

Nowadays the group is restricted to South America with only the armadillos found on the North American continent. In times past they were more widespread.

The aardvark of southern Africa, an anteater with the usual adaptations of a long tubular snout, a long feeding tongue, powerful claws for digging into termite nests, and a powerful tail to support it while digging, is nothing to do with this group. It developed all these features independently by convergent evolution, and probably evolved from early relatives of either the elephants or the hoofed animals.

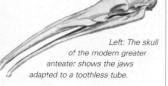

Left: The skull of the modern greater anteater shows the jaws adapted to a toothless tube.

Riostegotherium

Riostegotherium, and a related genus *Protostegotherium* are known from fossils of armoured ossicles found in late Palaeocene deposits of Brazil. These are so similar to the ossicles of the armour of modern armadillos that it is obvious that in South America as far back as the late Palaeocene armoured armadillos existed. Leg bones found in the area probably came from *Riostegotherium* as well.

Features: The bony scutes are of two types – one type consists of a fused mass forming a rigid armoured buckler, and the other consists of isolated and moveable individual pieces. This indicates that *Riostegotherium* had rigid armour over parts of its body separated by moveable armour that allowed the animal to articulate and possibly curl up. This is exactly the arrangement in the modern armadillo. The numbers of scutes found suggests that these early armadillos were less armoured than modern forms.

Distribution: Brazil.
Classification: Placentalia, Xenarthra.
Meaning of name: Roofed mammal from the river
Named by: Oliveira and Bergqvist, 1998.
Time: Late Paleocene epoch of the early Tertiary.
Size: 50cm (2ft).
Lifestyle: Anteater.
Species: R. yanei.

Right: The muscle attachments on the limb bones show that Riostegotherium *was a powerful digger, hunting for its prey below the surface of the soil.*

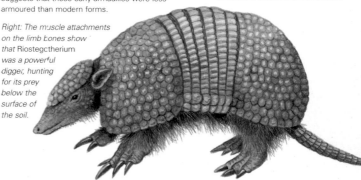

INSECTIVORES AND THEIR KIN

We often think of the insectivores as being the most primitive of mammals. With the dawn of the Tertiary they continued to flourish, mostly in their primitive shrew-like shapes, but a number developed into specialist forms. The classification of the insectivores is complex and changes often. Various orders are moved in and out of the superorder on a regular basis by different authorities.

Leptictidium

When *Leptictidium* was discovered in the fine-grained oil shales at Messel in Germany it was first reconstructed as a running animal – able to sprint, dinosaur-like on its hind legs balanced by its long tail. Apart from humans, no other mammal is known to run on hind legs. However, it is now believed that it was a jumping animal, like the long-tailed jerboas and desert rats of today.

Above: Leptictidium *hunted little animals through the undergrowth of the early Tertiary forests of Europe.*

Features: The skeleton of *Leptictidium* is similar to that of the modern elephant shrew, except for the extremely long lightweight hind legs with muscles concentrated around the thigh, and its long tail. The skull has anchor points for strong snout muscles indicating that it had a mobile trunk-like nose. Its body is short and plump and its front feet small and delicate and adapted to holding food. The stomach contents of good skeletons show that its diet consisted of insects, lizards, small mammals and plant material. Its body is very short, and the extremely long tail would have acted as a balancing organ as it leapt after food.

Distribution: Germany.
Classification: Placentalia, Leptiotida.
Meaning of name: Delicate weasel.
Named by: Tobien, 1962.
Time: Eocene epoch of the early Tertiary.
Size: 90cm (3ft).
Lifestyle: Omnivore, but mostly carnivorous.
Species: *L. tobieni*, *L. nasutum*, *L. auderiense*.

Icaronycteris

Bats evolved early in the age of mammals, and it seems that all the evolutionary developments – hand-supported wing, echolocation, upside-down roosting – appeared all at once. Modern-type bats are known from Palaeocene lake deposits in Wyoming, USA, and also from the oil shales of Messel, Germany – both places of exquisite preservation.

Features: Despite its early age, *Icaronycteris* would be difficult to distinguish from a typical modern bat. Analysis of the skull and the presence of moth scales in the stomach show that it hunted by echolocation. However, there are a few differences between it and modern bats. Bats and birds have rigid bodies, to help support the stresses of flight, but *Icaronycteris* does not show this to such an extent as modern bats. The jaws have more teeth, and these are arranged like those of an insectivore. The tail is long and not attached to the legs by a membrane. As well as a claw on the thumb there is also one on the forefinger.

Above: Winging its way over the Eocene lakes of North America, Icaronycteris *would have been indistinguishable from a modern bat in the gloom.*

Distribution: Wyoming, USA.
Classification: Placentalia, Chiroptera.
Meaning of name: Icarus night flyer (after a winged hero in Greek mythology).
Named by: Jepsen, 1966.
Time: Eocene epoch of the early Tertiary.
Size: 37cm- (14in-) wingspan.
Lifestyle: Night hunter.
Species: *I. index*.

Palaeoryctes

Distribution: North America; Africa.
Classification: Placentalia, Cimolesta.
Meaning of name: Early digger.
Named by: Matthew, 1913.
Time: Paleocene epoch of the early Tertiary.
Size: 12cm (5in).
Lifestyle: Insectivore.
Species: *P. puercensis*, *P. punctatus*, *P. minimus*, *P. cruoris*.

The palaeoryctid group consisted of mammals that had an insectivorous lifestyle but a very primitive tooth arrangement. The early forms, such as *Cymolesetes* of the Cretaceous, had very generalized teeth for piercing, shearing and grinding, but the later forms such as *Palaeoryctes* were more specialized for eating insects. It may have been close to the ancestry of the modern insectivores, or even that of the flesh-eating creodonts of the later early Tertiary.

Features: *Palaeoryctes* resembles the modern shrew, with its sleek body and pointed snout. The molar teeth are equipped with very high pointed cusps evolved for piercing the tough carapaces of insects. The strong forelimbs appear to be adapted for digging after prey. Several species of *Palaeoryctes* are known, ranging in size from that of the modern pygmy shrew – the smallest mammal alive today – to that of a squirrel.

Above: Modern shrews are similar to Palaeoryctes *(below).*

Eomys

The eomids have been known from their teeth for more than a century. They represent a group of rodents quite close to the modern pocket gophers, and lasted from late Eocene times until the Pliocene. Their peak was in the Oligocene. They were so widespread that their teeth are used as index fossils to date the rocks in which they are found, with the recognizable teeth of different species defining dates of the individual beds. The hardness of the teeth means that they were easily fossilized.

Features: Until 1996 the eomids, and the genus *Eomys* in particular, were known only from teeth. In that year a complete fossilized body was found, with furry outline and, surprisingly, a patagium – a flying membrane like that of a flying squirrel. A bony spur from the elbow supports this membrane, as it does in the flying squirrel. This makes the eomids the oldest-known family of gliding rodents.

Distribution: Germany.
Classification: Placentalia, Rodentia.
Meaning of name: Early mouse.
Named by: Schlosser, 1884.
Time: Oligocene epoch of the early Tertiary.
Size: 10cm (4in).
Lifestyle: Gnawer.
Species: *E. zitteli*, *E. major*, *E. abnatensis*, *E. quercyi*, *E. minimus*, *E. orientalis*, *E. craiky*, *E. orbicularis*, *E. intermedius*, *E. maximus*, *E. minutes*, *E. burkei*.

Right: Eomys *would have used its outspread patagia to glide from tree to tree.*

PANTODONTS

The first mammals that reached any real size were the members of the pantodont order. They had short stumpy feet, each with five toes, and massive pillar-like limbs. Their teeth usually had the full primitive compliment of 44, and were quite basic in structure. Their brain capacity was very small for the size of animal, but adequate for their lifestyle as browsers in the lush forests of the early Palaeocene.

Coryphodon

One of the most prolific of the pantodonts, *Coryphodon* are found throughout North America, Europe and Asia where they lived as waterside browsers and rooters. It was long-lived too, from the latest Palaeocene to the late Eocene. The pantodonts were replaced by the rhinoceroses that developed similar lifestyles in the Oligocene.

Features: *Coryphodon* is a hippopotamus-sized animal with a long skull and the canine teeth elongated into tusks, which are better developed in the male. The chewing teeth have prominent transverse crests suggesting that it browsed on soft forest vegetation. The brain to body ratio seems to have been the smallest of any mammal known.

Distribution: Widespread over the Northern Hemisphere.
Classification: Ungulata, Pantodonta.
Meaning of name: Peaked tooth.
Named by: Owen, 1845.
Time: Eocene epoch of the early Tertiary.
Size: 2m (6½ft).
Lifestyle: Semi-aquatic rooter.
Species: *C. anthracoideus, C. molestus, C. latidens.*

Left: Coryphodon *pursued a semi-aquatic lifestyle in the rivers and swamps of the northern continents.*

Barylambda

This animal seems to have anticipated the lifestyle of the giant ground sloths that were to come at the end of the Tertiary. Its forequarters were quite lightweight and its centre of gravity lay far back. It may have rested its great weight on its massive hind limbs and heavy tail and leaned on trees to reach the leaves of the higher branches.

Features: *Barylambda* has particularly heavy hind legs with plantigrade feet – that means that they rest with the soles on the ground like those of a bear, rather than on their toes, like those of a more active animal. The tail is heavy too, and the massive hindquarters may have allowed the animal to rear up on its hind legs to reach up. The head is quite small and the dentition primitive, with fairly long canines and unspecialized rear teeth.

Distribution: Western North America.
Classification: Ungulata, Pantodonta, Barylambdidae.
Meaning of name: Broad wave – referring to the grinding surface of the back teeth.
Named by: Patterson, 1937.
Time: Paleocene epoch of the early Tertiary.
Size: 2.5m (8ft).
Lifestyle: Browser.
Species: *B. faberi, B. churchilli.*

Left: Barylambda *fed on leaves that it reached by rearing up on its hindquarters.*

Pantolambda

Distribution: North America.
Classification: Ungulata, Pantodonta.
Meaning of name: Completely waved – referring to the grinding surface of the back teeth.
Named by: Cope, 1883.
Time: Early Paleocene epoch of the early Tertiary.
Size: 1.5m (5ft).
Lifestyle: Rooter.
Species: P. intermedium, P. cavirictis, P. bathmodon.

The earliest of the pantodonts was *Pantolambda*, from the beginning of the Palaeocene. The heavy limbs suggest that it may have spent much of its time wallowing, capybara-like, in the rivers and swamps. The claws were broad and weight-bearing and so *Pantolambda* may be regarded as the earliest hoofed mammal.

Features: Species of *Pantolambda* range from the size of a beaver to about the size of a sheep, with a long body, short stocky limbs and broad toes. Its skull is long and low and there are large canine teeth. The forelimbs, that are splayed to the side, the long tail and the primitive skull suggest that *Pantolambda* evolved from the primitive meat-eaters that went on to develop into the carnivorous creodonts.

Right: Pantolambda *spent its life rooting in the soil for roots and other food.*

PANTODONTS WORLDWIDE

The pantodonts were a relatively short-lived group of mammals. They evolved rapidly to fill the niche of the large herbivores that was left open with the extinction of the big plant-eating dinosaurs. They were quite widespread over the Northern Hemisphere. Those from Asia tended to be quite small, with some only the size of a rat. Those from Europe and America were large animals. One Asian form appears to have had a short trunk like that of a tapir. They all shared the primitive tooth arrangement, with a full compliment of 44 teeth, and they had a distinctive wavy pattern on the grinding surfaces of those at the back. The fossils of the teeth show little wear, suggesting that the vegetation they ate was relatively soft.
Archaeolambda – a small early form from Asia that may have been a tree-dweller.
Caenolambda – a generalized genus, rather like *Barylambda* but smaller.
Bemalambda – from the early Palaeocene of south China.

Below: The pantodonts were mostly large hippopotamus-like animals.

Titanoides

The big pantodonts such as *Titanoides* were adapted for tearing into the ground with their heavy claws on their broad front feet, looking for the roots and tubers on which they fed. Their big canines would have helped them in this, hooking them out of the soil and slicing them up. The modern equivalent in lifestyle would be the forest hog, but this was very much bigger.

Features *Titanoides* may have resembled a bear in appearance, with its heavy body, its strong forelimbs and its big claws on the five toes of its plantigrade feet. In build and bulk it is similar to the other big pantodonts, but has longer claws and has long canines on the upper jaw and hook-like canines on the lower.

Distribution: Montana, North Dakota, USA.
Classification: Ungulata, Pantodonta.
Meaning of name: Like a big one.
Named by: Patterson, 1932.
Time: Late Palaeocene epoch of the early Tertiary.
Size: 1.6m (5ft).
Lifestyle: Rooter.
Species: T. primaevus, T. gidleyi, T. looki, T. major, T. nanus.

Left: Titanoides *was the biggest mammal of the subtropical swamps of the Palaeocene American Mid-West.*

CONDYLARTHS

The most prolific group of the early Palaeocene mammals were the condylarths – apparently making up about 70 per cent of the fauna in North America. In body shape they were very primitive, not having evolved much from their insect-eating ancestors. As a result they tended to look like wolves or other carnivores, but they were omnivorous or plant-eaters.

Protungulatum

The sediments in which this small mammal was found also contained the bones and teeth of dinosaurs. The suggestion was that it lived at the end of the Cretaceous period, or that the dinosaurs survived into the Palaeocene. The modern interpretation is that the dinosaur material was eroded from older beds and redeposited in the early Palaeocene sediments.

Features: The teeth of *Protungulatum* are quite primitive but they anticipate the later evolution of true plant-eating teeth, being low-crowned with broad grinding areas. The skeleton gives the appearance of a once-insectivorous or carnivorous animal adapting to a vegetarian way of life. The limbs are shorter than expected for a plant-eating animal. The feet are five-toed with big claws – hence its name. However, little else is known of *Protungulatum* apart from the teeth, and so the restoration must be speculative.

Distribution: North America.
Classification: Ungulata, Condylarthra.
Meaning of name: First hoof.
Named by: Sloan and Van Valen, 1965.
Time: Early Palaeocene epoch of the early Tertiary.
Size: 20cm (8in).
Lifestyle: Omnivore.
Species: *P. donnae, P. gorgun, P. sloani, P. mckeeveri.*

Left: The appearance of Protungulatum *is uncertain, but this restoration gives an idea of what a small generalized omnivorous mammal looked like.*

Chriacus

Our understanding of the appearance of arctocyonids – one of the condylarth families – is based on an almost complete and articulated skeleton of *Chriacus* – perhaps the most complete early Tertiary mammal skeleton found. This shows that it was at home in the trees and on the ground, rather like a modern raccoon. Although the hind limbs were adapted to climbing, the front feet were probably capable of digging in the soil. It was a very versatile animal.

Features: The limbs are muscular, as shown by the prominent flanges on the bones that anchored the muscles, the joints are mobile and the feet have five digits with strong claws, probably used for balancing while running in branches, or it may have been prehensile. The teeth are plant-eater's teeth, but they seem to be versatile enough for tackling insects and small animals. The incisors form a comb-like structure for grooming.

Distribution: North America.
Classification: Ungulata, Condylarthra.
Meaning of name: Useful.
Named by: Cope, 1883.
Time: Early Palaeocene to early Eocene epochs of the early Tertiary
Size: 1m (about 3ft).
Lifestyle: Omnivore.
Species: *C. pelvidens, C. baldwini, C. metocemeti, C. punitor, C. calenancus, C. katrinae, C. oconostotaie.*

Left: The fine-comb structure of the front teeth was perhaps used for grooming the fur.

Phenacodus

Distribution: North America.
Classification: Ungulata, Condylartha.
Meaning of name: Obvious teeth.
Named by: Cope, 1881.
Time: Late Palaeocene to middle Eocene epochs of the early Tertiary.
Size: 1.5m (5ft).
Lifestyle: Browser.
Species: P. zuniensis, P. brachypternus, P. vortmani, P. intermedius, P. matthewi, P. grangeri, P. bisonensis, P. magnus.

Sheep-sized *Phenacodus* seems to have been close to the ancestry of the hoofed animals – particularly the odd-toed hoofed animals such as the horses and the rhinoceroses – and its appearance begins to reflect this, with its long straight legs. *Phenacodus* and its close relatives in the phenacodontids – a group defined by their large and obvious teeth – make up about half of the mammal specimens of the North American Palaeocene.

Right: Several species of Phenacodus are known; the larger ones existing during cooler climates and the smaller types flourishing when conditions warmed.

Features: The limbs of *Phenacodus* are straighter than those of other condylarths, and it has a digitigrade stance – walking on its toes rather than the soles of its feet. This usually indicates a running mode of life. The back is long and arched and the tail heavy. The feet are long, but are flexible and quite primitive in structure, retaining all five toes on each foot and each toe having the original number of bones – not reduced in number to cut down the weight as is usual in a running animal. As in all condylarths, the teeth show little specialization.

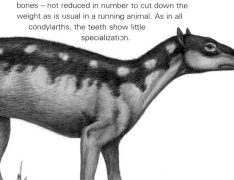

OTHER CODYLARTHS

Quettacyon – from the early Eocene of Pakistan, and is the oldest Tertiary land mammal from south Asia.
Stylidodon – with a digging ability similar to that of the modern aardvark.
Meniscotherium – having very advanced plant-grinding teeth for such an early mammal – almost like some of the early horses.
Tetraclaenodon – related to *Phenacondon* but with a more advanced foot structure for running.
Kopidodon – from a perfect skeleton from the Messel oil shales, this was a raccoon-sized climber with a very long tail, and with teeth that seem adapted for eating fruit.

Below: Skull of Meniscotherium.

Hyopsodus

Perhaps the smallest-known of the condylarths, *Hyopsodus* and its close relatives looked more like little insectivores than any relatives of the hoofed animals. In fact some forms have teeth that show them to have been specialist insect-eaters. The best-known genus is *Hyopsodus* itself, and it was an abundant animal ranging over the whole of the Northern Hemisphere in the late Palaeocene. It was probably a tree-living omnivore scampering about in tree branches like a modern squirrel – a very different lifestyle from that of its big, hoofed, running relatives.

Features: *Hyopsodus* has a long low body, like that of a weasel. The limbs are short and suited both for running on the ground and climbing in trees. The teeth are quite unspecialized, with the canines not being particularly big, but the molars are bunodont – they have rounded cusps that would be useful n tackling anything, and indicate an omnivorous diet.

Distribution: Northern Hemisphere.
Classification: Ungulata, Condylarthra.
Meaning of name: Like a pig.
Named by: Leidy, 1870.
Time: Late Paleocene to middle Eocene epochs of the early Tertiary.
Size: 30cm (1ft).
Lifestyle: Arboreal omnivore.
Species: H. gracilis, H. loomisi, H. wardi, H. paulus, H. marshi, H. despiciens, H. lepidus, H. minisculus, H. wortmani.

Right: Insects, both on the ground and in the trees, were hunted by little Hyopsodus.

ACREODI

Closely related to the condylarths – the early hoofed animals – the Acreodi were the first well-adapted hunting mammals. Despite their vegetarian connections, these tended to be big fierce predators, preying on their plant-eating relatives, and included some of the biggest carnivorous land-living animals known. Although the order is the Acreodi, it is often called by the name of its only family – the mesonychids.

Mesonyx

Although *Mesonyx* resembled a wolf in its appearance and habits, it seems to have lived near water, suggesting that it may have been amphibious. This seems to tie in with the suggestion that these animals may have been close to the ancestry of the whales.

Features: The teeth of *Mesonyx* are distinctly those of a carnivore, with killing incisors and canines and meat-shearing blades on the lower molars. The upper molars are triangular, like those of early whales. Unlike other meat-eating animals, *Mesonyx* does not have claws on its feet; instead it has little hooves. The tail is long and heavy, the body is slim and lithe and the feet are digitigrade, as befits a hunter.

Distribution: North America.
Classification: Ungulata, Mesonychia.
Meaning of name: Middle claw.
Named by: Cope, 1872.
Time: Early Eocene epoch of the early Tertiary
Size: 1.5m (5ft).
Lifestyle: Hunter.
Species: *M. obtusidens*.

Left: The wolf-like appearance of Mesonyx *belies its close relationship to hoofed herbivores and to whales.*

Harpagolestes

If *Mesonyx* was the wolf among the mesonychids, then *Harpagolestes* would have been the hyena. It was about the size of a bear and had a dentition that was evolved to rip up and crush carcasses; it must have been the major scavenger of the Eocene woodlands and forests.

Features: The position of the mandibular condyle – the hinge on which the jaw works – shows a strong leverage for producing a powerful bite. There is a high sagital crest running fore and aft along the top of the skull that in life would have held strong jaw muscles. The teeth are enormous, with the canines able to rip prey limb from limb and molars producing a broad bone-crushing pavement. These are all indicative of a scavenging diet. Evidence from a closely related genus, *Ankalagon*, suggests that only the males had the massive crushing teeth.

Distribution: Mongolia.
Classification: Ungulata, Mesonychia.
Meaning of name: Grapnel stealer.
Named by: Wortman, 1901.
Time: Middle Eocene epoch of the early Tertiary.
Size: 2m (6½ft).
Lifestyle: Omnivore or scavenger.
Species: *H. macrocephalus*, *H. orientalis*.

Right: The teeth of Harpagolestes *would have made short work of any carcass.*

Andrewsarchus

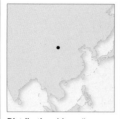

Distribution: Mongolia.
Classification: Ungulata, Mesonychia.
Meaning of name: Andrews' beast (after American palaeontologist Roy Chapman Andrews).
Named by: Pao, 1923.
Time: Middle Eocene epoch of the early Tertiary.
Size: 4m (13ft).
Lifestyle: Hunter or scavenger.
Species: A. mongoliensis. A. henanensis, A. crassum.

This huge, heavy bear-like relative of *Mesonyx* must have been one of the biggest meat-eating land-living mammals of all time. With the abundant vegetation giving rise to many big plant-eating animals, it was inevitable that very big meat-eating animals would develop to prey on them.

Features: The skull is long, narrowing from a broad rear that held the jaw muscles, forming an hour-glass restriction along the jaws and expanding once more at the snout to hold the big canine and incisor teeth. The back teeth are broad and adapted for crushing bones. Only the skull and lower jaw have been found, and the restoration of the rest of the body is speculative, but it must have been very powerful to back up the size of the skull.

Rght: Andrewsarchus was the biggest meat-eater of its time.

Most members of the Acreodi are known from deposits in the Mid-West of North America and across Asia.
Dissarcusium, **Hukoutherium** and **Yantanglestes** – known from fragmentary crania in Asia.
Hapalodectes – in shape and size like a modern giant otter, but with hooves on its toes. It was probably amphibious.
Sinonyx – a large wolf-like type from China with big canine teeth.
Synoplotherium – another wolf-sized animal, very similar to *Mesonyx*, that probably hunted in packs in North America.

Below: Wolf-like hunting mesonychid, Sinonyx.

Pachyaena

This wolf-sized running predator shows that the carnivorous mesonychid group evolved from the omnivorous and herbivorous arctocyonid condylarths – a group of the early heavy plant-eaters that superficially resembled bears, although they ranged from bear- to monkey-sized. The limbs seem to combine the features of modern odd-toed and even-toed ungulates and the carnivores. The teeth, which were modified for shearing and holding meat, also show that they were derived from the arctocyonids.

Features: The structure of the limbs shows *Pachyaena* to have been a running animal. It is digitigrade and the joints of the limbs would have restricted the movements to a fore-and-aft swing – a feature of running hoofed animals as well as carnivores – the toes are furnished with hoofs rather than claws. Despite the running specializations the leg bones are quite thick, and this indicates that *Pachyaena* was adapted for endurance rather than speed.

Distribution: Wyoming, USA.
Classification: Ungulata, Mesonychia.
Meaning of name: Thick pig.
Named by: Cope, 1874
Time: Early Eocene epoch of the early Tertiary.
Size: 1m (about 3ft).
Lifestyle: Hunter.
Species: P. ossifraga, P. gracilis, P. gigantean.

Right: Pachyaena had four toes on each foot, and was clearly a running animal.

DINOCERATA, TILLODONTIA AND EMBRITHOPODA

A common pattern for a large herbivorous mammal consists of a stocky body, carried on four pillar-like legs, and a smallish head bearing a spectacular array of horns. This rhinoceros-like appearance evolved independently in these three unrelated mammal orders in the early Tertiary.

Prodinoceras

The early members of the Dinocerata appeared in Asia in Palaeocene times. From there they migrated to the North American continent. *Prodinoceras* was about the size of a modern tapir and, although smaller than the later American forms, was still one of the biggest animals of the area at that time.

Features: An early member of the Dinocerata, *Prodinoceras* had the enlarged upper canines, protected by the pair of flanges on the lower jaw, that is one of the features of the group. These teeth are probably more pronounced in the males and would have been used for display. *Prodinoceras* does not yet show the development of the spectacular horns. The chewing teeth, each with two crests, are typical of all members of the group.

Left: Prodinoceras has a close relative in Probathyopsis from Western USA. Some scientists think it is the same genus.

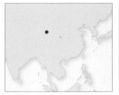

Distribution: Mongolia.
Classification: Ungulata, Dinocerata.
Meaning of name: Before Dinoceras (one of the early names of *Uintatherium*).
Named by: Matthew, Granger and Simpson, 1929.
Time: Late Paleocene epoch of the early Tertiary.
Size: 2.9m (9½ft).
Lifestyle: Rooter and browser.
Species: *P. martyr, P. turfanensis, P. primigenium, P. simplum, P. diconicus, P. lacustris, P. xinjiangensis.*

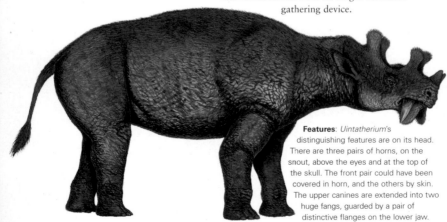

Uintatherium

The Eocene forests of western North America were home to many strange herbivorous animals, none so strange as the rhinoceros-like *Uintatherium* and its relatives. The horns and the fangs were first found as individual fragments and thought to have belonged to totally different animals. The sabre-like canines and the lack of front teeth suggest that *Uintatherium* used its tongue as a food-gathering device.

Distribution: Wyoming, USA.
Classification: Ungulata, Dinocerata.
Meaning of name: Uinta beast (after the mountain range and local native American tribe).
Named by: Leidy, 1872.
Time: Eocene epoch of the early Tertiary.
Size: 4m (13ft).
Lifestyle: Rooter and browser.
Species: *U. robustum.*

Features: *Uintatherium's* distinguishing features are on its head. There are three pairs of horns, on the snout, above the eyes and at the top of the skull. The front pair could have been covered in horn, and the others by skin. The upper canines are extended into two huge fangs, guarded by a pair of distinctive flanges on the lower jaw.

Left: From a distance Uintatherium *could have been mistaken for a rhinoceros.*

UINTATHERIUM

Leidy's assistants found the first remains of *Uintatherium* in Wyoming, USA, in 1872. Leidy named it and published the first description of it at that time. Mash and Cope were also working in the area and discovered other large mammal remains which they named *Tinoceras*, and *Dinoceras*. The more precise collecting techniques of Cope and Marsh meant that their specimens were more complete than those of Leidy. Almost all these specimens were later verified as *Uintatherium*, all except a skull found by Cope and named *Eobasileus*. This is still regarded as a closely related genus within the group.

The work done by Marsh on the subject is regarded as the major contribution to the understanding of this family of animals, and his name for the group, *Dinocerata*, is accepted today. However, because Leidy gave the name *Uintatherium* first, that is the name by which the original genus is known.

Trogosus

The Tillodontia represent a group of rooters that do not seem to fit into the grand scheme of mammal evolution. Their enlarged incisors seem to suggest that they belong to the ancestral line of the elephants, but equally they may be related to the rodents. The most well-known genus is *Trogosus*, but others include *Tillodon* from North America, *Kuanhuanius* from China and *Higotherium* from Japan.

Features: We can only guess at the appearance of the whole animal, as only the skulls and jawbones are known from the tillodont group. They have simple three-cusped upper molars. The second incisors are very big and rootless. These were probably used for rooting in the ground, and, this being the case, we can restore the rest of the body of *Trogosus* with a bear-like body and limbs to conform to this lifestyle.

Distribution: Wyoming, USA.
Classification: Ungulata (possibly), Tillodontia.
Meaning of name: Cave one.
Named by: Leidy, 1871.
Time: Eocene epoch of the early Tertiary.
Size: 1.5m (5ft).
Lifestyle: Rooter.
Species: *T. fodiens*, *T. latidens*, *T. castoroides*, *T. hyracoides*, *T. vetulus*.

Right: Trogosus would have resembled a bear with the head of a giant beaver.

Arsinoitherium

Distribution: Egypt.
Classification: Ungulata, Embrithopoda
Meaning of name: Arsinoe's beast (from an ancient queen of the area).
Named by: Beadnell, 1902.
Time: Early Oligocene epoch of the early Tertiary.
Size: 3.4m (11ft).
Lifestyle: Rooter or browser.
Species: *A. zitteli*, *A. andrewsii*, *A. giganteus*.

The African *Arsinoitherium* was once thought to be the only genus in the Embrithopoda – the horned, rhinoceros-like animals. Others have now been found in Turkey and Romania. It does not seem to be related to any other mammal group, and evolved its rhinoceros-like build independently of any of the others of the Tertiary.

Features: The most obvious feature is the pair of massive horns on the nose – hollow conical structures made of bone – and another pair above the eyes that are comparatively insignificant. The feet have five hoofed toes, and there are a 44 teeth in the mouth. The teeth are high-crowned and suited for chewing tough vegetation. The first-found species, *A. zitelli*, is large, with the others about a third of its size.

Left: There is a suggestion that Arsinoitherium may be close to the ancestors of the elephants.

HORSES AND THE
EARLY ODD-TOED UNGULATES

The evolutionary development of the horses is one of the most familiar examples of the evolutionary development of a particular group of animals. Unlike the fleet-footed plains-running animals that we know today, the earliest horses were small browsing animals that scampered through forest undergrowth.

Palaeotherium

The first specimens of *Palaeotherium* were found in the gypsum mines near Paris and described by Cuvier. Since then several dozen complete specimens of the close relative *Propalaeotherium* have come to light in the Messel oil shales in Germany, including some that were pregnant with single foals. Although not on the direct ancestral line to the horses, *Palaeotherium* was closely related.

Features: The body of *Palaeotherium* is rather like that of a lightweight tapir, although later members of the palaeothere group are as big as rhinoceroses. *Palaeotherium* itself has a long head and may have had a short trunk suitable for browsing from low bushes. There are four toes on the front feet and three on the hind, forming spreading feet that would have been ideal for walking on soft, boggy forest soil.

Distribution: Southern England; northern France.
Classification: Ungulata, Perissodactyla, Palaeotheriidae.
Meaning of name: Ancient mammal.
Named by: Cuvier, 1804.
Time: Eocene to early Oligocene epochs of the early Tertiary.
Size: 75cm (2½ft).
Lifestyle: Browser.
Species: *P. castrense, P. ruetimeyeri, P. magnum, P. crassum, P. medium, P. curtum, P. lautricense.*

Left: A distant cousin of the early horses, Palaeotherium *probably lived in small herds in the undergrowth.*

Hyracotherium

The earliest horse known is *Hyracotherium*. It, or a close relative, gave rise to all the subsequent horses, and also the side branch that led to *Palaeotherium*. In early books it is seen with the more evocative name *Eohippus* – "dawn horse" – but the name *Hyracotherium* was given earlier and so that takes priority.

Features: *Hyracotherium* has the build of an undergrowth dweller, like a modern mouse deer. There are four toes on the front feet and three on the rear, unlike the single toe of the modern horse. These had weight-bearing pads like those of a dog. The teeth are low-crowned and suitable for chewing soft leaves, rather than the high-crowned grass-grinders of modern horses. The brain is larger than that of other small mammals of the time.

Distribution: Widespread across Europe; Asia; and North America.
Classification: Ungulata, Perissodactyla, Hippomorpha.
Meaning of name: Hyrax mammal.
Named by: Owen, 1841.
Time: Eocene epoch of the early Tertiary.
Size: 60cm (2ft).
Lifestyle: Browser
Species: *H. angustidens, H. index.*

Left: The body of Hyracotherium *was long and the backbone arched, giving the animal a rather hunchbacked appearance.*

Epihippus

Distribution: Wyoming, USA.
Classification: Ungulata, Perissodactyla, Hippomorpha.
Meaning of name: Marginal horse.
Named by: Marsh, 1878.
Time: Late Eocene epoch of the early Tertiary.
Size: 60cm (2ft).
Lifestyle: Browser.
Species: *E. uintensis*, *E. gracilis*.

As the Eocene drew on and the landscapes became more open, the horses began to adapt to the changing conditions. The fossils of *Epihippus* have been found in strata that seem to have been deposited in upland meadows rather than riverside forests – the home of their predecessors.

Right: The fossils of Epihippus *were found in North America, so the early evolution of the horses took place there.*

Features: The only real difference between *Epihippus* and its predecessors is the fact that, whereas the latter had only three true grinding teeth in each jaw – the molars – in *Epihippus* the last two premolars became grinding teeth as well, increasing the tooth area and enabling the animal to tackle even tougher foods. The front feet still have four toes and the rear three, but now the middle toe is stronger, anticipating the reduction of the outer toes in later horses.

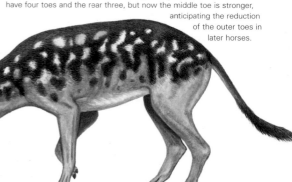

THE EVOLUTION OF THE HORSE

The story of the evolution of the horse is not a straight line from *Hyracotherium* through a series of middle forms until we get to the modern horse. Instead the line has all sorts of branches that blossomed and came to nothing, while one of those branches brought us to the stage we see today.

The early part of horse evolution, during the Eocene and the Oligocene of the early Tertiary, is among the forest-living types. The widespread grasslands that are the typical domain of modern horses did not develop until the Miocene, at the beginning of the late Tertiary, and that is when the really spectacular horse evolution took place. However, there were a number of interesting stages in early times. In the Oligocene the dense forests were giving way to more open country and the horses were beginning to develop accordingly – with more adaptations to running and to a diet of tougher vegetation – grass.

EARLY TERTIARY HORSES

Orohippus – very similar to *Hyracotherium* but slightly later and with slightly different teeth.

Duchesnehippus – possibly a subgenus or a species of *Epihippus* but with slightly more advanced teeth.

Haplohippus – representing a short-lived side branch.

Miohippus – slightly more advanced than *Mesohippus* at the very end of the Oligocene.

Mesohippus

Mesohippus marks the beginning of the change from the small browsing forest- and woodland-dwelling horses of the early Tertiary to the bigger grazing, running, plains-dwelling horses of the late Tertiary. From this stock the many types of running horse evolved. There were several species, each adapted to slightly different habitats and the foods found there.

Features: *Mesohippus* lacks the fourth toe on the front foot, making the front foot three-toed like the back. There is another tooth added to the grinding battery, and an elongation of the face means that a gap appears between the cropping incisor teeth at the front and the grinding molars at the back of the mouth. The eyes are rounder than in its ancestors, and set further back – a sign of low grazing. Despite this, its main food still seems to be twigs and fruit.

Distribution: Nebraska, North and South Dakota, Colorado, USA.
Classification: Ungulata, Perissodactyla, Hippomorpha.
Meaning of name: Middle horse.
Named by: Marsh, 1876.
Time: Late Oligocene epoch of the early Tertiary.
Size: 1m (about 3ft).
Lifestyle: Browser or grazer.
Species: *M. eulophus*, *M. obliquidens*, *M. bairdi*, *M. intermedius*, *M. validus*, *M. gridlei*.

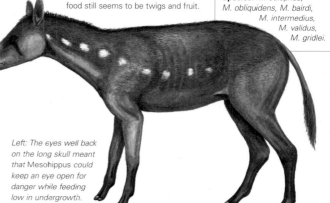

Left: The eyes well back on the long skull meant that Mesohippus *could keep an eye open for danger while feeding low in undergrowth.*

EOCENE BRONTOTHERES

The Brontotheriidae, sometimes called the Titanotheriidae, represents a family of odd-toed ungulates, distantly related to the tapirs and rhinoceroses but probably most closely related to the horses. They became the biggest animals on earth during the Oligocene, but the earlier Eocene forms were modestly sized animals.

Eotitanops

Like its distant cousin, *Hyracotherium*, *Eotitanops* was a small rabbit-sized animal that scampered through the undergrowth of the early Eocene forests. Although it does not seem to have been a common animal, many species of *Eotitanops* were named in the past, but only two are now recognized.

Features: As in other early perissodactyls – the odd-toed ungulates – there are four toes on the front feet and three on the back. In life these would have been equipped with pads, like those of a dog, and would have been suitable for walking on soft boggy forest soil. The teeth show that it browsed soft forest leaves. Although there is a vast difference in appearance and lifestyle between this little animal and the monsters that represent the group in Oligocene times, the shape of the teeth shows that they all belong to the same family.

Left: Early brontotheres, horses, rhinoceroses and tapirs all closely resembled one another.

Distribution: Wyoming and Colorado, USA.
Classification: Ungulata, Perissodactyla, Brontotheroidea.
Meaning of name: Like a dawn titan.
Named by: Osborn, 1907.
Time: Early to middle Eocene epoch of the early Tertiary.
Size: 75cm (2¾ft).
Lifestyle: Browser.
Species: *E. borealis*, *E. minimus*.

Dolichorhinus

Like the modern forest-dwelling rhinoceroses, *Dolichorhinus* probably inhabited deep shady woodlands and fed from low-growing foliage. It had a very long head, and shared its environment with another brontothere, *Manteoceras*, which was almost identical but had a broad head. Presumably the differences indicated a difference in diet and so there would have been no competition.

Features: Like *Eotitanops* before it, this much larger animal still has the four toes at the front and the three at the back. In appearance and lifestyle it would have resembled a small hornless rhinoceros. The skull is very long and shows no sign of the ornamental horns that came to characterize the later members of the family. The teeth are elongated and low-crowned, adapted for soft vegetation.

Left: Dolichorhinus fed on the shoots and leaves of the undergrowth of the lush forests that covered central North America at that time.

Distribution: Wyoming, USA.
Classification: Ungulata, Perissodactyla, Brontotheroidea.
Meaning of name: Narrow snout.
Named by: Douglass, 1924.
Time: Middle Eocene epoch of the early Tertiary.
Size: 2m (6½ft).
Lifestyle: Forest browser.
Species: *D. intermedius*, *D. hyognathus*.

BRONTOTHERE BEHAVIOUR

An insight into the behaviour of brontotheres came with the discovery of a bone bed in late Eocene rocks in Wyoming, USA. The partially articulated skeletons of about 25 horse-sized brontotheres of the genus *Mesatirhinus* were found in an area less than 100m (110yd) square. They consisted of individuals of all stages of growth from yearlings, accompanied by their mothers, to animals of about 15 years old.

The deposits in which they were found consisted of a sequence of river channels and flood deposits and the individual bed showed signs of rapid deposition as if by a sudden flood. The interpretation is that they were part of a migrating herd that was crossing a river, upstream when they were overwhelmed by a flash flood. Their bodies were washed along by the current and deposited together, to be picked over by scavengers before being buried by subsequent floods.

We can see from this that at least the medium-sized brontotheres travelled in herds or large family groups. However, it is not possible to tell whether the deposit represents the entire group or only part of a much larger herd.

Below: Mesatirhinus *herd.*

Nanotitanops

The brontotheres were found all over the Northern Hemisphere. They are known from most of North America as far north as the Canadian Arctic islands. They are also known across Europe and Asia, with a group of dwarf genera from the far east of China. The remains of these tiny brontotheres have been found in fissure deposits near Shanghai.

Features: *Nanotitanops* is by far the smallest brontothere known. Its teeth are much more advanced than the early small forms such as *Eotitanops*, and so it is not a primitive ancestral form. Other small brontotheres are known from the area, but they do not seem to be closely related. They appear to have been trapped in fissures, presumably while feeding on the vegetation that grows around cave mouths.

Above:
Nanotitanops *appeared quite late in the evolution of the brontotheres, although it resembled one of the small early forms.*

Distribution: Jiangsu province, China.
Classification: Ungulata, Perissodactyla, Brontotheroidea.
Meaning of name: Like a dwarf titan (the original name given was *Nanotitan*, but that had already been allocated to a fossil insect).
Named by: Qi and Beard, 1996 (renamed 1998).
Time: Middle Eocene epoch of the early Tertiary.
Size: 50cm (20in).
Lifestyle: Low browser.
Species: *N. shanghuangensis.*

Aktautitan

Distribution: Kazakhstan.
Classification: Ungulata, Perissodactyla, Brontotheroicea.
Meaning of name: Titan from Aktau Mountain.
Named by: Mihlbachler, Lucas, Emry and Bayshashov, 2004.
Time: Middle Eocene epoch of the early Tertiary.
Size: 2.5m (8ft).
Lifestyle: Grazer, possibly semi-aquatic.
Species: *A. hippopotamopus.*

With the hippopotamus-like build and proportions of the limbs it is tempting to imagine that this large early brontothere was semi-aquatic, wallowing in shallow lakes and feasting from the abundant lakeside vegetation. Indeed the remains have been found in lake deposits. This theory is entirely plausible but there is no other proof of such a lifestyle.

Features: The spectacular nasal horns that came to characterize the later brontotheres are apparent in *Aktautitan*, although they are nothing like as big as in the later forms. The body is large and the legs are particularly short and stumpy, giving the whole animal a rather hippopotamus-like appearance. The teeth are adapted for soft vegetation. Footprints have been found that were probably made by *Aktautitan*.

Right: The species name
A. hippopotamopus *means hippopotamus-foot, as the shape of the feet has given rise to the idea that it may have been semi-aquatic.*

LATER BRONTOTHERES

Like the uintatheres before them and the rhinoceroses after them, the brontotheres were big browsing horned animals. Like them, they started as small generalized beasts and quickly developed into the biggest animals of their day. Despite their superficially similar appearance, these groups were not closely related to one another and their similarity is a result of convergent evolution.

Palaeosyops

About the size of a tapir, *Palaeosyops* was considerably bigger than its predecessor *Eotitanops*, a trend that was to continue in the group. The earliest species, *P. fontinalis*, actually overlaps in time with the last species of *Eotitanops*, *E. minimus* in Wyoming, USA. The remains have been found in shallow lake and mudflat deposits.

Features: Although it is quite modest in size, *Palaeosyops* is the largest mammal found in the area at the time. It differs from *Eotitanops* by its size and by the more sophisticated arrangement of the teeth. The canines are more prominent than in other brontotheres, and may have been used for rooting around on the forest floor. However, the teeth were not adapted for tough foods like tubers or roots, but for soft leaves.

Distribution: Wyoming, Colorado, Montana, USA.
Classification: Ungulata, Perissodactyla, Brontotheroidea.
Meaning of name: Early curve.
Named by: Leidy, 1870.
Time: Early to middle Eocene epoch of the early Tertiary.
Size: 1.5m (5ft).
Lifestyle: Low browser.
Species: *P. paludosus, P. fontinalis, P. laevidens, P. robustus, P. laticeps.*

Left: Palaeosyops was an early brontothere that had yet to develop the nasal ornamentation.

Protitanotherium

Towards the end of the Eocene the brontotheres were developing into very large animals, some of them being larger than rhinoceroses and approaching the size of elephants. *Protitanotherium* was among the first of the really big brontotheres.

Features: With the appearance of *Protitanotherium* we see the beginnings of the "horn" structures on the nose. These are formed from an elongation of the nasal bones. In *Protitanotherium* they are small bumps, probably used as protection while head-butting with rivals. They would not have been sheathed in horn but covered in skin, rather like the horns of a giraffe. The development of this ornamentation gives rise to the saddle-shape of the skull, which is so typical of the later brontotheres.

Distribution: North America.
Classification: Ungulata, Perissodactyla, Brontotheroidea.
Meaning of name: Early titan beast.
Named by: Lucas, 1983.
Time: Late Eocene epoch of the early Tertiary.
Size: 4m (13ft).
Lifestyle: Low browser.
Species: *P. emarginatum, P. superbum, P. curryi.*

Right: The horns are taken as an indication that these animals lived in groups, and that there was rivalry between the males for mating rights.

Megacerops

Distribution: North Dakota, USA.
Classification: Ungulata, Perissodactyla, Brontotheroidea.
Meaning of name: Big horned head.
Named by: Leidy, 1870.
Time: Early Oligocene epoch of the early Tertiary.
Size: 5m (16ft).
Lifestyle: Browser.
Species: *M. tyleri, M. coloradensis, M. (Brontotherium) gigas, M. (Brontotherium) ingens, M. (Brontops) dispar, M. (Titanops) elatus, M. kuwagatarhinus.*

This is the biggest and best-known of the brontotheres. It has also been known as *Brontotherium*, *Brontops* and *Titanops* and these are all now regarded as the same genus, with the name *Megacerops* having priority.

Features: A veritable monster, *Megacerops* stands 2.5m (8ft) high at the shoulder. The massive forequarters support a skull that is saddle-shaped, being high at the rear, sweeping down over a tiny brain (no larger than your fist) and turning up into a huge Y-shaped bony ornament on the nose. In life this ornament would have been covered in skin rather than horn, and may have been used as a battering ram, or for display between the males of the herd. The female specimens do have the horns on the nose, but these are nowhere near as big as those of the males.

Below: A big Megacerops *may have had the appearance of an elephant when viewed from a distance – the size would have been about the same.*

THUNDER BEASTS

Across the Mid-West of North America the native peoples have the legend of the thunderbird. This mythical creature is responsible for the seasonal rains that are so important to the people of the plains. When it spreads its wings, the dark clouds gather, and when it shakes its feathers the rain falls.

When they found huge bones buried in the soil of their prairies, the local tribes took them for the bones of the great thunderbirds. This story was passed on to the pioneer palaeontologists, such as Cope, Marsh and Leidy when they came into the area. In recognition of the local tradition Marsh named the group of animals to which they belonged, the *Brontotheria*, the "thunder beasts."

Embolotherium

This was the Asian equivalent of the mighty *Megacerops* of North America. It roamed the plains of Oligocene Mongolia in big herds. There were a number of species, and each had a differently shaped horn ornament. As in *Megacerops* it is hollow, and does not seem to differ between sexes, suggesting that it was not used for any kind of mating behaviour.

Features: *Embolotherium* has the same huge build as *Megacerops*, but the main difference is in the nose ornament. Instead of a Y-shaped structure it has an undivided, almost spade-shaped ornament. It probably supported a nasal structure that would have been apparent as a fleshy area in front of it. It may have been used as a sounding box when bellowing or it may have been part of a system that cooled the breathed air. The eyes are tiny and set well forward in the skull.

Distribution: Mongolia.
Classification: Ungulata, Perissodactyla, Brontotheroidea.
Meaning of name: Battering ram animal.
Named by: Osborn, 1929
Time: Oligocene epoch of the early Tertiary.
Size: 5m (16ft).
Lifestyle: Low browser.
Species: *E. andrewsi, E. grangeri.*

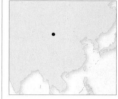

Left: There has not been a complete skeleton of Embolotherium *found, but we can restore it based on what we know of* Megacerops.

RHINOCEROSES AND TAPIRS

*Perhaps the most successful of the big horned plant-eaters of Tertiary times were the rhinoceroses –
successful in as much as they survive today. Their heyday came after that of the Dinocerata and the
Brontotheriidae, but their success likewise waned and we are down to only a handful of genera in
modern times. The tapirs seem more closely related to rhinoceroses than to other perissodactyla.*

Heptodon

The tapirs, of which there is only one modern genus, have
remained largely unchanged as forest dwellers throughout
the Tertiary. As in the rest of the perissodactyls, the earliest
tapir was an unspectacular scampering animal feeding on
forest undergrowth, and only distinguished from the early
forms of the other
perissodactyls by
the shape of the
teeth.

Features: The skull has a
complete complement of 44
teeth, showing how primitive the
whole group is. However, unlike
other primitive perissodactyls
there are large canines on the
upper jaw and a diastema – a
gap between the front
teeth and the back.
It is about half
the size of a
modern tapir,
and in life
would not have
had the short
trunk that characterizes
modern species.

Distribution: Wyoming, USA.
Classification: Ungulata,
Perissodactyla, Tapiroidea.
Meaning of name: Seven
teeth.
Named by: Cope, 1882.
Time: Early Eocene epoch of
the early Tertiary.
Size: 1m (about 3ft).
Lifestyle: Rooter and
browser.
Species: *H. niushanensis,
H. minimus, H. calciculus,
H. posticus.*

Left: Heptodon
*probably had a
muscular upper lip –
would have evolved
into the trunk in later
forms.*

Hyrachyus

The similarity of *Hyrachyus* to the most primitive
rhinoceroses has led some palaeontologists to suggest that it
is ancestral to this group rather than to the tapirs, or is even
part of the stock from which both groups evolved. However,
the consensus is that it is actually a tapiroid (a member
of the tapir group).

Features: *Hyrachyus* differs from
Heptodon in being slightly larger.
It also has slightly higher teeth –
a rhinoceros feature that has led
to the uncertainty in the
classification. The skull,
particularly the shape of the
nasal bones which are
quite small and
leave a large
area for
muscle
attachment
on the upper jaw, shows the
presence of a muscular upper lip
– a forerunner of the short trunk
that later genera of the tapir
group would find so useful in
food gathering.

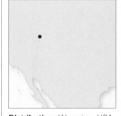

Distribution: Wyoming, USA.
Classification: Ungulata,
Perissodactyla, Tapiroidea.
Meaning of name: Like a
hyrax.
Named by: Leidy, 1870.
Time: Middle Eocene epoch
of the early Tertiary.
Size: 1m (about 3ft).
Lifestyle: Browser.
Species: *H. modestus.
H. eximius, H. asiaticus.*

Left: Hyrachyus *shows the
beginnings of the development of
the short trunk.*

THE DISTRIBUTION OF RHINOCEROSES AND TAPIRS

Although the broad grouping that encompasses the rhinoceroses and tapirs – the "tapiromorphs" in this context – evolved from the same animals as the broad grouping that encompasses *Palaeotherium* and the horses – the "hippomorphs" – their distribution in early Eocene times is quite different. There are examples found in all the northern continents, *Cymbalophus* being European, *Systemodon* North American and *Orientolophus* is Asian.

This implies that the origin of the group lay way back in Palaeocene times to allow them time to disperse over most of the Northern Hemisphere. They probably appeared in North America and spread from there.

The most primitive hippomorphs are found in Europe and in the USA, but it is in the USA alone that they continued to evolve and develop in later times.

Hyracodon

From an animal close to *Hyrachyus* there evolved the rhinoceros group. Their main difference was in the shape of the teeth, being taller and beginning to develop a shearing action rather than a mere grinding one. *Hyracodon* was one of the earliest rhinoceroses. It browsed the low vegetation of the forests that still covered much of North America in Oligocene times.

Features: The small stature and light build of *Hyracodon* gives the initial impression that this might have been a running animal. However, studies of the leg bones show that it would not have been any more capable of a running mode of life than, say, a modern forest hog, and certainly not well adapted to a life on the open plains that were beginning to develop at the end of the Oligocene.

Distribution: South Dakota, USA.
Classification: Ungulata, Perissodactyla, Ceratomorpha.
Meaning of name: Hyrax tooth.
Named by: Leidy, 1850.
Time: Oligocene epoch of the early Tertiary.
Size: 1.5m (5ft).
Lifestyle: Browser.
Species: *H. browni*, *H. nebraskensis*, *H. leidyanus*, *H. priscidens*, *H. petersodi*, *H. medius*.

Right: Hyracodon was one of the last of the forest-dwelling primitive rhinoceroses.

Metamynodon

No sooner was the rhinoceros line established than it began to diversify, producing animals that were adapted to all sorts of habitats. The majority of *Metamynodon* specimens have been found in river sandstones, and the stocky build of this genus suggests that it had a hippopotamus-like amphibious way of life.

Features: The tusks and broad muzzle are adaptations to feeding on aquatic vegetation, as in a modern hippopotamus. The eyes are high on the skull, allowing them to be used while most of the head was submerged. The short and powerful legs are ideal for pushing through river bed mud. The broad ribcage would have been buoyed up by the water. All these adaptations are absent in its close relative *Amynodon*, which was a purely dry land animal.

Distribution: Nebraska, South Dakota, North Dakota, USA.
Classification: Ungulata, Perissodactyla, Ceratomorpha.
Meaning of name: Beyond mynodon – a primitive rhinoceros group.
Named by: Scott and Osborn, 1887.
Time: Late Eocene to early Oligocene epochs of the early Tertiary.
Size: 4m (13ft).
Lifestyle: Aquatic rooter.
Species: *M. chadronensis*, *M. mckinneyi*, *M. planifrons*.

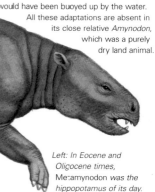

Left: In Eocene and Oligocene times, Metamynodon was the hippopotamus of its day.

THE DEVELOPMENT OF THE RHINOCEROSES

In Oligocene times the rhinoceroses flourished, adapting to the more open habitats that were spreading into all the northern continents and even into South America, and developing into the biggest land mammals the world has known. It was the Hyracodontidae, the family descended from Hyracodon, or something very like it, that included the largest land mammals of all time.

Indricotherium

Originally known as *Baluchitherium*, *Indricotherium* was also referred to as *Paraceratherium* and *Dzungariotherium*. These names may represent valid genera, but are regarded here as different species of the one genus. It is the biggest-known land mammal. It was a rhinoceros that stood twice the height of a modern elephant and reached a weight of 12 tonnes. Molecular studies of its teeth indicate that it fed in enclosed woodland – presumably the shady habitat was necessary to keep its great body cool. The first remains found consisted only of a gigantic skull. And from this the animal was restored as a massive elephantine animal. Later discoveries of leg bones indicated that it was more lightly built than originally imagined.

Features: As in the largest dinosaurs the vertebrae of the back and neck are sculpted into hollows in order to keep down the weight while retaining the strength. The legs are huge, but not as thick as first thought, and the weight is taken on only three toes per foot – the modern rhinoceros pattern. The teeth are distinctive with a pair of front teeth at the top pointing downwards like tusks and the lower pair pointing forward. There seems to have been a muscular upper lip that could have been used as a short trunk in conjunction with these. Contrary to what would be expected of such elephantine animals, the knee, elbow and ankle joints are quite flexible – more like those of modern rhinoceroses than the columnar weight-bearing limbs of elephants. This may mean a very active animal, with the ability to gallop despite its great size.

Right: Despite the fact that Indricotherium *was not as massive as originally thought, it is still the biggest land mammal.*

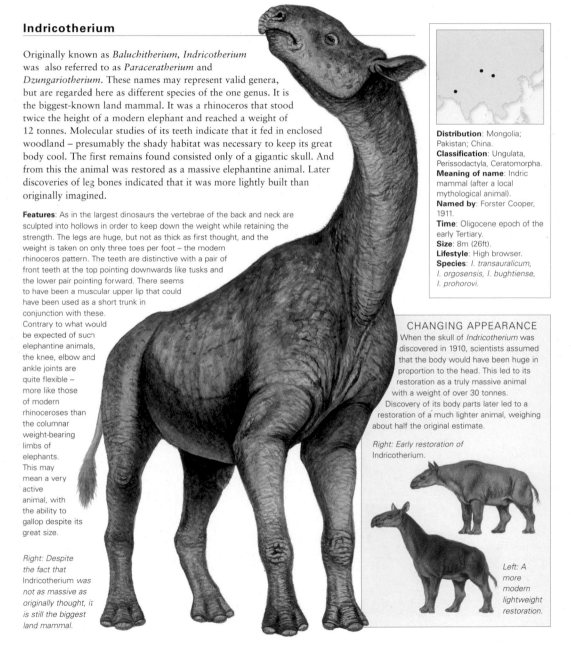

Distribution: Mongolia; Pakistan; China.
Classification: Ungulata, Perissodactyla, Ceratomorpha.
Meaning of name: Indric mammal (after a local mythological animal).
Named by: Forster Cooper, 1911.
Time: Oligocene epoch of the early Tertiary.
Size: 8m (26ft).
Lifestyle: High browser.
Species: *I. transauralicum, I. orgosensis, I. bughtiense, I. prohorovi.*

CHANGING APPEARANCE

When the skull of *Indricotherium* was discovered in 1910, scientists assumed that the body would have been huge in proportion to the head. This led to its restoration as a truly massive animal with a weight of over 30 tonnes. Discovery of its body parts later led to a restoration of a much lighter animal, weighing about half the original estimate.

Right: Early restoration of Indricotherium.

Left: A more modern lightweight restoration.

Trigonias

Distribution: Montana, USA; France.
Classification: Ungulata, Perissodactyla, Ceratomorpha.
Meaning of name: Three-point jaw articulation.
Named by: Lucas, 1900.
Time: Oligocene epoch of the early Tertiary.
Size: 2.5m (8ft).
Lifestyle: Browser.
Species: *T. osborni*, *T. wellsi*, *T. yoderensis*.

The Rhinocerotidae, the family to which modern rhinoceroses belong, appeared in early Oligocene times and spread through North America, Europe, Asia and Africa. However, by the end of the Miocene they were extinct on the North American continent. *Trigonias* was the earliest known form of the family.

Below: From a distance Trigonias *would have resembled a modern rhinoceros, but lacked the horn.*

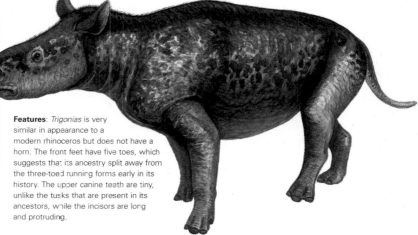

Features: *Trigonias* is very similar in appearance to a modern rhinoceros but does not have a horn. The front feet have five toes, which suggests that its ancestry split away from the three-toed running forms early in its history. The upper canine teeth are tiny, unlike the tusks that are present in its ancestors, while the incisors are long and protruding.

Cadurcodon

Distribution: Kazakhstan; Europe.
Classification: Ungulata, Perissodactyla, Ceratomorpha.
Meaning of name: Cadurco's tooth.
Named by: Kretzoi, 1942.
Time: Late Eocene to late Oligocene epoch of the early Tertiary.
Size: 2.5m (8ft).
Lifestyle: Browser.
Species: *C. ardynense*, *C. saisanensis*.

As we have seen, the early rhinoceroses lacked a horn or other head ornament. However, *Cadurcodon* had a head structure of a different sort. Its nostrils were on the end of a short trunk like that of a tapir. This would have been used for gathering food.

Features: The bones of the skull show that *Cadurcodon* had particularly large nasal cavities and anchor points for nose muscles. These features are identical to those of modern tapirs that sport a short prehensile trunk. The stout hippopotamus-like body suggests that this animal spent at least some of its time in the water, using its trunk to sniff out and gather the best of the waterside plants.

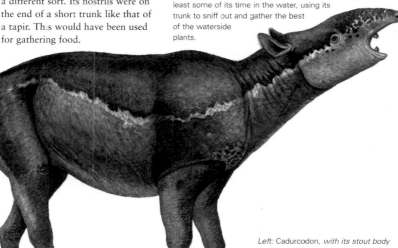

Left: Cadurcodon, *with its stout body and its short trunk, would have looked like a modern tapir. Its lifestyle was probably similar too.*

SOUTH AMERICAN MISCELLANY

South America has only the most tenuous link to North America through the Isthmus of Panama. For most of the Tertiary this link did not exist and South America was an island continent. Like the island continent of Australia today, it had a unique Tertiary fauna, with animals found nowhere else that had evolved in isolation from any other continent.

Didolodus

The didolodonts are an ill-defined family of mammals that encompasses most of the unique fauna of South America. They began as small insignificant animals like *Asmithwoodwardia*, but soon evolved into medium-sized creatures like hare-sized *Didolodus*. They were a quite long-lived family, starting in the Palaeocene and existing until the late Miocene.

Features: *Didolodus* is the best-known of the didolodonts. Its teeth resemble those of the early hoofed animals so it must have had a very similar lifestyle. In life it would have scampered through the undergrowth of the forests and fed on the shoots and leaves.

Distribution: Argentina.
Classification: Ungulata, Litopterna.
Meaning of name: (Obscure).
Named: Ameghino, 1879.
Time: Late Eocene epoch of the early Tertiary.
Size: 60cm (2ft).
Lifestyle: Fleet-footed browser.
Species: *D. mullticuspis*, *D. minor*.

Right: In lifestyle, Didolodus would have resembled a rabbit.

Pyrotherium

On any continent that is isolated from other landmasses we tend to find convergent evolution working – animals evolve the same shapes and features as totally unrelated animals elsewhere in response to evolutionary pressures allowing them to live similar lifestyles in similar habitats and environments. The pyrotheres were the South American equivalent of the elephants.

Features: The short neck and the head of *Pyrotherium* show similarity to unrelated elephants. The incisor teeth are adapted into forward pointing tusks, two pairs on the upper jaw and one on the lower. In life these would have worked along with a trunk that developed from the extended upper lip. The massive body is supported by stout pillar-like legs with plantigrade feet. The name derives from the fact that the remains were first found in sediments of volcanic ash.

Distribution: Argentina.
Classification: Ungulata, Pyrotheria.
Meaning of name: Fire mammal.
Named by: Ameghino, 1889.
Time: Early Oligocene epoch of the early Tertiary.
Size: 3m (10ft).
Lifestyle: Rooter.
Species: *P. romeroi*, *P. macfaddeni*.

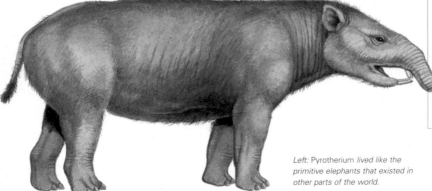

Left: Pyrotherium lived like the primitive elephants that existed in other parts of the world.

Notostylops

Distribution: Argentina.
Classification: Ungulata, Notoungulata.
Meaning of name: Like a southern stem.
Named by: Ameghino, 1897.
Time: Early Eocene epoch of the early Tertiary.
Size: 75cm (2½ ft).
Lifestyle: Low browser.
Species: *N. murinus*.

The Notostylopidae represents a group of South American mammals that show an early specialization of teeth. The front pair of incisors were big and chisel-shaped like those of rodents and the second pair and the canines reduced. The lower incisors, however, pointed forward, not upwards.

Features: *Notostylops* has a short deep face to accommodate the rodent-like dentition. It has a pair of big gnawing incisors at the front, separated by a gap from grinding teeth at the back. In this the dentition was similar to that of rodents but unlike in rodents the front teeth did not grow continually. This was probably an adaptation to nipping rather than gnawing. The rest of the body is rather rabbit-like.

Left: Notostylops *remains are so common that the rocks in which they are found are referred to as the* Notostylopus *beds.*

Thomashuxleya

When Ameghino named the vast number of South American Tertiary mammals whose study he pioneered, he often did so in honour of the great palaeontologists of the day whom he particularly admired. This peccary-like animal was named after the British zoologist Thomas Huxley.

Features: *Thomashuxleya* is a robust, sheep-sized unspecialized animal with a big head in relation to its body. All 44 teeth are present in the jaws – a primitive feature – but the canines are enlarged as prominent tusks for rooting in the ground. Otherwise it does not seem specialized for any particular way of life, and may have been a generalized feeder, in appearance and habits rather like the modern peccary.

Distribution: Argentina.
Classification: Ungulata, Notoungulata.
Meaning of name: From Thomas Huxley, the 19th-century British naturalist.
Named by: Ameghino, 1901
Time: Early Eocene epoch of the early Tertiary.
Size: 1.3m (4ft).
Lifestyle: General feeder.
Species: *T. externa*.

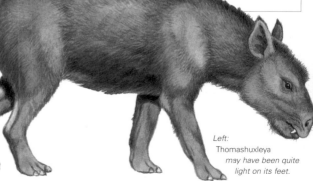

Left: Thomashuxleya *may have been quite light on its feet.*

SOUTH AMERICAN MISCELLANY II

During the early Tertiary the vast continent of South America had many climate zones and habitats. As a result all kinds of animals evolved there, occupying all ecological niches and exploiting all the food supplies presented. For tens of millions of years the continent of South America remained stationary, and the conditions were stable for a long period of time, and so the major groups of animals were long-lived.

Scarrittia

The leonitinids are a primitive group of South American mammals of uncertain affinities. They seem to be an early form of the toxodonts – a hippopotamus-like family that flourished later in the Tertiary. *Scarrittia* is the only one that is known from good remains.

Features: In build *Scarrittia* resembles a flat-footed rhinoceros – a heavy animal with a long body and neck and stout legs with three-toed, hoofed feet, and a short tail. The teeth are unspecialized, with a full complement of 44 squeezed into the short jaws and only the first tooth adapted to an incisor shape, and so it is not clear what kind of feeding style it had.

Left: The remains of Scarrittia were found in deposits formed in a crater lake. The animals may have been poisoned by volcanic fumes when coming to drink.

Distribution: Argentina.
Classification: Ungulata, Notoungulata.
Meaning of name: After H. S. Scarritt, the sponsor of the expedition that found it.
Named by: Simpson, 1934.
Time: Early Oligocene epoch of the early tertiary.
Size: 2m (about 6ft).
Lifestyle: General browser.
Species: *S. canquelensis, S. robusta.*

Rhynchippus

Although the name means "snout horse" and the family to which it belonged is called the notohipids, meaning the "southern horses", this animal is in no way related to the horses. However, it evolved the horse-type shape in response to the opening up of the environment and the spreading of open plains rather than enclosed forest.

Features: Many South American hoofed mammals developed horse-like forms, especially in the late Tertiary when the grasslands flourished. *Rhynchippus* has the deep compact body and the long legs, and the elongated face with cropping teeth that are found on typical grazing, plains-running animals like horses. The canines are not developed into tusks as in other South American hoofed mammals, but have the same cropping structure as the incisors, and the back teeth have convoluted enamel for grinding tough grasses. These features developed independently through convergent evolution.

Left: The rounded muzzle of Rhynchippus *allowed it to crop big mouthfuls of grass.*

Distribution: Argentina.
Classification: Mammal, Notoungulata.
Meaning of name: Snout horse.
Named by: Soria and Alvarenga, 1989.
Time: Early Oligocene epoch of the early Tertiary.
Size: 1m (about 3ft).
Lifestyle: Running grazer.
Species: *R. brasiliensis, R. equinus, R. pumulis.*

Pachyrukhos

Distribution: Argentina.
Classification: Ungulata, Notoungulata.
Meaning of name: (Obscure)
Named by: Ameghino. 1885.
Time: Late Oligocene to middle Miocene epochs of the Tertiary.
Size: 30cm (1ft).
Lifestyle: Low browser or grazer.
Species: *P. moyani*.

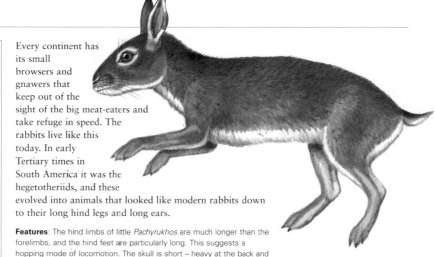

Every continent has its small browsers and gnawers that keep out of the sight of the big meat-eaters and take refuge in speed. The rabbits live like this today. In early Tertiary times in South America it was the hegetotheriids, and these evolved into animals that looked like modern rabbits down to their long hind legs and long ears.

Features: The hind limbs of little *Pachyrukhos* are much longer than the forelimbs, and the hind feet are particularly long. This suggests a hopping mode of locomotion. The skull is short – heavy at the back and narrowing to a pointed muzzle. The teeth are adapted for a diet of nuts and other tough plant material. The eyes are big and the ear structure suggests that the animal sported a pair of long ears in life.

Above: Pachyrukhos *was probably a nocturnal feeder.*

NOTOUNGULATES

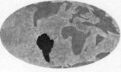

Above: South America as an island continent in the early Tertiary.

The notoungulates – the "hoofed animals from the south" – were the most abundant of the South American ungulates, comprising four suborders containing more than 100 genera. Many were small and looked more like rabbits and rodents than hoofed animals, while others were sheep- or rhinoceros-sized. They were all so different that it is from detailed anatomical evidence, mostly the arrangement of cusps on the teeth and the bones of the ear structure, that they are shown to be related to each other.

Some genera have been found in North America (*Arctostylops* from the Palaeocene and Eocene of Wyoming) and in Asia (*Palaeostylops* from Mongolia and China). They or their ancestors must have reached these continents at times when they were physically joined. However, they appear to have lasted only a very short period of time on these other landmasses. It was only in South America, away from the competition of other mammal types, that the notoungulates thrived from the Palaeocene to the Oligocene, declining and dying out only in the Miocene.

Archaeohyrax

Distribution: Argentina.
Classification: Ungulate, Notoungulata.
Meaning of name: Ancient hyrax.
Named by: Ameghino, 1897.
Time: Oligocene epoch of the early Tertiary.
Size: 45cm (1½ft).
Lifestyle: Generalist feeder.
Species: *A. patagonicus*.

The archaeohyracids were closely related to the hegetotheriids, and, like them, were small scurrying general feeders in the forest undergrowth. Apart from that information, very little is known about them, except for the fact that they are the earliest South American – indeed some of the earliest in the world – mammals to possess high-crowned cheek teeth. It is known from a single well-preserved skull found in the 1890s.

Features: The high-crowned cheek teeth suggest that *Archaeohyrax* and its relatives were adapted to chewing abrasive plant material such as grasses. Apart from this feature the skull is something like that of a hyrax with a tall blunt muzzle. Nothing is known of the rest of the skeleton and so the restoration here is rather speculative, and based on the modern hyrax, to which the animal is in no way related.

Right: Archaeohyrax *may have resembled the modern hyrax of Africa.*

EARLY EVEN-TOED UNGULATES

In modern times the main grazing animals belong to the order Artiodactyla – the even-toed ungulates.
They are characterized by the presence of the cloven hoof (meaning formed by the hoofs of two toes side
by side). The artiodactyls evolved about the same time as the perissodactyls – the odd-toed ungulates –
but were later in developing and did not really spread until the perissodactyls declined.

Diacodexis

The earliest artiodactyls would have been almost indistinguishable
from the earliest perissodactyls in general appearance and lifestyle,
but they had all five toes on each foot. These reduced to two toes as
the group evolved. One of the earliest families was the Dichobunidae,
which looked more like rabbits
than ungulates.

Features: With the
general build of a
modern muntjac deer,
Diacodexis has the
features of an animal
that lived by
scampering through
undergrowth. The
articulation of the legs
shows that it was able
to run, but not as
efficiently as others of
the group. The teeth
are low and have
rounded cusps – an
indication that it may
have been
omnivorous.

Distribution: France;
Germany; Wyoming, USA;
India; Pakistan.
Classification: Ungulata,
Artiodactyla, Dichobunidae.
Meaning of name: Across
the book.
Named by: Cope, 1882.
Time: Early Eocene epoch of
the early Tertiary.
Size: 5cm (2in).
Lifestyle: Low browser.
Species: *D. pakistanensis*,
D. secans, *D. metsiacus*.

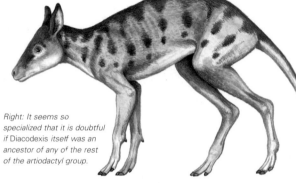

Right: It seems so
specialized that it is doubtful
if Diacodexis *itself was an*
ancestor of any of the rest
of the artiodactyl group.

Elomeryx

The anthracotheres were a family of early ungulates
that may have given rise to the hippopotamus and its
relatives. Certainly they seem to have had a
hippopotamus-like semi-aquatic lifestyle. Their name
means "coal mammals" – a reference to the fact that
they were first found in the brown coal deposits
of France.

Features: *Elomeryx* has a long body
and short stumpy legs. The broad
feet have five toes on the front and
four on the back. The head is long
and rather horse-like. The incisors are
spoon shaped – ideal for digging in
mud – and the canines form
tusks that could be
used for uprooting
underwater
vegetation.
The
premolars
are rather
whale-like, and
this, along with the possible aquatic
lifestyle, has led to the suggestion
that *Elomeryx* lies on the evolutionary
line to the whales.

Distribution: France; North
Dakota, USA.
Classification: Ungulata,
Artiodactyla,
Anthracotheriidae.
Meaning of name: Obscure.
Named: Marsh, 1894
Time: Late Eocene to late
Oligocene epochs of the
early Tertiary.
Size: 1.5m (5ft).
Lifestyle: Browser.
Species: *E. armatus*,
E. woodi, *E. crispus*.

Left: The wide heavy feet of
Elomeryx *would have been ideal for*
wallowing in soft mud, like the
modern hippopotamus.

Archaeotherium

The early artiodactyls soon evolved into pig-like shapes, and the entelodonts were a family that outwardly resembled the warthogs, with the bony processes on the lower jaw. The group originated in Asia in the late Eocene and subsequently spread into Europe and North America, becoming extinct in the Miocene. Some became very large animals indeed.

Features: *Archaeotherium* is one of the earlier members of the entelodont family. Imagine a big warthog with a crocodile-like head. This is the typical appearance of an entelodont. The shoulders are high, to hold the massive muscles needed to support the heavy head. The brain is tiny but the sense of smell is very well developed. Unlike the warthogs, *Archaeotherium* and the other entelodonts do not have the sensitive disc on the snout, but rather large nostrils on the side.

Distribution: Colorado, USA; China; Mongolia.
Classification: Ungulata, Artiodactyla, Entelodontidae.
Meaning of name: Ancient mammal.
Named by: Leidy, 1850.
Time: Early Oligocene epoch of the early Tertiary to early Miocene epoch of the late Tertiary.
Size: 1.2m (4ft).
Lifestyle: Hunter or scavenger.
Species: *A. mortoni*.

Right: The feet are more adapted for running than are those of warthogs and Archaeotherium may have been a hunter.

Dinohyus

Distribution: Nebraska and South Dakota, USA.
Classification: Ungulata, Artiodactyla, Entelodontidae.
Meaning of name: Terrible pig.
Named by: Peterson, 1907.
Time: Miocene epoch of the late Tertiary.
Size: 3m (10ft).
Lifestyle: Omnivore.
Species: *D. hollandi*.

Although *Dinohyus* is found in beds formed in the early part of the late Tertiary it is included here as a member of the entelodonts group that had its heyday in early Tertiary times. It was about the last of the line. It is also the biggest one known, being larger than a buffalo.

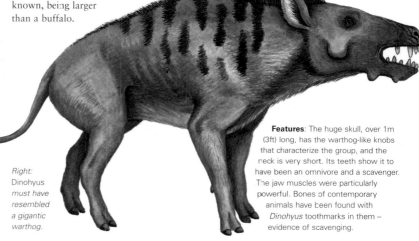

Right: Dinohyus must have resembled a gigantic warthog.

Features: The huge skull, over 1m (3ft) long, has the warthog-like knobs that characterize the group, and the neck is very short. Its teeth show it to have been an omnivore and a scavenger. The jaw muscles were particularly powerful. Bones of contemporary animals have been found with *Dinohyus* toothmarks in them – evidence of scavenging.

TYLOPODS

The tylopods were a group of artiodactyls that fell somewhere between the pig-like suines and the more advanced long-legged pecorans – the giraffes, cattle and deer. They appeared in the late Eocene and reached their peak in Oligocene and Miocene times. They survive today in reduced numbers as the camels, hence the name which means "padded foot".

Cainotherium

The cainotheres were the most primitive of the tylopods. They were common in the Oligocene of Europe and, like the most primitive representatives of any group, were small and unspecialized, browsing in the undergrowth of forests. Some members of the group, however, were quite specialized, living in isolation on the islands of early Tertiary Europe.

Features: *Cainotherium* is typical of the cainotheres. It is rather rabbit-like in appearance and lifestyle. The hind legs are longer than the front and the hind foot is particularly long. The feet are quite advanced in having two functioning toes with the others reduced to slivers. The shape of the brain shows that its senses of hearing and of smell were well developed. The teeth are long and low, and mostly adapted for eating soft leaves.

Distribution: France; Spain.
Classification: Ungulata, Artiodactyla, Tylopoda.
Meaning of name: Empty mammal.
Named: Bravard, 1835
Time: Late Oligocene epoch of the early Tertiary to the early Miocene epoch of the late Tertiary.
Size: 30cm (1ft).
Lifestyle: Low browser.
Species: *C. laticurvatum, C. commune, C. geoffroyi.*

Left: Cainotherium lolloped along through the undergrowth like a rabbit.

Agriochoerus

A tree-climbing camel! Not quite, but it was the only known representative of the agriochoeriids and was a specialist tree-living beast. It was built like a big squirrel, but was a member of the tylopods. It lived in the trees of the open woodlands and riverside forests of the Mid-West of North America in Oligocene times.

Features: The feet of *Agriochoerus* are quite different from those of all other tylopods. Both fore feet and hind feet have four functioning toes that are equipped with strong claws instead of the more usual hoofs. They seem to have been adapted for scrambling along branches although they would not have been very good for climbing up vertical trunks. The animal has a very long tail, which would have been used for balance.

Distribution: South Dakota, USA.
Classification: Ungulata, Artiodactyla, Tylopoda.
Meaning of name: Coarse pig.
Named by: Leidy, 1853.
Time: Oligocene epoch of the early Tertiary.
Size: 2m (6½ft).
Lifestyle: Arboreal browser.
Species: *A. maximus, A. latifrons, A. major.*

Left: Agriochoerus became extinct as the woodlands of the Oligocene gave way to the grasslands of the Miocene.

Merycoidodon

Distribution: South Dakota, USA.
Classification: Ungulata, Artiodactyla, Tylopoda.
Meaning of name: Ruminant-like teeth.
Named by: Leidy, 1848
Time: Early to late Oligocene epoch of the early Tertiary.
Size: 1.4m (4½ft).
Lifestyle: Browser.
Species: M. culbersonii.

The most successful of the early tylopods in North America were a group known as the oreodonts. They were pig-sized running animals with short legs and primitive teeth. Their remains are so common that they represent an index fossil for the Oligocene beds of South Dakota. They seem to have lived in big herds.

Features: Merycoidodon is a typical oreodont. Its body is long and its legs short. The upper and lower leg bones are about the same length, indicating that it was not a strong runner. There is a full complement of 44 teeth in the jaws – a primitive feature – but those at the front of the lower jaw are distinctly longer. This is a feature that would develop further in the later camels. A pit in the skull in front of the eye indicates where a scent gland would have been.

Right: Vast herds of Merycoidodon roamed the plains that were opening across North America in Oligocene times.

OTHER TYLOPODS

The various types of tylopods were abundant throughout the woodlands of the Northern Hemisphere in Eocene and Oligocene times.
Prodesmatochoerus – a small very generalized oreodont from the Oligocene of South Dakota.
Leptauchenia is very similar to *Merycoidodon* but rarer and with eyes situated on the top of the head. This has been interpreted as evidence of an amphibious lifestyle, but there is no other indication of it.
Montanatylopus – a large form from the early Oligocene of Montana.
Protoreodon – like *Agriochoerus* but even smaller.
Bathygenys – like *Merycoidodon*, but yielding specimens from which the cast of the brain can be studied. The cast shows that the brain was organized in the same way as that of modern artiodactyls, but only about a third of the relative size.

Brachycrus

Towards the end of the success story of the oreodonts, in the early part of the late Tertiary, a number of different forms evolved. Some, such as *Leptauchenia*, had the eyes and nostrils high on the head, suggesting a hippopotamus-like, semi-aquatic lifestyle. Others, like *Brachycrus*, had their nostrils elongated into short trunks. The low-crowned teeth suggest that the diet consisted of soft plant material, rather than the tough grasses that were spreading across the land.

Features: The head and jaws of *Brachycrus* are deep and there are big nasal openings that reach back along the top of the skull. This suggests the presence of a fleshy proboscis like that of a tapir. Like a tapir it would have used it for sniffing and rooting about on the forest floor or on the muddy beds of streams and lakes.

Distribution: North America.
Classification: Ungulata, Artiodactyla, Tylopoda.
Meaning of name: Short cross.
Named by: Merriam, 1919.
Time: Early to middle Miocene epoch of the late Tertiary.
Size: 1m (about 3ft).
Lifestyle: Browser.
Species: B. laticeps, B. siouense.

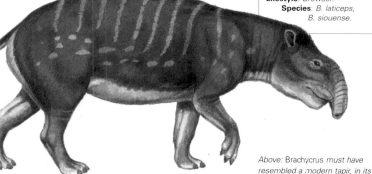

Above: Brachycrus must have resembled a modern tapir, in its appearance and lifestyle.

CAMELS

The camels were the first of the true ruminants – the animals that chew the cud. They developed in Eocene times, and their early history took place entirely in North America. They did not really become big successful running animals until the late Tertiary. Nowadays they have passed their peak, being reduced to a handful of specialized mountain- and desert-living species.

Protylopus

Again the earliest member of this group started off as a scampering rabbit-sized beast browsing forest undergrowth with teeth that were completely unspecialized. From such humble beginnings there developed a wide range of specialized animals.

Features: The feet of *Protylopus* have four functional toes. In the forelimbs they all reach the ground. On the longer hind limbs only the third and fourth are weight-bearers, the others being reduced to dew claws (claws that serve no function). The toes that touch the ground have small sharp hooves, rather than the pads possessed by the later camels. In the forelimb the radius and ulna are fused together in elderly specimens – a stage in the development of a simplified lightweight running leg.

Distribution: Utah, USA.
Classification: Ungulata, Artiodactyla, Tylopoda.
Meaning of name: Before the camels.
Named by: Wortman, 1898.
Time: Late Eocene epoch of the early Tertiary.
Size: 50cm (20in).
Lifestyle: Browser.
Species: *P. petersoni.*

Left: It would be difficult to tell this ancestral camel from many of the other ancestral plant-eaters of the Eocene.

Poebrotherium

The landscape of the Mid-West began to open up in Oligocene times, with the thick forest giving way to open woodland. The camels began to evolve into more modern-looking animals, with long running legs that could move them across the open ground quickly. *Poebrotherium* was one of these.

Features: *Poebrotherium* is about the size of a sheep. The shape of its skull would have made the head look rather like one of the modern llamas. It has two functioning toes on each foot – the first appearance of the traditional cloven hoof – and the two toes are spread, as a weight-bearing feature. The teeth are still primitive, but gaps are beginning to appear between different parts of the dentition, a precursor to more specialized feeding.

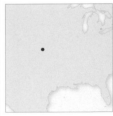

Distribution: South Dakota, USA.
Classification: Ungulata, Artiodactyla, Tylopoda.
Meaning of name: Something mammal.
Named by: Leidy, 1848.
Time: Oligocene epoch of the early Tertiary.
Size: 90cm (3ft).
Lifestyle: Browser.
Species: *P. wilsoni.*

Left: A skull of Poebrotherium *was the first fossil to have been sent to the great American palaeontologist Joseph Leidy from the Mid-West.*

Archaeomeryx

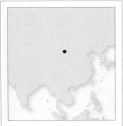

The leptomerycids represent another family that links the Tylopa with the later pecorans. They were small chevrotain-like animals that showed an adaptation of the foot bones to a running mode of life. This adaptation was similar to that of the camels but seems to have evolved independently.

Features: The skeleton of *Archaeomeryx* suggests that in life it must have resembled the modern chevrotain, with its nimble legs, its longish body and its low-slung shoulders. However, its long tail is quite unlike that of any modern hoofed animal. On the lower jaw the first pair of premolar teeth are elongated into canine-like tusks that would have been used for rooting about in vegetation.

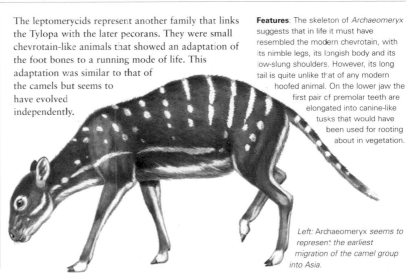

Distribution: China.
Classification: Ungulata, Artiodactyla, Tylopoda.
Meaning of name: Ancient hoof.
Named by: Matthew and Granger, 1925.
Time: Middle Eocene epoch of the early Tertiary.
Size: 75cm (2¼ft).
Lifestyle: Browser.
Species: *A. optatus*.

Left: Archaeomeryx seems to represent the earliest migration of the camel group into Asia.

A REPEATING STORY

There appears to be a pattern in the evolution of the major groups of plant-eating mammals.

They appear as small generalized animals, almost indistinguishable from the early members of the other groups. In appearance and lifestyle the earliest camels resemble the earliest horses, the earliest rhinoceroses, the earliest tapirs and so on. Then, if conditions are right, they expand to fill a niche and become very successful at occupying that niche for some time. Amongst the camels this occurred in late Tertiary times as they became the principal grazers of the open plains. As time goes on these animals are replaced by another group, in the camels' case the deer and antelope, and the only ultimate survivors are those that have become highly adapted to very specialized niches – niches that no other animals have evolved to exploit. The only camels that survive today are the desert-living camels of Africa and Asia and the mountain-living lamas and their relatives of South America – veritable grotesques compared with the range of beasts we find in their history.

This pattern – small and generalized to successful and specialized to restricted and grotesque – is seen in other groups, like the rhinoceroses and giraffes.

Below: A small and generalized Cainotherium, and (right) a specialized Bactrian camel.

Xiphodon

The xiphodonts represent another camel-like family that were common in late Eocene and early Oligocene times. These, however, were restricted to Europe, far from the focus of camel evolution in the middle of North America. It seems likely that their camel-like features evolved independently, in response to similar environmental conditions on the two continents.

Features: *Xiphodon* has only two functional toes on each foot – a feature that developed earlier in this group than in the camels proper. The side second and fifth toes are reduced to mere splints. The teeth are superficially camel-like but are really quite primitive, and there is a pair of strong canine tusks that give the group its name.

Distribution: Spain.
Classification: Ungulata, Artiodactyla, Tylopoda.
Meaning of name: Sword tooth.
Named by: Cuvier, 1825.
Time: Eocene epoch the early Tertiary.
Size: 1m (about 3ft).
Lifestyle: Browser.
Species: *X. gracile*.

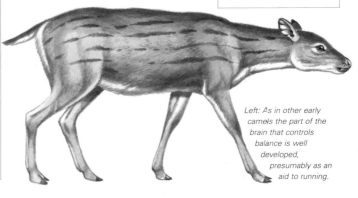

Left: As in other early camels the part of the brain that controls balance is well developed, presumably as an aid to running.

ELEPHANTS

The biggest land animals of today are the elephants. They had a long and varied history, with the large size, the tusks and the trunk evolving quite early. In Palaeocene times they seem to have sprung from a group of condylarths. The increase in body size went along with an increase in head size and the decrease in the length of the neck. They belong to the same group as the hyraxes, and the dugongs and manatees.

Phosphatherium

The oldest-known ancestor of the elephant was dog-sized *Phosphatherium*, uncovered in phosphate mines in Morocco in 1996. Only skulls are known, but it seems likely that the body was long and the legs short – resembling the shape of other plant-eaters of the time.

Features: In bodily form *Phosphatherium* must have been very similar to the ancestral condylarth. The skull is very primitive with a long facial part and a narrow snout. The front teeth protrude forward. The arrangement of skull bones along the jaw, and the teeth show that it is a very primitive member of the elephant line.

Distribution: Morocco.
Classification: Ungulata, Proboscidea.
Meaning of name: Phosphate mammal.
Named by: Gheerbrant, Sudre and Tassy, 1996.
Time: Early Eocene epoch of the early Tertiary.
Size: 1m (about 3ft).
Lifestyle: Rooter and browser.
Species: *P. escuilliei*.

Left: Phosphatherium *appeared a mere five million years after the dinosaurs disappeared – showing the rapid evolution of mammal types.*

Moeritherium

The best-known ancestral elephant was a little pig-sized animal, probably semi-aquatic. The eyes and ears were high on the skull, suggesting that it lived partly in the water like the modern hippopotamus. The presence of most of the primitive elephants in north Africa suggests that this is where the group evolved.

Features: The skeleton of *Moeritherium* is similar in build and in size to that of the modern pygmy hippopotamus, suggesting a similar lifestyle. There is no sign of a trunk but the upper lip and nose were probably long, as in modern tapirs. The incisors are longer than normal, but not quite forming the tusks that we see in later elephants. All the cheek teeth are present and adapted for grinding.

Distribution: Egypt; Mali; Senegal.
Classification: Ungulata, Proboscidea.
Meaning of name: Mammal from Moeris Oasis.
Named by: Andrews, 1904.
Time: Late Eocene to early Oligocene epochs of the early Tertiary.
Size: 1m (about 3ft).
Lifestyle: Rooter and browser.
Species: *M. trigodon*.

Left: Moeritherium *looked and lived more like a hippopotamus than an early elephant.*

Barytherium

Distribution: Egypt; Libya.
Classification: Ungulata, Proboscidea.
Meaning of name: Heavy mammal.
Named by: Andrews, 1901.
Time: Late Eocene and early Oligocene of the early Tertiary.
Size: 3m (10ft).
Lifestyle: Browser.
Species: *B. grave.*

The barytheres are known only from Africa and are restricted to Eocene and early Oligocene times. They were large animals, similar in size to modern elephants. They were a side branch of the elephant line, rather than direct ancestors of later types and were successful for some time in the Eocene and Oligocene epochs.

Features: *Barytherium* has two pairs of tusks in both the upper and lower jaw, and these are quite short and stumpy. In the upper jaw the outer pair are shorter than the inner, while in the lower jaw the inner pair is larger, and together with the second pair form a short shovel structure. This produced a shearing action of the teeth like that of a modern hippopotamus. The back teeth are grinding teeth and are separated from these front tusks by a gap.

Left: We do not know if Barytherium *had a trunk, but the arrangement of the teeth suggests that it did.*

ELEPHANT LINEAGE

Most of the earliest elephants have been found in north Africa. They did not spread into the other continents until Miocene times in the late Tertiary.

Anthracobune – a primitive form that may have given rise to *Moeritherium* and the rest of the elephants.

Jozaria – similar to *Anthrocobune*, from Pakistan.

Piligrimella – an early Eocene relative of the anthracobunids with massive cusps on the molars.

Daouitherium – a large elephant with primitive teeth from the early Eocene of Morocco.

Numidotherium – similar to *Barytherium* but from Libya.

Palaeomastodon – lower tusks flattened and rather long, working with a lower jaw bone to form a scoop for working plants out of mud. Upper tusks were short and curved outwards. It probably had a short trunk.

Phiomia

In the swamps that covered Egypt in Eocene times, both primitive *Moeritherium* and more elephant-like *Phiomia* lived together. They probably did not compete for food, as their adaptations were quite different. *Moeritherium* would have grazed the waterside vegetation, while *Phiomia* would have fed in the forests.

Features: In *Phiomia* we see the development of the tusks that were to be important tools for elephants in the future. There are no first incisors, but the second incisors on both upper and lower jaws are elongated and project forward with an outward curve. A short trunk would have been present in life to act with the tusks to pull food into the mouth. The grinding surfaces of the back teeth consist of low rounded cones – the typical mastodont pattern.

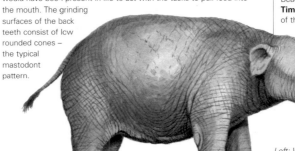

Distribution: Egypt; Ethiopia.
Classification: Ungulata, Proboscidea.
Meaning of name: From the Fayum area in Egypt.
Named by: Andrews and Beadnell, 1902.
Time: Early Oligocene epoch of the early Tertiary.
Size: 3m (10ft).
Lifestyle: Browser.
Species: *P. major, P. winton*

Left: With Phiomia *the typical elephant shape – bulky body, pillar-like legs, short head with a trunk and tusks – was established.*

CREODONTS – THE MEAT-EATING MAMMALS

*The blossoming of the wide range of plant-eating mammals that occurred at the beginning of
the Tertiary, to take up the ecological niches left vacant by the departed dinosaurs, was
accompanied inevitably by the development of meat-eating mammals that preyed on them.
The creodonts were an early group of meat-eaters.*

Oxyaena

Oxyaena was typical of the oxyaenid family
of creodonts. Their feet were not well
adapted to running and they tended to walk
with a flat-footed plantigrade stance rather
than on the tips of their toes as more speedy
animals do. The oxyaenids appeared in
various places from Mongolia to North
America at about the same period of time.
This suggests that they evolved
elsewhere, possibly Africa,
and spread out.
Remains have not
yet been found in
their place of
origin.

Features: About the size of a domestic cat, *Oxyaena*
has the build of a generalized carnivore and would
have fed on birds, small mammals, birds' eggs and
even insects. It has a long body, short limbs and a
very long tail. The mobile nature of the foot, a
feature that distinguishes the creodonts from the
carnivores proper, indicates that it may have been an
agile climber and hunted from ambush rather than a
fleet-footed animal that ran after its prey.

Distribution: North
America.
Classification: Creodonta.
Meaning of name: Sharp
hyena.
Named by: Cope, 1874.
Time: Eocene epoch of the
early Tertiary.
Size: 1m (about 3ft).
Lifestyle: Hunter.
Species: *O. gulo, O. lupine,
O. forcipata, O. intermidia.*

*Left: The various
species of* Oxyaena
*resembled one
another but
differed in size.*

Patriofelis

The shapes and sizes of many of the early creodonts anticipated
the variety of the true carnivores to come. *Patriofelis* was bear-
sized and bear-like in appearance, although it had a long heavy
tail. It would not have been a lithe hunter as its close relatives
were. It had the typical meat-eater's teeth, with strong killing
canines and meat-shearing teeth at the back. A feature of the
creodonts as a whole was the presence of a cleft in the claw
bone. The significance of this is unclear.

Features: *Patriofelis* is one of
the biggest and heaviest of the
oxyaenids. Its strong limbs,
especially about the forequarters, indicate that this was
an ambush predator, lying in
wait for prey and pouncing on
it, killing it with a quick bite to
the neck. The teeth are heavy
like those of a hyena and
indicate an ability to
crack bones.

Distribution: North America.
Classification: Creodonta.
Meaning of name: Father of
cats.
Named by: Leidy, 1870.
Time: Middle Eocene epoch
of the early Tertiary.
Size: 3m (10ft).
Lifestyle: Hunter.
Species: *P. ulta, P. ferox,
P. coloradensis.*

*Above: The jaws of
Patriofelis were first studied
by Leidy in 1870 and he saw
that it was a large carnivore.
Marsh found parts of the
rest of the skeleton two
years later.*

Sarkastodon

Distribution: Mongolia.
Classification: Creodonta.
Meaning of name: Flesh-tearing tooth.
Named by: Granger, 1938.
Time: Late Eocene epoch of the early Tertiary.
Size: 3m (10ft).
Lifestyle: Hunter or scavenger.
Species: *S. mongoliensis*.

Another of the very big creodonts that evolved to hunt the big plant-eaters of the Eocene and Oligocene central Asian plains was bear-like *Sarkastodon*. Like modern grizzly bears, *Sarkastodon* was probably omnivorous, not restricting itself to a meat diet but tackling a wide range of food types.

Features: *Sarkastodon* has hoof-like claws on all feet. Its teeth are heavy and strong, adapted for eating all kinds of food, from meat and bone, to leaves and roots. Unlike a bear, it has a thick, raccoon-like tail. It is one of the biggest of the oxaenids.

Left: Sarkastodon, which stood as high as a man at the shoulders, would have been powerful enough to kill a modern elephant.

GENERALIZED HUNTERS

Primitive mammals tend to have the same generalized anatomy, and there is very little difference between plant-eaters and meat-eaters. At one time the acreodi such as *Mesonyx* were regarded as creodonts, but they are now classed with the early hoofed animals – carnivorous hoofed animals! The more specialized forms were thought to have been ancestral to the Carnivora.

The creodonts were the main meat-eaters of the Northern Hemisphere during the early part of the Tertiary. They did not reach the southern continents, and the meat-eating niche there was taken by the marsupials. The main differences between the Creodonta and the later more advanced Carnivora lies in the fact that their brains are so much smaller. The ear is not as complex. Their feet are not as well adapted for running as in the Carnivora. Their toe bones have a cleft in them. The back teeth in both groups are adapted for a carnassial – meat-shearing – action, but the carnassial teeth of creodonts are formed from the first molars while those of the carnivores are formed from the second premolars.

The generalized build of the creodonts made them perfectly adapted to hunting the unspecialized plant-eaters of the early Tertiary. It was when the plant-eaters began to diversify into different forms in the early Oligocene that the creodonts found themselves at a disadvantage. The more versatile members of the Carnivora then took over.

Machaeroides

It is clear, when looking at *Machaeroides*, that the creodonts evolved into every niche subsequently occupied by the carnivores. Here we have the creodont version of the sabre-toothed tiger. Like the sabre-toothed tiger of the late Tertiary and Quaternary, *Machaeroides* evolved long stabbing canines to inflict deep wounds on prey larger than itself, so that it bled to death.

Features: *Machaeroides* is not so much a cat-like animal as the later sabre tooths – it is more dog-like in build as befits a member of the oxaenids. The front limbs are particularly strong around the humerus, and this would seem to suggest a digging specialization, which is puzzling considering the ambush hunting adaptation shown by the sabre teeth.

Distribution: North America.
Classification: Creodonta.
Meaning of name: Like a sabre-tooth.
Named by: Matthew, 1909.
Time: Eocene epoch of the early Tertiary.
Size: 2m (6½ft).
Lifestyle: Hunter.
Species: *M. eothen*, *M. simpsoni*.

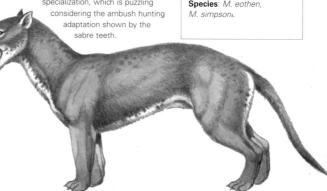

HYAENODONTID CREODONTS

The creodonts were divided into two main families. The oxyaenids tended to be rather cat-like, while the hyaenodontids had features that resembled those of dogs. The oxyaenids were the family that evolved first, while the hyaenodontids came later, some of them lasting into the Miocene, after which the whole group died out and were replaced by the more modern carnivores.

Sinopa

As with all meat-eating groups of animals, there was a variety of shapes and sizes among the creodonts, all adapted to a particular lifestyle and a particular type of prey. The mongoose-like build of *Sinopa* suggests that it hunted small prey such as birds, small mammals or reptiles, through the undergrowth of the early Tertiary forests.

Features: *Sinopa* is a long, slim, fox-sized creodont that would have hunted the smaller animals of the time. Its short feet show that it was more adapted for scampering than for running. The strong shoulders suggest that it may have hunted by ambush and sudden attack. The pointed muzzle would have given it a mongoose-like appearance. The skull is long and narrow, with killing teeth at the front and meat-shearing teeth at the back.

Distribution: North America.
Classification: Creodonta.
Meaning of name: Fused foot.
Named by: Liedy, 1871.
Time: Eocene epoch of the early Tertiary.
Size: 1m (about 3ft).
Lifestyle: Predator.
Species: *S. rapax, S. pungens, S. major, S. minor, S. grangeri.*

Right: There were species of Sinopa in North America and Europe, and possibly one in north Africa.

Hyaenodon

There are several species of *Hyaenodon* known, and these vary in size from that of a rat to that of a lion. The bigger ones probably hunted alone, while smaller forms would have hunted in packs, possibly at night. The discovery of coprolites near their prey suggests that, as in some modern predators, they defecated on their kill to keep rivals away.

Features: The teeth of *Hyaenodon* are big, and adapted for killing, for crushing bones and for shearing meat. The canines are long and sabre-like. The carnassial, or meat-shearing, teeth are set far back in the jaws. The opening of the mouth is long to accommodate this.

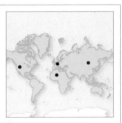

Distribution: Mongolia; North America; France; Africa.
Classification: Creodonta.
Meaning of name: Hyena tooth.
Named by: Leidy, 1869.
Time: Eocene and Oligocene epochs of the early Tertiary.
Size: From 0.6–3m (2–10ft).
Lifestyle: Hunter.
Species: *H. horridus, H. crucians, H. gigas, H. cruentus.*

Right: One species, H. cruentus, had fairly long legs and probably ran after its prey in packs, but most of the others were short-legged animals.

Tritemnodon

Agile and lithe, *Tritemnodon* would have been the civet of its day. Like the modern civet it would have spent as much time in the trees as on the ground. It would have fed on the smaller animals of the time, as well as birds' eggs and insects. Close relatives are known from northern India (*Paratritemnodon*), from Europe (*Prodissopsalis*) and from China (*Propterodon*).

Features: The limb proportions indicate that *Tritemnodon* was a climber, its powerful front legs allowing it to pull itself up trees. By middle Eocene times the variety of niches that could be occupied by meat-eating animals was great. Every type of living thing must have had its predator, very much like today.

Below: With its long body, short legs and long tail, Tritemnodon must have looked like the modern civet.

Distribution: Wyoming, USA.
Classification: Mammal, Creodonta.
Meaning of name: Three rejected teeth.
Named by: Matthew, 1909.
Time: Middle Eocene epoch of the early Tertiary.
Size: 1m (bout 3ft).
Lifestyle: Hunter.
Species: *T. strenuous*, *T. agilis*.

Megistotherium

Distribution: Egypt; Libya.
Classification: Creodonta.
Meaning of name: Satyr mammal.
Named by: Savage, 1973.
Time: Miocene epoch of the late Tertiary.
Size: 3m (10ft).
Lifestyle: Hunter or scavenger.
Species: *M. osteothlastes*.

The creodonts lasted into the Miocene, but were then replaced by the true carnivores. As usual, the last of a line became highly specialized allowing them to survive in niches that were still closed to the newcomers. *Megistotherium*, as well as being among the last of the creodonts, was also the biggest with a skull that was twice as long as that of a tiger – the biggest of our modern land carnivores.

Features: Like other creodonts, *Megistotherium* has clawed feet, huge jaws, sharp meat-shearing teeth and a tiny brain. However, it was so much bigger than the normal run of creodont, being one of the biggest land-living meat-eaters to have been found so far, as big as a bison and with a weight of around 817kg (1,800lbs). This makes it the same order of size as the mesonychid *Andrewsarchus*. It seems unlikely that such a big animal was able to survive by hunting, and so it was probably mostly a scavenger.

Right: As with most animals that we know only from the skull, the appearance of the body and the overall size of Megistotherium are speculative.

THE CARNIVORA

As the conditions changed at the end of the early Tertiary, the creodonts – the first specialized meat-eaters – began to die away. Their place was taken by the members of the order Carnivora, the group to which nearly all the meat-eating mammals today belong. Like the creodonts they evolved from small unspecialized animals into a number of particular niches, and evolved the shapes accordingly.

Miacis

The most primitive and earliest family of the carnivores were the miacids. They were small mammals with a long body and tail, flexible limbs and spreading paws. A thumb-like first toe suggests that they had the ability to climb. They were probably adapted to a tree-living existence as the ground would still have been patrolled by the creodonts.

Right: Miacis *is one of the small mammals found perfectly preserved in the oil shales at Messel.*

Features: *Miacis* is the typical miacid, looking and behaving like a modern pine marten although only about the size of a weasel. As in many primitive mammals, but unlike the later members of the carnivore group, it has a full set of 44 teeth in the jaws. It would have used them to catch birds and small mammals, and for breaking into birds' eggs as it hunted through the trees of the Palaeocene and Eocene forests.

Distribution: Germany.
Classification: Carnivora, Miacoidea.
Meaning of name: Mother animal.
Named by: Cope, 1872.
Time: Paleocene to the middle Eocene of the early Tertiary.
Size: 20cm (8in).
Lifestyle: Arboreal hunter.
Species: *M. parvivorous, M. sylvestris*.

Daphoenus

The amphicyonids were the so-called "bear dogs". The family gained its name because of the resemblance of the head and teeth to those of dogs, and the resemblance of the body and limbs to those of bears. They took over from the creodonts as the ground-dwelling predators in Eocene times and became very abundant, surviving well into the Miocene when they were replaced by the true dogs.

Features: *Daphoenus* is the most common North American amphicyonid, although the group spread across Europe as well. It is smaller than the rest of its group, some of which reach the size of tigers. Coprolites associated with *Daphoenus* shows that they killed and chewed their food as the modern coyote does. The skull is more badger-like than dog-like as in the rest of the group with broad crushing molar teeth, rather than meat-shearing teeth, and a crest on the top of the skull to hold powerful jaw muscles.

Distribution: North America.
Classification: Carnivora, Amphicyonidae.
Meaning of name: (Obscure)
Named by: Lartet, 1836.
Time: Oligocene epoch of the early Tertiary.
Size: 1.5m (5ft).
Lifestyle: Hunter.
Species: *D. nebrascensis, D. vetus, D. lambei*.

Left: The lightly built legs show that Daphoenus *must have been a running animal.*

Plesictis

Distribution: China, France, North America.
Classification: Carnivora, Miacoidea.
Meaning of name: Almost a marten.
Named by: Pomel, 1846.
Time: Early Oligocene to Miocene epochs of the early and late Tertiary.
Size: 75cm (2¼ft).
Lifestyle: Arboreal hunter.
Species: *P. vireti*, *P. mayeri*, *P. himilidens*, *P. genettoides*.

The mustelids form the family of carnivores that, today, incorporate the weasels, the stoats, the martens and the otters. They tend to be quite small, have long lithe bodies and often long tails. They evolved from the miacids in Oliogocene times. *Plesictis* is the most primitive known and almost represents a transitional form between the miacids and the mustelids. Their sinuous bodies allow them to wriggle through undergrowth, bound over open country, scamper up trees and swim in streams.

Features: The long and heavy tail would have acted as a balance while running through branches. The face is short and the brain case long and expanded – a characteristic of the mustelid group. The big eyes on the skull show that this must have been a crepuscular or nocturnal animal. The hind teeth are square in cross section and have blunt cusps, although meat-shearing structures are present, indicating that it probably had an omnivorous diet – feeding on small mammals, birds' eggs, insects and plant matter.

Left: A tree-living hunter with a long tail for balance, Plesictis may have resembled the modern cacomistle.

Hesperocyon

The canids are the dogs, familiar to us today as the wolf, fox, jackal and, of course, as the domesticated companion. The earliest dogs were not very dog-like and resembled mongooses more than anything else. They evolved in central North America and may have lived in underground colonies like modern meerkats.

Features: The skull is beginning to show the evolution of the advanced carnivores, with the teeth being more specialized, with meat-shearing blades on the first molars, and reduced in number to 42 – the last molar tooth being missing from each side of the upper jaw. The ear structure is much more complex than in other contemporary meat-eaters and is distinctly dog-like, being enclosed in bone rather than gristle. The body, however, is long and flexible and the legs short, with spreading five-toed paws. This is a very long-ranging genus, lasting more than 13 million years.

Distribution: North America.
Classification: Carnivora, Canidae.
Meaning of name: Western dog.
Named by: Scott, 1890.
Time: Early Oligocene epoch of the early Tertiary.
Size: 75cm (2¼ft).
Lifestyle: Omnivore but particularly a meat-eater.
Species: *H. gregarius*, *H. coloradensis*.

Right: Hesperocyon looked more like a mongoose than a dog, and may have lived in underground colonies like the modern meerkat.

EARLY CATS

Of the several main lines of carnivore evolution it was the felids – the cats – that set off with an early start. Appearing in the Eocene, they evolved along a number of successful lines in the Oligocene, some of them anticipating the sabre-toothed hunters to come. These early forms, however, were more modest in size.

Nimravus

The early Oligocene saw the appearance of a number of cat families, one of which was the Nimravidae, or the false sabretooths, featured on these pages. They tended to have long low bodies and relatively short legs and, although their canine teeth were longer than average for cats, they were not as well developed as those of the true sabretooths. The ear structure was simpler than that of more advanced cats, having a single, rather than a double, chamber.

Right: As in most cats, Nimravus had a stereoscopic picture of its prey.

Features: *Nimravus* is the animal that gives its name to the group. In build it is similar to the modern caracal of Africa and Asia, but with a longer neck.

It has rather dog-like feet, but the very thin sharp claws have a version of the retraction mechanism that is so typical of cats, preventing them from damage as the animal walked. Its legs are longer than those of the other nimravids, suggesting that it hunted by pursuit rather than ambush.

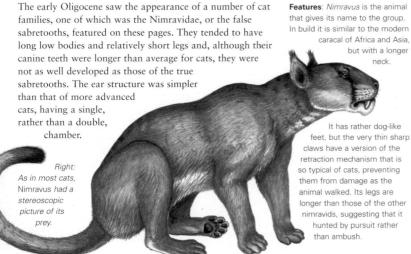

Distribution: North and South Dakota, Colorado, Nebraska and Wyoming, USA.
Classification: Carnivora, Feliformia.
Meaning of name: Ancestral hunter.
Named by: Cope, 1878.
Time: Early Oligocene to early Miocene epochs of the early Tertiary.
Size: 1.2m (4ft).
Lifestyle: Hunter.
Species: *N. brachyops, N. altidens, N. edwasrdsi, N. gomphodus, N. intermedius, N. spectator.*

Dinictis

Although the legs of *Dinictis* were longer than those of its relatives, its feet and those of the other nimravids were short, indicating that the animal walked with a flat-footed plantigrade stance like a bear, not a tiptoed digitigrade stance like a modern cat. Its claws were retractable, but not to such an extent as we find in modern cats.

Features: Despite its very primitive build, *Dinictis* has the characteristic features of a modern cat, with the short face and the forward pointing eyes that would have allowed stereoscopic vision – a valuable ability in a hunting animal. In life it probably also possessed the nictating membrane, the so-called third eyel d. The sabre teeth formed of the upper canines are thick and round in cross section, rather than thin as in other nimravids.

Distribution: South Dakota, USA.
Classification: Carnivora, Feliformia.
Meaning of name: Terrible cat.
Named by: Leidy, 1854.
Time: Oligocene epoch of the early Tertiary.
Size: 1.2m (4ft).
Lifestyle: Hunter.
Species: *D. felina, D. squalidens.*

Left: The cheek teeth of Dinictis were reduced to two pairs in the upper jaw, and highly specialized for shearing meat.

Hoplophoneus

Distribution: North and South Dakota, Nebraska, USA.
Classification: Carnivora, Feliformia.
Meaning of name: Murderer.
Named by: Le dy, 1857 (as Drepanodon), renamed by Cope, 1874.
Time: Late Oligocene epoch of the early Tertiary.
Size: 1.3m (4¼ft).
Lifestyle: Ambush hunter.
Species: *H. prmaevus, H. mentaiis, H. occidentalis, H. sicarius, H. dakotensis*

The nimravids were built as ambush predators rather than as pursuit hunters. *Hoplophoneus* shows these adaptations, with its short muscular legs, its plantigrade feet and its deep-stabbing canines. This nimravid was about the size of a modern jaguar and probably dropped upon its prey from overhanging branches.

Features: *Hoplophoneus* has shorter bulkier limbs than other nimravids. The whole skull is shaped to drive down the sabre teeth with a powerful blow. Its sabre teeth are much longer and thinner and work with a jaw articulation that gave the mouth a very wide gape allowing room for the sabres to be deployed efficiently. There is a prominent flange on the lower jaw that would have protected the sabres when the mouth was closed.

Above: When Leidy studied the skull in 1857 he thought that it was a species of the sabretooth Machairodus. It was Cope who named it Hoplophoneus *in 1874.*

CATS' CLAWS

A feature unique to cats is the presence of retractable claws. The joints of the toes are shaped to be able to lift the end bone with its claw up and off the ground, and the ligaments between them are adapted accordingly. This brings the claw back into a protective sheath. There is a fine balance of tendons between those that extend the claw and those that retract it. The result is that the cat's claws, essential for hunting, are protected when it is walking – the animal takes its weight on pads on its feet. This gives rise to the silent stalking movement that is so familiar. The early cats had this ability to some extent.

Eusmilus

Many different types of false sabretooths inhabited the world at the same time, presumably hunting different prey. However, they did not live peaceably together. A skull of *Nimravus* has been found that had been pierced by the sabre tooth of *Eusmilus*. The former survived the attack, as the wound had healed itself.

Features: The biggest sabre teeth among the nimravids are found in *Eusmilus*, where they are almost as long as the skull itself and very flattened. The jaw opens to more than 90 degrees to allow them to be deployed. Deep flanges on the lower jaw protect the sabres, as in *Hoplophoneus*. The lower canines, by contrast, are no bigger than those of any conventional cat. The jaw muscles are weak, but the neck muscles are very strong to drive the stabbing action. The teeth are reduced in number to 26 – a more extreme specialization than in its contemporaries, and even less than the 30 found in modern big cats.

Distribution: France; Colorado, Nebraska, North Dakota, South Dakota, Wyoming, USA.
Classification: Carnivora, Feliformia.
Meaning of name: True sabre.
Named by: Gervais, 1876.
Time: Late Oligocene epoch of the early Tertiary.
Size: 2.5m (8ft).
Lifestyle: Ambush hunter.
Species: *E. sicarius, E. bidentatus, E. cerebralis.*

Left: Eusmilus was a leopard-sized hunting cat, with sabres on the upper jaw guarded by bony flanges on the lower.

WHALES

The whale is perhaps the most impressive work of evolution, showing the transition of a land-living mammal to a totally aquatic one. Over the last decade or two, all kinds of transitional forms have come to light from the early part of the Tertiary, but even so, there is still a great deal of debate as to whether the whale evolved from a meat-eating mesonychian or a herbivorous ungulate.

Pakicetus

Most of the earliest whales have been found in north-west India and Pakistan. In the Tertiary, this area would have been one of shallow seas, formed when the continents of India and Asia came together as the Tethys Ocean closed. *Pakicetus* would have lived on land, venturing into the water to feed. In appearance it was a land-living animal, with only the details of the skull showing that it ventured into the water and that it was ancestral to the aquatic whales. The bones of the ankle suggest it evolved from the ungulates.

Features: *Pakicetus* has eye sockets on the top of the skull and nostrils at the tip of the snout – indicative of a partially aquatic existence. The shape of the molar teeth – three roots and a single triangular cusp – and the structure of the ear are very similar to those of the other primitive whales. However, the ear lacks the sinuses that a modern whale uses to hear underwater. Apart from these features, *Pakicetus* would appear to have been a very dog-like animal, feeding in shallow streams and ponds.

Left: Skeletons of Pakicetus have been found in river sediments, indicating that the species lived in river mouths.

Distribution: Pakistan.
Classification: Cetacea.
Meaning of name: Whale from Pakistan.
Named by: Gingerich and Russell, 1981.
Time: Early Eocene epoch of the early Tertiary.
Size: 1m (about 3ft).
Lifestyle: Semi-aquatic.
Species: *P. attocki, P. inachus.*

Ambulocetus

Ambulocetus was found in roughly the same region as *Pakicetus*, in marine rocks that were slightly younger. This was an animal that would have lived along the shoreline, hunting like a crocodile and swimming like an otter with up and down undulatory movements of the body and tail. Chemical analysis of the teeth shows that *Ambulocetus* lived in both salt and fresh water, as does the estuarine crocodile of modern day Australia.

Features: The limbs of *Ambulocetus* show evidence of an amphibious lifestyle, with forelimbs equipped with hooves and hind feet that were adapted for swimming. It would have moved about on land but clumsily, pushing itself along like a sea lion by its hind legs and muscular tail. Its hip bones are fused to the backbone, unlike that of later whales. The feet were probably webbed to help in swimming.

Below: The arrangement of nasal passages shows that Ambulocetus *could swallow underwater, and the ear structure was adapted to hearing while submerged.*

Distribution: Pakistan.
Classification: Cetacea.
Meaning of name: Walking whale.
Named by: Thewissen, Madar and Hussain, 1996.
Time: Eocene epoch of the early Tertiary.
Size: 3m (10ft).
Lifestyle: Aquatic hunter.
Species: *A. natans.*

Rodhocetus

Distribution: Pakistan.
Classification: Primates, Cetacea.
Meaning of name: Whale from Rodho.
Named by: Gingrich, 1994.
Time: Middle Eocene epoch of the early Tertiary.
Size: 3m (10ft).
Lifestyle: Aquatic hunter.
Species: *P. kasrani*, *R. balochistanensis*.

This animal shows further development towards an aquatic way of life, with its streamlined shape, its increased flexibility of the body and the strength of the back and tail rather than of the limbs. When on land it would have walked clumsily on the soles of its hind feet, so that the ankle took all the weight. Evidently it took the weight of its forequarters on the tips of the toes.

Features: The front and hind feet have splayed toes, and their structure suggests that they were webbed. The hip bones are not fused to the backbone – a better arrangement for the flexibility needed for swimming than the strength needed for walking. The vertebrae at the base of the tail have tall spines indicating that the tail was very muscular and may have carried a fluke. The neck vertebrae are shortened as in modern whales leading to a streamlining of the head and neck.

Left: When swimming Rodhocetus *probably held its limbs against its body for a more streamlined shape, while propelling itself along with its powerful tail.*

OTHER EARLY TERTIARY WHALES

Nalacetus – like a mesonychian, giving support for the idea of evolution from this group.
Ichthyolestes – very similar to *Pakicetus* but only about half the size.
Remingtonocetus – otter-sized animal with powerful swimming limbs, small eyes and a tapering snout.
Artiocetus – a whale with ungulate-like ankle bones that suggest that the whales evolved from the hoofed mammals rather than the meat-eating mesonychians.
Protocetus – early whale-shaped whale, with nostrils far back on the skull forming an elementary blowhole.
Natchitochia – only known from a few vertebrae that have a hip-bone attached to the backbone, unlike the other archaeocetes.
Durodon – like a very short *Basilosaurus*.
Pontogeneus – similar to *Basilosaurus*.
Zygorhiza – like *Basilosaurus* but much smaller.

Above: A very early, but unrelated, example of a mammal returning to water is Castorocauda *from the middle Jurassic of China.*

Basilosaurus

Basilosaurus is the whale with a name like a dinosaur. When Harlan discovered it in 1843 he thought it was a big reptile and named it accordingly. When he noticed his mistake he changed the name to *Zeuglodon*, meaning "yoked teeth", after the distinctive shape of the teeth with two roots and a high pointed crown. However, the scientific rules apply, and the first name given is regarded as the valid one. *Basilosaurus* has been adopted as the state fossil of Alabama.

Features: The great length of *Basilosaurus* is achieved by an increase in the length of the individual vertebrae. The head is relatively small and the snout pointed. There are two shapes of teeth, those at the front are conical and pointed for gripping prey and those at the back triangular, serrated and blade-like for shredding it. The forelimbs are paddles, but have an elbow joint like a seal. The hind legs are present but they are tiny, and the hip bones are not attached to the backbone.

Below: The obvious feature of this whale is its enormous length and sea-serpent-like form.

Distribution: Egypt; Australia; Louisiana, Mississippi and Alabama, USA.
Classification: Primates, Cetacea.
Meaning of name: Emperor lizard.
Named by: Harlan, 1843.
Time: Middle to late Eocene epoch of the early Tertiary.
Size: 2m (65ft).
Lifestyle: Fish- and squid-hunter.
Species: *B. cetoides*, *B. hussaini*, *B. isis*.

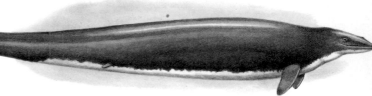

PRIMATES

The primates are the mammal group to which lemurs, monkeys, apes and humans belong. They first appeared in the Palaeocene forests and soon became adapted to tree-living life, with hands and feet evolved for gripping, eyes that could judge distances, and a reproductive strategy that involved a small number of offspring that were carefully looked after.

Plesiadapis

The plesiadapids were an early form of lemur. They were abundant in Palaeocene and Eocene times and spread throughout Europe and North America, implying that the two continents were close enough to share their faunas. In general they had lemur-like bodies but rodent-like heads with eyes at the side, and long snouts and jaws with a gap between the canines at the front and the grinding teeth at the back.

Features: The body is quite robust with fairly short limbs, long fingers and toes with claws that are compressed from side to side, and a long tail. In the mouth the teeth are rodent-like, with long gnawing incisors separated by a gap from grinding cheek teeth at the back. The hands and feet are articulated so that they could turn towards one another, evidently an adaptation for climbing vertical stems and trunks, although it probably walked on branches as well.

Distribution: France; Colorado, USA.
Classification: Primates, Adapiformes.
Meaning of name: Almost Adapis (another extinct lemur).
Named by: Gervais, 1877.
Time: Late Palaeocene and early Eocene epochs of the early Tertiary.
Size: 80cm (2½ft).
Lifestyle: Arboreal plant-eater.
Species: *P. tricuspidens*, *P. russelli*.

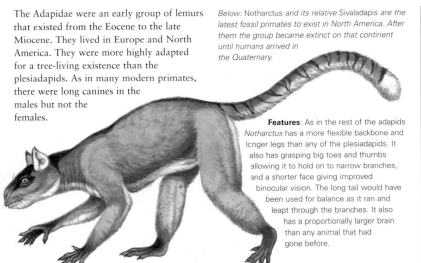

Left: Plesiadapis used its gnawing and grinding teeth to eat the fruits and seeds of the Palaeocene and Eocene forest trees.

Notharctus

The Adapidae were an early group of lemurs that existed from the Eocene to the late Miocene. They lived in Europe and North America. They were more highly adapted for a tree-living existence than the plesiadapids. As in many modern primates, there were long canines in the males but not the females.

Below: Notharctus and its relative Sivaladapis are the latest fossil primates to exist in North America. After them the group became extinct on that continent until humans arrived in the Quaternary.

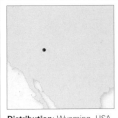

Features: As in the rest of the adapids *Notharctus* has a more flexible backbone and longer legs than any of the plesiadapids. It also has grasping big toes and thumbs allowing it to hold on to narrow branches, and a shorter face giving improved binocular vision. The long tail would have been used for balance as it ran and leapt through the branches. It also has a proportionally larger brain than any animal that had gone before.

Distribution: Wyoming, USA.
Classification: Primates, Adapiformes.
Meaning of name: False bear.
Named by: Leidy, 1870.
Time: Early to middle Eocene epoch of the early Tertiary.
Size: 40cm (16in).
Lifestyle: Arboreal leaf-eater.
Species: *N. tenebrosus*, *N. robinsoni*, *N. pubnax*, *N. robustior*.

Shoshonius

Distribution: Wyoming, USA.
Classification: Haplorhini, Primates.
Meaning of name: From Shoshone native American people.
Named by: Beard, Krishtalka and Stuky, 1991.
Time: Early Eocene epoch of the early Tertiary.
Size: 25cm (10in).
Lifestyle: Arboreal insectivore.
Species: *S. cooperi*.

The tarsiers are a group of prosimians that are highly specialized for night activity in trees. The modern forms have big eyes and ears, and have gripping pads on their fingers and toes. Nowadays there is only one species, *Tarsius*, but in the early Tertiary it was a very widespread and abundant group, and these seem to have diverged from the main primate line at least in Eocene times.

Right: With its short face and its very large eyes, Shoshonius must have had the facial appearance of the modern tarsier.

Features: *Shoshonius* is known from four fossil skulls and limb fragments found in the 1990s in Wyoming, USA. The ear structure and the eye sockets are tarsier-like, and suggest that in life it had the large ears and eyes that are so distinctive of the group. The few limb bones known indicate that it was not so well adapted for leaping as the tarsier is; the short femur in particular suggests that it moved about in branches in a manner similar to the modern bushbaby.

PRIMAL PRIMATES

The ancestors of the primates probably lie in the varied little mammal types of the Cretaceous. They would have been little tree-shrew-like animals, scampering in the branches above the heads of the dinosaurs.

They came into their own in Palaeocene times. The earliest known is *Altiatlasius* from the late Palaeocene of Morocco, which was rather like a squirrel in its lifestyle and appearance, but had hands that were capable of grasping.

The prosimian group – the lemurs, the tarsiers and their relatives – really got under way in Eocene times, and this was the time of their greatest diversity, ranging from North America, through Europe and Asia and into Africa. It was at this time that they reached Madagascar – their modern stronghold.

We know very little of primate evolution in the succeeding Oligocene, as few fossils have been found, but what we have discovered indicates that the anthropoids – the suborder that contains the two modern groups of monkeys and also the apes – evolved at this time. The platyrhines – or the Old World monkeys – are known from the early Oligocene of South America. These are characterized by broad noses and prehensile tails. *Branisella* and *Tremacebus* are Oligocene platyrhines that would have looked like the spider and howler monkeys of today.

The catarrhines – the narrow-nosed Old World monkeys from Africa and Asia – are common in rocks laid down in Miocene times. The apes also evolved from this group in the Miocene.

Eosimias

One of several tiny mouse-sized primates, *Eosimias* lived in China in Eocene times. It is known from bones that may have come from an owl pellet, showing that this tiny animal was small enough to have been the prey of the contemporary owls, but these bones are enough to show that it was a transitional form between the prosimians – the lemurs and tarsiers – and the monkeys.

Features: With only some teeth, parts of the jaw and the ankle bones to go on, it is difficult to get a full picture of *Eosimias*. The molar teeth show the shearing action associated with the prosimians but it also has the large canines of the monkeys. The foot bones indicate that *Eosimias* walked on the tops of branches, like monkeys do, rather than clinging to vertical stems as is usual among the prosimians.

Distribution: China.
Classification: Catarrhini, Primates.
Meaning of name: Early monkey.
Named by: Beard, 1994.
Time: Eocene epoch of the early Tertiary.
Size: 10cm (4in).
Lifestyle: Arboreal omnivore.
Species: *E. sinensis*, *E. centinnicus*.

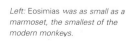

Left: Eosimias was as small as a marmoset, the smallest of the modern monkeys.

SOUTH AMERICAN MARSUPIALS

By the late Tertiary, Australia had broken away from Antarctica, being the last fragment of old Pangaea to separate. At this time South America was isolated, both from Antarctica and from North America. On both these island continents the marsupials – the pouch mammals – continued to be important, particularly in the roles of meat-eaters.

Cladosictis

The borhyaenids were the carnivorous marsupials of South America. They occupied the same niches as the placentals and so developed the same body patterns as placental mammals did elsewhere through the phenomenon of convergent evolution. They evolved from quite generalized stock, such as *Cladosictis*, which must have lived rather like the mongoose it resembled, hunting any small prey it could find and probably even raiding birds' nests or swimming after fish.

Features: *Cladosictis* has a long body, short limbs and a long muscular tail. The head is long and narrow. Its teeth are organized like those of a placental carnivore, with incisors at the front for nipping, a pair of canines at the side for stabbing, and blade-like premolars and molars at the back for shearing meat. Unlike placental meat-eaters, though, there are four pairs of incisors rather than three.

Distribution: Argentina.
Classification: Marsupialia.
Meaning of name: Branched weasel.
Named by: Amighino, 1887.
Time: Late Oligocene to Early Miocene epochs of the early and late Tertiary.
Size: 80cm (32in).
Lifestyle: Hunter.
Species: *C. centralis, C. bardus, C. patagonica.*

Left: Cladosictis may have raided the nests of the huge flightless birds that lived in contemporary South America.

Borhyaena

The largest of the borhyaenids were bear-like, with heavy bodies, and walked with a plantigrade action – on the flats of their feet. They were the main predator of the time, until they were replaced by giant ground-dwelling birds similar to those that existed on the northern continents in the earliest Tertiary. A closely related form, *Prothylacinus*, was very similar but ambushed its prey from overhead branches rather than hunting it on the ground as *Borhyaena* did.

Features: *Borhyaena*, the animal that gives its name to the group, is about the size of a wolf, has a body like a bear, a head like a hyena and a long tail that is unlike anything possessed by a placental carnivore. As well as the meat-shearing molars, it has huge premolars that are adapted for crushing bones. The jaw is very similar to that of the recently extinct thylacine of Australia, leading to the suggestion that the two were closely related. However, DNA studies show that the similarities are superficial and the result of convergent evolution. Being a heavy animal it would have hunted its prey by ambush rather than running it down – the prey in this instance being the strange ungulates that existed on the South American plains.

Distribution: Argentina.
Classification: Mammal, Marsupialia.
Meaning of name: Strong hyena.
Named by: Ameghino, 1894.
Time: Late Oligocene to early Miocene epochs of the early and late Tertiary.
Size: 1.5m (5ft).
Lifestyle: Hunter.
Species: *B. tuberata, B. excavata, B. macrodonta.*

Left: Ground hunters like Borhyaena probably evolved from tree-living ancestors.

Thylacosmilus

Distribution: South America.
Classification: Marsupialia.
Meaning of name: Pouched sabre.
Named by: Riggs, 1933.
Time: Late Miocene to Early Pliocene of the late Tertiary.
Size: 1.2m (4ft).
Lifestyle: Ambush hunter.
Species: *T. ferox*, *T. lentis*.

Here is the classic sabre-toothed tiger shape appearing among the marsupials. There is some evidence that *Thylacosmilus* had a sophisticated family life, with the ability to look after its young long after they were weaned and left the pouch. It survived for most of the late Tertiary, only becoming extinct when the placental carnivores invaded from North America.

Features: The sabre teeth of *Thylacosmilus* developed from a pair of incisors. These teeth were longer than in any conventional sabre-toothed cat, and they continued growing throughout life. The lower jaw carries a pair of flanges that act as supports for these teeth when the mouth is shut. The line of the head indicates that the neck muscles were massive, to power the lunge of the sabre teeth.

Left: A full skeleton of Thylacosmilus *is not known, and so the restoration is based on the body of other big marsupials.*

LOST MARSUPIALS

Marsupials hold sway in Australia today, as they did throughout the Tertiary despite the influx of animals from Asia that appeared as the continent drifted north and the island chains between the two landmasses formed.

South America also had impressive marsupials throughout the Tertiary, but these are all extinct, having been replaced by invaders from their northern neighbour. Why the difference?

American palaeontologist Steven J. Gould advanced a theory in the 1970s to account for this. Throughout its Tertiary history, Australia moved northwards from Antarctica at a considerable rate – faster than any other continental movement. As a result the landmass moved from polar climates, through temperate conditions, to the tropical desert and rainforest that we find there now. The animal life had to evolve quickly just to keep up with these changes in conditions, and so the modern fauna is tested and resilient.

South America, on the other hand, has remained in the same latitudes for the last 60 million years. Conditions have not changed. The indigenous fauna became adapted to these conditions and no others. As a result, when the land bridge of Central America was established to the northern continents, the inhabitants were unprepared for the competition that would ensue. Northern invaders had been evolving all this time and when they were able to move into the new continent they were able to adapt quickly and take the place of the South American fauna.

Necrolestes

This little animal is known only from the bones of the jaw and snout, and some leg bones, but from these we can deduce that it was a burrowing hunter. It is so unusual that it cannot be compared with any of the known groups of South American marsupials. We can, however, deduce that it was the marsupial equivalent of the moles of the Northern Hemisphere.

Features: The snout has an upturned tip, which seems to be the base for fleshy folds – possibly sensitive organs of touch like those of the modern star-nosed moles. These would have been used for feeling through the darkness of the soil in which it burrowed, hunting out worms and other underground invertebrates. The teeth are tiny and sharp, with five pairs of incisors on the upper jaw and four on the lower. The canines are sharp and the molars have triangular crowns. The front legs seem to be evolved for burrowing.

Distribution: Argentina.
Classification: Marsupialia.
Meaning of name: Grave robber.
Named by: Ameghino, 1891.
Time: Early Miocene epoch of the late Tertiary.
Size: 15cm (6in).
Lifestyle: Burrowing insectivore.
Species: *N. patagonensis*.

Left: The teeth of Necrolestes *were so similar to those of the modern golden mole that researchers have thought that it was a placental mammal rather than a marsupial.*

AUSTRALIAN MARSUPIALS

In Tertiary times, as today, Australia was the stronghold of the marsupials – the pouched mammals. They differ from the placental mammals that dominate the rest of the world, in giving birth to very immature young, often little more than two-limbed and naked foetuses that can just about manage to crawl to the pouch, where they are suckled until they are ready to make an appearance in the world.

Palorchestes

When part of the skull was first found, it was thought to have been that of some kind of giant kangaroo – hence the name. It was about 100 years after Owen's first description of this creature that more remains came to light, showing that it was more closely related to the diprotodonts. In structure, however, it must have been the Australian equivalent of the giant ground sloth of the Americas – a huge tree-ripper.

Features: Specimens of *Palorchestes* are rare, with many of them lying uncatalogued in Australian museums for decades until fuller remains came to light in the 1970s. The forelimbs are massive and made for grasping, and the front feet are equipped with long laterally compressed claws, ideal for ripping down trees. The broad shelves of bone around the nasal passages and the holes for the passage of nerves and blood vessels indicate that in life it supported a substantial trunk. It probably had a long tongue.

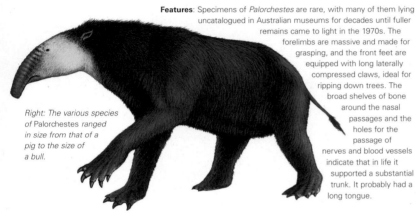

Right: The various species of Palorchestes *ranged in size from that of a pig to the size of a bull.*

Distribution: Australia.
Classification: Marsupialia.
Meaning of name: Ancient leaper.
Named by: Owen, 1874.
Time: Pliocene epoch of the late Tertiary to Pleistocene epoch of the Quaternary.
Size: 2m (6½ft).
Lifestyle: High browser.
Species: *P. azael, P. parvus, P. painei.*

Thylacoleo

Marsupial hunters in Australia have a long history. In Oligocene and Miocene times there was *Priscileo*, the smallest and most primitive of the group, the size of a cuscus. This was followed by late Miocene *Wakaleo*, the size of a large dog. The ultimate comes in *Thylacoleo* of the Pleistocene. Despite its obvious carnivorous adaptations, it has sometimes been suggested that it was a specialized fruit-eater.

Features: The cheek teeth consist of huge blade-like premolars, structured like the carnassial teeth of placental carnivores. The tiny canines are replaced by a pair of huge pointed incisors that were adapted for piercing, holding and lacerating. The jaw held muscles that must have given it the strongest bite known for that size of animal. The hand is equipped with a huge killing claw on an opposable thumb. The thumb swings over to a strong bone in the wrist to give a powerful grip.

Left: Leopard sized, with a killer claw, and meat-shearing teeth powered by strong neck muscles, Thylacoleo *was the top predator of the time.*

Distribution: Australia.
Classification: Marsupialia.
Meaning of name: Marsupial lion.
Named by: Owen, 1859.
Time: Pleistocene epoch of the Quaternary.
Size: 1.5m (5ft).
Lifestyle: Hunter, ambush predator.
Species: *T. carnifex.*

Neohelos

Distribution: Australia.
Classification: Marsupialia.
Meaning of name: New wart (referring to a structure on the tooth).
Named by: Stirton, 1967.
Time: Oligocene epoch of the early Tertiary to the Miocene epoch of the late Tertiary.
Size: 2m (6½ft).
Lifestyle: Forest browser.
Species: *N. tirarensis*, *N. stirtoni* and two others as yet unnamed.

The forests of Tertiary Australia had their big browsers, like the tapirs of today. *Neohelos* was a cow-sized member of the widespread diprotodon group – the marsupial group that resembled giant wombats and grew to gigantic sizes in the Quaternary. Many specimens of this big animal have been found in the Riversleigh fossil site in the Northern Territories. There have been so many remains found that it may have lived in herds.

Right: When it was discovered Neohelos *was identified as an intermediate form between the more primitive and the more highly evolved diprotodonts, such as* Zygomaturus.

Features: *Neohelos* is the big browser in the fossil-bearing beds of the period, and has the typical browser's shape of a heavy deep body supported on stout limbs. The skull is quite large, but the space for the brain is tiny – no bigger than an orange. The animal shows sexual dimorphism, with the males being larger than the females. *Neohelos* was identified and named in 1967 on the basis of isolated teeth. Since then almost complete remains of this animal have been found.

THE REVERSLEIGH FOSSILS

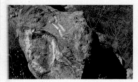

Left: The arid terrain around Riversleigh.

One of the best sites for Tertiary and Quaternary fossils in Australia is the area of Riversleigh in Queensland. As in the Naracoorte site in southern Australia, this site has been an arid plain of limestone since Oligocene times and the fossils have been preserved in fissure deposits and cave deposits at various periods of time. The limestone was laid down in shallow seas in the early Tertiary.

In the Miocene the area became dry land and was thickly forested, and the remains of forest-dwelling animals date from that time. Thereafter the area gradually became drier and was able to support fewer types of animals. The fossils in cave deposits dating from the later time are those of animals adapted to dry grasslands, like grass-eating kangaroos. At about this time we find the first remains of rats, indicating an influx of placental mammals due to the fact that the continent was beginning to close with the Asian continent to the north.

As well as ancient marsupials there are the fossils of the egg-laying monotremes – the other group of typically Australian mammals. These include *Zaglossus*, an echidna as big as a small sheep, and *Obdurodon*, a platypus with more teeth than the modern example.

Wynardia

Found sometime before the 1860s, *Wynardia* is the first of the Tertiary Australian land mammals to be found – everything found before then was from the relatively recent Quaternary. It lived at a time when Tasmania was low-lying and covered in rainforest. It seems to have been a slow mover, like a modern cuscus and would have lived both in the trees and on the ground.

Features: *Wynardia* is known from a single specimen that is almost complete except for the feet and the tail. It has a big head, prominent ears and eyes that point forward like those of a koala. Strong jaws show it to have lived on the tough plants of the rainforest, but it may have scavenged meat as well. Fossils of close relatives have been found in central Australia.

Right: The best-preserved skeleton of Wynardia *was found in marine sediments. It had evidently been washed out to sea from its forest habitat.*

Distribution: Tasmania.
Classification: Marsupialia.
Meaning of name: From Wyndard, where it was found.
Named by: Spencer, 1901.
Time: Miocene epoch of the late Tertiary.
Size: 1m (about 3ft).
Lifestyle: Arboreal browser or scavenger.
Species: *W. bassiana*.

XENARTHRA – THE TOOTHLESS ONES

The xenarthrans were once known as "edentates" meaning "lacking teeth". Toothlessness indeed characterizes the anteaters – the most prominent of the modern representatives of this group. Other living members have teeth, but they are usually small and rootless and often without enamel. Anteaters, sloths and armadillos are the modern examples.

Hapalops

The ground sloths evolved to become the big plant-eaters of Quaternary times, but, as usual, they began as quite small animals. *Hapalops*, the earliest known, was the size of a small sheep, and was probably partly arboreal and partly a ground dweller. They first evolved in South America.

Features: The stout body of *Hapalops* is supported on quite long and heavy legs. The front feet have long curved claws, used for food gathering, and these must have forced the animal to walk on its knuckles while on the ground, like a modern-day gorilla does. There are very few teeth in the mouth, with only four or five pairs of cheek teeth and these were quite small. It probably fed on soft leaves. Its brain size is larger, proportionally, than that of living sloths. It seems likely that fermentation of food took place in the gut to compensate for any inefficiencies in the chewing mechanism.

Left: The forelimbs of Hapalops were particularly important in locomotion and food-gathering.

Distribution: Argentina.
Classification: Placentalia, Xenarthra.
Meaning of name: Gentle appearance.
Named by: Ameghino, 1887.
Time: Early to middle Miocene epoch of the late Tertiary.
Size: 1.3m (4ft).
Lifestyle: Semi-arboreal leaf-eater.
Species: *H. indifferens, H. longiceis, H. cadens.*

Peltephilus

The biggest of today's edentates, the giant armadillo of Brazil, is as heavy as a human and may dig up graves to scavenge the corpses. *Peltephilus* was an armadillo about the same size, and may have had a similar lifestyle. Along with its relative *Macroeuphractus*, it appears to have had adaptations for a carnivorous mode of life as well as a burrowing one, although recent work suggests that it may have been a specialist herbivore like the later glyptodonts.

Right: The horns on the face may have helped to protect the eyes when Peltephilus *borrowed in the soil after food.*

Features: The skull of *Peltephilus* has a pair, or possibly two pairs, of horns which in life would have been covered with skin or horn and used for defence or display. The teeth, which form a continuous series from the front to the back, appear to be adapted for meat-shearing and bone-crushing. The scutes of the back and tail are formed in transverse rows, loosely overlapping.

Distribution: Argentina.
Classification: Placentalia, Xenarthra
Meaning of name: Armour lover.
Named by: Ameghino, 1887.
Time: Oligocene to Miocene epochs of the early and late Tertiary.
Size: 1.5m (5ft).
Lifestyle: Hunter or scavenger, or possibly specialized herbivore.
Species: *P. ferox.*

Thalassocnus

Distribution: Peru.
Classification: Placentalia, Xenarthra.
Meaning of name: Sea sloth.
Named by: Muizon and McDonald, 1995.
Time: Late Miocene to late Pliocene epochs of the late Tertiary.
Size: 2m (6½ft).
Lifestyle: Aquatic or shoreline scavenger.
Species: T. natans, T. littoralis, T. carolomartini, T. antiquus, T. yaucensis.

We do not associate sloths with water, but a genus of giant ground sloths particularly adapted to a shoreline habitat and showing an ability for foraging underwater existed in late Tertiary times on the west coast of South America. The abrasion on their hind teeth shows that they fed underwater on sea grass, chewing mouthfuls of sand at the same time. The earliest species of *Thalassocnus* probably grazed on stranded seaweed or on sea grasses growing in very shallow waters, but the later species seem to have been specialized diving underwater feeders.

Right: The hinterland of the time was barren desert. Thalassocnus evolved to exploit the only local food source – the sea.

Features: The skull is longer than that of other ground sloths, and is expanded at the downturned snout, probably supporting thick lips for manipulating the food as in the modern dugong. The hands have strong claws, which were probably used like those of the modern marine iguana, for holding on to the sea floor while it fed. The arm bones are expanded, rather like those of a seal, and adapted for underwater locomotion. The tail is flattened and muscular, indicating its use as a swimming organ.

MODERN XENARTHRA

The xenarthrans are restricted to the Americas, mostly South America. The original classification "edentate" encompassed the animals that we now class as the xenarthrans – the anteaters, armadillos and sloths – and also the aardvark and the pangolin of the Old World. These have since proved to be derived from other ancestors.

Right: A modern tamandua anteater.

Right: A three-toed sloth.

Below: A nine-banded armadillo.

Stegotherium

The history of the armadillos starts in Eocene times with, as usual, very unspecialized animals. By the Miocene they had evolved into a number of different shapes, but were mostly specialized as insect-eaters. In *Stegotherium* this adaptation was taken to an extreme, and it could have subsisted only on ants and termites.

Features: The body structure of *Stegotherium* is very similar to that of a modern armadillo, but the head is quite different, having an elongated snout that resembles that of an anteater. Like an anteater the teeth are reduced, and consist of a few pairs of weak cylindrical lumps in the back of the jaw. The armour is like that of modern armadillos in being formed of plates that lie in bands across the back, allowing a degree of articulation.

Distribution: Argentina.
Classification: Placentalia, Xenarthra.
Meaning of name: Roofed mammal.
Named by: Ameghino, 1887.
Time: Early Miocene epoch of the late Tertiary.
Size: 1m (about 3ft).
Lifestyle: Anteater.
Species: S. simplex, S. tessalatum, S. variegatum.

Left: An anteater beneath the shell of an armadillo – that would describe the general appearance of Stegotherium.

INSECTIVORES

The most abundant group of mammals, and the one that contains the smallest individuals, is the Insectivora. These are the little insect-eating mammals of today. In build they are similar to the primitive mammals that existed throughout the Mesozoic before the Age of Mammals, and during their history they have developed into a number of different forms, not all of which restricted their diet to insects.

Dimylus

The dimylids are a group of insectivores that seem to have been related to the shrews and the moles. They are known from a handful of genera, mostly represented by fossil teeth. A single skull with attached jaws has been found and from this we can see that it was a semi-aquatic animal like the modern desman, and may have fed on freshwater shellfish. Central and northern Europe at that time was a low-lying area crossed by rivers and streams and dotted with lakes and swamps – an ideal habitat for such an animal.

Features: The strong jaw action needed for tackling hard-shelled prey is indicated by the teeth, in which the enamel is particularly thick. The canines have deep double roots, and the cheek teeth are low and broad, adapted for crushing. These features are like those of shellfish-eating animals such as sea otters. From these details we can build a picture of a small mammal that has adaptations to an amphibious mode of life – with a flattened swimming tail and webbed feet. The only known skull shows that the group is related to the shrews and the moles, probably more closely to the latter.

Left: Judging by the teeth and the jaws, and the inferred lifestyle, the appearance of Dimylus was probably similar to that of other small aquatic insectivores.

Distribution: Germany.
Classification: Insectivora.
Meaning of name: Double millstone.
Named by: Meyer and Jahrb, 1846.
Time: Early Miocene epoch of the late Tertiary.
Size: 15cm (6in).
Lifestyle: Aquatic shellfish-eater.
Species: *D. paradoxus.*

Deinogalerix

The hedgehogs are an old group of insectivores, evolving in Europe in the Eocene and expanding into Asia and Africa by the Miocene, and also into North America, but they became extinct there by the end of the epoch. *Deinogalerix*, found in fissure deposits in southern Italy, is the biggest-known of the group. Other animals of the time and place were also very large, suggesting that in late Miocene times the area was an island archipelago. Small animals often have very large species evolving in island habitats.

Features: Unlike many modern hedgehogs, *Deinogalerix* in life does not have the body hairs modified to spines – it is more closely related to the modern hairy hedgehog. It also has a long tail. However, like the smaller modern forms it has a long snout, sharp insect-grabbing teeth and short legs. As well as being an insectivore it probably scavenged meat from dead animals.

Left: Deinogalerix – the largest-known of the hedgehogs – was probably restricted of a few islands along the edge of the closing Tethys Ocean.

Distribution: Italy.
Classification: Insectivora.
Meaning of name: Terrible polecat.
Named by: Fruedenthal, 1973.
Time: Late Miocene epoch of the late Tertiary.
Size: 60cm (2ft).
Lifestyle: Insectivore or scavenger.
Species: *D. koeingswaldi.*

Proterix

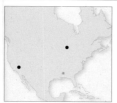

Distribution: South Dakota, California, Nebraska, USA.
Classification: Insectivora.
Meaning of name: First hedgehog.
Named by: Matthew, 1903.
Time: Late Oligocene to early Miocene epochs of the early and late Tertiary.
Size: 20cm (8in).
Lifestyle: Burrowing insectivore.
Species: *P. bicuspis, P. loomisi.*

The hedgehogs that penetrated North America became extreme in their specializations before becoming extinct. *Lanthanotherium* was as tiny as the smallest shrew. *Proterix* was an armour-headed burrowing form that seems to have lacked limbs. It would have burrowed its way through soil, using spade-like actions of its head and undulations of its body, pursuing worms and burrowing invertebrates as a mole does.

Features: *Proterix* is known from a number of skulls and jaws, the skulls extending backwards in a pair of bony plates. There are associated vertebrae, strong and compressed in the neck region, but there has never been a discovery of limb bones or of shoulder or hip girdles. The implication is that the limbs had become atrophied in their extreme adaptation to an underground existence. Should this be true, then it would be the only example known of a land mammal that has lost its limbs.

Below: The strangest of all insectivores, Proterix must have resembled a furry cylinder with a spade-like head – extreme adaptations for a worm-like existence.

INSECTIVORA

The Insectivora is a classification that covers a number of widespread and differing groups. As well as those illustrated here, the major groups are as follows:

Adapicoricidae – an early group only known from teeth, which are rather like those of a hedgehog. *Adapisorex* is the Palaeocene typical example.

Heterosoricinae – a primitive shrew group that existed from Eocene to Miocene, known only from teeth until a full skeleton complete with soft tissues of *Lusorex* was found in Miocene lake deposits in China, showing that it was very similar in appearance and build to the modern shrews.

Talpidae – these are the moles, specifically adapted to a burrowing way of life. Pliocene *Hesperoscalops* was similar to today's moles.

Chrysochloridae – the golden moles, unrelated to the above but having the same adaptations. Miocene *Prochrysochloris* had the body of a golden mole but the teeth of a tenrec.

Tenrecidae – restricted to Madagascar, the Seychelles and parts of Africa, this hedgehog-like group dates back to the early Miocene with *Parageogale* from Kenya.

Soricidae – the shrews. These are by far the most abundant of today's insectivores and include the world's smallest mammals. Pliocene *Allosorex* was a big carnivorous example of this group.

Scandentia – the tree shrews. Uncommon both today and in the fossil record, they may be ancestral to the primates. *Adapisoryculus* was a very early form, from the Palaeocene.

Myohyrax

The elephant shrews, or macroscelids, are an unusual group of insectivores – so different that they are now not regarded as belonging to the order Insectivora. They have long snouts – hence the name – and tend to have big eyes although they are active by day. Fossil forms are unlike the modern types, being adapted to all sorts of environments and lifestyles – *Mylomygale* was like a rodent, while *Myohyrax* looked just like a hyrax. This range in forms has led to uncertainty in identification and classification in the past.

Features: *Myohyrax* is about the size and the build of a modern hyrax, and had big eyes and a long snout. It would have scampered over rocks like the hyrax rather than jumping along on hind legs as modern members of the group do. Although it looks like an insectivore, *Myohyrax* has self-sharpening teeth more like those of a rodent, and may have fed on seeds and nuts.

Distribution: Africa.
Classification: Macroscelidea.
Meaning of name: Mouse hyrax.
Named by: Andrews, 1914.
Time: Miocene epoch of the late Tertiary.
Size: 15cm (6in).
Lifestyle: Insectivore.
Species: *M. oswaldi, M. boederieini, M.osborni.*

Right: All members of the elephant shrew group are restricted to Africa.

RODENTS

Along with the insectivores, the rodents are among the most abundant mammals on earth. This is due to their generally small size – small animals are naturally more abundant than large ones, as their small size means that they exploit much less of the resource of the environment than large animals. However, in their early history there were some very large rodents, as well as the more familiar small types.

Ceratogaulus

The protogomorphs are regarded as the earliest true rodents, being known from the squirrel-like paramyids, like *Paramys*, from the Eocene. In Oligocene times they diversified into mouse-like and beaver-like forms, and in the Miocene they died out as conditions changed from forest habitats to open grasslands. *Ceratogaulus* was one of the last and most spectacular.

Features: The most obvious feature of this animal is the pair of horns on the snout. Head and neck muscles show that the head could be lowered so that the horns stick out at the front – the same stance produced by a rhinoceros in defensive mode. This indicates that the horns were used for defence rather than for display. The front legs are stout and powerful and the toes are long and carry very big laterally compressed claws. These, and the small eyes, indicate that this was a burrowing animal spending most of its time underground.

Left: Ceratogaulus *is often illustrated under its old name* Epigaulus. *Apart from the armadillo-relative* Peltephilus, *it is the only known horned burrowing mammal.*

Distribution: Colorado, USA.
Classification: Glires, Rodentia.
Meaning of name: Horned marten.
Named by: Matthew, 1902.
Time: Miocene epoch of the late Tertiary.
Size: 30cm (1ft).
Lifestyle: Burrowing plant-eater.
Species: *C. rhinoceros, C. anecdotus, C. minor, C. hatcheri.*

Phoberomys

The biggest rodent known resembled a guinea pig – but was as big as a cow. Rodents probably arrived in South America in Eocene times. There they found open plains that were not yet exploited by hoofed animals, and the very few meat-eaters present were inefficient hunting marsupials. *Phoberomys* evolved to occupy the plant-eating niches that were present. Big rodents persist in South America in the form of the capybara – the largest rodent of the present day.

Features: *Phoberomys* is more than ten times the weight of the capybara – the biggest rodent known today. It lived by river banks and fed on grasses, and may have been, like its relative the capybara, partially aquatic. The huge body would have held a voluminous gut that was able to process large quantities of otherwise indigestible grass. The structure of the hips and tail show that it could have sat on its haunches feeding itself with its front paws. The back teeth are tall and adapted for grinding tough substances like grass.

Left: On its haunches Phoberomys *resembled a giant ground sloth. On all fours it walked on straight legs – more like a cow than a guinea pig.*

Distribution: Venezuela.
Classification: Glires, Rodentia.
Meaning of name: Terror mouse.
Named by: Sánchez-Villagra, Aguilera and Horovitz, 2003.
Time: Late Miocene epoch of the late Tertiary.
Size: 4.5m (15ft).
Lifestyle: Grazer.
Species: *P. pattersoni.*

Sivacanthion

The modern porcupines exist in two separate groups. The New World porcupines are part of the caviomorph group, and are closely related to the capybaras and the guinea pigs of that hemisphere and the Old World porcupines belong to a separate group called the phiomorphs. The latter are the older group, with representatives being found in the Oligocene of Egypt. The oldest recognizable porcupine is *Sivacanthion* from the Miocene of Pakistan.

Below: With its stout body and its armour of quills, Sivacanthion would have had the same lifestyle as the modern-day porcupines.

Distribution: Pakistan.
Classification: Glires, Rodentia.
Meaning of name: Siva's spiked one (after the local deity).
Named by: Colbert, 1933.
Time: Middle Miocene epoch of the late Tertiary.
Size: 50cm (20in).
Lifestyle: Gnawer.
Species: *S. complicatus*.

Features: The build of *Sivacanthion* is very like that of a modern porcupine, although details of the anatomy suggest that it is not a direct ancestor but a side branch of the family tree. The spines are modified hairs that act as a defensive mechanism. It belongs to a small group that are known in the Indian sub-continent – all the other phiomorphs are restricted to Africa.

JAWS AND TEETH

The jaws and the teeth of rodents and lagomorphs are quite distinctive. At the front of the mouth is a pair of incisors on the upper and lower jaws – two pairs in the case of the lagomorphs. These continue to grow throughout life, being worn away at the chisel-shaped tips as the animal feeds.

The back teeth are tall and complex, with transverse ridges, and used for grinding. Between the incisors and the back teeth is a diastema – a gap in the tooth row. This gives room for the food to be processed while it is being gnawed at the front and chewed at the back.

The jaw muscles are quite complex, allowing the jaw to move up and down, sideways, and to and fro, in order to grind up tough food and extract as much nutrition as possible.

Below A skull of a rabbit (lagomorph).

Right: A skull of a rat (rodent).

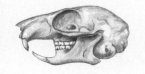

Palaeolagus

The rabbits and hares are not strictly rodents, but belong to the lagomorphs, a related group. With the rodents they share the continually growing incisors and the cheek teeth with the transverse ridges. However, there are two pairs of upper incisors rather than the single pair as in rodents. The history of the group goes back to late Palaeocene times. Most fossils found are of fragments of animals eaten by predators.

Features: The best Tertiary age lagomorph skeleton found so far is that of *Palaeolagus*. This shows an animal very similar in build and adaptations to the modern rabbits. The cheek teeth are shorter; it does not have the long thumping feet of modern representatives and the legs are less well adapted for leaping. These are features that are seen to evolve as the Tertiary develops. Like modern lagomorphs *Palaeolagus* fed on grass.

Right: Apart from the short feet and the longer tail, Palaeolagus would have looked just like a modern rabbit.

Distribution: Nebraska, USA.
Classification: Glires, Lagomorpha.
Meaning of name: Ancient rabbit.
Named by: Leidy, 1856.
Time: Oligocene epoch of the early Tertiary.
Size: 25cm (10in).
Lifestyle: Grazer.
Species: *P. philoi, P. primus, P. burkei, P. haydeni, P. hemirhizis, P. intermedius, P. temnodon.*

RUNNING HORSES

With the demise of the forests that had formed so great a proportion of the landscape of the early Tertiary, and the increasing area of open landscapes, animals that were adapted to a running lifestyle appeared. The horses that had, up to this point, been small undergrowth animals started to develop into the elegant running animals that we know today.

Parahippus

In *Parahippus* the running leg that was to be a feature of horses to come had developed. In this, the lightweight foot was worked by tendons from muscles concentrated close to the body, and the movement produced a fore-and-aft motion.

Right: The transitional form between the undergrowth horses and the plains-living horses appears to be Parahippus.

Features: *Parahippus* still has three toes as in the more primitive horses, but now the toes at the side are smaller, and most of its weight is borne on the middle toes. It was an efficient runner. The head is longer, reflecting a grazing habit, with the eye socket well back in the skull. The teeth are taller than in the preceding *Mesohippus* and are strengthened by hard-wearing cement crests on the enamel. This enabled *Parahippus* to chew the new tough grasses.

Distribution: Florida, USA.
Classification: Perissodactyla, Hippomorpha.
Meaning of name: Beside a horse.
Named by: Leidy, 1858.
Time: Miocene epoch of the late Tertiary.
Size: 1.5m (5ft or 10 hands).
Lifestyle: Grazer.
Species: *P. leonensis*, *P. mourningi*, *P. pristinus*, *P. pawniensis*.

Merychippus

Right: Merychippus appears to have lived in herds on the grasslands.

With *Merychippus* the horse line really spread on to the developing grasslands. It seems to have given rise to a number of lines of horse evolution, all evolved for living on the plains. The meaning of the name is something of a misnomer. Horses do not ruminate as cows do. The way that horses deal with the indigestible grasses is to hold them for long periods in their hind gut.

Features: The later species of *Parahippus* are difficult to distinguish from their successor, *Merychippus*. In the latter the middle toe is very pronounced and the side toes do not reach the ground. The head is very like that of a modern horse, with the strong grass-ripping incisor teeth at the front, then a diastema, and a set of tall grinding teeth at the back.

Distribution: Nebraska, Oregon, USA.
Classification: Perissodactyla, Hippomorpha.
Meaning of name: Ruminant horse (a misnomer).
Named by: Leidy, 1857.
Time: Late Miocene epoch of the late Tertiary.
Size: 1.5m (5ft or 10 hands).
Lifestyle: Grazer.
Species: *M. insignis*, *M. californicus*.

Pliohippus

Distribution: Colorado, North and South Dakota, Nebraska, USA; Canada.
Classification: Perissodactyla, Hippomorpha.
Meaning of name: Horse of the Pliocene (although it is now thought that it may have been restricted to the Miocene).
Named by: Marsh 1874.
Time: Late Miocene and possibly into the Pliocene epochs of the late Tertiary.
Size: 2m (6½ft or 12 hands).
Lifestyle: Running grazer.
Species: See caption.

This was the predecessor of the modern running horse. The ability to run is a good survival adaptation. Over the open plains danger can be seen from far off, and the best way to escape distant danger is to run further away. It is possible that *Pliohippus* actually encompasses two genera, the other being *Dinohippus* with smaller depressions in the skull in front of the eyes.

Features: *Pliohippus* is a truly one-toed horse, with a large hoof on the middle toe and no sign of any bones of the other toes. It has taller grinding teeth and a smaller brain than the modern horse, *Equus*, but the only significant difference is the presence of a deep depression in the skull in front of the eyes. This may have held additional muscles for thick and mobile lips or it may have been part of the nasal structure allowing it to produce a different sound from the modern horse's neigh.

Below: Many species of Pliohippus *are known to have lived: P. mirabilis, P. campestris, P. pernix, P. robustus, P. fossulatus, P. pachyops, P. lullianus, P. nobilis, P. osborni, P. mexicanus. Also Dinohippus spectans, D. leidyanus, D. leardi, D interpolatus, D. mexicanus.*

Anchitherium

HORSE EVOLUTION

The late Miocene represents the heyday of horse evolution in abundance and diversity. Some were browsers, feeding on the soft leaves of trees and bushes; others were grazers, feeding on the tough grasses of the plains.
Hypohippus – a big browsing horse similar to *Anchitherium* but from North America.
Megahippus – a big browser from North America, later than *Hypohippus*.
Sinohippus – a browsing horse from China.
Archaeohippus – a member of a short-lived group of dwarf horses that browsed the thickets of North America.
Parahippus – the last of the browsing horses on the main evolutionary line, with features intermediate between *Parahippus* and *Merychippus*.
Kalobatippus – closely related to *Parahippus*.
Protohippus – a small North American grazer.
Calippus – another dwarf grazer from North America.
Hipparion – a group of grazers that migrated from North America to Asia.
Astrohippus – a short-lived descendant of *Pliohippus*.
Hippidion – one of a group that migrated to South America when the Isthmus of Panama was established.

The evolution of the horse is often shown as a straight line or a family tree. In fact it has all sorts of side branches that became highly successful but died out with no issue. *Anchitherium* was one of these, branching away from the main line in the Oligocene times and crossing from North America into Europe and Asia where it continued into the Miocene as a browsing animal like its ancestors.

Features: As befits a leaf-eating animal rather than a grazer, the cheek teeth are much lower than those of the contemporary plains-living horses. It retains the three functional toes of its forest-dwelling ancestors, and is not such a specialized running animal as the others of the time.

Right: Anchitherium's browsing lifestyle probably evolved in the open woodland of Miocene Spain.

Distribution: France; Florida, USA; Germany; Spain.
Classification: Perissodactyla, Hippomorpha.
Meaning of name: Near to a mammal (probably "near to a horse").
Named by: Meyer, 1844.
Time: Miocene epoch of the late Tertiary.
Size: 1m (about 3ft or 6 hands).
Lifestyle: Browser.
Species: A. clarenci, A. aurelianense, A. corcolense, A. paraquinum, A. castellanum, A. cursor, A. procerum, A. hippodides, A. matritense, A. alberdiae.

CHALICOTHERES

The chalicotheres represent a strange group of perissodactyls that resemble nothing more than horses with big claws. They existed from Eocene to Pleistocene times so they were quite long-lived. However they never formed an important part of the ecosystem. Their remains are first found in Eocene rocks, but it is only in the late Tertiary that they become spectacular.

Protomoropus

This creature was discovered in the 1970s and regarded as a species of *Hyracotherium*. However, later studies published in 2004 show that it was actually an early representative of the chalicothere line. It appears that the group arose in Asia in the early Eocene, and migrated to North America and from there to Europe in the late Eocene where they then evolved into later genera and species.

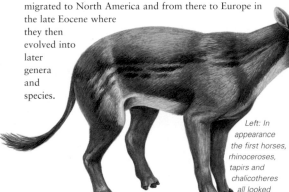

Left: In appearance the first horses, rhinoceroses, tapirs and chalicotheres all looked very similar.

Features: As with many mammal groups, the earliest forms are not well known. In this case *Protomoropus* is known only from a set of cheek teeth found in Mongolia. The details of the teeth show great similarities to those of the later members of the group, and show that they evolved from the same ancestors as the horses, the rhinoceroses and the brontotheres. It must have resembled its close relative *Litolophus*.

Distribution: Mongolia.
Classification: Perissodactyla, Chalicotheroidea.
Meaning of name: Before *Moropus* (a better-known later chalicothere).
Named by: Hooker and Dashzeveg, 2004.
Time: Early Eocene epoch of the early Tertiary.
Size: 45cm (18in).
Lifestyle: Browser.
Species: *P. gabuniai*.

Chalicotherium

Chalicotherium gives its name to the whole group. The hind legs were very short when compared with the very long forelimbs, and this must have given the animal a gorilla-like stance in life. Most of the weight would have been taken on the stout hind legs. It would have walked on its knuckles, again like a gorilla, with the claws curled away for safety.

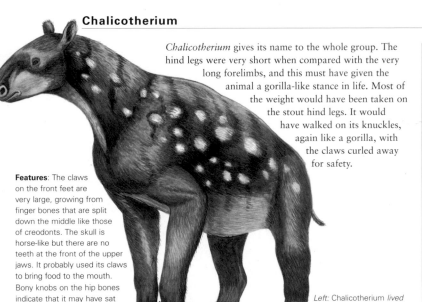

Features: The claws on the front feet are very large, growing from finger bones that are split down the middle like those of creodonts. The skull is horse-like but there are no teeth at the front of the upper jaws. It probably used its claws to bring food to the mouth. Bony knobs on the hip bones indicate that it may have sat down to eat, as the modern giant panda does.

Left: Chalicotherium lived in forests and browsed on bushes and trees.

Distribution: Kazakhstan.
Classification: Perissodactyla, Chalicotheroidea.
Meaning of name: Pebble mammal.
Named by: Kaup, 1833.
Time: Miocene and Pliocene epochs of the late Tertiary.
Size: 2.75m (9ft).
Lifestyle: Browser.
Species: *C. gigantium, C. rusingense, C. pilgrimi, C. wetzleri, C. salinum.*

Moropus

Distribution: North America.
Classification: Perissodactyla, Chalicotheroidea.
Meaning of name: Sloth foot.
Named by: Marsh, 1877.
Time: Early Miocene epoch of the late Tertiary.
Size: 3m (10ft).
Lifestyle: Browser.
Species: *M. oregonensis, M. elatus, M. hollandi, M. matthewi, M. merriami.*

When palaeontologists first found the spectacular claws they thought that they were from some kind of giant sloth, but since then articulated skeletons of *Moropus* have been uncovered. The claws were probably used for digging up roots, or even for defence. It did not walk on its knuckles like *Chalicotherium*, but the claws could be lifted clear of the ground when the forelimbs took the weight. The strong hind legs show that it could stand upright to reach food high in trees.

Features: There are three big claws on each foot, those on the front being much bigger than those on the back. The bones of the second finger are fused to form a structure that would strengthen the inside edge of the hand. The front legs are longer than the hind, but not as markedly as in *Chalicotherium*. The joints of the neck vertebrae show that it could hold its head up, as well as reach the ground.

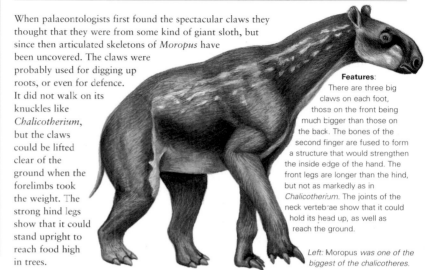

Left: Moropus was one of the biggest of the chalicotheres.

Kalimantsia

Several genera of chalicotheres have been found in southern Europe, including *Chalicotherium* itself. These have been found in areas that would have been quite open habitat at the time, and are accompanied by the remains of horses, early deer and various carnivorous mammals. The discovery of *Kalimantsia* and related animals suggests that the chalicotheres of Europe can be divided into two families – those with long horse-like heads, and those with the short muzzles.

Features: The skull of *Kalimantsia*, the only part so far found, has a much shorter muzzle than the horse-like shapes of the rest of the group, and is very high and domed at the back. This would give a head that resembled that of one of the boneheaded dinosaurs such as *Pachycephalosaurus*. On other chalicothere genera from Wyoming, USA, domed skulls have been interpreted as being used by males in sparring contests. The teeth of *Kalimantsia* are long and low, and adapted for eating leaves.

Distribution: Bulgaria.
Classification: Perissodactyla, Chalicotheroidea.
Meaning of name: From Kalimants in Bulgaria.
Named by: Geraads, Spassov and Kovachev, 2001.
Time: Late Miocene epoch of the late Tertiary.
Size: 3m (10ft).
Lifestyle: Browser.
Species: *K. bulgarica.*

Left: Kalimantsia had a tall crest along its head. In related animals this was expanded into a dome.

RHINOCEROSES

The rhinoceroses and their relatives the tapirs continued to expand during the late Tertiary. As with the horses they occupied the open plains. However, they did not adopt the lightweight running mode of life to any great extent and retained their three-toed feet. The expansion and success of the rhinoceroses led to the development of all kinds of shapes as adaptations to all kinds of environments.

Teleoceras

This creature was the rhinoceros equivalent of the hippopotamus, and must have spent its time wallowing in mud and water. Some researchers suggest that body shape is not enough to indicate an aquatic lifestyle, but isotope studies on the teeth seem to confirm it, showing that in life they absorbed the isotopes of oxygen that would be abundant in standing water.

Features: *Teleoceras* has the long, barrel-shaped body and stumpy legs that we would associate with a hippopotamus. There was a single small horn on the nose. The build is well known as a whole herd was found in 1980 buried in a volcanic ashfall in Nebraska, USA.

Distribution: Nebraska, Florida, USA.
Classification: Ungulata, Perissodactyla, Ceratomorpha.
Meaning of name: Long one with a horn.
Named by: Hatcher, 1894.
Time: Middle to late Miocene epoch of the late Teriary.
Size: 4m (13ft).
Lifestyle: Aquatic rooter or browser.
Species: *T. meridianum, T. guymonense, T. major, T. medicornutum, T. hicksi, T. proterum, T. fossiger.*

Left: Teleoceras lived in herds where there was conflict between adult males, a social structure more similar to that of the modern hippopotamus than the rhinoceros.

Diceratherium

Diceratherium was a small rhinoceros, about the size of a Shetland pony, which lived in big herds, moving fleet-footed across the Mid-West plains like bison along with vast herds of small gazelle-like camels called *Oxydactylus*. An interesting fossil of this animal was found in 1935 – its body had lain in shallow water that was covered by an advancing lava flow. A hollow shape of the animal was left in the solidified lava, with scraps of teeth and bones inside.

Features: *Diceratherium* takes its name from the two horns mounted on the snout. These are mounted side by side as in the brontotheres of previous times, not one behind the other as in modern rhinoceroses. It is the earliest-known horned rhinoceros.

Left: In an area of Nebraska, USA, the fossils of Diceratherium are so common that the rock sequences in which they are found are named the Diceratherium beds.

Distribution: Nebraska, USA.
Classification: Ungulata, Perissodactyla, Ceratomorpha.
Meaning of name: Two-horned mammal.
Named by: Marsh, 1875.
Time: Miocene epoch of the late Tertiary.
Size: 2m (6½ft).
Lifestyle: Browser.
Species: *D. niobrarense, D. armatum, D. annectens, D. gregorii, D. tridactylum.*

Menoceras

Distribution: Florida, Nebraska, USA.
Classification: Ungulata, Perissodactyla, Ceratomorpha.
Meaning of name: Crescent horn.
Named by: Troxell, 1921.
Time: Miocene epoch of the late Tertiary.
Size: 1.5m (5ft).
Lifestyle: Browser.
Species: *M. arikarensis*, *M. barbouri*.

Of the herd-living rhinoceroses that roamed the plains of the Mid-West of North America in early Miocene times, pig-sized *Menoceras* was one of the smallest. The large number of bones of this animal found in once place at the famous Agate Springs locality in Nebraska, USA, famous for the remains of all kinds of Tertiary mammals indicates that it lived in large herds. Its remains have been found in Florida, USA, and it is the earliest rhinoceros to appear as far south as this. A related rhinoceros, *Floridaceras*, appeared there later in the Miocene.

Features: Like its close relative *Diceratherium*, *Menoceras* has a pair of horns side by side on the snout. As in modern rhinoceroses these would have been formed of keratin and so have not fossilized, but the pair of rough areas on the bones of the snout show where they would have been attached. These horns were carried only by the males and used in sparring.

Left: The slim legs of Menoceras are unusual for a rhinoceros, but it was one of the smallest rhinoceroses of the time.

MODERN RHINOS

The five species of modern rhinoceros that live now represent the end of a long line. They look similar to one another, but belong to several different evolutionary lines that diverged in the middle Oligocene. One line kept to the forests and wooded areas and continued browsing, eating leaves from trees and bushes. This line gives us the Sumatran rhino, which has not changed much from the Miocene forms. The second gave rise to the forest-dwelling one-horned rhinos of Asia. The white rhino of Africa split from the Asian rhino stock in Pliocene times and adapted to a grazing lifestyle on the African plains. It has developed a broad mouth with thick lips and no cutting teeth to allow it to crop grass. The name "white rhino" is a corruption of the term "wide rhino" and refers to the mouth shape.

Above: A Sumatran rhino.

Above: A white rhino.

Right: An Indian rhino.

Miotapirus

The close relatives of the rhinoceroses, the tapirs, appeared early in the Oligocene. They have been a conservative group, with the modern forms being almost identical to those of the past. *Miotapirus* was typical, and it seems to have been quite an adaptable animal, living from sea level to heights of 4,500m (15,000ft).

Features: *Miotapirus* has a short heavy body, short legs and tail, rounded back, a short neck and large head with short flexible trunk that we associate with a modern tapir. Like its modern counterpart, *Miotapirus* was probably nocturnal, rooting about in the thickets and undergrowth of forests, using its trunk to grasp the forest plants.

Distribution: Texas, Nebraska, South Dakota, Wyoming, possibly Oregon, USA.
Classification: Ungulata, Perissodactyla, Tapiroidea.
Meaning of name: Miocene tapir.
Named by: Schlaikjer, 1937.
Time: Early Miocene epoch of the late Tertiary.
Size: 2m (6½ft).
Lifestyle: Rooter and browser.
Species: *M. harrisonensis*, *M. marslandensis*.

Left: From the small Hyracotherium-like ancestors, the tapirs, including Miotapirus, have evolved little in bodily form.

LITOPTERNS

Throughout the late Tertiary South America continued to be an island continent, separated from all other landmasses. Accordingly the animals there evolved independently. In many instances the indigenous animals developed forms similar to those found in the rest of the world in response to similar environmental conditions. The litopterns were the South American equivalents of the horses and camels.

Thoatherium

Convergent evolution was taken to an extreme with *Thoatherium*. While the horses, such as *Pliohippus*, of the northern continents moved into open grasslands and developed body shapes and feeding strategies accordingly, the hoofed litopterns of South America did so too, and developed almost exactly the same shapes. *Thoatherium* was the South American equivalent of the horse.

Features: To look at, we would think this creature was a horse, albeit a small one, with its short body, long running legs, and long neck and head. Like the horse, its toes are reduced to a single hoofed digit, and the rest are atrophied to an even greater extent than in the horses. It is the smallest of the litoptern group. However, the teeth remain simple and so it probably fed on leaves rather than grasses.

Distribution: Argentina.
Classification: Ungulata, Litopterna.
Meaning of name: Swift mammal.
Named by: Ameghino, 1887.
Time: Early Miocene epoch of the late Tertiary.
Size: 70cm (2ft 4in).
Lifestyle: Running browser.
Species: *T. minisculum.*

Left: Long thin legs, powered by muscles close to the shoulders and hips, show that Thoatherium *was a fast-running animal.*

Diadiaphorus

Another horse-like litoptern was *Diadiaphorus*. This was larger than *Thoatherium* and lived a little later, but it had the same running legs. However, it had not completely lost the toes that were used for running. The horse-like head contained quite a large brain for its size.

Features: In horse-like *Diadiaphorus* the lower limb bones – the ulna and radius in the forelimb and the tibia and fibula in the hind – are not fused together as they are in the skeletons of the true horses. Nevertheless this is essentially a running animal with a large middle toe on which the full weight was carried. The toes to each side have atrophied but they are still present. The teeth are low-crowned and adapted for browsing rather than grazing.

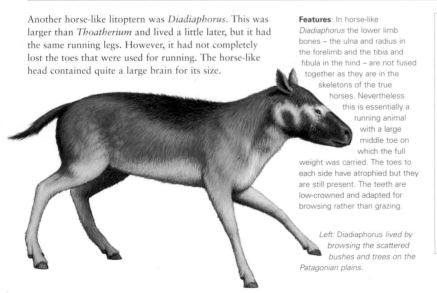

Distribution: Argentina.
Classification: Ungulata, Litopterna.
Meaning of name: One that carries through.
Named by: Ameghino, 1887.
Time: Early Miocene epoch of the late Tertiary.
Size: 1.2 m (4ft).
Lifestyle: Running browser.
Species: *D. majusculus, D. caniadensis, D. zamius.*

Left: Diadiaphorus *lived by browsing the scattered bushes and trees on the Patagonian plains.*

Theosodon

Distribution: Argentina and
Chile.
Classification: Ungulata,
Litopterna.
Meaning of name: God
tooth.
Named by: Ameghino, 1887.
Time: Early Miocene epoch
of the late Tertiary.
Size: 2m (6½ft).
Lifestyle: Running browser.
Species: *T. tontanus,
T. garretorum, T. gracilis,
T. lydekkeri.*

A deer-sized, long-necked animal,
Theosodon resembled a llama or a guanaco
that lives in the same area today.
Its biggest superficial
difference would
have been the
three-toed, rather
than two-toed, feet and its tapir-
like trunk. *Theosodon* and its
relatives were successful right up until
they were replaced by other mammals
that migrated southwards from North
America once the land bridge of
Central America was established.

*Right: Theosodon would have resembled the
modern guanaco except for the heavier feet that
sported three toes.*

Features: The shape of the skull around the nostrils
indicates that in life *Thoesodon* probably had a short
trunk, perhaps like that of a saiga antelope of today –
probably used for the same purpose, to filter and
moisten the dry and dusty air of the plains. The
mouth has a full set of 44 teeth, which is unusual for
such a late placental mammal, and the lower jaw
is narrow.

EVOLUTIONARY WITNESS

Specimens of litopterns were excavated by
Charles Darwin in 1834 while he visited
Patagonia and Argentina on his voyage on *HMS
Beagle. Macrauchenia* in particular struck him as
strange. He noted its similarity to the artiodactyl
llamas and guanacos of the region, but
acknowledged that it must have been more
closely related to the perissodactyls like the
tapirs and rhinoceroses because of the number
of toes. He was mystified that such a large
quadruped herbivore could have existed on the
arid plains of Patagonia, but argued that it must
have had the same adaptations as the modern
guanaco that lived there.

This observation, along with the similarities he
noted between living and fossilized South
American species, helped to concentrate his
mind on how animals could change in order to
live in particular environments. However, it was
decades before
he could put all
this into a
coherent theory.

*Left: The three-
toed foot of
a litoptern.*

Macrauchenia

The macraucheniid line, known as far back
as the early Miocene, reached its climax
with *Macrauchenia* itself, an animal that
appeared to combine the build of a
perissodactyl with the adaptations of an
artiodactyl. It was a tall animal with a long
neck and a trunk, but its feet were more like
those of a rhinoceros. These features
evolved independently of similar features on
animals from other parts of the world.

Distribution: Argentina.
Classification: Ungulata,
Litopterna.
Meaning of name: Long
neck.
Named by: Owen, 1838.
Time: Pliocene epoch of the
late Tertiary to the
Pleistocene epoch of the
Quaternary.
Size: 3m (10ft).
Lifestyle: High browser.
Species: *M. patachoneica.*

*Below: The front limbs are
those of a running animal,
but the hind limbs are not,
and so it is unclear just
how it moved.*

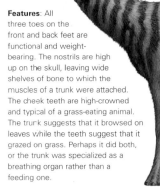

Features: All
three toes on the
front and back feet are
functional and weight-
bearing. The nostrils are high
up on the skull, leaving wide
shelves of bone to which the
muscles of a trunk were attached.
The cheek teeth are high-crowned
and typical of a grass-eating animal.
The trunk suggests that it browsed on
leaves while the teeth suggest that it
grazed on grass. Perhaps it did both,
or the trunk was specialized as a
breathing organ rather than a
feeding one.

NOTOUNGULATES AND OTHERS

If the litopterns were the South American equivalents of the light-footed running mammals, then the notoungulates, literally "the southern hoofed animals", were the non-running versions. They ranged from small rodent-like animals to heavy hippopotamus-sized beasts. One or two made it to North America in early Tertiary times, but they did not last long there. South America was their bastion.

Protypotherium

Small and rabbit-like, *Protypotherium* was a typical representative of the interatheriid family – the group that resembled rodents – of the notoungulates. They were a long-lived rodent-like group, surviving from the Palaeocene and not becoming extinct until the late Miocene.

Features: The rodent-like skull of *Protypotherium* has a rat-like snout, but the teeth are completely unspecialized. The neck is short and the tail long, and the small rabbit-like body is carried on long legs, furnished with stout claws. These were probably used for digging in the ground or for fighting with other animals for control over their burrows. It would have been a leaf-eater, but may have scavenged carrion as well.

Left: A rabbit-like body and rat-like head and tail – that is what Protypotherium *must have looked like.*

Distribution: Argentina.
Classification: Ungulata, Notoungulata.
Meaning of name: Early type of mammal.
Named by: Ameghino, 1882.
Time: Early Miocene epoch of the late Tertiary.
Size: 40cm (16in).
Lifestyle: Browser or scavenger.
Species: *P. australe, P. minitum, P. anomopodus.*

Homalodotherium

The claw-toed chalicotheres of the northern continent were a strange group of perissodactyls – horse-like but with claws instead of hooves. Strange though they were, even they had their South American equivalents, in the homalodothere group of the notoungulates. *Homalodotherium* is the only well-known genus of this group.

Features: *Homalodotherium* has longer front than hind legs. The hind feet are plantigrade, with the whole foot on the ground, but the front legs are digitigrade, supported on the tips of its fingers. On the front feet the hooves are replaced by strong curved claws. The front limbs are very versatile, with anchor points for strong muscles and flexible wrist joints. It may have been able to stand on its hind legs to feed, pulling down leaves and twigs with its claws. As in most notoungulates, the teeth of *Homalodotherium* are primitive and unspecialized.

Left: Homalodotherium *was the largest of the Miocene South American mammals, although it was small compared with those on other continents.*

Distribution: Argentina.
Classification: Ungulata, Notoungulata.
Meaning of name: Even-toothed mammal.
Named by: Riggs, 1930.
Time: Early and middle Miocene epochs of the late Tertiary.
Size: 2m (6½ft).
Lifestyle: High browser.
Species: *H. cunninghami, H. crassum, H. excursum, H. segoviae.*

Trigodon

Distribution: Argentina.
Classification: Ungulata, Notoungulata.
Meaning of name: Three-part tooth.
Named by: Ameghino, 1882.
Time: Pliocene epoch of the late Tertiary.
Size: 2.7m (9ft).
Lifestyle: Slow-moving grazer.
Species: *T. gaudryi*.

The toxodonts were rather rhinoceros-like or hippopotamus-like notoungulates. They had high-crowned teeth and so they were adapted to grazing on the spreading pampas of Miocene South America. *Toxodon* itself, a hippopotamus-sized animal, had a broad mouth and a flexible lip for pulling in grasses. *Adinotherium* was sheep sized and more elegant. *Trigodon* was horned like a rhinoceros.

Features: Along with *Toxodon*, *Trigodon* is the biggest of the notoungulates. There are four toes on each foot, spread to take the great weight of the animal. The head is deep, and surmounted by a horn above the eyes. Presumably this horn was used like that of a rhinoceros for defence. The main enemies of the time would have been giant flightless birds and carnivorous marsupials.

Below: In late Tertiary times Trigodon *was the South American equivalent of the rhinoceroses – the heavy horned herbivores.*

A DIVERSE GROUP

The notoungulates were the most diverse and abundant of the South American mammals, comprising more than 150 genera in about 13 families – far too many to be illustrated in a representative selection. They are all related as shown by the structure of the ear and the tooth pattern – with a distinctive crown on the upper molars. They can be divided into two main groups – the heavy Toxondontia and the more lightweight Typotheria. The former include:
Adinotherium – like *Trigodon* but sheep-sized and with a much smaller horn.
Nesodon – with a big head and huge grass-grinding teeth.
Toxodon – like a hippopotamus.
Mixotoxodon – the only notoungulate to leave South America, living in Guatemala in Central America.
Among the Typotheria were:
Eohyrax – like a hyrax
Hegethotherium – even more like a rabbit than *Protoypotherium* was.

Astrapotherium

The astrapotheres were closely related to the notoungulates. Like the pyrotheres before them, they were elephant-like, with their big heads, their prehensile trunks and their strong tusks. However, the neck was not shortened as we usually see in animals with heavy trunks.

Features: The skeleton of *Astrapotherium* has a long neck, short legs and small feet. The nasal bones are very high, indicating that in life it supported an elephant-like trunk. The canine teeth on the upper jaw are enormous, and work with a shearing action against those on the lower jaw, just like those of a hippopotamus. There are no upper incisors, and the lower ones stick forward and act against the bony pad on the top of the jaws as in the grass-gathering mechanism of modern cows.

Distribution: Argentina.
Classification: Ungulata, Astrapotheria.
Meaning of name: Star mammal.
Named: Ameghino, 1887.
Time: Early Miocene epoch of the late Tertiary.
Size: 2.5m (8ft).
Lifestyle: Semi-aquatic rooter.
Species: *A. magnum, A. granda*.

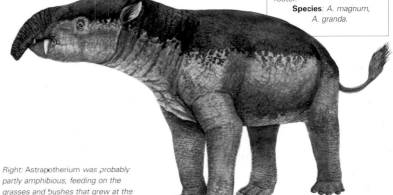

Right: Astrapotherium *was probably partly amphibious, feeding on the grasses and bushes that grew at the water side.*

CAMELS

The heyday of the camels came with the spreading grasslands in the late Tertiary. They expanded into different groups on the North American plains, and at the Tertiary/Quaternary boundary they crossed land bridges west into Asia and Africa and south into South America. Nowadays they are extinct in North America and exist as specialized desert and mountain types on other continents.

Procamelus

Below: Procamelus has all the features of a modern camel, from the shape of the skull and arrangement of teeth, to the structure of the foot bones.

This is the first genus that can be really recognized as a camel. At this time the characteristic pacing movement of the camel developed – with both legs at the same side moving at the same time. This gives a rather ungainly gait, but it prevents the very long legs from becoming entangled with one another at speed.

Features: The skull is long, like that of a modern camel. The working incisors are restricted to the lower jaw, working against a pad on the upper. Behind these, on both jaws, is a widely spaced set of little teeth, until the batteries of tall grass-grinding cheek teeth at the back. The wrist bones of the two functioning toes are fused into a single canon bone, and the toe tips are splayed, with weight-bearing pads on the tips. In other words *Procamelus* is essentially a modern camel.

Distribution: Montana, North Dakota, Idaho, USA.
Classification: Ungulata, Artiodactyla, Tylopoda.
Meaning of name: Early camel.
Named by: Leidy, 1858.
Time: Late Miocene epoch of the late Tertiary.
Size: 1.5m (5ft).
Lifestyle: Running grazer.
Species: *P. occidentalis*, *P. coartatus*, *P. grandis*, *P. gracilis*, *P. leptocolon*, *P. gazini*.

Aepycamelus

Below: Like the modern giraffe, Aepycamelus was able to browse the branches that were way above the height of the other animals of the time.

Also referred to as *Alticamelus*, *Aepycamelus* belongs to a Miocene group that are known as the "giraffe camels". These lived at the same time and place as other members of the camel group – all adapted to different feeding strategies in the environment. The giraffe camels branched away from the main camel line back in Oligocene times. In the early Miocene they included a very common form called *Oxydactylus* which had long legs and a long S-shaped neck. The group reached its climax in the tall and elegant *Aepycamelus* later in the epoch.

Features: As the nickname of the group suggests *Aepycamelus* has particularly long legs and a very long neck, and has a body shape that is adapted to a lifestyle like that of modern giraffes, pacing across the grasslands from one stand of tall trees to the next. Unlike other contemporary camels it has upper incisor teeth at the front, and low-crowned cheek teeth, indicating that it fed on leaves rather than grasses.

Distribution: Colorado, Nevada, Texas, Nebraska, Montana, Florida, California, USA.
Classification: Ungulata, Artiodactyla, Tylopoda.
Meaning of name: Tall camel.
Named by: Leidy, 1886.
Time: Middle to late Miocene epoch of the late Tertiary.
Size: 5.5m (18ft).
Lifestyle: High browser.
Species: *A. major*, *A. giraffinus*, *A. alexandrae*, *A. bradyi*, *A. elrodi*, *A. priscus*, *A. proceras*, *A. robustus*, *A. stocki*.

Titanotylopus

Distribution: Nebraska, USA; possibly Ukraine.
Classification: Ungulata, Artiodactyla, Tylopoda.
Meaning of name: Giant knobbly foot.
Named by: Barbour and Schultz, 1934.
Time: Pliocene epoch of the late Tertiary to Pleistocene epoch of the Quaternary.
Size: 4m (13ft).
Lifestyle: Grazer.
Species: T. nebraskenis, T. spatulus.

At the end of the Pliocene there was another evolutionary burst amongst the camels, and several new genera came on to the scene. Some of these migrated southwards and became the modern llamas and their relatives in South America. Another group became truly massive and were the giants of the camel line. They lived in the northern reaches of the North American continent and migrated, via the landmass that bridged the Bering Strait, into northern Asia. A Ukrainian genus *Gigantotylopus* is thought by some palaeontologists to be a species of *Titanotylopus*.

Features: This gigantic camel, standing 3.5m (11½ft) tall at the shoulder and weighing about a tonne, has long massive limbs, a small braincase and tall spines on the backbone indicating the presence of a hump. The male has a more robust skull and longer canine teeth that the female. It was the biggest animal around, taller than the elephants that lived in the same area at that time.

Right: The broad feet of the camels, including Titanotylopus, *may counteract the instability brought about by the pacing form of locomotion.*

AGATE FOSSIL BEDS

The best place for fossil camels is the Agate Fossil Beds National Monument in Nebraska. In Miocene times this was an area of seasonal rainfall and grassland. The nutrition of the soil was constantly replenished by volcanic ash falling from eruptions taking place in the newly arisen Rocky Mountains to the west. The area was subjected to periods of drought, and when this happened the animal life of the area congregated around dwindling water holes. During serious dry spells they perished here and their bodies were eventually covered by the steady fall of ash, and fossilized.

As well as the various forms of camels, the area has remains of rhinoceroses such as *Menoceras*, chalicotheres like *Moropus*, entelodonts like *Dinohyus* and many smaller animals.

A common geological structure is the "devil's corkscrew" – a vertical twisted column cutting through the strata and representing a filled burrow. The animal that formed it was a beaver-like rodent called *Palaeocastor*.

Stenomylus

Some of the Miocene camels were small and graceful, resembling antelope, and moving about on the North American plains in herds. Skeletons of dozens of *Stenomylus* have been found in the Agate Fossil Beds National Monument in Nebraska, seemingly all killed suddenly and buried quickly.

Below: In appearance and habit, Stenomylus *resembled a modern antelope.*

Features: The structure of the backbone, and the musculature that would have been attached, shows that *Stenomylus* would have moved by high speed running and short, choppy leaps. The toes are not splayed as in other camels, and in life they were furnished with hard hooves for sure-footed movement on hard rough terrain.

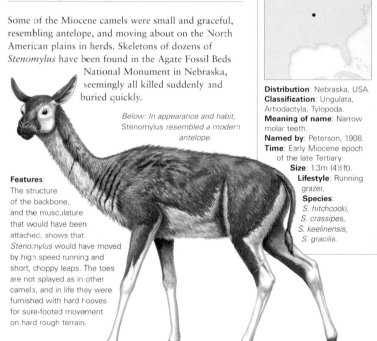

Distribution: Nebraska, USA.
Classification: Ungulata, Artiodactyla, Tylopoda.
Meaning of name: Narrow molar teeth.
Named by: Peterson, 1908.
Time: Early Miocene epoch of the late Tertiary.
Size: 1.3m (4½ft).
Lifestyle: Running grazer.
Species: S. hitchcocki, S. crassipes, S. keelinensis, S. gracilis.

PROCERATIDS

The proceratids represent a group of deer-sized artiodactyls that existed from late Eocene to early Pliocene, mostly in the southern parts of the North American continent. Since their discovery in the 1890s they were thought to have been closely related to the camels, but work done in the 1990s cast doubt on this. It is not now clear to what branch of the artiodactyls they were related.

Protoceras

Othniel Charles Marsh established this proceratid group on the basis of his discovery of *Protoceras* in 1891, and since then about 20 genera have been recognized. *Protoceras* was one of the earliest and most primitive. It probably lived in the wooded hills above the plains.

Features: *Protoceras* is an early member of the group and, as such, has many primitive features. The earliest forms still had incisor teeth in the front of the upper jaw but these had been lost by the time *Protoceras* evolved. Each foot of *Protoceras* has four toes unlike the later members which had two toes. The most obvious feature, however, is the head ornamentation, which consists of three pairs of bony knobs – a pair over the nose, a pair over the eyes and a pair at the rear. In life these were probably covered in skin like the ossicones of a giraffe.

Left: Male Protoceras *had the full three pairs of horns. Females only had the topmost pair.*

Distribution: South Dakota, USA.
Classification: Ungulata, Artiodactyla.
Meaning of name: First horn.
Named by: Marsh, 1891.
Time: Late Oligocene epoch of the early Tertiary to the early Miocene of the late Tertiary.
Size: 1m (about 3ft).
Lifestyle: Browser or grazer.
Species: *P. celer*.

Syndioceras

Syndioceras was a more deer-like animal than its predecessor *Protoceras*. As forest dwellers *Syndioceras* and the advanced protoceratids may have resembled the forest deer and antelope of today, such as the bongo, rather than the swift-footed grassland gazelles. They now had two toes on each foot, but the ankle bones were not fused into a canon bone.

Features: The head ornamentation of *Syndioceras* consists of a pair of horns on the snout and another above the eyes. Those on the snout grow from a single base but then diverge and curl back. Those above the eyes curve upwards and inwards like those of cattle. They are well developed in males but very much reduced in females, indicating that they were used in sexual display. This indicates that they were woodland animals, as the males of many medium-sized ungulates in forest habitats today have horns for this purpose.

Right: Like the horns of the giraffe, those of Syndioceras *and the other proceratids would have been covered in skin.*

Distribution: Nebraska, USA.
Classification: Ungulata, Artiodactyla.
Meaning of name: Together horn.
Named by: Barbour, 1905.
Time: Early Miocene epoch of the late Tertiary.
Size: 1.5m (5ft).
Lifestyle: Browser or grazer.
Species: *S. cooki*.

Synthetoceras

Distribution: Texas, USA.
Classification: Ungulata, Artiodactyla.
Meaning of name: Combined horn.
Named by: Stirton, 1932.
Time: Late Miocene to early Pliocene epochs of the late Tertiary.
Size: 2m (6½ ft).
Lifestyle: Grazer.
Species: S. tricornatus.

Although the stubby horns of the early protoceratids, such as *Protoceras*, were arranged so that they would be seen best from the side, those of the later forms, such as *Syndioceras* and *Synthetoceras*, with their sideways sweeps and their diverging prongs, would have most impact when viewed from the front. This may indicate the development of a more aggressive display strategy. The tall grinding teeth indicate that *Synthetoceras* was a grazer, feeding on grass, rather than a browser like the earlier forms.

Right: Synthetoceras *was probably a herd-living animal, with the males using the horn arrangement for displays of dominance.*

Features: The skull of *Synthetoceras* is particularly long and narrow. Like *Syndioceras*, it has a pair of horns above the eyes and ornamentation on the nose. The horns above the eyes are curved and cow-like, the same as those of *Syndioceras*, but the structure on the nose consists of a very large Y-shaped horn that curves forwards and upwards and spreads into two prongs at the tip.

OTHER PROTOCERATIDS

Paratoceras – from the early Miocene of southern Mexico – the southernmost found of the group.
Prosynthetoceras – from Florida where it was found in association with a rich subtropical flora indicating that it was a forest dweller.
Lambdoceras – like *Synthetoceras*, mostly from Texas but also found further north.

Right: Paratoceras.

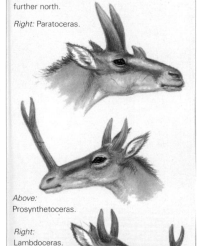

Above: Prosynthetoceras.

Right: Lambdoceras.

Kyptoceras

One of the last and the largest of the protoceratids was *Kyptoceras*. Being restricted to the southern states, its distribution suggests that towards the end of their existence they were driven south as the grasslands expanded in the north.

Features: The cheek teeth are much higher than those of other *Protoceratids*, indicating that they ate tougher food. They probably inhabited the margins of the forests and grazed from the surrounding grasslands as well as browsing from the trees and bushes. The head ornamentation consists of a pair of inward curving horns on the snout and a huge pair of forward curving horns, up to half a metre long, over the eyes.

Distribution: Florida, North Carolina, USA.
Classification: Ungulata, Artiodactyla.
Meaning of name: Curved horn.
Named by: Webb, 1981.
Time: Pliocene epoch of the late Tertiary.
Size: 2m (6½ ft).
Lifestyle: Browser or grazer.
Species: K. amatorum.

Right: The remains of Kyptoceras *were found by fossil collector Frank Garcia, who expressed a "love" of fossils. It was given the species name* amatorum, *meaning "lover", in honour of this.*

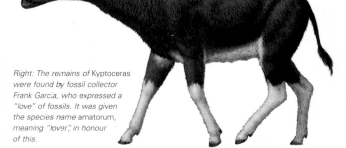

GIRAFFES

Giraffes, like the camels, represent a group that is relatively unimportant today but have had a distinguished history throughout the Age of Mammals. Modern representatives consist of the characteristic long-necked giraffe of the African grasslands, of which there is a single species and several subspecies, and the single more compact species of okapi of the African rainforests.

Giraffokeryx

Below: The early discovery of horned skulls of Giraffokeryx may have given rise to legends of dragons among the people of northern India and Pakistan.

Cattle, deer and giraffes are included in the broad group, the pecorans. Their resemblance is in the way they process their food, by chewing the cud. They also share the feature of paired ornaments on their heads, be they horns, antlers or ossicones – skin-covered bony structures. The pecorans also have a highly modified foot structure, in which the joints are simplified by the fusion of some of the leg and foot bones. This restricts the foot movement and turns the limbs into fast running structures. The palaeotragines were the earliest of the giraffe line among the pecorans.

Features: *Giraffokeryx* resembles the modern okapi, except that the skull has a quite different arrangement of ossicones. There are two well-developed pairs, projecting upwards and sideways and curving back. Those on top of the head are much larger than those above the eyes. Like the okapi, to which it is closely related, *Giraffokeryx* would have spent its time in deep woodlands, browsing from bushes.

Distribution: Border of India and Pakistan.
Classification: Ungulata, Artiodactyla, Ruminantia.
Meaning of name: Giraffe and Keryx – an ancient Greek priest.
Named by: Pilgrim, 1910.
Time: Middle Miocene epoch of the late Tertiary.
Size: 2m (6½ft).
Lifestyle: Forest browser.
Species: *G. punjabienses*.

Climacoceras

Below: The colour scheme of Climacoceras, as well as that of the other mammals, is entirely hypothetical, and based on that of modern animals from similar environments.

Climacoceras looked like an okapi with the horns of a deer. The hooves, the tail and the general skeletal structure show that it is an early giraffe. What's more, the resemblance in the horns is superficial. Like modern giraffes it probably lived in small herds, with males developing bigger horns for display and aggression. A recent theory about the evolution of the giraffe's long neck suggests that it might be a display structure with which males contest against each other for the females.

Features: At first glance this animal would appear to be a deer, with a magnificent set of antlers. However, there is no sign of a burr, where an antler would begin to grow seasonally and subsequently drop off at the end of the breeding period. The branched outgrowths of *Climacoceras* are not antlers but true ossicones, permanent bony structures on the animal's head.

Distribution: East Africa.
Classification: Ungulata, Artiodactyla, Ruminantia.
Meaning of name: Ladder horn.
Named by: MacInnes, 1936.
Time: Early Miocene epoch of the late Tertiary.
Size: 2m (6½ft).
Lifestyle: Browser.
Species: *C. africanus, C. gentryi*.

Prolybitherium

Distribution: Libya; Egypt.
Classification: Ungulata, Artiodactyla, Fuminantia.
Meaning of name: Early mammal from Libya.
Named by: Arambourg, 1961.
Time: Early Miocene epoch of the late Tertiary.
Size: 1.5m (5ft).
Lifestyle Forest browser.
Species: *P. magnieri*.

Before the Quaternary, the northern reaches of the Sahara Desert were not the arid wastes known today. There were thick forests in these regions inhabited by typical forest-type animals. The largest forest browsers were primitive giraffes that resembled the modern okapi. *Prolybitherium* was one of these, and, like the okapi, it probably had a solitary existence. The broad flat ossicones may have been used in sparring. Rival males could have encountered each other head-on, and pushed until one or the other gave way. It is possible that the skin covering was shed every year.

Features: Low flat horns are usually the sign of a forest-living animal. Look at the modern duikers and serows. Such structures are less likely to become entangled in branches and thick vegetation than more flamboyant structures. The ossicones of *Prolybitherium* consist of broad flat plates, reaching a span of about 14cm (5½ in).

Right: Apart from the unique head ornamentation, Prolybitherium *would have looked very much like the modern okapi.*

Sivatherium

Apart from the modern long-necked giraffe of the African plains, perhaps the most spectacular of the giraffe line was *Sivatherium*, looking more like a moose than a giraffe. It lived alongside the modern giraffe and many other members of the modern fauna in the late Tertiary and the Quaternary.

Features: The moose-like appearance of *Sivatherium* is due to its heavy body, its stoutly muscled forequarters, its powerful legs and its thick broad trunk-like snout. The big head carries a small pair of conical ossicones above the eyes, and another massive pair that sweep outwards from the top of the head. The cheek teeth are high and strong, indicating that, unlike the rest of the primitive giraffes, *Sivatherium* was a grass-eater.

Distribution: India; Morocco; Kenya; South Africa.
Classification: Ungulata, Artiodactyla, Ruminantia.
Meaning of name: Siva's mammal (from the Hindu god of beasts).
Named by: Falconer and Cautley, 1832.
Time: Late Pliocene epoch of the late Tertiary to Pleistocene epoch of the Quaternary.
Size: 4m (13ft).
Lifestyle: Grazer.
Species: *S. maurusium*, *S. giganteum*, *S. hendeyi*.

Left: Massive Sivatherium *was one of the most widespread of the fossil giraffes.*

DEER

The cervids, or the deer, are the even-toed ungulates that have antlers. Antlers differ from other horny growths by the fact that they are shed every year and grow again. They can be distinguished by a bony lump – a burr – at the base representing the point at which they are shed. The deer were quite late to evolve, but have gone on to become the principal browsers of the Northern Hemisphere.

Dicrocerus

The first true deer seem to have evolved in Europe and then spread to the rest of the northern continents. The elegant little animal *Dicrocerus* is regarded as the first true deer known. It is the first appearance in the fossil record of antlers that were shed seasonally.

Features: The antlers are short, simply branched, with a long tine sweeping backwards and a shorter one pointing to the front, and carried on long pedicles with an oval section. The antlers are thinner and sweep backwards more in the juveniles. The teeth show a greater development of enamel than in earlier artiodactyls and the structure is more convoluted, indicating an adaptation to tough foods like grass. There is still a pair of distinctive canines on the upper jaw.

Left: Dicrocerus *is the first ruminant known with deciduous antlers – those that could be shed and regrown every year.*

Distribution: Europe; Asia.
Classification: Ungulata, Artiodactyla, Ruminantia.
Meaning of name: Two horned.
Named by: Lartet, 1837.
Time: Early Miocene epoch of the late Tertiary.
Size: 60cm (2ft).
Lifestyle: Browser or grazer.
Species: *D. elegans, D. teres, D. grangeri.*

Procranioceras

Features: The woodland-dwelling dromomerycids are larger than the later grass-eating types, and *Procranioceras* is a member of the larger group. Its teeth are low-crowned, as befits a leaf-eater. In build it resembles any deer, but the head is distinctive – having a simply branched pair of antlers growing from a very long pair of pedicles above the eyes, and a single one curving upwards from the back of the skull.

During the Miocene there was a widespread change of climate and vegetation across the Northern Hemisphere. The climate dried and the forests gave way to grasslands. This change is marked in the change of animal life and can be plotted in the dromomerycid family of primitive deer. Early forms such as *Drepanomeryx* were definitely leaf-eaters, while later types like *Cranioceras* were both leaf- and grass-eaters. *Procranioceras* was one of the last of the specialized leaf-eaters.

Left: Procranioceras *is perhaps unique in being a deer with three horns – a pair in the usual position and a central one sweeping backwards.*

Distribution: Nebraska, USA.
Classification: Ungulata, Artiodactyla, Ruminantia.
Meaning of name: Before the horned skull.
Named by: Frick, 1937.
Time: Middle Miocene epoch of the late Tertiary.
Size: 1.5m (5ft).
Lifestyle: Running browser.
Species: *P. skinneri.*

Euprox

One of the earliest of the cervids was *Euprox*, and it was related to the modern mutjac. Like the modern equivalent it was probably a shy little animal, living a solitary existence except in the mating season, scampering through the forest undergrowth and eating leaves and ground plants. It was abundant throughout middle Miocene Europe. As in modern deer it would only have been the males that supported the antlers, the females being bare-headed.

Below: The canine teeth would have been used both for display and for digging in the forest soil for food.

Features: The pedicles – the bony stumps on which the antlers grew – are long, occupying about half the length of the fully grown antler. There is a pair of long canine tusks on the upper jaw, particularly prominent in the male. These would have been used in combat and display.

Distribution: Europe; China.
Classification: Ungulata, Artiodactyla, Ruminantia.
Meaning of name: Truly nearest.
Named by: Kaup, 1839.
Time: Middle Miocene epoch of the late Tertiary.
Size: 45cm (18in).
Lifestyle: Forest browser.
Species: *E. dicranocerus, E. furcatus, E. robustus.*

Eucladoceros

The deer *Eucladoceros* appeared in the late Pliocene and existed well into the succeeding Pleistocene ice age, when, as the species *E. ctenoides*, it became the most common deer species in certain areas of central France. Several deer species shared their grassy domain with herds of bison that were the most significant grass-eaters of the time.

Below: Eucladoceros had the most flamboyant and intricate set of antlers known of any deer.

Features: *Eucladoceros* is a conventional looking deer, but with an amazing pair of antlers in the appropriate season. Each one has up to 12 tines, and the branches seem to be irregular, following no observable pattern. On the open plains where they lived, such a head ornament would have been visible from a long way away, and used for display over long distances.

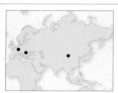

Distribution: Europe; mid-latitude Asia.
Classification: Ungulata, Artiodactyla, Ruminantia.
Meaning of name: Well-branched horn.
Named by: Falconer, 1868.
Time: Pliocene epoch of the late Tertiary to Pleistocene epoch of the Quaternary.
Size: 2.5m (8ft).
Lifestyle: Grazer.
Species: *E. dicranios, E. tetraceros, E. ctenoides.*

A RUMINANT'S STOMACH

The adaptation to grass-eating as seen in the ruminants lies in the structure of the stomach. In other herbivorous mammals the stomach serves to mash up the food before transferring it to the intestines where the nutrients are absorbed. In a ruminant the stomach is divided into chambers, the first of which contains symbiotic bacteria. These work on the cellulose of grasses, breaking it down into fatty acids that can be absorbed by the animal. After this breakdown, the remaining components pass into the intestines where they deliver their nutrients in the normal way.

The treatment of the cellulose takes some time, and often the half-digested material is returned to the mouth for further chewing, to break it down into even smaller pieces for the bacteria to process. The resulting action is known as "chewing the cud" and gives the animal a rather relaxed and reflective appearance. This is the origin of the term "ruminant" which is from the Greek, meaning "thoughtful".

PRONGHORNS AND CATTLE

*The antilocaprids represent a North American family of artiodactyls that appeared in the early Miocene,
flourished during the middle and late Miocene, and declined during the Pliocene and Pleistocene leaving
only the pronghorn antelope of today. The bovids are closely related and represent the cows in the
modern landscape – probably the most important grass-eaters of today.*

Merycodus

The antilocaprids differ from the deer in that their horns
have bony cores that are never shed. The horny
coverings that form the antler-like sheath are shed
annually, like the antlers of deer. *Merycodus* was
one of the first of the group to appear.

Features: With its short
horns, paired and branched
in a Y-shape, *Merycodus*
resembles the modern
pronghorn, except it is only
about a quarter of the
size. As in the
modern pronghorn it
kept the bony cores of
the horn but grew new horny
sheaths every year. Between the
bony core and the horny sheath
there is a layer of hair. The rest
of the body seems to be
intermediate between the deer
and the pronghorns.

Left: Little Merycodus *may have
been the ancestor of the modern
pronghorn. All the physical
features are there, but on a much
smaller scale.*

Distribution: Nevada,
Colorado, California,
Nebraska, USA; Mexico.
Classification: Ungulata,
Artiodactyla, Ruminantia.
Meaning of name: Like an
antelope.
Named by: Leidy, 1854.
Time: Miocene epoch of the
late Tertiary.
Size: 80cm (32ft).
Lifestyle: Browser or grazer.
Species: *M. nevadensis*.

Tsaidamotherium

Tsaidamotherium seems to be related to the modern musk-
ox, which is itself intermediate between the goats and the
cattle. Unlike the modern forms, the head ornamentation
was distinctly asymmetrical. Presumably it was used for
display over the wide Asian plains on which it lived.

Features: The remarkable feature
of *Tsaidamotherium* is the
arrangement of the horns. From
the usual pair of conical hollow-
cored horns found in the cattle,
one of the pair – the right – has
developed into a broad conical
structure and has taken a central
place on the top of
the head like a
unicorn. The left
horn, meanwhile,
has atrophied and
is present but very
small.

Distribution: Mongolia.
Classification: Ungulata,
Artiodactyla, Ruminantia.
Meaning of name: Mammal
from Tsaida.
Named by: Bohlin, 1935.
Time: Late Miocene epoch of
the late Tertiary.
Size: 2m (6½ft).
Lifestyle: Grazer.
Species: *T. hedini*.

Left: Tsaidamotherium *must
have resembled a unicorn, with
the prominent central horn on
its forehead.*

Hoplitomeryx

Distribution: Southern Italy.
Classification: Ungulata, Artiodactyla, Ruminantia.
Meaning of name: Helmeted antelope.
Named by: Leinders, 1983.
Time: Late Miocene epoch of the late Tertiary.
Size: 1m (about 3ft).
Lifestyle: Browser or grazer of rocky outcrop plants.
Species: *H. matthei*.

In the early Miocene, the southern part of Italy was cut off from the lands round about and formed an island. The mammal life there evolved in isolation into strange forms. These included the giant hedgehog *Deinogalerix*, and the peculiar horned ruminant *Hoplitomeryx*. It was so specialized that there is uncertainty as to what branch of the ruminants it belonged.

Right: It is thought that the horns of Hoplitomeryx *evolved to fight off* Garganoaetus, *an eagle-sized bird of prey that lived in the area.*

Features: The feet of *Hoplitomeryx* are not adapted to running, and it probably spent its time goat-like among the limestone crags of the region. It has two sabre-like canine teeth in its upper jaw. Its head ornamentation is spectacular, with a single brow horn and two pairs of smaller horns behind it. It probably evolved from hornless ancestors that were stranded in the area as the sea rose and the island was cut off.

THE RANGE OF PRONGHORNS

In Miocene and Pliocene times there was a vast range of pronghorns, each with a different head ornament and, presumably, adaptations to slightly different niches.

Hayoceros – with a straight horn pointing upwards and a branched tine reaching forwards.
Ilingoceros – with a pair of horns, each comprising two branches twisted tightly together.
Osbornoceros – with a pair of slightly twisted unbranched horns.
Ramoceros – carrying a pair of broad hand-like horns.

Hayoceros.

Illingoceros.

Ramoceros

Osbornoceros.

Hexameryx

In the latter part of the Tertiary *Hexameryx* was one of the vast range of the pronghorns that roamed North America. The variety of horn structures would have been used to distinguish species from one another as they roamed the plains in herds. Rarely were the horns as complex as those found on *Hexameryx*.

Features: As the name suggests, the distinctive feature of *Hexameryx* is the presence of three pairs of horns on the head – one pair straight up, another pair curving forward and up, and the third curving backwards and inwards. Apart from that the body of *Hexameryx* is typical of the group.

Right: Science acknowledges two species of Hexameryx – H. elmori *and* H. simpsoni. *However, the differences may be sexual, and one may be the male and the other the female of the same species.*

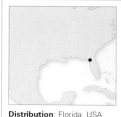

Distribution: Florida, USA.
Classification: Ungulata, Artiodactyla, Ruminantia.
Meaning of name: Antelope with six.
Named by: White, 1941.
Time: Late Miocene epoch of the late Tertiary.
Size: 1.2m (4ft).
Lifestyle: Grazer.
Species: *H. elmori*, *H. simpsoni*.

ELEPHANTS

By the late Tertiary, the elephants had developed into forms that we would recognize today – big-bodied animals on columnar legs, with short necks and big heads furnished with tusks and a trunk. The arrangements of tusks varied, and indicated a variety of feeding strategies and lifestyles from scraping in mud to browsing from high trees.

Gomphotherium

The mastodons were a group of elephants characterized by the shape of the teeth – having cusps that were rounded and conical. The earliest are known from Africa and India in the early Miocene, but by mid Miocene times they had migrated to North America.

Features: This large elephant, as big as a modern Indian elephant, has four tusks – a pair extending from the lower jaw straight forward, and a pair on the upper jaw, curved slightly outward. Although there is no direct evidence for a trunk, the head is long and low and the neck is so short that the head would not have been able to reach the ground, and a trunk would have been necessary to work along with the tusks for feeding.

Left: Gomphotherium is the earliest-known mastodon from North America.

Distribution: Japan; France; Kenya; Pakistan; Nebraska, USA.
Classification: Ungulata, Probiscidea.
Meaning of name: Bolted mammal.
Named by: Cuvier, 1806.
Time: Early Miocene to early Pliocene epoch of the late Tertiary.
Size: 4m (13ft).
Lifestyle: Rooter and browser.
Species: *G. angustidens*, and about a dozen others.

Deinotherium

The deinotheres were a side branch of the elephant line. Diverging from the main evolutionary sequence in early Miocene times, they quickly reached a very large size and were a very successful group, becoming the contemporaries of primitive humans before they became extinct. The purpose of the distinctive tusks is still unclear.

Features: The obvious difference between *Deinotherium* and the other elephants is the fact that the tusks are confined to the lower jaw and curve downwards and backwards, anchored by the two lower jaw bones that are strongly bound together. Apart from this the skull is very short and the brain capacity small. The nasal opening is large and situated well up on the skull, indicating that it had a big strong trunk.

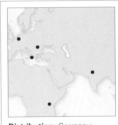

Above: The unusual nature of the tusks prompted Victorian scientists to reconstruct Deinotherium with its jaw upside down.

Distribution: Germany; Bohemia; Crete; India; Kenya.
Classification: Ungulata, Probiscidea.
Meaning of name: Terrible mammal.
Named by: Kaup, 1829.
Time: Miocene epoch of the late Tertiary to Pleistocene epoch of the Quaternary.
Size: 5m (16ft).
Lifestyle: Rooter.
Species: *D. giganteum*.

Amebelodon

The grassy plains that spread across central North America during the Miocene were crossed by winding rivers in which luxuriant water plants flourished. These were exploited by all kinds of specialist animals including the shovel-tusked elephants. *Amebelodon* was typical of these. A close relative from Europe and Asia was *Platybelodon* – very similar but with a shorter and broader shovel jaw.

Features: The tusks of the lower jaw of *Amebelodon* are elongated, flattened and bound together in a shovel structure. This would have dug into the soft muds of the river banks, severing the roots of the water plants which would then have been drawn up the shovel by the broad muscular trunk. The shovel tuskers were so specialized for a particular way of life that they could not survive when conditions changed.

Distribution: Colorado, Nebraska, USA.
Classification: Ungulata, Probiscidea.
Meaning of name: Shovel tooth.
Named by: Barbour, 1927.
Time: Late Miocene epoch of the late Tertiary.
Size: 4m (13ft)
Lifestyle: Semi-aquatic rooter.
Species: *A. britti*, *A. floridarus*.

Right: Wear on the lower tusks suggests that the shovel structure was used for scraping bark from trees as well as for dredging in mud.

THE ELEPHANT'S TRUNK

The trunk of an elephant is one of the most versatile tools in the animal kingdom. It is used as a hand, for collecting food from the ground or from high in a tree; as a suction pump for drawing up drinking water and transferring it to its mouth, or washing water and spraying it over itself. It is also used for dispensing dust as a guard against flies, and as a sounding box for varying the timbre of the voice. Finally, it is used as a snorkel when swimming.

The trunk is formed from the muscles of the nose and upper lip, and consists of two tubes, which represent the nostrils. These are surrounded by muscles – about 40,000 of them. At the tip there are structures like fingers – two on the modern African elephant and one on the Indian elephant. Since the trunk is all muscle, it leaves no direct fossils. We can deduce from muscle scars on the skull whether or not a fossil elephant had a trunk, but we cannot be certain about the details.

The trunk may have evolved as a snorkel when the early elephants were partly aquatic. This is in keeping with the presence of relict aquatic structures in the elephant's body, such as a pressure resistant wall of tissue flanking the lungs and the shape of the kidney ducts in the elephant embryo that is similar to that of aquatic animals. As the group evolved into large animals, and the jaws and teeth elongated, the trunk continued to evolve as a manipulatory organ to work with the tusks in food gathering.

Anancus

Anancus was very common over most of Europe and Asia in the early Pliocene and reached Africa in the succeeding Pleistocene. The tusks of this elephant were more than 4m (13ft) long, as long as the rest of the animal, and may have been used to root around in forest soil.

Features: With *Anancus*, the modern shape of the elephant appears – with the short head and the single pair of tusks confined to the upper jaw. However, the legs are shorter than those found on a modern elephant. The tusks are straight and very long, and the teeth are the typical mastodont shape. It was an animal of moist tropical forest – its broad feet seem to be well adapted to soft soil.

Right: Bones of Anancus appear to have been studied by ancient Greeks who misinterpreted them as the bones of giant humans.

Distribution: Europe; Africa; Asia.
Classification: Ungulata, Probiscidea.
Meaning of name: After the Roman king.
Named by: Aymard, 1855.
Time: Late Miocene epoch of the late Tertiary to the early Pleistocene epoch of the Quaternary.
Size: 3m (10ft).
Lifestyle: High browser.
Species: *A. arvernensis*, *A. jebtebsus*.

PINNIPEDS

Apart from the whales, which have adapted totally to the marine lifestyle, there are a number of mammals that show aquatic adaptations but have not totally severed their links with the land. The most abundant of these are in a family of the Carnivora called the pinnipeds. These comprise the seals, the sea lions and the walruses. Evidence from anatomy and genetics shows that these three are closely related.

Enaliarctos

A splendid example of an animal that shows a partly terrestrial and a partly marine lifestyle, *Enaliarctos* probably lived rather like the modern sea otter. It lived on land but hunted in the water. Judging by the teeth, which have both fish-seizing and meat-shearing adaptations, it was quite a generalist feeder. The enaliarctids seems to have been directly ancestral to the sea lions and the walruses, and also to a short-lived middle Miocene group, the desmatophocids, that resembled sea lions with tails.

Features: *Enaliarctos* is almost half way between an otter and a sea lion. It has a sleek streamlined body and, although it has a distinct tail and legs, the feet are modified into paddles. The skull is quite like that of a sea lion, with the big eyes and nasal openings. These senses associated with sensory whiskers are quite well developed and the inner ear is constructed to hear sound underwater.

Distribution: Pacific coast of North America.
Classification: Carnivora, Pinnipedia.
Meaning of name: Sea bear.
Named by: Mitchell and Telford, 1973.
Time: Early Miocene epoch of the late Tertiary.
Size: 1.5m (5ft).
Lifestyle: Fish-hunter.
Species: *E. mealsi*, *E. mitchelli*.

Left: Enaliarctos, *as well as having fish-snatching front teeth, also had crushing molars for shellfish and meat-shearing carnassials for larger prey.*

Imagotaria

The earliest walrus known was *Neotherium*, a bear-like animal from the middle Miocene of the Pacific Rim. It probably evolved from *Enaliarctos* or something very close to it. Through the Miocene we can follow an evolutionary progression from this through *Proneotherium* to *Imagotaria*, in which the chewing dentition is being replaced by a dentition specialized for holding prey. Of the modern groups of pinnipeds, it appears that the walruses reached their greatest diversity first. A wide variety of walrus types existed at the boundary of the Miocene and Pliocene. Since then they have declined and only a single species exists today, with two subspecies – the Atlantic and Pacific walruses respectively.

Features: By the late Miocene the distinctive walrus-like features and adaptations to exploiting a shellfish diet had evolved in *Imagotaria*, although it probably preyed on fish as well. In build *Imagotaria* is more like a sea lion than a walrus. However, the canine teeth are larger than those of its ancestors and were probably used for digging up shells.

Distribution: Pacific coast.
Classification: Carnivora, Pinnipedia.
Meaning of name: Image of a sea lion.
Named by: Mitchell, 1968.
Time: Later Miocene epoch of the late Tertiary.
Size: 1.5m (5ft).
Lifestyle: Shellfish-eater.
Species: *I. downsi*.

Left: Although Imagotaria *is classed as a walrus, in life it would have looked more like a sea lion.*

PINNIPED ANATOMY

The pinnipeds – or "winged feet" – make up a distinct group of animals comprising the modern seals, sea lions, fur seals and walruses. They represent a very widespread modern group ranging from the frigid Arctic and Antarctic Oceans to the tropical Galapagos Islands. They are all adapted as aquatic hunters and thrive purely marine diet such fish, crabs and shellfish. Because they all share the same aquatic features – streamlined shape, hind feet evolved into a swimming tail-like structure, front feet evolved into wing-like paddles – it is assumed that they are all closely related. However, there are various scientific views on this.

One theory is that the walruses and the sea lions evolved from dog-like ancestors and then spread throughout the Pacific ocean, not penetrating the Atlantic until late Miocene times, and that the seals evolved from otter-like ancestors and occupied the Atlantic first and then the Pacific. The former are known as the otarioids and the latter the phocoids. Under this scheme, their resemblances would have come about by convergent evolution.

The other theory is that they all evolved from the same bear-like ancestor in the late Oligocene or early Miocene and the divergence into the various groups came later in the Miocene.

Both schemes acknowledge that they are all part of the Carnivora. Genetic studies on the modern types suggest that they are more closely related to one another than originally thought and this leans towards the second evolutionary view.

Aivukus

The modern walrus shape had arrived with *Aivukus*. It was a shellfish-eater of the American Pacific coastline. In late Miocene time the walrus family had spread through the gap that separated North and South America and had entered the Atlantic. They reached Europe by the early Pliocene. Shortly afterwards they became extinct in the Pacific and colonized the Arctic from the European side.

Features: The tusks that we associate with the modern walrus had begun to appear in *Aivukus*, although they were still quite short. The cheek teeth are broad, like those of the modern walrus and adapted for crushing hard shells. The lower jaw is deep, to hold the muscles. The whole animal shows that it was adapted to feed from the sea bottom in shallow waters – just like modern types.

Below: Aivukus and its relatives spread across the north Pacific Ocean in the Miocene.

Distribution: California, USA; Mexico.
Classification: Carnivora, Pinnipedia.
Meaning of name: From Aivu.
Named by: Repenning and Tedford, 1977.
Time: Miocene epoch of the late Tertiary.
Size: 10m (33ft).
Lifestyle: Shellfish-eater.
Species: *A. cedrosensis*.

Acrophoca

The main difference between the seals and the sea lions is the ability of the latter to turn their hind feet forward to enable them to move on land. The seals, on the other hand, cannot do this and have to make do with a clumsy wriggling action for movement while out of the water. This is made up for in the elegance of the swimming technique.

Features: *Acrophoca* is essentially a modern seal in build, but it has a rather longer neck and its snout is quite pointed. It has all the adaptations of a modern seal but not quite so marked, and it probably spent most of its time close to the shore. The length of the neck vertebrae – greater than in modern seals – and the long snout give *Acrophoca* a serpentine appearance, unlike the distinctive bullet shape of its modern relatives.

Distribution: Peru.
Classification: Carnivora, Pinnipedia.
Meaning of name: Extreme seal.
Named by: de Muizon, 1981.
Time: Early Pliocene epoch of the late Tertiary.
Size: 1.5m (5ft).
Lifestyle: Fish-hunter.
Species: *A. longirostris*.

Left: The long neck of Acrophoca has given it the reputation of a "swan-necked seal" and as the seal equivalent of a plesiosaur.

MARINE GRAZERS

The rocky coastlines of the temperate oceans have always produced a prolific crop of seaweeds. And, as we have seen, whenever there is a potential food supply something will evolve to exploit it. The late Tertiary saw the development of several groups of seaweed-eating marine mammals, in particular the desmostylans and the sirenians.

Desmostylus

The Miocene saw the brief flourishing of an order of animals called the desmostylans. These were built like hippopotamuses with thick set bodies and stumpy legs with broad feet. They had an amphibious lifestyle paddling around the coastal waters feeding on kelp. Sometimes referred to as "seahorses", their evolution and their relationship to other mammals is something of a mystery, but they could be related to elephants or manatees.

Features: The lower jaw of *Desmostylus* is broad and shovel-like, the front teeth jutting outwards from the front edge – a similar arrangement to the shovel jaw of the shovel-tusked elephants. The back teeth are tall and cylindrical and arranged like the links in a chain, hence the name.

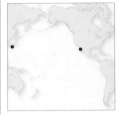

Distribution: North Pacific coastline.
Classification: Ungulata, Desmostyla.
Meaning of name: Chain pillar.
Named by: Marsh, 1888.
Time: Miocene epoch of the late Tertiary.
Size: 1.8m (6ft).
Lifestyle: Possibly seaweed-eater.
Species: *D. Hesperus*.

Left: The heavy body and thick legs would have made Desmostylus clumsy on land.

Paleoparadoxia

The ribs of *Paleoparadoxia* and *Desmostylus* show pachystasis – a thickening that would decrease the animal's buoyancy – to keep it on the bottom while foraging underwater. These animals are typical of the coasts of the northern Pacific, but some desmostylan remains have been found in Florida, indicating that the group lived in the Atlantic as well. There are only two complete skeletons known, one from Japan and the other from California.

Features: The skull of *Paleoparadoxia* is similar to that of a horse, with a deep jaw bone, an extended snout and high cheek teeth. However, there the resemblance ends. The incisors and the canines project forward, but not in such a regular manner as those of *Desmostylus*. The limbs are short and stout, furnished with broad four-toed feet with hoof-like nails. The lower arm bones are fused together, so the forefoot could not be turned. The result is a permanent stance with the foot turned inwards.

Distribution: North Pacific coast, particularly Japan; Alaska, California, USA.
Classification: Ungulata, Desmostyla.
Meaning of name: Ancient puzzle.
Named by: Tokunaga, 1939.
Time: Miocene epoch of the late Tertiary.
Size: 10m (33ft).
Lifestyle: Fish-hunter or scavenger.
Species: *P. tabadai*.

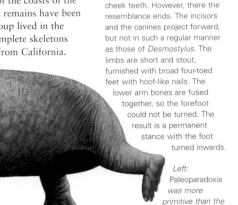

Left: Paleoparadoxia was more primitive than the similar Desmostylus.

HYDRODAMALIS

The climax of the sirenian story comes with the arrival of *Hydrodamalis*. This was the giant of the group, at 8m (26ft) long, and it existed from the Pliocene to historical times. Most of our knowledge of this beast comes from the studies of the German biologist and explorer Georg Wilhelm Steller in the 1750s when its herds were restricted to the islands of the Bering Sea. It was subsequently known as Steller's sea cow. The body was enormous, and the creature spent most of its time on the surface of the water. Its neck was quite flexible for such a large animal, allowing it to graze quite widely without moving its great body too far. The single pair of flippers were held well beneath the body, and were used for "walking" through the shallows while feeding.

Hydrodamalis was hunted by humans for its meat, which was similar to beef in texture, and for its milk, which was consumed directly, or made into butter. Its skin was used for shoe leather and for making kayaks. Having had no history of contact with humans, the Steller's sea cows were placid and fearless animals. They were hunted to extinction by 1768, less than 20 years after their discovery.

Above: The gigantic Steller's sea cow.

Prorastomus

Another group of aquatic herbivores are the sirenians, the sea cows. Today they are represented by the manatee and the dugong, but the family was in existence since the early Eocene. Like the desmostylans, they may share common ancestry with the elephants. The earliest-known is middle Eocene *Prorastomus*, which is included here among the later Tertiary animals because of its importance.

Features: *Prorastomus* was probably a land dweller but took to the sea to feed. Its thick snout and the arrangement of the teeth, which is similar to that of modern sirenians, suggest that it was adapted to feeding on floating seaweeds and on growing seagrass. *Prorastomus* is known only from a skull and vertebrae, but the more complete skeleton of the slightly younger *Pezosiren*, also from Jamaica, with its strong hind limbs and its thick ribs, gives us a good idea of what the animal must have looked like.

Distribution: Jamaica.
Classification: Ungulata, Sirenia.
Meaning of name: Broad jaws at the front.
Named by: Owen, 1855.
Time: Middle Eocene epoch of the early Tertiary.
Size: 1.5m (5ft).
Lifestyle: Seaweed grazer.
Species: *P. sirenoides*.

Below: This restoration is based on the more complete remains of Pezosiren.

Rytiodus

Distribution: France; Libya.
Classification: Ungulata, Sirenia.
Meaning of name: From *Rytina* – wrinkled – an old name for Steller's sea cow.
Named by: Lartet, 1866.
Time: Late Oligocene epoch of the early Tertiary to Miocene epoch of the late Tertiary.
Size: 6m (20ft).
Lifestyle: Aquatic grazer.
Species: *R. capgrandi*.

By the late Oligocene a group called the rytiodids had evolved. These were so highly adapted that, essentially, they looked like modern manatees. The group consists of four or five genera that were mostly confined to the Caribbean region, but the best-known, *Rytiodus*, had migrated across to the Old World. By the end of the Pliocene the whole group seems to have become extinct in the Caribbean and western Atlantic. The modern descendants are the dugongs, which are found around the Indian Ocean.

Features: *Rytiodus* has a massive but streamlined body, a single pair of flippers at the front, and a broad swimming tail. The snout is downturned to help it to graze on the sea bottom and the mouth is equipped with a massive pair of tusks. The tusks have knife-like edges that would have been self-sharpening, and ideal for cutting the tough stems of seagrasses – the staple diet of the modern sirenians. Smaller sirenians, like *Metaxytherium*, existed at the same time, but these were adapted to feeding on the leaves of the seagrasses rather than the stems.

Left: Rytiodus *was twice the size of modern sea cows, but very much smaller than the gigantic recently extinct* Hydrodamalis.

MUSTELLIDS, PROCYONIDS AND BEARS

Today the mustellids are known as the weasels, stoats and badgers. They are generally small, long-bodied hunters. Procyonids are the modern raccoons, pandas and coatis. The bears were the last carnivore group to evolve. All of these have a distinguished fossil record. Victorian naturalists recognized that these groups were closely related.

Potamotherium

Potamotherium is the only fossil mustellid that is well known. This is possibly because it lived near the water and so its remains had a better chance of being fossilized than those of any other of the group. It is possible that the seals evolved from *Potamotherium* or something very close to it. Certainly the articulation of the backbone shows that it swam with an up and down undulating movement, very much like that of a seal. The teeth show it to have been a mustellid.

Right: Like most water-hunting mammals Potamotherium had a poor sense of smell, made up for by its eyesight and hearing.

Features: Otter-like *Potamotherium* has a long sinuous body and short legs. The head is flat and the tail is long and powerful. On land it would have moved with a bounding action with its head close to the ground. In the water it would have swum like its modern counterpart. The senses of sight and hearing would have been its most important ones.

Distribution: France; Germany.
Classification: Carnivora, Musteloidea.
Meaning of name: River mammal.
Named by: Geoffroy, 1833.
Time: Early Miocene epoch of the late Tertiary.
Size: 1.5m (5ft).
Lifestyle: Fish-hunter.
Species: *P. valletoni.*

Megalictis

Below: Megalictis would have been one of the main predators of the many types of running grass-eaters from the Agate Fossil Beds in Nebraska.

We usually regard the mustelids as the small and adaptable members of the Carnivora. *Megalictis*, however, is the largest mustelid known. A similar giant mustelid, *Aelurocyon*, is now thought to have been the same animal.

Features: *Megalictis* is an enormous mustelid, a bit like a wolverine but the size of a black bear. Two sizes of the single species are known, and it has been interpreted that the larger size represents the males and the smaller size the females. The joints of the forelimb suggest that *Megalictis* was an ambush hunter, springing on prey and grappling it to the ground rather than chasing it.

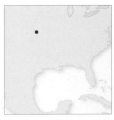

Distribution: Nebraska, USA.
Classification: Carnivora, Musteloidea.
Meaning of name: Giant marten.
Named: Matthew, 1907
Time: Early Miocene epoch of the late Tertiary
Size: 2m (6½ft).
Lifestyle: Hunter or scavenger.
Species: *M. ferox.*

Chapalmalania

Distribution: Argentina.
Classification: Carnivora, Procyonidae.
Meaning of name: Scrubland ripper.
Named by: Ameghino, 1908.
Time: Pliocene epoch of the late Tertiary.
Size: 1.5m (5ft).
Lifestyle: Omnivore.
Species: C. altaefrontis.

The raccoons are closely related to the bears, and when the remains of *Chapalmalania* were first found they were thought to be those of a bear, based on the size. In fact, it was a raccoon, but about the size of a modern giant panda. The raccoons crossed to South America from the north, during one of the few times that this was possible, where the genus diversified. *Chapalmalania* evolved as a mountain-living omnivore.

Right: Chapalmalania had killing teeth at the front of the mouth and grinding teeth at the back.

Features: The resemblance between *Chapalmalania* and the modern giant panda is one of convergent evolution. Its teeth show that it could cope with a wide variety of foods, and probably ate plants, eggs, fish, fruits, nuts, insects and carrion. Like most of the largest carnivores, it would not have restricted its diet to meat.

COPE'S RULE

Many animals were spectacularly huge. They seem to be the result of a biological rule that has been accepted since the 1870s, called Cope's Rule. It is named after the 19th-century American palaeontologist Edward Drinker Cope who noted from the fossil record that groups of animals tend to have larger and larger species through the passage of geological time. He detected the trend in every group of animal from mammals to corals. The development of the big plains-living horse of today from the scampering undergrowth dweller of the early Tertiary was his prime example. The development of the giant weasels and pandas seen here could be another.

Since the 1970s, however, this idea has been falling out of favour. Several reasons are cited. Firstly, after a mass extinction, it would be the small adaptable animals that would survive. These would evolve quickly to an optimum size, giving the impression of a trend towards gigantism. Then there is the fact that bigger animals, or the parts of them, fossilize more easily than smaller, more delicate, ones, meaning that in any population of fossil animals the smaller ones may have been overlooked. There are also many branches of the evolutionary tree on which animals become smaller over time, because of environmental factors, such as dwarfism on islands.

Of course, it is the big animals that get most of the attention, and often it may seem that these are the ultimate evolutionary versions of that particular group of animals.

All in all, Cope's Rule held sway for more than 100 years. It is no longer accepted as valid.

Agriotherium

There are no bears in Africa today. However, they did once exist on that continent. The ancestors of *Agriotherium* reached Africa from Asia in Miocene times, and its remains have been found in the south, indicating that it did have a history in the continent in between.

Features: *Agriotherium* is a very large bear, with particularly long legs. In build it is dog-like and primitive. However, its teeth are very similar to those of modern bears, indicating that it was omnivorous. Wear on the teeth indicates that there was a large proportion of plant matter in its diet. Its meat ration was probably obtained by scavenging.

Distribution: Spain; India; Namibia; USA.
Classification: Carnivora, Ursidae.
Meaning of name: Sour mammal.
Named by: Wagner, 1837.
Time: Late Miocene to Pleistocene.
Size: 2m (6½ft).
Lifestyle: Hunter or scavenger.
Species: A. africanum, A. schneideri, A. gregoryi, A. sivalensis.

Left: The shape of the body and limbs show that Agriotherium was not adapted for hunting, either by ambush or pursuit.

DOGS

The dogs evolved and developed on the North American continent, and did not spread to other parts of the world until the end of the Miocene. Their teeth are distinctive, and led to their success. Their canines are very prominent (which is why these teeth are called "canines" in all mammal groups) and the cheek teeth are adapted for shearing meat. Their ability to hunt in packs was also a factor in their success.

Cynodesmus

By Miocene times there were carnivores that actually looked like the dogs we would recognize today. The opening of the grasslands across the plains of North America encouraged the evolution of long-legged running grass-eaters which, in turn led to the evolution of the running hunters, and the shapes of the modern big carnivores. *Cynodesmus* may have been ancestral to modern dogs.

Features: The body and legs of *Cynodesmus* are distinctly dog-like, but they do not have quite the sophisticated running mechanisms that we would see today. There are still five toes on each foot, as in its ancestors, but the first is small and atrophied, forming a "dew claw". The claws themselves are narrow and sharp, rather than the blunt weight-bearing structures of modern dogs. The face, too, is different, having a shorter muzzle. Despite all that, *Cynodesmus* could have been mistaken for a modern coyote from a distance.

Left: A longer back and heavier tail are all that would have distinguished Cynodesmus *in appearance from a coyote.*

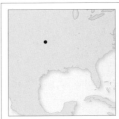

Distribution: Nebraska, USA.
Classification: Carnivora, Caniformia.
Meaning of name: Musk dog.
Named by: Scott, 1893.
Time: Late Oligocene epoch of the early Tertiary to early Miocene epoch of the late Tertiary.
Size: about 1m (3ft).
Lifestyle: Hunter.
Species: *C. brachypus, C. martini, C. thooides.*

Amphicyon

The amphicyonids are known as the bear dogs – a group of heavy carnivores that existed from the Eocene to the Miocene. They were the hunters that replaced the creodonts when they declined, and were themselves replaced by the true dogs in the latter part of the Tertiary. The amphicyonids came in a wide variety of forms. *Borophagus* – big and bear-like. *Daphoenodon* – smaller, about the size of a wolf, and adapted for running rather than ambush. *Cynelos* – the size and shape of a coyote.

Features: The head of a dog on the body of a bear, would be a good general description of *Amphicyon* – the typical amphicyonid. As in a bear its short limbs are plantigrade – the animal walked on the soles of its feet rather than its finger tips. The front legs are powerful and it would have been able to kill with a blow from its paws. However, it would not have been purely carnivorous. Like modern bears it would have eaten a variety of foods.

Left: Footprints of Amphicyon show that it paced with the feet on the same side moving at once, like a camel.

Distribution: France; Germany; Turkey; Nebraska, USA.
Classification: Carnivora, Caniformia.
Meaning of name: Ambiguous dog.
Named by: Lartet, 1836.
Time: Middle Oligocene epoch of the early Tertiary to early Miocene epoch of the late Tertiary.
Size: 2m (6½ft).
Lifestyle: Hunter or scavenger.
Species: *A. major, A. giganteus, A. longiramus.*

Osteoborus

The borophagines were scavenging dogs that came to prominence in the late Miocene. They were primarily scavengers and had the same lifestyle in North America as the hyenas have on the plains of Africa today – mostly existing on the kills of other animals. The teeth were adapted to bone-crushing like those of modern hyenas, but, unlike hyenas, they had no meat-shearing teeth. Borophagines, like *Osteoborus*, were more adapted to scavenging than hyenas are.

Distribution: Nebraska, USA; Honduras.
Classification: Carnivora, Caniformia.
Meaning of name: Bone crusher.
Named by: Martin, 1928.
Time: Late Miocene epoch of the late Tertiary to early Pleistocene of the Quaternary.
Size: 80cm (2½ ft).
Lifestyle: Scavenger.
Species: *O. cynoides*, *O. hilli*, *O. galushai*, *O. secundus*, *O. direptor*.

Features: The premolar teeth of *Osteoborus* are massive and covered in thick enamel, and adapted for crushing bone. The muscles of the jaw are huge to work this action. The skull is short to accommodate all this. This would have given the head a distinctly hyena look.

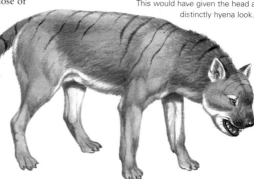

Right: Throughout evolution there were animals that were specifically evolved to scavenge the carcasses of dead beasts. Osteoborus represented the late Tertiary example.

HUNTER OR SCAVENGER

The borophagines, the dog group with the huge bone-crushing jaws and teeth, have always been regarded as the hyenas of the late Tertiary plains – big animals that concentrated on scavenging corpses, rather than hunting for themselves. However, there are lines of evidence that suggest that they may have been pack hunters as well.

The teeth were adapted for grinding bones, but some genera were also equipped with meat-shearing, carnassial, blades as well. This in itself suggests the animals had a hunting mode of life for part of the time. Research into modern members of the Carnivora shows that a carnivore weighing more than about 21kg (46lbs) is usually a hunter, and always hunts prey larger than itself. Many of the borophagines fall into this category. The fact that they did not have the sharp claws of the cats and their relatives shows that it was difficult for them to make a quick kill. Modern animals overcome this by hunting in packs, subduing their prey by cooperative action. It may be that the large borophagines were pack-hunting animals, but not all scientists agree.

Epicyon

This was the biggest of the dog tribe, being comparable in size to a large bear. It spread across North America replacing the big amphicyonids when they became extinct. It is possible that their large size came about during the general cooling of the climate of the time. There is an ecological rule – Bergman's rule – that states that large-bodied animals tend to evolve in cooler conditions. The simple reason is that a bigger body can keep the heat in better than a small body.

Distribution: Texas, Florida, Nebraska, Oklahoma, USA.
Classification: Carnivora, Caniformia.
Meaning of name: More than a dog.
Named by: Leidy, 1858.
Time: Late Miocene epoch of the late Tertiary.
Size: 1.5m (5ft).
Lifestyle: Hunter or scavenger.
Species: *E. haydeni*, *E. validus*, *E. saevus*.

Features: The jaws are powerful, the palate broad and the teeth huge. This shows a different kind of killing and feeding mechanism from any modern dog. It may have existed only on carrion or it may have hunted the other large mammals around – singly or in packs. The fossil evidence is unclear on this point.

Left: Epicyon was probably the largest dog that ever lived.

HYENAS AND CATS

Hyenas and cats, along with the mongooses, all fall into the broad carnivore classification of the feliformes as distinct from the major group that encompasses the dogs, the bears, the weasels and the pandas. It is the bony structure of the ear that unites them. By the late tertiary they had all developed their distinctive shapes and lifestyles.

Ictitherium

The hyenas evolved in the early Miocene as civet-like carnivores. The earliest-known is *Protictitherium* which was mostly arboreal and would have been indistinguishable from a civet. Only later did they evolve into the specialist bone-crushers that we know today. *Ictitherium* was one of the most common of the early generalist types, with abundant remains known from various sites around the Mediterranean.

Features: In its dentition, *Ictitherium* is very much like a modern civet, but the bodily shape is more jackal-like than its ancestors. The structure of the ear shows that it is quite separate from the viverroids to which the civets belong. It is possible that the concentration of their remains means that they hunted in packs. Their remains have been found as bonebeds in flood deposits, indicating where a pack had been overcome.

Left: At times in the Pliocene the bones of Ictitherium *outnumber those of all other carnivores put together.*

Distribution: Morocco; Libya; Greece; Slovakia; Turkey; China.
Classification: Carnivora, Feliformia.
Meaning of name: Marten mammal.
Named by: Zdansky, 1924.
Time: Middle Miocene to early Pliocene epoch of the late Tertiary.
Size: 1.2m (4ft).
Lifestyle: Omnivore.
Species: *I. intuberculatum, I. viverrinum, I. pannonicum, I. hipparionum, I. tauricum, I. ibericum, I. kurteni, I. arkesilai.*

Percrocuta

The bone-crushing hyenas quickly evolved from their more generalized ancestors in Miocene times. *Percrocuta* and its close relative *Pachycrocuta* from the later Pleistocene epoch were probably the biggest hyenas that ever lived – about the size of a modern lion. They probably evolved in Europe and since spread throughout the continents of Asia and Africa.

Features: Like the modern hyena, *Percrocuta* has the sloping back, brought about by the long front legs. The head is large and the jaws are extremely powerful, and furnished with enormous bone-crushing teeth. However, the rear teeth are meat-shearing and rather cat-like, indicating a more varied diet than the bone fragments and carrion that we would normally expect.

Left: As big as a modern lion, Percrocuta *must have been the most fearsome meat-eater of its day.*

Distribution: Africa; Asia; Europe.
Classification: Carnivora, Feliformia.
Meaning of name: Thoroughly a hyena.
Named by: Schlosser, 1903.
Time: Middle to late Miocene epoch of the late Tertiary.
Size: 1.5m (5ft).
Lifestyle: Hunter or scavenger.
Species: *P. gigantea, P. tobieni, P. australis, P. tungurensis, P. hebiensis, P. miocenica.*

Machairodus

Distribution: Africa; Europe; Asia; Arizona, USA.
Classification: Carnivora, Feliformia.
Meaning of name: Knife tooth.
Named by: Kaup, 1833.
Time: Late Miocene to Pliocene epoch of the late Tertiary.
Size: 1.5m (5ft).
Lifestyle: Hunter.
Species: M. coloradensis, M. giganteus, M. aphanistus, M. africanus, M. oradensis and about a dozen others.

By the late Tertiary the cats had developed into several different lines of evolution. There were those that pursued their prey at speed and killed by a quick bite to the neck, and there were those that were ambush hunters. Perhaps the most famous of the ambush hunters were the sabre tooths – those with the canine teeth enlarged into killing fangs.

Features: Machairodus is one of the earliest of the true sabre-toothed cats. It is as big as a modern lion. The legs are very stout and obviously not used for running very far. The sabre teeth are big, although not as big as those of other sabretooths to come. The incisors project forward, unlike those of modern cats. Machairodus must have hunted by ambushing its prey and wrestling it to the ground.

Right: There was a large variation in size between the many species of Machairodus. Each may have had a different colour and pattern.

SABRE TEETH

What was the purpose of the sabre teeth of the sabre-toothed cats? There are a number of ideas.

It was once thought that the sabre teeth were used like daggers. They were plunged deep into the body of the prey until they hit a vital organ. However, the points are too blunt for this and they are too thin – any movement of the prey would break the teeth. What is more, if the tooth hit bone during such an attack it would shatter.

Another idea is that they were used to inflict deep bleeding wounds on the flanks of the prey and the cat would then wait for it to bleed to death. The problem with that idea is that there were so many other predators that there would be too much competition for such a dying animal.

A third idea is that the sabre-toothed cats used their forward-pointing incisors to grip their prey, and used the sabre canines to tear at the arteries and veins of the neck or even at the windpipe. This seems an unnecessarily complex way of hunting.

The debate continues.

Megantereon

One of the ambush-hunting cat groups was the dirk tooths. These had blade-like canine fangs, but they were not as well developed as those that we find in the sabre tooths. It seems likely that these big teeth were used for killing the big grass-eating mammals that inhabited the plains of late Tertiary times. In Dmanisi, Georgia, the skull of an early human, *Australopithecus*, has been found with tooth punctures that match the dirk teeth of *Megantereon*, so clearly the stabbing teeth were powerful enough to penetrate bone.

Features: The upper canines are large, but not as large as in the sabre tooths. There is a flange on the lower jaw to protect them when the mouth is closed. The lower canines are not nearly as enlarged. The strength of the forequarters shows that this cat wrestled its prey to the ground before killing it.

Distribution: South Africa; Kenya; India; France; Spain; China; Greece; Florida, USA.
Classification: Carnivora, Feliformia.
Meaning of name: Big animal.
Named by: Cuvier.
Time: Late Miocene epoch of the late Tertiary to early Pleistocene epoch of the Quaternary.
Size: 1.2m (4ft).
Lifestyle: Hunter.
Species: M. whitei, M. cultridens, M. falconeri, M. inexpectatus.

Right: There is one good skeleton of Megantereon, found in France, though other bones are known.

PRIMATES

During the Miocene there were about 100 species of apes and monkeys that lived in Africa, Europe and Asia. Some of these were the ancestors of modern primates, including humans, while others were evolutionary offshoots. They were mostly arboreal animals, and all had the full set of five fingers and toes, and had brains that were proportionally large.

Oreopithecus

Found in the coal mines of Italy, *Oreopithecus* has become known as the "abominable coalman". It lived on islands that were covered in forested swamps – the origin of the lignite, or brown coal, in which its fossils were discovered. It fed on the leaves and shoots of these plants. The lack of predators on the islands allowed it to exploit habitats on the ground as well as in the trees. It is known from over 50 specimens that have been found over 130 years, and so its features are very well known, compared with other fossil apes.

Features: *Oreopithecus* is regarded as being on the evolutionary line to humans by its mixture of features. The feet are like those of a monkey. Its face is like that of an ape. The teeth are like those of primitive humans. The hip bones and the backbone, as well as the vertically oriented knee, suggest that it could walk on the ground, but the arms are longer than the legs, suggesting that it spent much of its time in trees where it moved by brachiation – swinging by the hands. The hands are capable of delicate grasping – more so than in modern apes.

Left: Oreopithecus *probably gained its upright walking ability independently of the later australopithicines of Africa.*

Distribution: Tuscany and Sardinia, Italy.
Classification: Primates, Catarrhini.
Meaning of name: Mountain ape.
Named by: Gervais, 1872.
Time: Late Miocene epoch of the late Tertiary.
Size: 1.2m (4ft) tall.
Lifestyle: Forest-living vegetarian.
Species: *O. bambolii*.

Pliopethecus

The pliopithecoids were among the first fossil primates to be discovered and studied. Gibbon-like *Pliopethecus* was typical of the group and gives it its name. They were known mostly from fossil teeth found in the 1840s and it was not until the 1960s that enough of the skeleton was found to enable scientists to build proper restorations of what it was like in life. The pliopethecoid group evolved in Africa and moved northwards across the Tethys and colonized the forests of late Tertiary Europe and Asia. At this time there were about 30 distinct types of ape, most of which are now extinct, and it is not entirely clear how they were related to each other or to modern apes.

Left: Pliopithecus, *with its strong arms and curved fingers, would have swung through the trees like a modern gibbon.*

Features: *Pliopithecus* was believed to have been ancestral to the gibbon line, with its long arms and brachiating ability, but not all scientists agree, thinking the similarities to be superficial and convergent. It certainly resembles a gibbon, with its short face, its large eyes and its prominent canine teeth. The arms, although long, are about the same length as the legs and there is a short tail. It is now thought that *Pliopithecus* is more primitive than originally believed and that it predates the evolutionary split of the apes from the Old World monkeys.

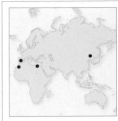

Distribution: Egypt; France; Spain; China.
Classification: Primates, Catarrhini.
Meaning of name: Pliocene ape.
Named by: Lartet, 1849.
Time: Middle to late Miocene epoch of the late Tertiary.
Size: about 1m (3ft) tall.
Lifestyle: Arboreal leaf-eater.
Species: *P. piveteaui, P. antiquus, P. platydon, P. zhanxiangi.*

Proconsul

Distribution: Africa.
Classification: Primates,
Catarrhini.
Meaning of name: Before
Consul (the name of a
chimpanzee in London Zoo at
the time of discovery).
Named by: Hopwood, 1948.
Time: Miocene epoch of the
late Tertiary.
Size: 1m (about 3ft) tall.
Lifestyle: Arboreal omnivore.
Species: *P. africanus*,
P. heseloni, *P. nyanzae*.

Proconsul is regarded as the first ape, or at least the earliest ape to have been found. The brain size is comparable to that of a modern Old World monkey. No tail bones are known but the attachment of the spinal column to the hips indicates that one may have been present. The various species range in size from 15–50kg (33–110lb). *Proconsul* lived in the branches of tropical rainforest, as modern apes do.

Features: *Proconsul* has more flexible hips, shoulders, hands and feet than the monkeys – preadaptations to the brachiating locomotion in apes and the prehensitility of human hands. However, it retains several monkey-like features in the backbone and pelvis, and so it was more suited to walking along branch tops than to swinging beneath them. The thin enamel on the teeth suggests that it probably subsisted on soft fruit.

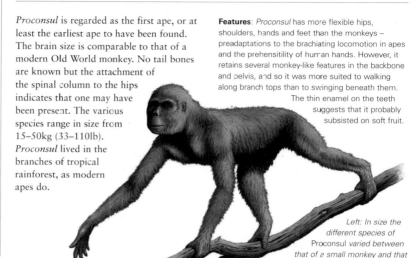

Left: In size the different species of Proconsul *varied between that of a small monkey and that of a female gorilla.*

OUR ANCESTOR

The evolution of humans, and the exact point at which they diverged from the line that led to the apes, has always been a controversial study. Almost every year there is a new discovery that changes our understanding of the subject. At the time of writing, perhaps the oldest primate that is accepted as our ancestor is a middle Miocene genus called *Sahelanthropus* from northern Chad. This is known from a single crushed skull, but the distortion has been straightened out by computer imagery. This shows that its features were more human than ape-like, it had small teeth and the joint between the backbone and the skull indicates an upright posture. It appears to have existed immediately after the split between the ape line and the human line and shows that bipedalism evolved very early.

Other early hominids, showing very similar adaptations, from about the same time include *Orrorin*, from Kenya, and *Ardipithecus* from Ethiopia.

Right: The face of Sahelanthropus.

Sivapithecus

Another animal that combines features of monkeys and apes was *Sivapithecus*. It had chimpanzee-like feet, an orangutan-like face and flexible wrists. Despite this mixture, it is classed in the pongidae – the family that contains the apes – but as a side branch. It lived on the Indian subcontinent but became extinct when the environments changed. *Ramapithecus*, a Miocene ape from the Himalayan foothills, is now regarded as a species of *Sivapithecus*.

Distribution: India.
Classification:
Primates, Catarrhini.
Meaning of name:
Ape of Siva (the Hindu
god of animals).
Named by: Gregory,
Hellman and Lewis, 1938.
Time: Miocene epoch of the
late Tertiary.
Size: 1m (about 3ft).
Lifestyle: Vegetarian.
Species: *S. sivalensis*,
S. parvada, *S. indicus*.

Left: For a century, some canine teeth from a fossil pig were thought to have come from Sivapithecus, *resulting in the erroneous idea that the males had bigger teeth than females.*

Features: A combination of physical features seems to indicate that Sivapithecus was at home on the ground as well as in the trees. The teeth, with the big canines and thickly enamelled molars, are more adapted to the coarse plants found on the open plains than the soft leaves of the forests.

THE
QUATERNARY
PERIOD

With the dawn of the Quaternary period we come to what are essentially modern times. In the last 1.8 million years the configuration of the continents has not changed very much. However, the climates have changed a great deal, for this is the time known as the Ice Age. During the Pleistocene epoch, the ice sheets spread out from the poles and down from the mountains to produce the glacial intervals that were separated by "interglacials" in which the climate was warmer than it is today. The Holocene epoch – the present day – has only lasted about 12,000 years, and may represent another interglacial.

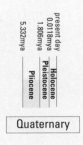

present day
0.0118mya — Holocene
1.806mya — Pleistocene
5.332mya — Pliocene

Quaternary

Above: The timeline shows the epochs that make up the Quaternary period of Earth's history to date.

1 Dire wolves
2 Woolly mammoths
3 Cave lion
4 Woolly rhinocerous
5 Cave bear
6 Bison
7 Geese

REPTILES

Though the Age of Reptiles is long over, there are plenty around today. The Quaternary shows the range of reptiles that is familiar nowadays. However, earlier in the period there were a number of spectacular versions, particularly in Australia, that would dwarf their contemporary relatives but that are now extinct.

Megalania

The largest lizard today is the Komodo dragon of the East Indies – a very large member of the monitor lizard group. *Megalania* was a close relative but weighed about four times as much. It had the body size of a lion, but it was shorter in proportion, due to the comparative shortness of the tail.

Features: No complete skeletons are known, but there are enough individual bones to give us a good idea of what this giant lizard looked like. The teeth are serrated like those of the other monitors, but they are more widely spaced and more curved. The claws on its feet are very large, and used in ambush hunting. It would have preyed on the giant marsupial megafauna that was typical of Pleistocene Australia.

Distribution: South Australia, Queensland, New South Wales.
Classification: Lepidosauria, Squamata.
Meaning of name: Greater ripper.
Named by: Owen, 1859.
Time: Pleistocene epoch of the epoch of the Quaternary.
Size: 5.5m (18ft).
Lifestyle: Hunter or scavenger.
Species: *M. prisca*.

Left: The diet of Megalania would have included Diprotodon, Zygomaturus *and the other giant marsupials of the time.*

Wonambi

Known from the famous Naracoorte caves in South Australia, where many of the megafauna fell to their deaths down swallow holes in dry limestone terrain, *Wonambi* was the giant serpent that lurked in the caves and fed on the hapless animals that fell in. No full skeleton is known, and the restoration is based on numerous disarticulated bone fragments.

Features: *Wonambi* was first described from eight vertebrae and a fragment of upper jaw bone with teeth. From these is it was obvious that the animal was a gigantic snake that, like the python, killed by constriction. The head is quite small for the size of the body, and the jaw could not be fully disarticulated suggesting it fed on medium-sized prey.

Distribution: South and western Australia.
Classification: Lepidosauria, Squamata.
Meaning of name: From an Aboriginal legendary giant snake.
Named by: Smith, 1975.
Time: Pleistocene epoch of the Quaternary.
Size: 5m (16ft).
Lifestyle: Cave lurker and hunter.
Species: *W. naracoortensis*.

Left: Wonambi was the last of an ancient snake family that can be traced back to the Cretaceous period.

Meiolania

Distribution: Lord Howe Island, Australia.
Classification: Chelonia, Meiolaniidae.
Meaning of name: Lesser ripper.
Named by: Owen, 1886.
Time: Pleistocene epoch of the Quaternary.
Size: 2m (6½ft).
Lifestyle: Not known.
Species: *M. platyceps*.

The first specimens found – mostly the skull – were so unlike those of a turtle that they were thought to have been from some kind of lizard. This is the derivation of the name – the "lesser ripper" as compared with the giant lizard *Megalania*, the "greater ripper". It is the biggest land-living turtle known.

Right: The group to which Meiolania *belonged – the* Meiolaniidae *– is found in South America suggesting that its origin dates to Gondwana.*

Features: The structure of the skull is so unusual that it is difficult to see to what group of modern turtle *Meiolania* belongs. On balance it seems related to the cryptodires – those that pull their head back into the shell – as opposed to the pleurodires – those that swing their necks round sideways. The spikes and bumps on the skull, particularly the cow-like horns – show that in life *Meiolania* did neither. The club on the tail is reminiscent of that of one of the ankylosaur dinosaurs.

THE AGE OF REPTILES CONTINUES

Reptiles are much more successful than they would appear. In some habitats, such as deserts, they are the dominant vertebrate group. Being cold-blooded they do not need to expend so much energy maintaining a constant body temperature. They can control their body heat by moving from warm to cool areas and by preferential circulation – by pumping surface-warmed blood into the tissues in the depths of the body, and pushing cold blood from the body's core to the outside. As a result they can exist on far less food than birds or mammals. This makes them ideal inhabitants of areas where food is scarce.

Although we regard the Age of Reptiles as being historical, today's reptiles are still a very robust group of animals. They are likely to exist well into the future, occupying specialized niches for as long as evolution continues.

Quinkana

We normally think of crocodiles as being aquatic hunters. That is because all modern species fall into this description. However, the group has a long history, and many of the ancient forms were land dwellers. A group of land-living crocodiles survived into the Tertiary and mostly lived in the Northern Hemisphere. *Quinkana* was a late-surviving Australian form that hunted in the limestone highlands of Queensland.

Features: *Quinkana* is known from several specimens including two skulls, and these show the typical features of a land-living crocodile. The skull is deeper than it is broad – more like that of a meat-eating dinosaur. The teeth are laterally compressed, blade-like and edged with serrations – adapted for slashing and tearing rather than for gripping and drowning. The rest of the body would be adapted for running and scrambling over rocks rather than for swimming.

Distribution: Queensland, Australia.
Classification: Crocodylomorpha, Mekosuchinae.
Meaning of name: From Quinkana – ghostly beings in Aboriginal myths.
Named by: Molnar, 1981.
Time: Miocene to Pleistocene.
Size: 3m (10ft).
Lifestyle: Hunter.
Species: *Q. babarra, Q. fortirostrum, Q. meboldi, Q. timera*.

Right: In appearance Quinkana *would have resembled the hunting crocodiles from pre-dinosaur times.*

BIRDS

As with the reptiles, all the modern bird groups had become established by Quaternary times. The birds that have existed during the last two million years would be very recognizable today. However, extreme specialized conditions particularly in remote areas such as islands have led to a few specialist examples that have since become extinct.

Aepyornis

Right: The discovery of fresh bones of Aepyornis *by Arab sailors may have given rise to the Arabian Nights myth of the giant bird, the roc.*

The heaviest bird known to exist was *Aepyornis* of Madagascar. It survived there until the 17th century, when it died out, presumably through overhunting by humans. There is still much debate over how it came to be in Madagascar. One theory is that its ancestors had always been there since the island broke away from Africa in Cretaceous times. This is supported by the presence of related birds in the Tertiary of Egypt, but the lack of fossil evidence on Madagascar argues against it. The other theory, not largely supported, is that they evolved quite late from winged ancestors that flew from Africa. Whatever its origin, *Aepyornis* flourished on the island in the absence of ground-living hunting animals that would have discouraged the evolution of such slow-moving animals.

Features: *Aepyornis* has the typical skeleton of a huge flightless bird. There is no breastbone and the wings are mere splinters of bone. The legs are massive and elephantine, too heavy for the bird to have been a runner.

Distribution: Madagascar.
Classification: Bird, Struthiornithiformes.
Meaning of name: Tall bird.
Named by: Geoffroy St. Hilaire, 1851.
Time: Pleistocene to recent epochs of the Quaternary.
Size: 3m (10ft) high.
Lifestyle: Grazer.
Species: *A. titan*.

Titanis

A whole group of flightless hunting birds, the Phorusrachidae, existed on the island continent of South America during the Tertiary. The best-known was *Phorusrachus* itself. When the continent became joined to North America in the Pleistocene, they spread northwards just before the extinction of the entire group. *Titanis* was one of the last of these, from Florida and Texas, USA.

Features: Recently found specimens of *Titanis* show that it actually had hands equipped with stubby, versatile fingers, like those of a *Tyrannosaurus*, rather than the stubby useless wings that the phorusrachids were thought to have possessed. Like one of the big meat-eating dinosaurs it would have chased down its prey and killed it with its huge jaws, which in this instance were equipped with a massive hooked beak.

Right: The arms and hands of Titanis *gave it the appearance of a theropod dinosaur, an animal it would have emulated in its lifestyle.*

Distribution: Florida, Texas, USA.
Classification: Bird, Gruiformes.
Meaning of name: Giant bird.
Named by: Brodkorp, 1963.
Time: Pleistocene epoch of the Quaternary.
Size: 2.8m (9ft) high.
Lifestyle: Hunter.
Species: *T. walleri*.

Genyornis

Distribution: Australia.
Classification: Bird,
Anseriformes.
Meaning of name: Jaw bird.
Named by: Stirling and Zietz,
1896.
Time: Pleistocene epoch of
the Quaternary.
Size: 2.2m (7ft) tall.
Lifestyle: Browser.
Species: G. newtoni,
G. australis, G. stirtoni.

Australia was host to a group of so-called "thunder birds" – huge flightless birds that were thought to have belonged to the local emu line. However, it now seems that they were more closely related to ducks. They existed from Eocene times to relatively recently and ranged in size from *Barawertornis*, which was the size of a cassowary, to *Dromornis* that stood about 3m (10ft) tall. *Genyornis* was the last of the line, and it co-existed with the first aborigines on the dry open plains.

Features: *Genyornis* stood as tall as an ostrich but was about twice the weight. The huge beak was probably used for cracking nuts and seeds. The three-toed feet have hooves rather than claws. Many of its relatives, such as *Bullockornis*, were probably fierce hunters but *Genyornis* seems to have been a vegetarian.

Left: Stomach stones have been found with Genyornis *skeletons, suggesting it had a herbivorous diet.*

VULNERABLE GROUND DWELLERS

Where there are no ground-dwelling predators, there is little need for flight. The evolution of flightless birds in island habitats is an evolutionary development that we see time and time again. The problem is that flightless birds are very vulnerable, and when outside pressures impinge, such as human settlers with their accompanying cats, dogs and goats, they tend to be wiped out very quickly.

Raphus – the dodo. This large-headed turkey-sized relative of the pigeon existed on Mauritius until the 1680s.

Pezaphaps – the Reunion and Rodrigues solitaires. These were similar to the dodo, living on the nearby islands of Reunion and Rodrigues, and were wiped out by the 1790s.

Pinquinus – the great auk. A flightless seabird from the islands of Canada, Greenland, Iceland, Norway, Ireland and the British Isles. It was the Northern Hemisphere's equivalent of the penguin, and was hunted to extinction by the 1840s.

*Right: The dodo –
the symbol of
extinction.*

Dinornis

The birds of New Zealand were the dominant animal group before humans arrived there. Until recently there were no mammals on the islands except for bats, and the birds had evolved into all the ecological niches that mammals had occupied throughout the world. The moas were the tall browsers, like giraffes, and, as such, were the tallest birds known. They did not become extinct until the 19th century.

Distribution: New Zealand.
Classification: Bird,
Struthioformes.
Meaning of name: Terrible
bird.
Named by: Owen, 1843.
Time: Pleistocene to Recent
epochs of the Quaternary.
Size: 3.5m (12ft) tall.
Lifestyle: High browser.
Species: D. giganteus,
D. maximus and about
20 others.

Features: A number of species of *Dinornis* and close relatives have been found, ranging from the short, thick-legged *Euryapteryx* to the famously tall *D. maximus*. They all have bulky bodies and a stone-filled gizzard to process their vegetable food. The skull is very small, the beak broad and flattened and the eyes tiny. The nasal bones indicate that it had a good sense of smell. There are no traces of a wing structure – even the wishbone has disappeared from the skeleton.

Left: We know about Dinornis *not just from the bones, but also from mummified skin and preserved feathers.*

MARSUPIALS

Since Mesozoic times Australia has been moving north at a rate greater than the movement of any other continent. In this time the original marsupial inhabitants continued to evolve as conditions altered with the changing latitudes. Only in the Quaternary, when Australia approached the Asian mainland, did the distinctive giant animals die out.

Procoptodon

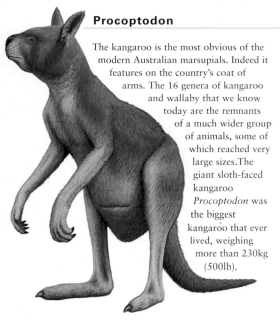

The kangaroo is the most obvious of the modern Australian marsupials. Indeed it features on the country's coat of arms. The 16 genera of kangaroo and wallaby that we know today are the remnants of a much wider group of animals, some of which reached very large sizes.The giant sloth-faced kangaroo *Procoptodon* was the biggest kangaroo that ever lived, weighing more than 230kg (500lb).

Features: *Procoptodon* is a member of the single-toed kangaroos. Apart from the number of toes, and the evident size, the most important difference between *Procoptodon* and modern kangaroos is in the articulation of the arms. This would have allowed the animal to reach high above its head and pull down branches with its extra-long fingers and long curved claws – something the modern grass-eaters cannot do. It probably could reach leaves 3m (10ft) above the ground. The face is short, with massive jaws, unlike the deer-like heads of the modern types.

Left: Procoptodon *was twice as tall as the red kangaroo – the biggest of the modern forms.*

Distribution: Australia.
Classification: Marsupialia.
Meaning of name: Forward hill tooth.
Named by: Owen, 1974.
Time: Pleistocene epoch of the Quaternary.
Size: 3m (10ft).
Lifestyle: Browser.
Species: *P. goliah, P. rapha, P. pusio, P. oreas.*

Diprotodon

This giant wombat-like animal, as big as a modern rhinoceros, was the first of the extinct Australian megafauna to be studied and named. It subsisted on a particular species of salt-bush which it could scrape up from the ground with its paws. Complete skeletons have been found preserved in the muds of salt lakes, where the heavy animals broke through the surface crust of salt and became entombed beneath.

Features: A big four-footed animal, *Diprotodon* has a massive head, neck and body. It walked with the feet turned inwards and in a plantigrade manner – on the soles of its feet, like a bear. The front claws are long, as if for digging. An odd feature is that the outer toe is the longest, unlike in other mammals. Skeletons of youngsters have been found in adult skeletons at places where the pouch would have been. It appears that the pouch opened rearwards, as in modern wombats.

Distribution: Australia.
Classification: Marsupialia.
Meaning of name: Two forward teeth.
Named by: Owen, 1838.
Time: Pleistocene epoch of the Quaternary.
Size: 3m (10ft).
Lifestyle: Browser or gazer.
Species: *D. optatum, D. australis, D. minor.*

Left: Diprotodon *is the biggest marsupial known.*

Zygomaturus

Distribution: South-eastern and south-western Australia.
Classification: Marsupialia.
Meaning of name: Big cheek bones.
Named by: Stirton, 1967.
Time: Pleistocene epoch of the Quaternary.
Size: 1.5m (5ft).
Lifestyle: Semi-aquatic browser.
Species: Z. trilobus, Z. keani.

With its raised nostrils and its general build that resembled a pygmy hippopotamus, *Zygomaturus* was probably a semi-aquatic swamp dweller, feeding in reed-beds at the sides of waterways – unlike the other diprotodonts that preferred the dry plains. It may have lived there in small herds. Its remains have been found mostly in areas of coastal swamp, but it is also found in river deposits well into the dry interior of the continent.

Right: Zygomaturus probably lived on vegetated river banks while its relative Diprotodon inhabited dry grasslands.

Features: The front of its lower jaw is narrow and extended forwards by a pair of incisor tusks that produce a structure rather like a garden fork. The hind teeth are broad for grinding tough plants. There is a raised structure over the snout that has led some researchers to suggest the presence of a rhinoceros-like horn, and others to suggest a short trunk.

THE NARACOORTE CAVES

During the early Tertiary, southern Australia was covered in shallow sea, depositing thick layers of limestone. The sea receded in the Pliocene and Pleistocene leaving the landscape as an arid limestone plain. Infrequent rains dissolved out caves and swallow holes across this area.

Nowadays, cattle can be seen congregating precariously at the lips of these swallow holes, feeding on the slightly richer vegetation encouraged by the moist air that rises from the caverns below. Inevitably some lose their footing and tumble to their deaths. It was the same in Pleistocene times, and much of what we know of the Pleistocene megafauna is from the remains of animals that have died in this way.

The best example is the Naracoorte cave system in the south-eastern corner of South Australia. It is now a national park, and has yielded remains of *Thylacoleo, Zygomaturus, Diprotodon, Procoptodon* and *Palorchestes*, as well as the giant snake *Wonambi* that fed on them. Remains of all these animals are on show at a visitor centre.

Left: The treacherous interior of a swallow hole.

Propleopus

The first piece of bone of this animal to be uncovered was a jaw bone very much like that of the modern rat-kangaroo, but as large as that of the biggest kangaroo today. Its remains are very rare, and mostly consist of teeth, and so any restoration must be regarded as highly speculative. These teeth are more like those of an insectivore than a plant-eater, but nothing that size would subsist on only insects. It was probably a carnivore or an omnivore.

Features: The lower incisors are stout and sharp and lack the edges that are used in other kangaroos for nipping off vegetation. They would be more useful as stabbing weapons. What is more, the wear on the teeth is similar to that found on those of meat-eaters.

Distribution: South-eastern Australia.
Classification: Mammal, Marsupialia.
Meaning of name: Before the lion-feet.
Named by: De Vis, 1888.
Time: Pleistocene epoch of the Quaternary.
Size: 1m (about 3ft).
Lifestyle: Hunter or scavenger.
Species: P. oscillans, P. chillagoensis.

Right: It seems reasonable to suppose that Propleopus was a killer kangaroo.

XENARTHRANS

Nowadays the xenarthrans consist only of the anteaters, the armadillos and the sloths. But in Pleistocene times, there were large and spectacular examples. The ancestors of these were found on the island continent of South America, but, when the land bridge was established in the Pleistocene, the group migrated to North America where several genera have been found.

Megatherium

The classic giant ground sloth is a familiar figure in many museums throughout the world. This was the ultimate in the evolution of a group that had existed throughout the Tertiary. It has always been regarded as an enormous slow tree ripper, but modern studies indicate that its great claws may also have been used for stabbing and killing, making it the big predator of the South American plains.

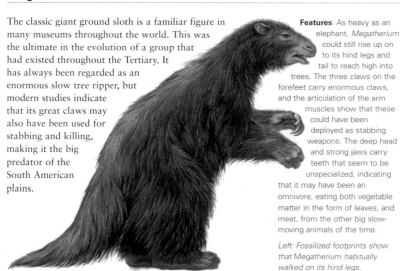

Features: As heavy as an elephant, *Megatherium* could still rise up on to its hind legs and tail to reach high into trees. The three claws on the forefeet carry enormous claws, and the articulation of the arm muscles show that these could have been deployed as stabbing weapons. The deep head and strong jaws carry teeth that seem to be unspecialized, indicating that it may have been an omnivore, eating both vegetable matter in the form of leaves, and meat, from the other big slow-moving animals of the time.

Left: Fossilized footprints show that Megatherium *habitually walked on its hind legs.*

Distribution: Argentina; Bolivia; Peru; Ecuador.
Classification: Mammal, Xenarthra.
Meaning of name: Giant mammal.
Named by: Cuvier, 1796.
Time: Pleistocene epoch of the Quaternary.
Size: 6m (20ft).
Lifestyle: Browser and possibly a hunter.
Species: *M. americanum*, and more than 20 others.

Glossotherium

Glossotherium was one of the giant ground sloths that settled in North America. It is famous for the numbers of its skeletons found in the Rancho la Brea tar pits in Los Angeles. In this area oil from underground reservoirs seeps to the surface and dries out forming pools of sticky tar, and since Pleistocene times this has trapped all kinds of animals, preserving the skeletons. *Glossotherium* was one of the largest of these.

Features: *Glossotherium* has a bulky skeleton with a heavy head and tail. The front feet were curled under while walking to protect the claws. The soles of the hind feet were permanently turned inward so that it walked on the edges of its feet. The hips are broad, and took the weight as the animal raised itself on its hindquarters to reach the tops of the trees on which it fed. Coprolites formed from its droppings showed that it subsisted on the dry desert shrubs of the area and the time.

Left: Glossotherium *was almost certainly hunted by the first American humans.*

Distribution: California, Florida, Texas, USA; Mexico; Chile; Uruguay; Argentina.
Classification: Mammal, Xenarthra.
Meaning of name: Tongue mammal.
Named by: Owen, 1842.
Time: Pliocene epoch of the late Tertiary to the Pleistocene and possibly the recent epochs of the Quaternary.
Size: 4m (13ft).
Lifestyle: Browser.
Species: *G. robustum* and about ten others.

Glyptodon

Distribution: Argentina.
Classification: Mammal, Xenarthra.
Meaning of name: Carved tooth.
Named by: Owen, 1839.
Time: Pleistocene epoch of the Quaternary.
Size: 3m (10ft).
Lifestyle: Grazer.
Species: G. clavipes.

The glyptodonts were, essentially, giant armadillos. Unlike the armadillos the armour carapace over the back was solid, and not articulated along certain bands. They evolved in South America in the Pliocene and become extinct in the Pleistocene, although not before some examples, such as *Glyptotherium*, had migrated to North America. *Glyptodon* itself was one of the biggest, its car-sized carapace being used as shelters by early natives of South America.

Right: Early humans in South America appear to have used the armour of Glyptodon *as shelters.*

Features: As in all members of the group, *Glyptodon* has no teeth in the front of the mouth but powerful grinding teeth at the back. The massive, deep jaws with their muscles attached to downward-pointing cheek bones show that they had a very strong bite, for chewing grasses and other tough vegetation. The continuous armour over the back is backed up with a helmet above the skull and rings of armour forming a tube around the tail.

OTHER GLYPTODONTS

Asterostemma – an early form probably dating from middle Miocene times in Venezuela.
Glyptatelus – from the Miocene of Argentina.
Eleutherocercus – like *Doedicrurus* from the Pliocene of Uruguay.
Glyptotherium – one of the later North American forms. A long-tailed type that existed in the Pleistocene of Arizona. .

OTHER GIANT GROUND SLOTHS

Mylodon – an ox-sized sloth found in Pleistocene cave deposits in Argentina. The dryness of the cave environment preserved patches of skin and hair, showing that the skin had little buttons of bone embedded in it.
Nothrotheriops – from the Pleistocene of Oklahoma, New Mexico and Texas, USA. Coprolites from this sloth, or one closely related, show that it fed on the stems and roots of coarse desert plants, as well as ferns and flowers.
Neomylodon – from the Pleistocene of Argentina. Coprolites show that, unlike *Nothrotheriops*, it fed mostly on grasses and sedges, with no sign of woody stems or roots.

Doedicrurus

The remains of glyptodonts are found associated with those of capybaras – a swamp-dwelling rodent. This may indicate that they fed on swamp plants. The short face has been interpreted by some scientists as the base for a short proboscis, like a tapir. Among the remains of *Doedicrurus* dented carapaces show that males battled for supremacy, apparently battering each other with their tails. Blood vessels in the shell show that the shell was covered in skin and may have been hairy.

Features: The shell consists of thick polygonal plates fused together in a rigid covering. It may have been a little more flexible around the edge. All in all the armour must have accounted for about 20 per cent of the animal's total weight. The big difference between *Doedicrurus* and *Glyptodon* is in its tail. As well as being armoured, it is very long and ends in a rigid shaft and a club, like a mediaeval mace. Undoubtedly this was used in defence.

Below: The horny spikes on the tail club are speculative. it may just have had a knobbly surface.

Distribution: Uruguay; Argentina.
Classification: Mammal, Xenarthra.
Meaning of name: Pestle tail.
Named by: Owen, 1847.
Time: Pleistocene epoch of the Quaternary.
Size: 4m (13ft).
Lifestyle: Grazer.
Species: D. clavicaudatus, D. patagonicus.

PERISSODACTYLS

With the coming of the Pleistocene Ice Age, the climates over most of the world changed, bringing colder weather to the higher latitudes and drier weather to more equatorial areas. The animal life adapted to this, evolving big sizes and furry coats, the better to keep the heat in. We see this among the rhinoceroses.

Coelodonta

The famous woolly rhinoceros of the Ice Age is *Coelodonta*. It evolved in Pliocene times in eastern Asia but then migrated to Europe during the Pleistocene, where it lived on the tundra and steppe grassland that bordered the glacier sheets that covered much of the Northern Hemisphere. Remains of its woolly coat have been recovered from frozen gravels in Siberia, and we know of its appearance from illustrations drawn by contemporary humans in Europe.

Features: *Coelodonta* is a large rhinoceros with two long horns formed from matted hair on the nose. The front one is up to 1m (3ft) long and flattened from side to side. Scuff marks on the horn suggest that *Coelodonta* used it for clearing snow from its grazing patches.

Distribution: Asia and Europe from Korea to Scotland.
Classification: Ungulata, Perissodactyla, Ceratomorpha.
Meaning of name: Hollow tooth.
Named by: Blumenbach, 1807.
Time: Pliocene epoch of the late Tertiary to the Pleistocene epoch of the Quaternary.
Size: 3.5m (11ft).
Lifestyle: Grazer.
Species: *C. antiquitatis*.

Right:
The woolly pelt is a grey-brown colour, possibly with a dark band around the middle.

Elasmotherium

The largest of the woolly rhinoceroses, *Elasmotherium*, almost the size of a modern elephant, lived on the steppes of southern Siberia. Although some earlier branches of the rhinoceros line produced much bigger animals, this is the largest known of the true rhinoceroses. Its horn, although built of hair like that of a modern rhinoceros, sprouted from its forehead rather than its nose.

Features: The tall-crowned teeth with the convoluted enamel show that this was a grass-eater. They had no roots but continued growing, resharpening their grinding surfaces as they wore away, like those of a horse. It has no incisor teeth and would have used strong lips to pluck grass. The massive horn base on the nose must have supported a huge structure, and from this a horn length of 2m (6½ft) is estimated.

Left: There is a theory that the survival of Elasmotherium *into historical times may have given rise to the legend of the unicorn.*

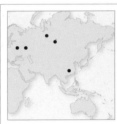

Distribution: Siberia.
Classification: Ungulata, Perissodactyla, Ceratomorpha.
Meaning of name: Plate mammal.
Named by: Fischer, 1808.
Time: Pleistocene epoch of the Quaternary.
Size: 5m (16ft).
Lifestyle: Grazer.
Species: *E. caucasicum, E. sibiricum, E. inexpectatum*.

Hippidion

Distribution: Bolivia; Argentina.
Classification: Ungulata, Perissodactyla, Hippomorpha.
Meaning of name: Like a horse.
Named by: Owen, 1869.
Time: Pleistocene epoch of the Quaternary.
Size: 2m (6½ft) long, 14 hands.
Lifestyle: Running grazer.
Species: H. saldiasi.

By the Pleistocene the modern horse *Equus* had evolved. Having developed through its long lineage on the North American continent and then spread out over the Northern Hemisphere, it then became extinct there between 13,000 and 11,000 years ago and did not return until introduced by Europeans. During the Pleistocene a line of the horses crossed into South America and existed for a time there. These consisted of the genera *Hippidion* and *Onohippidium*.

Features: *Hippidion* is a donkey-sized horse. The strange feature about it is the long domed nasal bone on the skull. It seems possible that in life this would have supported deep tubular nasal passages like those on the modern saiga antelope. As in the antelope this feature would have warmed and moistened the air it breathed in from the cold dry plains of Pleistocene South America.

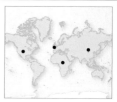

Right: The horse-like litopterns of South America were replaced by the true horses in Pliocene and Pleistocene times.

Hipparion

The wide range of horse types was whittled down during the Pleistocene, with *Equus* becoming the principal example. However, there were specialized genera as well. *Hipparion* had been widespread during the Pliocene and survived into the Pleistocene Ice Age in Africa. The running walk of the modern horse was thought to have been a result of training by human riders. However, footprints of *Hipparion* in African volcanic ash deposits show that this is a natural gait.

Distribution: North America, Europe, Asia, Africa.
Classification: Ungulata, Perissodactyla, Hippomorpha.
Meaning of name: Like a horse.
Named by: de Christol, 1832.
Time: Miocene epoch of the late Tertiary to the Pleistocene epoch of the Quaternary.
Size: 2m (6½ft) long, 14 hands.
Lifestyle Running grazer.
Species: *H. platystyle,
H. huangheense, H gettyi,
H. mohavense,
H. concudense,
H. anthonyi, H. prostylum,
H. campbelli, H. dietrichi.*

Features: In appearance *Hipparion* is essentially a modern horse. However, it still retains two side toes as well as the big hoofed middle toe. These were very small and did not reach the ground, but they had not disappeared as they had done on the main horse line.

Left: Hippar on left North America and spread to Europe and Asia in the late Miocene.

ARTIODACTYLS

It is the even-toed ungulates – those with the cloven hooves – that tend to be the main grass-eating mammals of today. The severe climates of the Pleistocene Ice Age brought about a number of the adaptations, and there were several spectacular examples that were contemporaries of early humans and have only recently become extinct.

Megaceros

Right: Megaceros had the biggest antlers of any deer.

The giant Irish elk (more closely related to the fallow deer than the elk) is known for its spectacular skull and antlers, excavated from the peat bogs of Ireland and mounted in many museums. One bog yielded specimens of more than 80 individuals. It is also well known from its frequent appearance on the walls of caves, mostly in central France, where Ice Age hunters recorded the animals they saw around them.

Features: The antlers could span 3.6m (12ft) and weigh around 40kg (88lb) which was approximately a seventh of the weight of the whole animal. They were confined to the males and were shed and regrown every year. Cave paintings suggest that in life it carried a small fatty hump over the shoulders, like some varieties of modern cattle, possibly a food store to help in the frequent harsh conditions during the Ice Age.

Distribution: Europe and Asia from Japan to Ireland.
Classification: Ungulata, Artiodactyla, Ruminantia.
Meaning of name: Big horn.
Named by: Blumenbach, 1799.
Time: Pleistocene epoch of the Quaternary.
Size: 2.5m (8ft).
Lifestyle: Grazer.
Species: *M. giganteus.*

Cervalces

The environment occupied by this North American deer is similar to that of the modern moose – swamps, bogs and wetlands surrounded by spruce forests and tundra. This suggests that it probably had the same lifestyle, wading in the chill waters and feeding on the lakeside vegetation. It may have been displaced by the true moose that crossed from Asia via the Bering Strait land bridge as the last Ice Age ended.

Features: *Cervalces* is built like a mixture of a moose and an elk. In appearance it would have looked like a moose with very long legs and the head like that of an elk. The antlers are broad and flat like those of a modern moose. In size the animal is larger than the modern moose.

Left: The antlers of Cervalces *were much more complex than those of the moose of the present day.*

Distribution: Central North America.
Classification: Ungulata, Artiodactyla, Ruminantia.
Meaning of name: Stag moose.
Named by: Scott, 1885.
Time: Pleistocene epoch of the Quaternary.
Size: 2m (6½ft).
Lifestyle: Semi-aquatic browser.
Species: *C. scotti.*

During the Pleistocene there was a wide range of very large mammals. These included the cold-adapted elephants and rhinoceroses and the herds of running animals and big carnivores that were widespread in Europe, Asia and North America, and also the giant xenarthrans of South America, and the huge marsupials of Australia. This assemblage is referred to as the Pleistocene megafauna. It only became extinct quite recently, and the reason is something of a puzzle.

The conflicting theories have been summarized as "kill, chill or ill".

The "kill" theory assumes that the spread of people across every habitable continent wiped out the big animals. The needs of the expanding human population, in terms of food, raw materials and living space drove them to extinction, either by hunting or by simply driving them out of living areas.

The "chill" theory relies on the fluctuating temperatures and other climatic conditions at the end of the Ice Age. A sudden cold snap as the world gradually came out of the Ice Age could have weakened the populations too much for them to continue.

The "ill" theory involves the spread of disease. Where the changing conditions allow previously separated populations to mingle, then it is likely that diseases will spread from one population to another that has no immunity. The spread of humans may have been partly responsible.

Where several plausible theories are proposed, the truth may lie in a combination of them all.

Metridiochoerus

This giant warthog the size of a buffalo was the contemporary of the early humans of East Africa. It spread as the grassy plains spread over that part of the continent in response to the drying of the climate. The conditions brought about the development of a unique fauna of large mammals that co-existed with the earliest humans, and have been found in association with their remains in the famous site of the Olduvai Gorge in Tanzania.

Features: The massive skull carries upper and lower canine teeth that are modified into tusks, both sets curving outwards and upwards. The back teeth are high crowned and bear complex enamel patterns that we usually see on specialized grass-eaters like horses. Although similar in appearance to modern warthogs, it was more than twice the size.

Distribution: East Africa.
Classification: Ungulata, Artiodactyla, Suidae.
Meaning of name: Frightful pig
Named by: Hopwood, 1926.
Time: Late Pliocene epoch of the late Tertiary to early Pleistocene epoch of the Quaternary.
Size: 2.5m (8ft).
Lifestyle: Rooter and omnivore.
Species: *M. andrewsi*, *M. jacksoni*, *M. modestus*, *M. nyanzae*, *M. compactus*.

Left: The gigantic warthog Metridiochoerus *sported a set of enormous tusks that curved upwards from the jaws.*

Pelorovis

Distribution: East Africa.
Classification: Ungulata, Artiodactyla.
Meaning of name: Giant (from Greek mythology) sheep.
Named by: Reck, 1925.
Time: Middle Pleistocene to recent epochs of the Quaternary.
Size: 3m (10ft).
Lifestyle: Grazer.
Species: *P. antiquus*, *P. oldowayensis*, *P. capensis*.

This giant relative of the modern water buffalo moved north from the southern part of the African continent as the dry conditions of the Ice Age spread, and grasslands reached into equatorial regions. Remains have been found with stone weapon points embedded in the backbone showing that it was hunted by early humans. Bushman rock engravings in South Africa show that it was alive and observed as recently as 6,000 years ago.

Below: Some restorations show the horns of Pelorovis *curving downwards, but contemporary rock drawings show them curling up.*

Features: The obvious feature of *Pelorovis*, apart from the great size, is the fantastic sweep of the horns. The bony core of the horns are each 2m (6½ft) long, and it seems likely that the horns themselves would have reached twice that size.

ELEPHANTS

Among the large animals of the Ice Age, the elephants are quite notable – especially those that were adapted to the cold conditions by their hairy coats and their deposits of fat. The best-known of the recently extinct elephants are the mammoths and the mastodons that roamed the cold plains of the northern continents in Pleistocene times.

Mammuthus

The mammoths were a wide-ranging group that spread from Africa to Europe and North America in early Pleistocene times. The typical woolly mammoth was the relatively small *M. premigenius*. We know the appearance of the woolly mammoth by engravings, sculptures and paintings done by Pleistocene humans, and by the perfectly preserved bodies found in Alaska, Canada and Siberia in mud frozen since the Pleistocene.

Right: Although Mammuthus *is usually shown with a red coat, the hair may actually have been black.*

Features: The hump on the shoulders gives it a back that slopes to the hips. The long shaggy hair covers the whole body. The tusks reach forward and down, and must have been used as snowploughs to scrape away the snowing covering of the ground plants on which it lived. The massive grinding teeth have the same distinctive ridges as in modern elephants, showing their close relationship.

Distribution: Widespread over the northern continents.
Classification: Ungulata, Proboscidea.
Meaning of name: Earth burrower.
Named by: Brookes, 1828.
Time: Pleistocene epoch of the Quaternary.
Size: 4m (13ft) long.
Lifestyle: Herd-living grazer.
Species: *M. primigenius* and see box.

Mammut

Confusingly, *Mammut* is not the scientific name for the mammoth, but for the other distinctive Ice Age elephant, the mastodon. The name, meaning "earth burrower", in both cases, comes from the mistaken belief that these were the remains of burrowing animals because they had been found frozen in the soil. The informal term "mastodon" means "breast-tooth" and refers to the molar teeth that have paired conical cusps rather than the more wrinkled appearance of other elephant teeth. It had a diet of leaves rather than of grasses.

Features: Compared with other elephants, the mastodon is quite squat and long in the body. The tusks of the upper jaw are long and curved, and sometimes there are vestigial tusks in the lower jaws. The hair is coarse and reddish-brown in colour.

Left; Mammut skeletons have been found in situ 300km (190 miles) off the north-east coast of North America, showing that the continental shelves were above sea level during the Ice Age.

Distribution: North America.
Classification: Ungulata, Proboscidea.
Meaning of name: Earth burrower.
Named by: Peale, 1801.
Time: Pleistocene epoch of the Quaternary.
Size: 4m (13ft) long.
Lifestyle: Grazer.
Species: *M. americanum*, *M. obscurus*.

Cuvieronius

Distribution: Arizona, Florida, USA; Mexico; Argentina.
Classification: Ungulata, Proboscidea.
Meaning of name: From Cuvier.
Named by: Osborn, 1923 (although specimens had been known and identified as other things for half a century before that).
Time: Pliocene epoch of the late Tertiary to recent.
Size: 3m (10ft).
Lifestyle: Forest browser.
Species: *C. hyodon*, *C. tropicus*.

This is unusual in being one of only two elephants known from South America. The other is *Stegomastodon*. They migrated here when the land bridge was established in the Pliocene, *Cuvieronus* moving down the valleys of the Andes, and *Stegomastodon* staying in the lower lands to the east. *Cuvieronus* may have existed on the pampas of Argentina into historical times. *Cuvieronus* was the one that was better adapted to cold conditions, and its remains have been found in mountain areas and the cooler areas to the south.

Right:
Cuvieronius
was a fairly
versatile feeder, eating the grass of the pampas as well as the leaves of the forests.

Features: *Cuvieronius* is a fairly small member of the elephant group that include the four-tusked gomphotheres of earlier times. Its notable feature is the tusks, which are long and spirally twisted like the tusk of a narwhal.

Archidiskodon

MAMMOTH SPECIES
Mammuthus primigenius is the best-known of the mammoth species, but it was not the only one.
M. meridionalis – the earliest-known species. It lived in open woodland in southern Europe but was not well adapted to the cold.
M. trogontherii – the first of the cold-adapted mammoths with woolly coats. It roamed the cold grasslands of central Europe feeding on the coarse grasses.
M. columbi – the biggest known, towering 4m (13ft) at the shoulder. It was one of the first to cross the land bridge into North America.
M. imperator – the typical and most widespread of the North American mammoths.
M. exilis – a small species from islands off California.
M. jeffersoni – an American species
M. lamarmorae – a dwarf species from Sardinia.
   The last mammoths lived 4,000 years ago on Wrangel Island in the Arctic Ocean. They were small, about 1.8m (6ft) high as a result of their island habitat.

This massive elephant, a relative of the modern Indian elephant *Elephas*, lived in the forested areas far from the edges of the Pleistocene ice caps, migrating there from Africa. It weighed twice as much as any modern elephant. There have been about 20 species of *Archidiskodon* identified, but recently many of these have been reassigned as species of the modern elephants, *Elephas* and *Loxodonta* and the mammoth *Mammuthus*.

Features: Although related to the Indian elephant, *Archidiskodon* has a body structure that is more like that of an African elephant. The high head with the long skull gives it a very high-fronted stance. The tusks are curved upwards and resemble those of a mammoth. They are crossed in front in some elderly specimens.

Distribution: Africa; Europe; Asia.
Classification: Ungulata, Proboscidea.
Meaning of name: Ancient disk tooth.
Named by: Nesti, 1825.
Time: Late Pliocene epoch of the late Tertiary to the Pleistocene epoch of the Quaternary.
Size: 6.5m (20ft) long.
Lifestyle: Forest browser.
   Species: *A. meridionalis*,
      A. gromovi,
         A. planifrons,
            A. imperator.

Left: Archidiskodon *is thought by some palaeontologists to have been the ancestor of the mammoths.*

CARNIVORES

With the presence of a Pleistocene megafauna, consisting of specialized plant-eating animals, it was inevitable that specialized meat-eating mammals would evolve to exploit this bonanza. Particular species of modern genera, such as the wolf Canis, *the lion* Panthera *and the bear* Ursus, *appeared and disappeared during the Pleistocene, but there were also unique genera, particularly among the cats.*

Smilodon

The largest and the most famous of the sabre-toothed cats was *Smilodon*. It is well-known from the many skeletons found in the Rancho La Brea tar pits in Los Angeles, where it fed on the animals such as giant sloths and mammoths that had become trapped there. However, it is known from all over North America, and there is a species, *S. populator*, from Argentina.

Features: The body is powerfully built with the muscles of the shoulders and neck arranged to drive a powerful downward lunge of its huge head. The lines of force in the skull are aligned with those of the sabre teeth. The jaws can open at more than 95 degrees. The teeth are oval in cross section and serrated like steak knives along their rear edges. Its short legs and bobbed tail show that it was not a fast runner, but an ambush hunter. The hyoid bones in the throat show that it roared like a lion.

Left: Smilodon has been adopted as the state fossil of California, because of its abundance in the Rancho La Brea tar pits.

Distribution: North America; Argentina; Brazil.
Classification: Carnivora, Feliformia.
Meaning of name: Sabre tooth.
Named by: Plieninger, 1846.
Time: Late Pleistocene epoch of the Quaternary.
Size: 1.2m (4ft).
Lifestyle: Hunter.
Species: *S. fatalis, S. gracilis, S. populator*. Also *S. floridus* and *S. californicus* which may be subspecies of *S. fatalis*.

Dinofelis

The "false sabretooths" that evolved in the late Tertiary continued into the Quaternary. They did not have true sabre teeth, but their killing canines seemed somewhere between the long flat teeth of the true sabretooths and the round-section conical teeth of the modern cats. They would have fed on smaller prey than the true sabretooths – possibly baboons and early humans.

Features: *Dinofelis* has a stocky build, about the size of a lion but more powerful, with massive forequarters showing that it was not a running hunter. Rather it was an ambusher, leaping upon its prey and holding it down with the strong front limbs while it killed using its big canines. The early species, *D. barlowi* and *D. cristata* were quite panther-like, but the later species were built more like the true sabre tooths.

Left: Remains of Dinofelis are not common, but they are mostly found where there are remains of hominids.

Distribution: Europe; Asia; Africa; North America.
Classification: Carnivora, Feliformia.
Meaning of name: Terrible cat.
Named by: Zdasky, 1924.
Time: Late Pliocene epoch of the late Tertiary to the early Pleistocene epoch of the Quaternary.
Size: 1.2m (4ft).
Lifestyle: Hunter.
Species: *D. abeli, D. barlowi, D. diastemata, D. paleoonco, D. piveteaui, D. therailurus, D.cristata.*

MODERN GENERA WITH PLEISTOCENE SPECIES

Panthera – the lion (right). The cave lion *P. leo spelaea* was probably the biggest cat that ever lived, existing in much of Europe

during the Ice Age. *P. leo atrox* lived in America at the same time.

Canis – the dog (right). *C. dirus* was the dire-wolf, larger than any modern wolf with more massive teeth and jaws. However, its legs were much shorter and it was not a fast sprinter.

Ursus – The giant cave bear (below) *U. spelaeus* lived in the mountains of Europe at the height of the Ice Age, and was known to early people of the area. Despite its formidable appearance it was a herbivore.

Homotherium

Between the sabretooths and the dirktooths there were the scimitar-tooths. Scimitar-tooths, like *Homotherium*, had backward-curved canine teeth, shorter and flatter than those of sabretooths, and they did not protrude beyond the deep chin. Specimens from the North Sea, dated at the height of the last Ice Age, show that *Homotherium* was tolerant of cold.

Features: The very long front legs and the long neck of *Homotherium* show that it had a totally different hunting style from contemporary cats. Its foot bones are short suggesting slow movement, but otherwise the legs seem to be built for running. This fact, along with the big nasal opening for fast and efficient breathing, suggests a sprinting hunter like a cheetah. Its remains have been found with those of young elephants, suggesting that these were its prey.

Distribution: Africa; Asia; Europe; North America.
Classification: Carnivora, Feliformia.
Meaning of name: The same mammal.
Named by: Fabrini, 1890.
Time: Pleistocene epoch of the Quaternary.
Size: 1.2m (4ft).
Lifestyle: Hunter.
Species *H. ethiopicum*, and four others.

Left: The visual part of the skull cavity indicates that it hunted by day rather than by night.

Miracinonyx

Distribution: Texas, Nevada, Wyoming, USA.
Classification: Carnivora, Feliformia.
Meaning of name: Amazing cheetah.
Named by: Adams, 1979.
Time: Pleistocene epoch of the Quaternary.
Size: 2.2m (7ft) including tail.
Lifestyle: Running hunter.
Species: *M. trumani*, *M. inexpectatus*.

The cheetah, the fastest mammal on Earth is confined to Africa. An American version, *Miracinonyx* sprinted across the prairies of Texas, Nevada and Wyoming in Pleistocene times, hunting the fleet-footed grass-eaters such as pronghorns. It was originally thought to have been a species of puma, with the similarities to the cheetah arising from convergent evolution. There is now debate as to whether *Miracinonyx* was more closely related to the puma or the cheetah.

Features: The long legs, the flexible spine and the heavy steering tail of the cheetah are evident in the build of *Miracinonyx* – so much so that some scientists regard it as a species of *Aconyx*, the modern cheetah. The head is small and the killing teeth are short and sharp, unlike the blade-like fangs of most of its contemporaries. A second species *M. inexpectatus* seems to have lived in mountain areas.

Left: The speed of Miracinonyx probably matched the 96km ph (60 mph) achieved by modern pronghorns.

PRIMATES

The Recent epoch is regarded as the Age of Humans, since overwhelmingly we represent the most significant species alive today. The differences between humans and their ape-like immediate ancestors are more to do with cultural than physical development. As with all evolutionary developments, our evolution was not a straight line process, but involved all sorts of side branches.

Megaladapis

While the monkeys and the apes were evolving in Africa, the lemurs in nearby isolated Madagascar continued to develop into all kinds of forms. The biggest was *Megaladapis*, which must have been about the size of a modern gorilla. As with many big specialized island animals, it was wiped out by human settlers, before it was studied scientifically.

Features: The skull is very long, more like that of a pig than a lemur. Its nose may have supported a short proboscis. Its lower incisors work against a toothless pad on the roof of the mouth, as in a cow. Its legs are short but its hands and feet are long, suggesting that it was a good climber especially on vertical trunks. This indicates a locomotion and a feeding habit like the modern koala.

Left: A gigantic koala with the head like a pig and teeth like a cow – a description of the appearance of Megaladapis.

Distribution: Madagascar.
Classification: Primates, Prosimia.
Meaning of name: Great lemur.
Named by: Forsythe Major, 1894.
Time: Pleistocene to Recent epochs of the Quaternary.
Size: 2m (6½ft).
Lifestyle: Arboreal leaf-eater.
Species: *M. edwardsi*, *M. gradidieri*, *M. madagascasriensis*.

Gigantopithecus

German palaeontologist Ralph von Koenigswald first found *Gigantopithecus* as a set of teeth in a Chinese apothecary shop in the 1930s, and the early collection of *Gigantopithecus* remains took place from such shops throughout China. Eventually the source was traced to a cave in Guangxi in southern China, which has since yielded several teeth and jaws.

Right: Size difference between male and female Gigantopithecus *was much greater than in any modern primate.*

Features: The largest primate so far found, a veritable King Kong, *Gigantopithecus* seems to have the body of a gorilla and the face of an orang-utan. Microscopic plant fragments in the teeth show a diet of leaves and bamboo. There is a theory that it survives in the mountains of South-east Asia, and is the source of the observations of the yeti.

Distribution: China.
Classification: Primates, Catarrhini.
Meaning of name: Giant ape.
Named by: von Koenigswald, 1935.
Time: Pliocene epoch of the late Tertiary to the mid-Pleistocene epoch of the Quaternary.
Size: 3m (10ft) high.
Lifestyle: Browser.
Species: *G. blacki*, *G. giganteus*.

Australopithecus

Australopithecus is generally regarded as the immediate ancestor of the genus *Homo*, to which modern humans *Homo sapiens* belong. It shows a transition from a tree-living habit to a ground-dwelling one, reflecting the spread of more open country in Africa at the beginning of the Pleistocene. Other genera on the ancestral human line include *Ardipithecus*, *Orrorin*, *Sahelanthropus* and *Kenanthropus* from the preceding Pliocene. Of this range of animals, all became extinct with only one leading onwards to the present genus *Homo*.

Features: Several species of *Australopithecus* are recognized. They all have a rather chimpanzee-like face with prominent brow-ridges, but have a larger brain capacity and the brain is arranged like that of a human. They are bipedal, but can still climb trees. The molar teeth with the thick enamel indicate that they were mainly vegetarian. The angle of the head of the femur and the way the skull attaches to the backbone indicates an upright stance, with the head at the top of the vertebral column rather than at the end.

Distribution: East Africa.
Classification: Primates, Catarrhini.
Meaning of name: Southern ape.
Named by: Dart, 1925.
Time: Pliocene epoch to the Pleistocene epoch of the Quaternary.
Size: 1.2m (4ft) tall.
Lifestyle: Browser.
Species: *A. africanus*, *A. afarensis*, *A. anamensis*, *A. bahrelghazali*, *A. garhi*, *A. ramidus*.

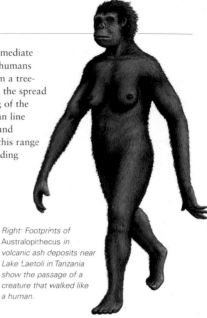

Right: Footprints of Australopithecus in volcanic ash deposits near Lake Laetoli in Tanzania show the passage of a creature that walked like a human.

THE GENUS HOMO

Our own genus, *Homo*, is naturally a topic of interest and debate. Several species are currently recognized, though controversy continues about precise nomenclature.

H. rudolfensis – 2.5mya to 1.8mya, from the Pliocene to early Pleistocene. Found in east Africa and used simple stone tools.

H. habilis – 2mya to 1.6mya, from the early Pleistocene of Tanzania. Possibly able to make single-faced stone tools. Probably a scavenger.

H. ergaster – 2mya to 1.6mya, from east Africa and Eurasia. Double-faced stone tools.

H. erectus – 2mya to 400,000 years. East Africa and Eurasia from Spain to Indonesia. Similar to *H. ergaster* but as tall as a modern human.

H. antecessor – 1mya to 800,000 years. southern Europe and northern Africa. Relatively modern face. Possibly complex funeral rituals.

H. heidelbergensis – 600,000 years. Precursor to the Neanderthals. Hunters of large game. Social rituals.

H. neanderthalensis – 230,000 years to 29,000 years, central Europe. Neanderthal Man. Complex stone tools with points and borers but no blades. Wooden spears. Possible musical instruments. Possibly ritual cannibalism.

H. floresiensis – up to 13,000 years ago. Similar to *H. erectus* but a dwarf island-dwelling form, only about a metre high.

H. sapiens – 200,000 years until today. Ourselves.

Paranthropus

Paranthropus used to be regarded as a robust form of *Australopithecus*, and sometimes still is. The jaws and teeth are more massive and powered by muscles attached to a skull ridge, indicating a diet of coarse plants, and earned the first species to be found the name "nutcracker man". Many remains have been found gnawed by carnivorous animals suggesting that it was a principal prey item at the time. It seems to have been a tool user. It branched from the *Australopithecus* line about 2.7 million years ago, at the end of the Pliocene.

Features: The heavy jaws and the big grinding teeth are the main feature of this hominid. The largest have a saggital crest running fore and aft over the top of the skull – sign of strong jaw muscles and diet of coarse plant material. The nasal passages have a large sinus system, presumably a system for cooling the large brain.

Distribution: East Africa.
Classification: Mammal, Primates.
Meaning of name: Almost an ape.
Named by: Broom, 1938.
Time: Pliocene epoch of the late Tertiary to Pleistocene epoch of the Quaternary.
Size: 1.6m (5ft) tall.
Lifestyle: Browser.
Species: *P. aethiopicus*, *P. boisei*, *P. robustus*.

Left: Heavy-jawed Paranthropus shared the African plains with the earliest species of the genus Homo.

GLOSSARY

Abelisaurid A group of theropods of the late Cretaceous particularly from the Southern Hemisphere.

Alvarezsaurid Long-legged running dinosaurs with diminutive forelimbs, often classed as primitive birds.

Amphibious A creature able to survive on land or in the water.

Ankylosaur Quadrupedal herbivorous ornithischian dinosaurs from the late Cretaceous, making up the suborder Ankylosauria.

Ankylosaurid A member of the Ankylosauridae, a family of the suborder Ankylosauria.

Antorbital fossa A hole in the skull between the snout and the eye socket.

Archosaur A member of the diapsid group of reptiles that includes the crocodiles, the pterosaurs and the dinosaurs – the so-called "ruling reptiles".

Arthropod A member of the invertebrate group with chitinous shells and jointed legs, including the crustaceans, insects, arachnids and centipedes.

Atrophy Wasting away of an organ that is no longer important, as a result of evolutionary development.

Belemnite A common Mesozoic cephalopod resembling a squid but having a bullet-shaped internal shell.

Bipedal Two-footed animal.

Caenagnathid A group of theropods related to the oviraptorids.

Carapace A thick, hard shell or shield that covers the body of some animals.

Carnosaur In old terminology, any big theropod, but in more modern terms a theropod belonging to the group that contains *Allosaurus* and its relatives.

Cartilaginous Referring to a skeleton composed entirely of cartilage, a tough, elastic tissue.

Cassowary A large flightless bird with a horny head crest and black plumage, from northern Australia.

Cenomanian A stage of the late Cretaceous period lasting from about 97 to 90 million years ago.

Cephalopod A mollusc with the limbs very close to the head, such as a snail or an octopus.

Ceratopsian Horned dinosaur.

Clade A group with common ancestry.

Cladogram A diagram illustrating the development of a clade.

Cololite A trace fossil consisting of the fossilized remains of the contents of an animal's digestive system.

Convergent evolution The evolutionary development of similar features on different animals that share a similar environment.

Coprolite A trace fossil consisting of the fossilized remains of an animal's droppings.

Crepuscular Active at twilight or dawn.

Crest A tuft of fur, feathers or skin or a ridge of bone along the top of the head.

Cretaceous The last period of time in the Mesozoic era.

Cycad A tropical or subtropical plant with unbranched stalk and fern-like leaves crowded together at the top.

Denticle A small tooth or tooth-like part.

Dermal Relating to the skin.

Diapsid A member of a major group of the reptiles, classed by the presence of two holes in the skull behind the eye socket, and comprising the majority of modern reptiles including the lizards, snakes and crocodiles.

Digitigrade Walking so that only the toes touch the ground.

Diplodocid A herbivorous quadrupedal dinosaur from the late Jurassic or early Cretaceous periods, with a long neck and tail.

Dorsal Relating to the back or spine

Fibula The outer thin bone from the knee to the ankle.

Fluke A blade-like projection at the end of the tail used for swimming, as in whales.

Gastralia A set of extra ribs covering the stomach area, as seen in some dinosaurs.

Gastrolith A stone in the stomach, deliberately swallowed to aid in digestion or in buoyancy.

Gavial A type of fish-eating crocodile from South-east Asia.

Genus (genera pl.) A taxonomic group into which a family is divided and containing one or more species, all with a common characteristic.

Gizzard The thick-walled part of the stomach in which food is broken up by muscles and possibly gastroliths.

Gondwana (sometimes called Gondwanaland) The southern of two ancient continents, comprising modern-day Africa, Australia, South America, Antarctica and the Indian subcontinent. It was formed from the break-up of the supercontinent Pangaea 200 million years ago.

Groundmass A matrix of rock in which larger crystals are found.

Homeothermic "Warm blooded", having the ability to keep the body at an almost constant temperature despite changes in the environment, as in mammals and birds and probably some of the dinosaurs.

Humanoid As an adjective, human-like in appearance, or as a noun, a member of the group to which humans belong.

Humerus The bone from the shoulder to the elbow.

Ichnogenus A genus based only on fossil footprints.

Ichnology The study of fossil footprints

Ichnospecies A species based only on fossil footprints.

Ichthyologist A scientist who studies fish.

Ischium A section of the hip bone which, in reptiles, sweeps backwards.

Isotope A form of the atom of a chemical element in which the atomic number is different from that of other atoms of the same element.

Jurassic The second period of the Mesozoic era and lasting for approximately 45 million years.

Keeled In a scale or an armour plate or a bone, having a ridge running along its length.

Keratinous Made up of keratin – a horny substance similar to fingernails.

Laurasia One of the two supercontinents formed by the break up of Pangaea 200 million years ago. It comprises modern North America, Greenland, Europe and Asia.

Lias bed The lowest series of rocks in the Jurassic system.

Maastrichtian The last age of the Cretaceous period, from 74 to 65 million years ago.

Megalosaur Large Jurassic or Cretaceous carnivorous bipedal dinosaur.

Mesozoic The era of geological time lasting from 208 to 65 million years ago and consisting of the Triassic, Jurassic and Cretaceous periods.

Mosasaur Cretaceous giant marine lizards with paddle-like limbs.

Nodule A small knot or lump – as a piece of armour bone embedded in the skin of an animal or as a mineral occurrence embedded in rock.

Nomen dubium Literally dubious name – a name given to an animal that is not fully supported by scientific study.

Olecranon Bony projection behind the elbow joint.

Olfactory Relating to the sense of smell.

Olfactory bulb The point from which the nerves concerned with the sense of smell originate.

Ornithischian An order of dinosaurs that includes the ornithopods, stegosaurs, ankylosaurs and marginocephalians - characterized by the hip bones, which are arranged like those of a bird.

Ornithomimid Bird-like, ostrich-mimic.

Ornithopod A herbivorous bipedal ornithischian dinosaur.

Ossicle A small bone.

Pachystasis A thickening of the bones, particularly the ribs, in some aquatic animals, such as the modern walrus, that helps the buoyancy of the animal in water.

Palaeogeography The study of what the geography was like in the past – the arrangement of the continents, the distribution of land and sea, and the climatic zones.

Palaeozoic The era of geological time that began 600 million years ago and lasted for 375 million years.

Paleontologist A scientist who studies fossils and the life of the past.

Pangaea The ancient supercontinent comprising all the present continents before they broke up 200 million years ago.

Pangolin A mammal from tropical Africa, southern Asia and Indonesia, with a long snout and a body covered in overlapping horny scales.

Patagium A web of skin between the neck, limbs or tail in gliding animals, that assists flight in place of a wing.

Petrifaction A process of forming fossils, particularly the process in which the organic matter of each cell of the creature is replaced by mineral.

Phalange A bone in the finger or toe.

Plantigrade Walking with the entire sole of the foot in contact with the ground.

Plastron Bony plate forming the underpart of the shell of a turtle or similar animal.

Plate A thin sheet that forms an overlapping layer of protection.

Plesiosaur A marine reptile with a long neck, short tail and paddle like limbs form Jurassic and Cretaceous times.

Polydactyly With more than the usual number of digits.

Polyphylangy With more than the usual number of bones in the digit.

Preadaptation The evolution of a particular feature for a function, that is useful for another.

Prehensile Adapted for grasping.

Premaxilla The front bone of the upper jaw of dinosaurs.

Pterodactyl The popular name for a member of the pterosaur group.

Pterosaur A flying reptile from Jurassic and Cretaceous times.

Pygidium The tail shield of an arthropod.

Pygostyle The bony structure formed from fused tail bones that is used as the base for the tail feathers of a bird.

Quadrupedal An animal that walks on all four limbs.

Refugium A geographical region that has remained unaltered by climate change

Rostral Beak- or snout-like.

Saurischian An order of dinosaurs that includes the theropods, therizinosaurs, prosauropods and sauropods – characterized by the arrangement of hip bones, similar to those of a lizard.

Sauropod A herbivorous quadrupedal saurischian dinosaur including *Apatosaurus*, *Diplodocus* and *Brachiosaurus*. Small heads and long necks and tails characterize the group.

Sclerotic A ring of bone inside the eye, used for adjusting focus or for adjusting pressure in a swimming animal.

Scute A horny plate that makes up part of an armour.

Seismic Relating to earthquakes or earth tremors.

Silica The oxide of the element silicon, which is a major constituent of the minerals of the Earth's crust.

Species A taxonomic group into which a genus is divided.

Stegosaurid A quadrupedal herbivorous ornithischian dinosaur, with bony plates and armour.

Stomach stone A gastrolith.

Swallow hole A depression in limestone terrain, usually into which a river or a stream disappears underground.

Syanapsid A member of a major group of the reptiles, classed by the presence of a single combined hole in the skull behind the eye socket, and comprising the mammal-like reptiles from which the mammals evolved.

Symphysis A growing together of parts joined by an intermediary layer, particularly the join at the front of the lower jaw.

Taphonomy The study of what happens to a dead organism before it becomes buried and fossilized.

Taxonomy A system of classification of organisms.

Thecodont A reptile of Triassic times with teeth set in sockets. They gave rise to dinosaurs, crocodiles, pterodactyls and birds.

Theropod A bipedal carnivorous saurischian dinosaur

Titanosaur A herbivorous quadrupedal dinosaur.

Triassic The first period of the Mesozoic era, which lasted 45 million years.

Turbinal A folded bone inside the nose of some animals, supporting a membrane used to adjust the temperature or humidity of breathed air.

Ulna The inner and longer bone of the forearm.

Viviparous Giving birth to live offspring.

Wastebasket genus A genus to which many dubious fossils are attributed.

INDEX

This edition is published by Hermes House
an imprint of Anness Publishing Ltd
info@anness.com
www.annesspublishing.com

If you like the images in this book and would like to investigate using them for publishing, promotions or advertising, please visit our website www.practicalpictures.com for more information.

A CIP catalogue record for this book is available from the British Library.

Publisher: Joanna Lorenz
Editorial Director: Helen Sudell
Editor: Simona Hill
Designer: Nigel Partridge
Illustrators: Anthony Duke, Julius Csotonyi, Stuart Jackson Carter, Peter Barrett, Andrey Atuchin, Anthony Duke, Robert Nicholls, Denys Ovenden, Alain Beneteau

PUBLISHER'S NOTE
Although the advice and information in this book are believed to be accurate and true at the time of going to press, neither the author nor the publisher can accept any legal responsibility or liability for any errors or omissions that may have been made.

ACKNOWLEDGEMENTS

The Publisher would like to thank the following picture agencies for granting permission to use their photographs in this book:
Key: l = left, r = right, t = top, c = centre, b = bottom

Alamy p24t, p26bl, p28, p39, p49, p51, p55, p121, p124, p181, p203, p205, p281, p491. Ardea p8, p24bl, p24br, p26t,p33tl, p33tr, p56, p60t, p60b, p61b, p207. Corbis p33br, p41, p77, p235, p252, p253, p386, p397, p443. The Natural History Museum p9, p29, p33bl, p61t, p249, p287, p457. David Varrichio p30.

The author would like to acknowledge the debt owed to the work of the scientific community, without whose dedication none of this fascinating stuff would be known. In particular he would thank the following for their direct help with this book, and he offers his sincere apologies to anyone he has overlooked.
Mike Benton, Ken Carpenter, Jenny Clack, Sue Evans, Peter Galton, Steve Hutt, Max Cardoso Langer, Don Lessem, Greg McDonald, Matthew C. Mihlbachler, Nate Murphy, Darren Naish, Tom Rich, Bryan Small.